LIBRARY

FLORIDA KEYS COMMUNITY COLLEGE

5901 W. JUNIOR COLLEGE ROAD

KEY WEST, FL 33040

UEMCU

INDUSTRIAL ELECTRICITY

INDUSTRIAL ELECTRICITY
Principles and Practices

THIRD EDITION

James E. Adams

Late Director
School of Industrial Electricity
Madisonville, Kentucky

Gordon Rockmaker

Editor in Chief, Electricity and Electronics
McGraw-Hill Book Co.
Former Instructor of Electricity
Institute of Design and Construction
Brooklyn, New York

McGRAW-HILL BOOK COMPANY
NEW YORK · ATLANTA · DALLAS · ST. LOUIS
SAN FRANCISCO · AUCKLAND · BOGOTÁ
GUATEMALA · HAMBURG · JOHANNESBURG
LISBON · LONDON · MADRID · MEXICO
MONTREAL · NEW DELHI · PANAMA · PARIS
SAN JUAN · SÃO PAULO · SINGAPORE
SYDNEY · TOKYO · TORONTO

Sponsoring Editor: Paul Berk
Editing Supervisor: Evelyn Belov
Design and Art Supervisor: Patricia Lowy
Production Supervisor: Priscilla Taguer

Text Designer: Levavi & Levavi
Cover Photographer: Ken Karp
Cover Prop: Relay courtesy of Dependable Industrial Supply Co., Inc., Brooklyn, New York, and Siemens-Allis, Inc.

Library of Congress Cataloging in Publication Data

Adams, James E.
  Industrial electricity.

  Rev. ed. of: Electrical principles and practices.
2nd ed. 1973.
  Includes index.
  1. Electric engineering. I. Rockmaker, Gordon.
II. Adams, James E.
Electrical principles and practices.
III. Title.
TK146.A3 1984   621.3   83-25552

Industrial Electricity: Principles and Practices
Third Edition

1 2 3 4 5 6 7 8 9 0 SEMBKP 8 9 1 0 9 8 7 6 5 4

ISBN 0-07-000327-0

CONTENTS

PREFACE

Industrial Electricity: Principles and Practices, Third Edition, is designed to train industrial electricians. That, as suggested in the title, is the straightforward objective of this text. But the modern industrial electrician is expected to have a depth of knowledge and a diversity of skills in both electricity and electronics. To be equipped with the principles and practices of the field requires the study of many subject areas. Thus the objective of training industrial electricians must assume a broad diversified aspect.

To avoid a fragmented, disjointed approach, this book focuses on just those principles and practices of industrial electricity that will provide a solid foundation to the entry-level student. The technology, equipment, tools, and methods will certainly change (very rapidly in some cases), so concentration on any one technique or equipment is avoided. What may appear to be an omission of some device or skill is usually a conscious choice—in effect, a judgment—of the author that the progress of technology will soon bypass that particular device or skill.

The modern industrial electrician must be able to install, maintain, and troubleshoot very sophisticated equipment. In dealing with these tasks, the electrician is often called upon to use mathematical formulas to solve electric circuit problems for loads, wire size, equipment specifications, costs, and the like. The student using this book must have the arithmetic skills needed for such calculations. Although an understanding of and facility in using basic algebra would be helpful, a knowledge of algebra is not a prerequisite for the use of this text.

The organization of chapters in this book is based on a logical sequence which provides a concise overview of the subject in Chapter 1, but without the details and definitions that are given in the ensuing 23 chapters and appendix. In general, the traditional direct current–alternating current order of topics has been followed. Since the basic machine of industrial electrical systems is the motor, emphasis in both the dc and ac chapters is placed on that piece of equipment. The final chapter covers what is certainly not the least important in the electricians' storehouse of knowledge, the solid-state electronic devices that are now rapidly replacing the electromagnetic and electromechanical devices which have dominated electrical installations.

Throughout the text, citations and references are given to the latest codes and industry standards such as the National Electrical Code (NEC), National Electrical Manufacturers Association (NEMA) standards, and the standards of the American National Standard Institute (ANSI). While these codes and standards may change periodically, it is essential that the student understand their applications and general approach.

No text is the product of a single person, and *Industrial Electricity: Principles and Practices,* Third Edition, is no exception. The late James E. Adams, author of the first two editions of *Electrical Principles and Practices,* will be remembered for his innovations in teaching, his dedication to his profession and to his students, and his love of the subject. This third edition leans heavily on his past efforts.

The author wishes to gratefully acknowledge the assistance of the numerous manufacturers and organizations that have contributed photographs, drawings, literature, suggestions, and moral support. Their names are cited throughout this book.

Thanks are also due to the reviewers of the manuscript for this book, Spencer Clark, Ronald Custer, and Eric David. Their comments and advice are most helpful.

Finally, a special and personal note of thanks and appreciation to my wife, Ellen, who, in addition to preparing the many drafts and the final version of the manuscript, corrected errors, suggested changes, spurred the author on to greater productivity, kept Kathryn, Linda, Marsha, and Morris occupied, and in every way made this book possible.

Gordon Rockmaker

ACKNOWLEDGMENTS

The author gratefully acknowledges the assistance, suggestions, contributions, and information extended by the following companies and organizations: Airpax Corp., Alflex Corporation, Allen-Bradley Company, Appleton Electric Company, Armature Coil Equipment, Inc., Baldor Electric Company, James G. Biddle Co., Bodine Electric Co., Caterpillar Tractor Co., Central Moloney Inc., CertainTeed Corp., Consolidated Edison Co. of New York, Inc., Crown Industrial Products Co., Eagle Picher Bearings, Eastern Air Devices, Eaton Corp. (Cutler-Hammer), General Cable Co., General Electric Company, Gould Inc., Greenlee Tool Div. Ex-Cell-O Corp., Hampton Products Co., Inc., Hearlihy & Co., Hurst Mfg. Corp., Ideal Industries Inc., Kirkwood Commutator Co., Klein Tools, Inc., Leviton Manufacturing Co., Inc., The Martindale Electric Co., McGraw-Hill Book Co., National Electric Coil Div. McGraw-Edison Service, National Fire Protection Association, Onan Corp. Subsidiary of McGraw-Edison Company, OTC Power Team Industrial Div. Owatonna Tool Co., Pyle-National Co., H. H. Robertson Company, Siemens-Allis, Inc., Simpson Electric Co., The L. S. Starrett Co., The Superior Electric Co., Union Carbide Corp., Wagner Div. McGraw-Edison Co., Walker Div. Butler Manufacturing Co., Ward-Leonard Electric Co., Inc., Westinghouse Electric Co., Weston Instruments, Wheatland Tube Company, and Whitefield Electric Company.

1
INTRODUCTION

This chapter presents an overview of what electricity is, how it is produced, and the principles and practices involved in its use. Many technical terms and principles are introduced without being defined. This is intentional; the main purpose of these brief preliminary discussions of electrical fundamentals is to provide a preview, with the more detailed and complete presentations of information (including definitions) to follow in later chapters.

1-1 ATOMIC THEORY AND STRUCTURE

The story of electricity begins with atomic theory and structure. All matter is made up of atoms or combinations of atoms. There are as many types of atoms as there are chemical elements. To date slightly more than 100 elements have been discovered. A few of these exist only in the laboratory.

An atom consists of a "core" or nucleus that contains protons, surrounded by an equal number of electrons in one or more "rings." Electricity may be thought of as a phenomenon involving atoms and their electrons. A complete atom has the same number of electrons and protons. Such an atom is electrically neutral. If an atom loses some of its electrons, the unbalance in the atom will favor the protons. Protons are electrically positive. Thus the atom that is missing the electrons is said to have a positive charge. The missing electrons, called *free* electrons, are said to have a negative charge.

1-2 DISPLACEMENT OF ELECTRONS

Electrons can be displaced from their atoms by mechanical friction, such as combing hair, rubbing on a plastic seat cover, and turbulence created by wind in clouds; by use of chemicals under certain conditions, such as in a storage battery; or by causing conductors to move through a magnetic field (or the magnetic field to move across a group of conductors) as in generators, alternators, and transformers. Other methods that produce a displacement of electrons from their atoms are heat, light, and solar action. The magnetic-field method is used to produce the large amounts of electric energy required by electric utilities and industrial power plants.

Producing free electrons is not enough if electricity is to be put to useful work. If a complete path is provided, the charged particles will move through that path. The movement of these charged particles is called *electric current*, and the particles are often referred to as *current carriers*. The charged particles could be either positive or negative. The negatively charged carriers are composed of the free electrons.

The electrons of some atoms can be more easily displaced than others. The materials of these types of atoms—such as silver, copper, and aluminum—are used as conductors of electricity. Some types of atoms contain electrons that are more difficult to displace. Materials composed of these atoms—such as nichrome (an alloy of nickel and chromium), tungsten, iron, and steel alloys—are used to limit current or produce heat. Other types of atoms contain electrons that can be displaced only under the most extremely severe conditions. Materials such as wood, mica, glass, porcelain, and rubber, for example, are used as electrical insulators. There are also materials which can behave either as conductors or as insulators depending on the electrical conditions in which they are used. These materials are called *semiconductors*. Carbon, silicon, and germanium are three common materials that display this property. They are used in transistors, diodes, integrated circuits, and other *solid-state* devices.

1-3 MOVING CHARGES

When charges move in a circuit, they produce magnetic, thermal (heat), and chemical action. When current (that is, a moving charge) flows in a wire, a magnetic field is formed around the wire. In practically every application in which a coil of wire is used in electrical equipment, the function of the coil is to produce a magnetic field. Much of our industrial electrical equipment operates on magnetic principles.

When negative charges are forced through a resistance, they produce heat. This effect is used in ranges and ovens, soldering irons, furnaces, and other heating appliances.

The chemical effect of electricity is used in refining ores, electroplating with chromium, copper, and cadmium, charging batteries, and producing light, to name just a few applications.

1-4 AUTOMOBILE ELECTRICAL SYSTEM

The basic electrical system of the modern automobile employs most of the main principles of electricity, magnetism, and solid-state electronics discussed in this text.

The automobile battery produces electricity by chemical means. When two dissimilar metals are immersed in an acid, electrons from both metals are displaced. This is the principle of the lead-acid battery which operates a starting motor that in turn helps start the engine. The battery also supplies energy to all the automobile accessories such as horns, radio, lights, windshield wipers, and other electrically operated devices when the engine is not running. An alternator [which is another name for an alternating-current (ac) generator] furnishes electric energy when the engine is running. The alternator also keeps the battery in a charged condition.

1-5 CONDUCTORS AND INSULATORS

Electrical paths to the various pieces of equipment operated by electricity are usually provided by copper wire. Copper contains atoms with loosely held electrons, which make it an efficient conductor of electricity. (Copper is second only to silver in conductivity, or the ability to conduct electric current.) The electron path is insulated throughout to keep electrons from straying without accomplishing useful work. Materials composed of atoms whose electrons are held extremely tight (such as rubber, thermoplastics, mica, vulcanized fiber, Bakelite plastic compounds, vinyl, varnish, and cotton, to name just a few) are used to insulate electric circuits. If electrons get out of their intended path, they may produce *leakage* current or a *short circuit*.

Light bulbs contain a filament (a thin wire) made of tungsten. Electrons in tungsten atoms are tightly held, and the friction produced by moving these electrons produces heat. The heat thus produced raises the temperature of the filament high enough so that some of its energy is emitted in the form of light.

1-6 ELECTROMAGNETISM

When current flows in a conductor, a magnetic field composed of magnetic lines of force is created at right angles to the conductor. If the wire is wound in the form of a coil, the field will pass through the center of the coil. If an iron core is placed in the coil, the magnetic field will be concentrated in the core and the effective magnetic strength will be greatly increased. An energized coil of wire with an open iron core is known as an *electromagnet*. An open iron core is one that does not provide a complete magnetic path for the flux. Thus poles will form on the open faces of the core. Magnetism produced by electricity is known as *electromagnetism*, but all magnetism, no matter how it is produced, is the same in characteristics and effects. Magnetism is conveniently thought of as consisting of imaginary lines of force having certain physical properties. When magnetic lines of force are made to travel through space, they tend to shorten themselves as much as possible, behaving somewhat like stretched rubber bands. Magnetic lines of force are always continuous and travel in closed paths. If they flow through two pieces of separated iron, they tend to draw the pieces together. Magnetic lines of force never cross one another.

An automobile starter relay uses these principles. A coil of wire connected to the battery is energized by turning the ignition key. The energized coil produces magnetism, which attracts an iron core, called an *armature*, connected to a switch. This closes the switch, which completes the electrical path from the battery to the starter motor.

An electromagnet is also used in the car's horn. It is arranged to draw a diaphragm or metal disk to it when energized. The disk is designed to break the circuit to the coil, which releases it. On release, the disk returns to its normal position and completes the circuit to the coil, which draws it back again. This back-and-forth movement of the disk sets up vibrations in the air which produce the characteristic sound of the horn.

Electromagnetism is used to open and close switches, which can ring bells, operate valves, and perform numerous mechanical functions. It is used in thousands of applications in the home, the factory, and in transportation systems. Modern solid-state electronics, however, is rapidly replacing many of these electromagnetic applications.

1-7 GENERATION OF ELECTRICITY

When a conductor moves across a magnetic field or when a moving magnetic field crosses a conductor, electrons in the atoms of the conductor are displaced and a voltage or *potential difference* results. If the conductor is in the form of a coil, voltage is produced or induced in each turn of the wire. The sum of the voltages in each turn of wire is the total voltage across the entire coil.

1-8 TRANSFORMATION OF ELECTRICITY

The principles of electromagnetism are used in a device called a *transformer*. In the home a transformer is used to reduce the house voltage of about 115 volts to a low voltage—about 12 volts—to operate a doorbell or chime. In this case the transformer is said to *step down* the voltage. The automobile ignition coil is another example of a transformer. The primary of the ignition coil is connected to a 12-volt supply. The voltage at the output of the transformer (called the *secondary*) is around 20,000 volts. This is enough to produce a strong spark across the points of a spark plug. The ignition coil is an example of a *step-up* transformer.

1-9 ELECTRIC MOTORS

The modern automobile has at least three electric motors as part of its electrical equipment—the starter motor, the heater fan motor, and the windshield wiper motor. These motors contain field pole pieces made of iron around which are wound coils of wire. When current flows through the coils, a strong magnetic field is produced at the faces of the pole pieces. An armature wound with several coils of wire is placed between the pole pieces and is free to rotate in bear-

ings. Current is supplied to the armature winding through brushes that are in contact with a commutator. Both ends of each coil are so connected to bars of the commutator that current can flow from a brush through a commutator bar to the winding, then through the winding to another commutator bar and brush and back to the source.

In flowing through the armature winding, the current produces magnetic fields in the armature. These armature fields are repelled by the magnetic field produced by the field pieces. The condition of repulsion between the two fields produces a turning force, called *torque*, against the armature and causes it to rotate.

1-10 MOTOR CONTROL

Since the action of two opposing magnetic fields produces rotation in a motor, controlling the strengths and polarity of the fields can control the operation of the motor. Basically *control* of a motor involves changing the speed of the motor (including stopping and starting) and the direction of rotation of the motor shaft (either clockwise or counterclockwise). A wide variety of controls are used in modern motor control, but almost all of them do the job by varying the current entering the field coils. In general, speed is controlled by weakening or strengthening the field. Rotation is controlled by the relative *direction* of current flow through the armature and field windings. The starting and stopping of motors are done by switches. Again, a wide variety of switches are used in modern motor control circuits. These range from the familiar on-off switches used in the home to complex interlocked relay and solid-state (electronic) systems used in industrial plants.

1-11 MEASURING INSTRUMENTS

Electrical quantities are measured and observed by instruments and meters. Many types of meters use a magnetic field to operate. Others use the flow of electrons to measure electric quantities. In the past most meters used moving pointers and dials. These are called *analog meters*. Most modern instruments display their numerical values directly. Meters that record quantities directly in numbers are called *digital meters*. The ammeter in your car may use a moving pointer to indicate whether the battery is being charged or discharged. The current flowing through the meter produces a magnetic field, and the pointer moves very much like the shaft of a motor. The information indicated by the pointer is read from a scale across which the pointer passes. Digital meters use solid-state electronic circuitry to monitor the effect of electrons flowing in a circuit. The information obtained is then applied to read-out devices such as light-emitting diodes (LEDs) or liquid crystal displays (LCDs) which display the information directly as numbers.

Common electrical meters are the *ammeter,* used for measuring current; the *voltmeter,* used for measuring voltage; the *ohmmeter,* used for measuring resistance; the *wattmeter,* used for measuring power; and the *kilowatthour meter,* used for measuring energy.

1-12 DIRECT AND ALTERNATING CURRENT

Direct current (dc) is used in such equipment as automobiles, mine locomotives, fork-lift trucks, elevators, metal refining, rolling mills, computers, and communication transmitters and receivers. Direct current is sometimes used in the long-distance transmission of large amounts of high-voltage power though it is converted back to alternating current before being distributed to the power company's customers.

Direct current and voltage do not vary with time. At any instant, if we were to measure the dc voltage and current in a circuit, their values would be the same as they were the instant before and the instant after—provided, of course, that the conditions of the circuit itself haven't changed. In addition, the polarity of the terminals (+ and −) would remain constant. The automobile battery delivers about 12 volts. Unless it is discharging, this value will remain constant at every instant in time. The terminal marked positive (+) will always be positive, and the terminal marked negative (−) will always be negative. These conditions are not true in alternating-current (ac) systems.

Alternating current is used in most industrial, commercial, and residential power systems. It is used where a wide range of voltages are required in the power company's system. For example, the power may be generated at 12,300 volts, but the different customers in the system may require 4160, 480, 208, or 120 volts. In alternating-current systems, transformers are used to change from one value of voltage to another. The transformer does this job with a relatively low loss of power. There are also no moving parts in the transformer to wear out or require constant lubrication. Transformers operate only with alternating current.

1-13 ALTERNATING-CURRENT CHARACTERISTICS

As discussed before, if a wire is placed in a moving or changing magnetic field, a voltage will be generated

in the wire. In an ac system, the voltage and current in the system are constantly changing both in value and in direction. Thus the magnetic field produced by alternating current is constantly changing both in strength and direction. It is this constant change that makes possible the use of transformers.

The changing values of voltage and current form a pattern called a *sine wave.* In this wave the voltage and current values rise from zero to some maximum value, then drop in a similar way to zero. The polarity of the voltage and current then reverses, and the voltage and current values rise to a maximum value. The maximum is the same as that reached previously *but with opposite polarity.* The wave then returns to zero and the pattern of the wave repeats. These waves continue as long as voltage and current are present. The pattern that goes from zero-to-maximum-to-zero-to-opposite-maximum-back-to-zero is called *a cycle.* The number of cycles that occur in 1 second is called the *frequency* of the system. Most electric power companies in the United States produce 60 cycles each second. The unit for frequency is the hertz. Therefore, the most common frequency in the United States is 60 hertz. There are some 50-hertz systems in this country and many outside this country. Transportation systems (electric railways) use 25 hertz. Solid-state (electronic) devices have also made the use of higher frequencies (for example, 4000 hertz) practical in certain power and lighting systems.

1-14 MOTORS AND GENERATORS

Alternating-current equipment does not operate as satisfactorily as direct current equipment in many respects, but the voltage-transformation features of alternating current and its ability to induce voltage in secondary windings more than offset its disadvantages.

The dc motor and the generator are practically the same in every respect of construction. The two principal electrical parts are the *field,* which is part of the stationary frame of the machine, and the *armature,* which is mounted on bearings and rotates freely in the center of the machine.

The ac motor is usually quite different from the ac generator. In both machines, however, the stationary part is called the *stator* and the rotating part is called the *rotor.* The rotating part of most ac motors does not receive its current directly from the line through brushes as in the case of dc motors. The current in the rotor windings of an ac induction motor is induced from the stator windings much like the primary and secondary windings of a transformer. The stator windings are connected to the line often through some starting device or control. Thus the need for brushes is eliminated from many ac motors, and the troubles and maintenance problems connected with brushes are absent.

There are three general classifications of dc motors and generators—series, shunt, and compound. Each has a field and an armature, the difference being in the way the field is connected in relation to the armature. The way in which they are connected influences the operating characteristics of the motors. There are two types of armature windings, *lap* and *wave*. The mechanical difference lies in the way the armature coil leads are connected to the commutator—in a lap winding they connect to adjacent commutator bars, while in a wave winding they connect several bars apart. A wave winding requires higher voltage for operation than a lap. Direct-current motor and generator armatures are identical in every respect. Where used in ac motors, armature windings are identical with dc armature windings.

Alternating-current motors are classified according to the system to which they are connected—that is, single phase or polyphase (generally three-phase). Single-phase motors are usually used at the lower voltages—120 and 208 volts. Since single-phase motors are not self-starting, they are often specified by the means used to start them: split phase, capacitor, repulsion, and shaded pole. The universal motor, a series motor, can be used on either alternating or direct current. This type of motor comes in small sizes and is self-starting. The stator of the three-phase motor is connected either as a star (wye) or delta. The stator windings themselves are identical. The only difference is the way their ends are connected together ·and then connected to the line. Alternating-current motors can be asynchronous or synchronous. In the asynchronous motor the speed changes with load. In the synchronous motor the motor speed is constant. If the load becomes excessive, the motor will simply stop. Alternating-current motors are generally specified as single speed, although various means can be used to give continuous speed control. Multiwinding motors are available where more than one, but usually not more than three, specific speeds are required. In effect these motors are actually two or three separate motors using the same stator and rotor.

1-15 HOME-STUDY TECHNIQUES

This chapter is intended as an introduction and quick overview of some of the main principles and practices involved in the generation and use of electricity. Although these subjects are within the grasp of students interested in them, several factors are important for efficient learning.

Anyone engaged in a course of study, either at home or in organized classes, should plan and follow a systematic study program. To begin with, good health practices are important. Efficient learning is aided considerably by the proper proportioning of work, recreation, and rest.

If you intend to study this textbook on your own outside a regular classroom setting, a schedule is indispensable. A place to study, such as a desk with provisions for containing study materials, and proper lighting should be arranged.

A recommended way of studying from a book is to thumb through it to gain a general knowledge of the nature and scope of the subject matter covered. Then each section or paragraph should first be read casually for a general idea of the information it contains. Next, each section should be studied thoroughly and the words, phrases, or sentences containing important information noted on a separate sheet or lightly underscored in pencil. These identified portions should then be studied until the subject is thoroughly mastered. This method of jotting down or underlining important facts also aids in review work in the future.

Any worthwhile undertaking requires three steps:

· setting a goal
· determination to achieve the goal
· action necessary to achieve the goal

SUMMARY

1. Electricity involves the atomic structure of materials.
2. Atoms contain electrons and protons.
3. Electrons can be displaced from atoms, but protons cannot ordinarily be displaced.
4. If an atom loses one or more of its electrons, it has more protons than electrons and is positively charged.
5. An accumulation of displaced electrons is called negative charge.
6. Movement of charge is called current.
7. Electrons can be displaced from their atoms by the use of magnetism, chemicals, or heat.
8. When electrons move, they create thermal, magnetic, and chemical effects.

9. Most industrial electricity is generated by the relative motion between conductors and magnetic lines of force.

10. Materials whose atoms contain loosely held electrons are conductors of electricity.

11. Materials whose atoms contain electrons that cannot be easily displaced are used as resistors and insulators.

12. Direct current of very high voltage is sometimes used in place of alternating current for long-distance transmission, though most transmission and distribution uses alternating current.

13. Alternating current is widely used because it can be easily transformed to higher or lower voltages and currents required in different applications.

14. In general, dc motors are easier to control than ac motors.

15. There are three types of dc motors or generators—series, shunt, and compound.

16. There are two types of armature windings—lap and wave.

17. There are two types of three-phase motor windings—star (wye) and delta.

QUESTIONS

1-1 Name two parts of an atom.

1-2 If an atom has some of its electrons displaced, is it positively or negatively charged?

1-3 List three ways in which electrons can be displaced from their atoms.

1-4 What is the movement of charged particles called?

1-5 Name three good conductors of electricity.

1-6 Name three poor conductors of electricity.

1-7 What type of field is produced by current in a wire?

1-8 What item in an automobile produces electricity by chemical action?

1-9 List five good electrical insulating materials.

1-10 What type of material can be used to concentrate or strengthen a magnetic field?

1-11 What two actions of a motor are involved in motor control?

1-12 What class of electric meters uses dials and pointers to indicate values?

1-13 What class of electric meters uses a direct numerical display to indicate values?

1-14 Name five common types of electric meters.

1-15 Does the automobile battery produce direct or alternating current?

1-16 What type of current is commonly used in business, industrial, and residential electrical systems?

1-17 What is the varying pattern of alternating voltage and current called?

1-18 What is the moving part of an ac motor called?

1-19 List the three general classes of dc generators.

1-20 What type of ac motor maintains a constant speed despite changes in load?

2

FUNDAMENTAL ELECTRICAL QUANTITIES

Electricity can be thought of as a condition. It is not something that occupies space or can be seen or touched. It is a condition that exists when electrons are displaced from atoms. Energy is required to displace electrons from their atoms; therefore displaced electrons contain energy. This energy can be made to do useful work. Electricity, then, can be defined as a form of energy.

In industrial electricity we are more interested in what electricity does than in what it is. However, to work with electricity safely and effectively, it is helpful to know something about the structure and nature of atoms.

2-1 ATOMS

Everything is composed of atoms. Atoms, in various combinations, form molecules, and matter is composed of molecules. For instance, a molecule of water is composed of 2 hydrogen atoms and 1 oxygen atom. Water is denoted chemically as H_2O since its molecules are 2 parts hydrogen (symbolized by H) and 1 part oxygen (symbolized by O). All known matter can be identified according to its molecular composition and atomic structure.

An *atom* is defined as the smallest part of an element that can exist and still maintain its chemical identification with that element. Infinitesimally small, an atom is a particle of matter composed of a core or nucleus which, among other things, contains one or more protons. *Protons* are described as positively charged particles of matter. An atom also has outside its core one or more negatively charged particles of matter called *electrons*. For an atom to be complete and balanced, or neutral, it must have an equal number of protons and electrons, or positive and negative charges, to neutralize each other. When an atom is complete with all its electrons, it is neutral, and no electrical condition exists.

The electrons of some atoms can be easily displaced from the atoms. When this takes place, a strained or unbalanced condition exists. The displaced electrons can form static electricity, or, if the proper conditions are present, they may produce the electricity we use in our homes, factories, offices, and schools. Thus, electricity can be said to result from an unbalanced atomic condition.

2-2 CHARGES

The words *positive* and *negative* are used to describe the electrical state of protons and electrons. Electrons are said to be negatively charged. They repel each other and are attracted by protons. Negative charge is indicated by the minus (−) sign. Protons are said to be positively charged. They repel each other and attract electrons. Positive charge is indicated by the plus (+) sign. The physical law regarding charges says: *Like charges repel each other; unlike charges attract each other*.

In order to have actual or possible current flow, there must be a charge. A charge is created when there is a surplus of protons or electrons in a given area. This area is charged positively if a surplus of protons exists or negatively if a surplus of electrons exists. If positively charged, the area is subject to a flow of electrons into it. If negatively charged, the area is subject to a flow of electrons out of it. When a charge

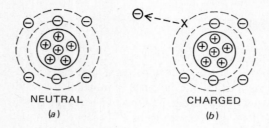

NEUTRAL
(a)

CHARGED
(b)

Fig. 2-1 Simplified diagram of the structure of a carbon atom. (a) The neutral state. (b) A positively charged atom.

exists in one area, an unlike charge exists in another area somewhere because electrons gained or lost in one area are moved to or from another area. Under this condition, an electron flow between areas is possible if a suitable path exists.

Figure 2-1 shows a simplified diagram of the structure of a carbon atom. A neutral state exists in Fig. 2-1(a), since there are six electrons and six protons in the atom. With one electron displaced from position X, as shown in Fig. 2-1(b), the atom now has five electrons and six protons. Since it has more positive protons than negative electrons, the atom is said to be positively charged.

2-3 FUNDAMENTAL QUANTITIES

The four electrical quantities commonly involved in electricity are current, voltage, resistance, and power. Electricity is often likened to water flowing in a pipe. Pressure is necessary to make water flow. Water must overcome the resistance of the walls of the pipe in its flow. As a result of water pressure, a certain quantity of water passes through the pipe in a given period of time. Similar relationships exist with electricity. The movement of charge is called *current*. Current requires pressure to overcome resistance and to flow in an electric circuit, and a certain quantity will flow as long as there is pressure and a suitable path.

There is a direct relationship between the factors of voltage (pressure), current (flow), resistance, and power in a circuit. In order to calculate conditions in a circuit, it is necessary to know how these factors are related and the units in which these factors are measured and expressed.

2-4 CURRENT

When the bundle or mass of displaced electrons (sometimes called "free" electrons) moves along a conductor, we say that *current* is flowing in the con-

ductor. Although the electrons do not actually flow through the conductor in the same sense as water flows through a pipe, they do transmit their *energy* through the conductor. In the field of industrial electricity current is often referred to as *amperage*.

2-5 THE AMPERE

Current is measured in *amperes* (abbreviated A). The term ampere is used in honor of André Marie Ampère, a French physicist who pioneered in the science of electricity. The letter symbol for current in amperes is I. Current is measured with an instrument called an *ammeter*.

When a charge of 6,280,000,000,000 (six trillion, two hundred eighty billion!) electrons passes a given point in a conductor in one second, we say that one ampere is "flowing." For smaller units of measurement the terms *milliampere* and *microampere* are used. A milliampere is one-thousandth (1/1000) of an ampere and a microampere is one-millionth (1/1,000,000) of an ampere. For a large unit of measurement kiloampere (kA) is sometimes used; it is 1000 A.

2-6 VOLTAGE

An atom that has lost some of its electrons has a surplus of protons and is positively charged. Protons are locked in the nucleus or core and cannot be displaced by ordinary means. Having a surplus of protons, the atom is in need of negative electrons to neutralize the positive charge. This results in a force of attraction for negatively charged electrons to neutralize the strain. Because other nearby electrons exert a force of repulsion against them, a two-way force is exerted on the displaced electrons. They are repelled toward the positively charged atom and are also attracted by it. The total force exerted on the electron is called *electromotive force* or simply *voltage*.

2-7 THE VOLT

Voltage is measured in *volts* (abbreviated V). The term volt is used in honor of Alessandro Volta, an Italian physicist who invented the electric cell (an ancestor to the common flashlight dry cell). The letter symbol for voltage is V. Voltage is also referred to as potential, potential difference (pd), or electromotive force (emf). Voltage is measured with a voltmeter. Smaller units of measurement are the millivolt (mV), which is one-thousandth (1/1000) of a volt, and microvolt (μV), which is one-millionth (1/1,000,000) of a volt. A larger unit of measurement is the kilovolt (kV), which is 1000 V.

2-8 RESISTANCE

The unit of measurement of electric resistance is the *ohm*. To avoid confusion with the numeral zero, ohm is abbreviated by the Greek capital letter omega (Ω). It is named in honor of Georg Ohm, a German physicist who originated Ohm's law. The letter symbol for resistance is R. Resistance may be measured with an ohmmeter. An ohm is that amount of resistance that will allow one ampere to flow when connected across one volt. A smaller unit, the microhm, is one-millionth (1/1,000,000) of an ohm. Larger units are the megohm, equal to one million ohms (1,000,000 Ω), and the kilohm, which is equal to one thousand ohms (1000 Ω).

2-9 ELECTRIC POWER

Electric power is measured in *watts* (abbreviated W). The name watt is used in honor of James Watt, Scottish engineer and inventor, who originated the term *horsepower* and defined it as a unit of mechanical power. The letter symbol for electric power is P. Power may be measured with a wattmeter. A larger unit of measurement of electric power is the kilowatt, abbreviated kW, which is 1000 W.

2-10 ELECTRIC ENERGY

Watts are used to measure the rate at which electric power is being used in a given amount of time. This measurement does not indicate how much electric energy has been used. Time is a factor that must be considered in determining the amount of energy used during a given period. Usually this is done by multiplying watts by hours. The result is watthours, abbre-

viated Wh. If power is measured in kilowatts and multiplied by hours, the result is *kilowatthours*, abbreviated kWh. The unit kilowatthour is used to measure a definite amount of electric energy. It is the unit used for calculating most electric bills. Users of electricity pay a flat rate or a sliding-scale rate based on a certain charge per kilowatthour of energy they have used.

2-11 OHM'S LAW

Georg Ohm, in the early days of electrical discoveries, suspected there was a direct relationship between voltage, resistance, and current in a circuit. He proved by experiments that such a relationship did exist. He found that for a given resistance, if he divided the value of voltage across the resistance by the value of current flowing through the resistance at that time, the answer would be the same *no matter how he changed the voltage*. Simply stated, he found

$$R = \frac{V}{I}$$

This can also be written as

$$V = I \times R$$

where V = voltage
I = current
R = resistance

This relationship is expressed in Ohm's law. It is simply a statement of relationship between volts, amperes, and ohms in an electric circuit.

Ohm's law can easily be memorized by the use of a simple diagram as shown in Fig. 2-2(a). The units are shown in the form of a formula. Assume a problem in which the current I is 10 A and the resistance R is 5 Ω and we want to know the voltage. Cover the symbol for voltage V with a finger as in Fig. 2-2(b), and the multiplication of I and R is indicated by the remaining symbols. In this case, I is 10 and R is 5; thus

$$V = 10 \text{ amperes (A)} \times 5 \text{ ohms } (\Omega) = 50 \text{ volts (V)}$$

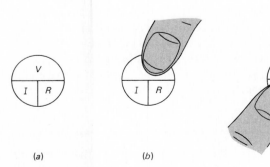

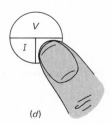

Fig. 2-2 A practical way to memorize Ohm's law.

(a)   (b)   (c)   (d)

The value of the covered symbol V is 50, and the answer is 50 V.

If we want to know the current in a circuit with 50 V and 5 Ω, cover the I as in Fig. 2-2(c), and the remainder of the formula indicates that V is to be divided by R. (The horizontal line of a formula means to divide the value above the line by the value below the line.) Thus the voltage V (50 V) is divided by the resistance R (5 Ω) or

$$I = 50 \div 5 = 10 \text{ A}$$

The value of I, or the current of the circuit, is therefore 10 A.

If we want to know the resistance in a circuit with 50 V and 10 A, cover R, as in Fig. 2-2(d), and V divided by I is indicated; thus

$$R = 50 \div 10 = 5 \ \Omega$$

Once this little diagram is memorized, it can minimize confusion in solving problems involving Ohm's law.

To show the relationship between the various factors in an electric circuit and how Ohm's law is used in determining unknown values, a simple electric circuit (Fig. 2-3) will be examined. In this circuit a 100-V dc generator is connected by 1000 ft of No. 10 wire to a 20-Ω heating resistor located 500 ft from the generator.

Problem 1

How many amperes will flow through the heating resistor, neglecting the resistance of the line wires?

Solution

By use of Ohm's law

$$I = \frac{V}{R}$$

where V = voltage (volts)
I = current (amperes)
R = resistance (ohms)

In this case $V = 100$ V and $R = 20$ Ω.

$$I = \frac{100 \text{ V}}{20 \ \Omega} = 5 \text{ A}$$

Problem 2

A meter placed in the heating resistor circuit shows that 5 A is flowing through the resistor. If the resistor is 20 Ω, what is the voltage across it?

Solution

The voltage across a circuit, or *any part of a circuit,* can be found by Ohm's law

$$V = I \times R$$

In this problem $I = 5$ A and $R = 20$ Ω.

$$V = 5 \text{ A} \times 20 \ \Omega$$
$$= 100 \text{ V}$$

The voltage across the resistor is 100 V.

Problem 3

The voltage across a heating resistor is measured at 100 V at the same time that 5 A is flowing through the resistor. What is the resistance of the heating resistor?

Solution

By Ohm's law

$$R = \frac{V}{I}$$

In this problem $V = 100$ V and $I = 5$ A.

$$R = \frac{100 \text{ V}}{5 \text{ A}} = 20 \ \Omega$$

The resistance of the heating resistor is 20 Ω.

2-12 VOLTAGE DROP

In any electric circuit the total supply voltage is distributed throughout the circuit. The voltage found by multiplying the current through a resistance by the value of the resistance itself is actually the voltage applied across that resistor. It is also called the *voltage drop* across the resistance. Obviously in this case the applied voltage equals the voltage drop.

Since voltage is equal to $I \times R$, voltage drop is often referred to as *IR drop*. Every part of a circuit that has resistance and carries current produces an *IR* drop.

In many circuit calculations we can ignore the relatively low resistance of the line wires or conductors. These wires do have resistance, however, and they will affect the amount of current that flows in a circuit. They will also affect the voltage across equipment connected to the circuit.

Problem 4

In Fig. 2-3 the current is actually measured and found to be 4.76 A instead of 5 A as previously

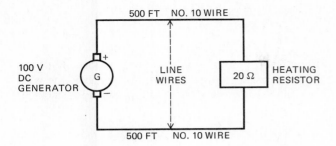

Fig. 2-3 A simple electric circuit.

calculated. What would account for this differ-ence? Would this difference affect any voltages in the circuit?

Solution

Ohm's law for resistance is

$$R = \frac{V}{I}$$

We are given that the generator voltage is 100 V and $I = 4.76$ A.
Therefore

$$R = \frac{100}{4.76} = 21.0 \ \Omega$$

But the heating resistor is only 20 Ω. Therefore the resistance of the line wires must be 1 Ω. (Pre-viously we omitted this small resistance in our calculations.) With 4.76 A flowing through the lines, the voltage *drop* in the lines, also called the *line drop*, is

$$V = I \times R$$
$$= 4.76 \times 1 = 4.76 \ V$$

The total voltage applied to the circuit is 100 V. With 4.76 V applied (or dropped) across the line wires, only 100 − 4.76 or 95.24 V is available to be applied across the heating resistor.

If the manufacturer of the resistor had specified that to operate the resistor properly no less then 97.5 V must be applied to it, the circuit of Fig. 2-3 would not be suitable. The only ways to solve this problem would be to increase the supply voltage slightly (not usually a practical solution) or to choose a wire size with less resistance. If the wire had less resistance, the *IR* drop of the line would be lower and more voltage would be available across the heating resistor.

Voltage drop in any part of a circuit is in proportion to the resistance of that part. In Fig. 2-4 three electri-cal loads (resistors) are connected to a voltage source. Notice that in this type of circuit current has only one path to follow from the + terminal of the generator to the − terminal.

The total resistance in this circuit is the sum of the resistances around the circuit. In the circuit of Fig. 2-4 the total resistance is $R1 + R2 + R3$ or $2 + 3 + 5 = 10 \ \Omega$.

Since the voltage drops are in proportion to the re-sistance, we need to know what part of the total resist-ance each of the load resistors makes up.

The total resistance of the circuit is 10 Ω. Resistor $R1$ is 2 Ω or 2/10 of the entire resistance. Resistor $R2$ is 3 Ω or 3/10 of the entire resistance. Resistor $R3$ is 5 Ω or 5/10 of the entire resistance.

The total voltage across the three resistors is the applied voltage, 100 V. The voltage drop across each resistor is proportional to its resistance.

$R1$ is 2/10 of the total resistance. Therefore, $R1$ has a voltage drop of

$$\frac{2}{10} \times 100 = 20 \ V$$

$R2$ is 3/10 of the total resistance. Therefore, $R2$ has a voltage drop of

$$\frac{3}{10} \times 100 = 30 \ V$$

$R3$ is 5/10 of the total resistance. Therefore, $R3$ has a voltage drop of

$$\frac{5}{10} \times 100 = 50 \ V$$

The total voltage drops must equal the applied voltage

$$20 \ V + 30 \ V + 50 \ V = 100 \ V$$

which checks with the generator voltage of 100 V.

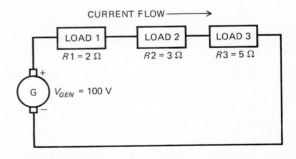

Fig. 2-4 Three resistors connected to a 100-V gen-erator.

2-13 VOLTAGE DROP PRACTICE

Most large electrical systems have voltage drop problems that must be considered in planning a wiring job. The cost of voltage drop must be weighed against the cost of wire. Selection of wire sizes on a job is governed by several factors. There are standards to go by, but each job is an individual case in itself, some falling within a standard and some not.

Because voltage that is lost in the line reduces the voltage available to the operating equipment, equipment must sometimes operate on *undervoltage*. Undervoltage is the word used to describe a condition where the voltage applied to a piece of equipment is less than that needed to make the equipment operate properly. Motors should be operated as near to their rated (or nameplate) voltage as practicable. Motors generally are capable of operating on voltages varying 10 percent under or over the amount shown on the motor nameplate, but any change from the nameplate rating affects the motor's efficiency. In general practice it is recommended that the line drop for motor conductors should never exceed 5 percent of the line voltage. Voltage drop in lines supplying lighting loads should be held to less than 3 percent of the voltage required by the lamps used. Good design practice will allow not over 1 percent voltage drop in main conductors and not over 1 percent drop in circuits supplying outlets and individual equipment. This limits voltage drop to equipment to 2 percent of the supply voltage.

The National Electrical Code is often used as a guide to minimums and maximums of electrical work. The 1984 issue of the Code makes the following statements concerning voltage drop:

> Conductors for branch circuits . . . sized to prevent a voltage drop exceeding 3 percent at the farthest outlet of power, heating, and lighting loads, or combinations of such loads, and where the maximum total voltage drop on both feeders and branch circuits to the farthest outlet does not exceed 5 percent, will provide reasonable efficiency of operation. (Art. 210-19).
>
> Conductors for feeders . . . sized to prevent a voltage drop exceeding 3 percent at the farthest outlet of power, heating, and lighting loads, or combinations of such loads and where the maximum total voltage drop on both feeders and branch circuits to the farthest outlet does not exceed 5 percent, will provide reasonable efficiency of operation. (Art. 215-2c).

The Code is described in Sec. 3-10 and cited freely throughout the balance of this textbook.

2-14 ELECTRIC POWER FORMULAS

Electric power can be calculated by multiplying voltage by current

$$P = V \times I$$

where P = power, W
V = voltage, V
I = current, A

If only the voltage and resistance of the circuit are known, power can be calculated by squaring the voltage (that is, $V \times V$) and dividing by the resistance.

$$P = \frac{V \times V}{R} = \frac{V^2}{R}$$

where R equals resistance in ohms.

If only the current and the resistance are known, power can be calculated by multiplying the square of the current (that is, $I \times I$) by the resistance.

$$P = I \times I \times R = I^2R$$

If voltage, current, and resistance are known, then any one of the above formulas can be used. The answers will be exactly the same.

If the amount of power used by a piece of equipment is multiplied by the time the equipment is used, the amount of electric energy used can be found. Most utilities base their rates on kilowatthours of energy. They will charge so many cents per kilowatthour of electricity.

If power is given in watts, it can be converted to kilowatts merely by dividing watts by 1000.

$$\text{Kilowatts} = \frac{\text{watts}}{1000}$$

To find the cost of energy, multiply kilowatthours by the cost per kilowatthour.

Problem 5

What is the monthly cost of operating a 20-Ω heater, drawing 6 A from a 120-V line, if the heater is used 8 hours (h) a day, 20 days a month, and the utility charges 10 cents per kilowatthour?

Solution

Any one of the power formulas can be used to find the power consumed by the heater.

$$P = V \times I = 120 \times 6 = 720 \text{ W}$$
$$= \frac{V^2}{R} = \frac{120 \times 120}{20} = 720 \text{ W}$$
$$= I^2R = 6 \times 6 \times 20 = 720 \text{ W}$$

To change to kilowatts

$$\text{Kilowatts} = \frac{\text{watts}}{1000} = \frac{720}{1000} = 0.72\text{kW}$$

The heater is used 8 h a day for 20 days or 8 × 20 = 160 h a month.

Kilowatthours = 0.72 kW × 160 h = 115.2 kWh

Since each kilowatthour costs 10 cents, the total charge for 115.2 kWh is

Cost = kilowatthours × cost per kilowatthour
= 115.2 × 0.10 = $11.52

Thus, it will cost $11.52 a month to operate the heater in this problem.

Note that the cost per kilowatthour varies widely throughout the country and depends on many factors including the type of fuel used to produce electricity, the complexity of the electrical system, and the amount of energy used by the customer. In addition, the amount the customer is billed usually includes a fuel adjustment charge and taxes.

2-15 POWER LOSS

Whenever current passes through a resistance, power is consumed. Often this power is consumed in the form of heat. The heat could be useful as in a space heater or hot water system or in an oven. The heat could also be wasted as in an incandescent lamp or the coils of a motor. In some cases reducing the heat produced even requires additional energy. Power consumed in a conductor by current passing through that conductor is usually wasted. It is power lost. Since $P = I^2R$, this power loss is often called I^2R loss.

Power is also lost by resistance in a circuit where it is not desired, as in poor wire connections, loose fuse clips, and rough contact surfaces on switches. Regardless of the form of resistance, heat will be produced when electricity flows through it. Heating at

any of these points indicates resistance that should not be there. If a poor connection is suspected in such places, an *IR* drop test can be made with a low-reading voltmeter. Touch the voltmeter test leads on each side of the suspected connection while it is under load, and if there is a voltage drop the voltmeter will show a certain number of volts. With the power off the line, an ohmmeter can be used to measure the actual resistance in ohms.

Figure 2-5 illustrates a knife switch and fuse assembly containing a fuse of the renewable-link type. In this assembly, 10 possible locations for poor contact or loose connections are shown and numbered. Trouble is no more likely to occur in an assembly of this type than in any other part of an electric circuit, but this assembly is used to illustrate the numerous times current travels through a connection from one part to another in a circuit.

Tracing current flow from left to right in the illustration, we see that the current has to transfer from the conductor to the connecting lug through a pressure connection 1; the lug is connected to the live side of the switch by a cap screw 2; the cap screw is an integral part of the switch clip which makes sliding contact with the switch blade 3; the switch blade is hinged on a rivet joint 4; the dead end of the switch is connected to a fuse clip which holds the ferrule end of a link fuse by compression 5; the fuse link is connected to the ferrule by compression also 6; the connections at 7 and 8 are the same as those at 6 and 5, respectively; the second ferrule is an integral part of the cap screw assembly 9 that holds the cable lug; the cable lug is connected to the cable by pressure (crimping) 10. Trouble is possible from a poor connection at any one of these 10 points. A low-reading voltmeter is shown with the test prods contacting the line wires at each end of the assembly. If no reading on the voltmeter is obtained here, with the equipment under heavy load, it can be considered in good condition.

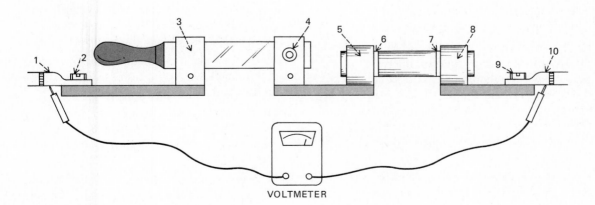

Fig. 2-5 Ten possible places of poor connection that produce power loss.

If a reading is obtained, trouble exists somewhere in the assembly and can be found by further testing between the numbered sections. For example, if fuse clip 5 is loose, a reading will be obtained between ferrule cap 6 and clip 5.

In the early stages of trouble from loose connections, erratic operation of equipment will be experienced, possibly for some time, before the condition develops to the point where trouble is apparent. A preventive measure is to check all connections periodically and keep them tight.

Heating due to I^2R can cause numerous troubles in a circuit. If a faulty fuseholder causes heating, it can cause fuses to open or "blow" with only a normal load on the line. Poor connection of cartridge fuse clips also leads to premature opening of fuses, charring of the fuse body, and lowering temper or "springiness" of the fuse clips. Such a condition can be noticed in the early stages by excessive heat and discoloration of metal parts. If the temper of the clips has not been destroyed, this condition can be remedied by polishing the contact areas of the clips and bending them to restore sufficient tension or pressure on the ferrule of the fuse for good contact.

The I^2R heating also causes premature opening of thermally operated control devices. Loose connections anywhere in a system can cause static in radios and interference in television pictures. The use of an electrical system normally causes constant warming and cooling of electric circuits, which cause expansion and contraction of the circuit metals, and this in turn leads to loosening of connections. Accordingly, all electrical systems should be checked periodically for loose connections. Systems containing aluminum conductors must be checked more often since aluminum connections have a tendency to "creep," or loosen, more than copper, especially when proper connectors are not used.

SUMMARY

1. Current is measured in amperes. The symbol for current is I. The abbreviation for ampere is A. Current is sometimes referred to as amperage.

2. Voltage is measured in volts. The abbreviation for volts is the letter V. The symbol for voltage is V. Voltage is also called electromotive force, emf, potential, potential difference, and pd.

3. Resistance is measured in ohms. The Greek capital letter omega, Ω, is used as an abbreviation for ohms. The symbol for resistance is R.

4. Electric power is measured in watts. The abbreviation for watt is W. The symbol for power is P.

5. One volt will cause one ampere to flow through a resistance of one ohm. The power consumed under those conditions will be one watt.

6. Electric energy is measured in watthours, abbreviated Wh, or kilowatthours, abbreviated kWh.

7. Prefixes to electrical units of measurement change the values of the units as follows: micro-, one-millionth; milli-, one-thousandth; kilo-, one thousand times; mega-, one million times. For example, a microvolt is one-millionth of a volt; a millivolt is one-thousandth of a volt; a kilovolt is one thousand volts; and a megavolt is one million volts.

8. Ohm's law is a statement of the relationship between voltage, current, and resistance in a circuit.

9. Ohm's law stated simply is

Voltage (volts) equals current (amperes) times resistance (ohms)

$$V = I \times R$$

or current (amperes) equals voltage (volts) divided by resistance (ohms)

$$I = \frac{V}{R}$$

or resistance (ohms) equals voltage (volts) divided by current (amperes)

$$R = \frac{V}{I}$$

10. Voltage drop is the voltage developed across a load or resistance due to current flowing through that part. The applied voltage across a load is always equal to the voltage drop across the load.

11. Because all conductors contain some resistance, when current flows, voltage drop is always present in a circuit.

12. To find the voltage drop in a circuit or part of a circuit, multiply the amperes flowing by the resistance in ohms. Voltage drop is also known as IR drop.

13. Good design practice in planning circuits requires voltage drop in branch circuits or feeders be held to 1 percent of the system voltage or less.

14. Motors are supposed to operate fairly satisfactorily with a voltage drop up to 10 percent of their rated voltage. Any deviation from rated voltage lowers motor efficiency.

15. Actual watts dissipated in resistance can be determined by squaring the amperes flowing and multiplying by the resistance in ohms. Thus any resistance—a heater, loose connection, poor contact, motor winding—or any part of an electric circuit with resistance produces heat in proportion to the resistance times the square of the amperes. That is, $P = I^2R$.

16. Loose connections and poor contacts are the principal causes of trouble in electrical equipment, causing voltage drop, heat, and erratic operation.

17. Constant warming and cooling of electrical equipment in the cycles of operation expand and contract metal parts, eventually causing loose connections to develop.

18. All electrical connections should be checked and tightened periodically, especially connections in enclosed control panels.

QUESTIONS

2-1 When is an atom neutral?

2-2 What particle has a negative charge?

2-3 What particle has a positive charge?

2-4 True or false: Like charges attract; unlike charges repel.

2-5 Name the three basic electrical quantities used in Ohm's law.

2-6 If current and voltage are known, what formula is used to find R?

2-7 If current and resistance are known, what formula is used to find V?

2-8 If voltage and resistance are known, what formula is used to find I?

2-9 What is the movement of charge called?

2-10 What formula is used to find power when voltage and current are known?

2-11 What electrical unit is used to measure electric energy usage in the home?

2-12 True or false: 1 million V is the same as 1 microvolt.

2-13 In the circuit of Fig 2-6: (*a*) find the current through the load; (*b*) find the power used by

the load; (*c*) if the circuit is on for 8 h, what is the energy consumption in terms of kilowatt-hours?

2-14 Figure 2-7 shows a flashlight circuit. Because of corrosion high resistance occurs at the points shown. What voltage will be applied across the lamp?

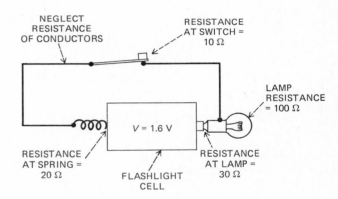

Fig. 2-7 Flashlight circuit for question 2-14.

2-15 Which of the following circuits will carry the most current? (*A*) $V = 120$ V, $R = 120$ Ω. (*B*) $V = 120$ V, $P = 120$ W. (*C*) $R = 120$ Ω, $P = 120$ W.

2-16 What is another name for voltage drop?

2-17 True or false: In general, line drop should be kept as low as possible. Explain your answer.

2-18 What name is given to the voltage drop that occurs across the conductors connected to a load?

2-19 Three loads are connected as shown in Fig. 2-8. Across which load will the greatest voltage drop occur?

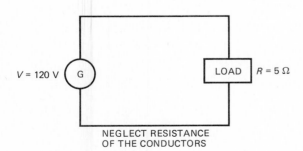

Fig. 2-6 Circuit for question 2-13.

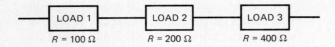

Fig. 2-8 Three loads for questions 2-19 and 2-20.

2-20 Referring to Fig. 2-8, which load will have twice the voltage drop of Load 2?

2-21 Will the current in each load in question 20 be different? Explain your answer.

2-22 If the supply voltage serving a lighting load is 120 V, what is the lowest voltage that should be connected across the load itself?

2-23 If a 480-V supply voltage is connected to a motor, what is the largest voltage drop that should occur in the lines connected to the motor?

2-24 The voltage across an electric space heater is 120 V. If 10 A is being drawn by the heater, what is the power being used?

2-25 What would it cost to operate the heater in question 2-24 for 5 h if the cost of electricity is 13 cents per kilowatthour?

3
ELECTRIC CONDUCTORS

Because electricians are constantly working with various kinds of electric conductors, they should "know their wires." The greater part of their on-the-job calculations involves wire sizes and capacities as well as a thorough knowledge of elementary circuit theory. Hence, the properties of wire conductors constitute an important part of their technical knowledge.

3-1 COPPER CONDUCTORS

Copper is used for conductors in practically all electric equipment. It contains electrons that can easily be moved and therefore affords little resistance to electron flow. Next to silver, it is the best conductor of electricity. Copper has many other characteristics that make it popular for electrical work. It bends easily, has good mechanical strength and a high degree of resistance to corrosion, and its size is not affected greatly by changes in temperature (we say it has a *low coefficient of expansion*). Another of its favorable characteristics is its ability to "take," or hold, solder, which is an alloy of tin and lead. A good soldered connection is one of the best electrical connections that can be made, though it is rarely used today in electrical power and lighting installations.

3-2 ANNEALED COPPER

Copper hardens when it is rolled, drawn, or forged, and resistance increases slightly (about 2.5 percent) when it hardens. The tensile strength (that is, its resistance to breaking due to stretching) is increased, and in the hardened state it is more difficult to bend. Hardness due to manufacture is known as *work-hardening,* and this can be reduced by annealing. The annealing process involves heating the copper and then allowing it to cool very slowly. This relieves the internal stresses in the metal. Nearly all wire used for winding electrical equipment is *soft-drawn* wire, that is, softened by annealing to make it easier to wind in coils. Construction wire, or *hard-drawn* wire, is usually left in a hardened condition, although some is annealed before the last few drawings and is classed as *semihard* wire. Hard-drawn or semihard wire will stretch less and is better for construction purposes.

Round copper wire is used in nearly all construction and winding work. Round wire is easier to manufacture than square or rectangular wire and will withstand more abuse without injury to insulation, which would later result in shorts or grounds.

3-3 STANDARD GAGE SIZES

The standard gage in the United States for measuring round copper wire is the *American Wire Gage* (AWG). This gage is the same as the *Brown and Sharpe* (B & S) gage used by machinists.

The AWG numbers run from 0000 (called "four-ought," or 4/0), which is the largest size, to No. 50, the smallest. The wire size decreases as the numbers increase; that is, the lower the number, the larger the

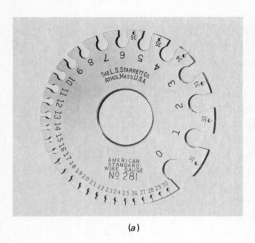
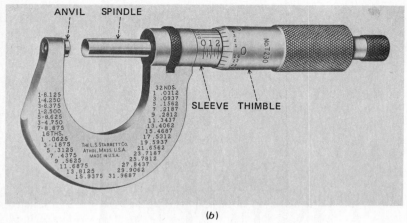

(a)   (b)

Fig. 3-1 Instruments used for measuring wire sizes. (a) American standard wire gage. The numbers indicate the wire size that will slide snuggly through the hole. (b) Micrometer. The markings (calibrations) on the thimble and sleeve indicate the distance between the spindle and anvil in decimal fractions of an inch. (*The L. S. Starrett Company.*)

diameter of the wire. Wire size can be measured using a standard gage or a micrometer. Figure 3-1(a) is an American Wire Gage showing the front side of the gage containing numbered slots for copper wire sizes. Figure 3-1(b) is a micrometer caliper with decimal equivalents of fractions of an inch engraved on its surface.

Wires or cables larger than 4/0 are measured in circular mils (CM). The circular mil is a unit of area. Machinists use the mil to indicate 0.001 inch (in.). One circular mil is the area of a circle whose diameter is one mil. The letters k and M are used as abbreviations for 1000.

The next larger wire size than 4/0 is 250,000 CM. This is commonly written as 250 MCM, though some wire tables and books may use the notation 250 kCM. Both notations mean the same, 250,000 CM.

When two or more wires are contained in a cable assembly, the size is referred to by naming the size of the wire first and then the number of wires in the assembly. For example, a cable containing three No. 10 conductors would be referred to as a 10-3 or 10/3 cable.

3-4 CIRCULAR AND SQUARE MILS

Wire calculations are based on the cross-sectional area of the wire measured in circular mils and square mils (SM). The cross-sectional area of a conductor is the area of the end of a conductor when cut at right angles to its length. The square-mil area of a square or rectangular conductor is the product of the two sides of the conductor measured in mils. The circular mil area

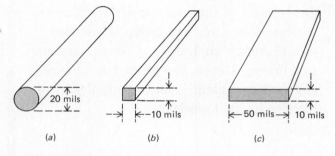

(a)   (b)   (c)

Fig. 3-2 Circular mils and square mils.

of a round wire is found by squaring its diameter in mils. Figure 3-2(a) illustrates a No. 24 wire with a diameter of 20 mils or a cross-sectional area of 400 CM. Figure 3-2(b) illustrates a square conductor with dimensions of 10 mils. It has a cross-sectional area of 100 SM. The 10 × 50 mils rectangular conductor in Figure 3-2(c) has a cross-sectional area of 500 SM.

3-5 WIRE MEASUREMENTS

Square or rectangular conductors are measured with a micrometer caliper, or "mike" for short. The micrometer is marked, or calibrated, in thousandths of an inch. Since a mil is one-thousandth of an inch, a micrometer can be read directly in mils. Round wire can also be measured with a micrometer instead of a gage. The reading in mils is then compared with a wire chart to get the gage size. Use of a gage is sometimes difficult because wires are not always drawn to an exact size. Plus and minus tolerances are allowed in wire drawing because of wear of the dies. A tolerance of plus or minus of about 1 percent is generally allowed for sizes 4/0 through 29, and 0.0001 in. for smaller wires. A new die starts with a minus tolerance and is used until it wears to the plus tolerance. The result of this is variation in sizes of wires of the same gage number. Thus one wire may crowd a gage slot when it should go into that slot. If wire is measured with a micrometer, its size can be determined by comparing the reading with a wire chart or the diameter of the wire shown on the back of the gage under the slot size in most wire gages.

For accurate measurement it is absolutely essential that the wire be clean. All insulation, such as the synthetic enamels on magnet wire, must be removed. When using a sharp edge or sandpaper, take care to avoid reducing the actual diameter of the wire.

Stranded wire is made by wrapping many small bare uninsulated wires together to form a single conductor. The size of a stranded wire is that of a solid wire having the same circular-mil area. To determine the size of a standard wire, measure the wire size of a single strand. The area of the strand multiplied by the number of strands making up the wire will be the total circular mil area of the stranded wire. By referring to a wire table, the corresponding solid wire size having the same area can be determined.

3-6 READING A MICROMETER

Micrometers are used in numerous places in electrical work such as measuring round, rectangular, and square wire and insulation thickness, shafts, and bearings. A micrometer, or mike, is easy to read when the principle of operation is understood. See Fig. 3-1(b) for names of parts of a micrometer caliper. Lines and numbers on the sleeve and thimble indicate in thousandths of an inch or mils the distance of the opening between the anvil and spindle where measurements are made. The values of the lines are illustrated on Figure 3-3(a). The long numbered lines on the sleeve above the starting line indicate 100 thousandths, or hundreds of mils. The shorter lines between the numbered lines indicate 25 thousandths, or 25 mils. The numbered lines on the thimble indicate thousandths, or 1 mil. In measuring, the values of the three readings are added for the complete total.

In use, the material to be measured is placed between the anvil and the spindle, and the thimble is turned until the anvil and the spindle are in light contact with the material. Then the values of the 100-mil lines, the 25-mil lines, and the 1-mil lines are added for the total thickness of the material in mils, or thousandths of an inch. All visible lines on the sleeve, and the thimble line that coincides with the starting line, are read.

In Fig. 3-3(b), the reading is 0.000. The sleeve lines are all zero at the starting line and the thimble line is zero. In Fig. 3-3(c), the thimble has been moved to 2, and with the sleeve lines zero, the reading is 0.002 in. In Fig. 3-3(d), the thimble has been moved to its zero line, and one 25-mil line is visible, so the reading is zero 100-mil lines, one 25-mil line, and zero on the thimble, making a total of 25 mils, or 0.025 in. In Fig. 3-3(e), the thimble has been ad-

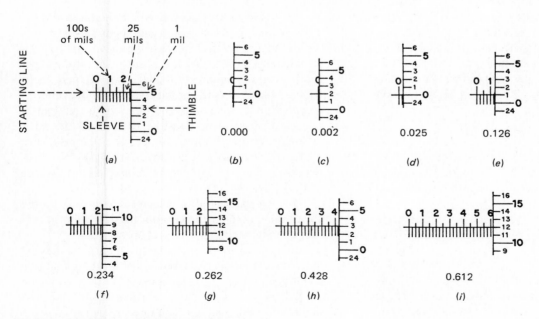

Fig. 3-3 Examples of micrometer readings.

vanced to where one 100-mil line and one 25-mil line are visible, and to the 1-mil line on the thimble, for a total of 126 mils, or 0.126 in. In Fig. 3-3(*f*), two 100-mil lines and one 25-mil line plus 9 mils on the thimble total 0.234 in. In Fig. 3-3(*g*), two 100-mil lines and two 25-mil lines plus 12-mils total 0.262 in. In Fig. 3-3(*h*), four 100-mil lines and one 25-mil line plus 3 mils on the thimble total 0.428 in. In Fig. 3-3(*i*), six 100-mil lines and no 25-mil line plus 12 mils on the thimble total 0.612 in. The reading shown in Fig. 3-3(*a*) is 0.255 in.

3-7 WIRES PARALLEL

When a job requires a wire that is too large to work easily, two or more wires with an equivalent cross-sectional area can be substituted. If a motor winding requires the cross-sectional area of copper of a No. 10 wire, and No. 10 is too stiff to work easily, two No. 12 or four No. 16 wires can be used. By connecting the ends together, the parallel wires contain an approximate equivalent cross-sectional area. This is frequently done in motor winding. When wires are paralleled in this manner, the winding is said to be two-in-hand or four-in-hand, as the case may be. In some cases six to eight wires are wound in-hand instead of using one larger wire. This is also referred to as *wires parallel* in a winding.

3-8 WIRE AREA CONVERSION

Occasionally it is necessary to change from round to square or rectangular wire, such as in making a change in types of wire in rewinding motors and transformers or in determining current-carrying capacities.

Square mils can be converted to circular mils by dividing by 0.7854. Circular mils can be converted to square mils by multiplying by 0.7854. Stated simply

$$SM \div 0.7854 = CM$$
$$CM \times 0.7854 = SM$$

In most cases, it is accurate enough to use 0.79. Figure 3-2(*a*) with 400 CM contains 400 × 0.79 or 314 SM. Figure 3-2(*c*) with 500 square mils contains 633 CM.

3-9 WIRE DATA

Table 3-1 is a wire data table showing American Wire Gage, or AWG, sizes, cross sectional area in circular mils, ohms per 1000 ft at 75°C (167°F) and other properties of copper and aluminum wire for each size from 18 through 2000 MCM. The table is adapted from Table 8 of the 1984 National Electrical Code. This table contains some valuable information regarding wire. For example, No. 10 wire is 0.102 in. or 102 mils in diameter, has 10,380 CM, and 1.21-Ω resistance per 1000 ft. For all practical purposes, we can say No. 10 has a diameter of 100 mils, 10,000 CM area, and 1-Ω resistance per 1000 ft. This should be memorized. For every three gage numbers down or up the wire sizes are either doubled or halved. By using No. 10 wire as a reference point, quick mental calculations of other wire sizes are possible.

Three numbers down from No. 10 is No. 7. Since this is not a common building wire size it is not given in the table. However, by the previous rule No. 7 has twice the circular-mil area and one-half the resistance in ohms of No. 10. Therefore, No. 7 is 20,000 CM and has 0.5-Ω resistance per 1000 ft. Three numbers up from No. 10 is No. 13. Again, this is not a common building wire so that it is omitted from the table. But it has one-half the circular-mil area and twice the resistance in ohms of No. 10. Three numbers up from No. 13 is No. 16. This is a common building wire size and is listed in Table 3-1. It has one-half the circular mils and twice the resistance of No. 13, or one-fourth the circular mils and four times the resistance of No. 10. If the data of No. 10 are remembered, the data of other sizes can be easily calculated from memory. Thus No. 16 has an area of 2500 CM and 4.0 Ω per 1000 ft resistance (actually 4 × 1.21 = 4.84 Ω). Compare these shortcut figures with the actual table values of 2580 CM and 4.89 Ω.

A decrease in wire size of 10 numbers reduces the cross-sectional area to about one-tenth and increases the resistance about 10 times. Thus No. 10 has approximately 10,000 CM and about 1-Ω resistance per 1000 ft, and 10 numbers higher, No. 20, has about 1000 CM and 10.0-Ω resistance per 1000 ft, which is a change in circular mils and ohms by a multiple of 10. An increase in wire size of 10 numbers increases the diameter by 10 times and decreases the resistance 10 times. Ten sizes larger than No. 10 is No. 1/0, which is 105,600 CM in diameter and has a resistance of 0.122 Ω per 1000 ft.

Table 3-1 values for resistance are valid only for the conditions noted in the column headings for stranding, coating, and especially temperature.

The column marked "Quantity" under Stranding indicates the number of strands that make up that size conductor. The "1" indicates solid or unstranded wire. Note that sizes No. 6 and larger are not commonly available as solid conductors.

TABLE 3-1
Conductor Properties

Size AWG/ MCM	Area Cir. Mils	Stranding		Overall		Copper		Aluminum
		Quantity	Diam. In.	Diam. In.	Area In.²	Uncoated ohm/MFT	Coated ohm/MFT	ohm/ MFT
18	1620	1	—	0.040	0.001	7.77	8.08	2.8
18	1620	7	0.015	0.046	0.002	7.95	8.45	3.1
16	2580	1	—	0.051	0.002	4.89	5.08	8.05
16	2580	7	0.019	0.058	0.003	4.99	5.29	8.2
14	4110	1	—	0.064	0.003	3.07	3.19	5.06
14	4110	7	0.024	0.073	0.004	3.14	3.26	5.17
12	6530	1	—	0.081	0.005	1.93	2.01	3.18
12	6530	7	0.030	0.092	0.006	1.98	2.05	3.25
10	10380	1	—	0.102	0.008	1.21	1.26	2.00
10	10380	7	0.038	0.116	0.011	1.24	1.29	2.04
8	16510	1	—	0.128	0.013	0.764	0.786	1.26
8	16510	7	0.049	0.146	0.017	0.778	0.809	1.28
6	26240	7	0.061	0.184	0.027	0.491	0.510	0.808
4	41740	7	0.077	0.232	0.042	0.308	0.321	0.508
3	52620	7	0.087	0.260	0.053	0.245	0.254	0.403
2	66360	7	0.097	0.292	0.067	0.194	0.201	0.319
1	83690	19	0.066	0.332	0.087	0.154	0.160	0.253
1/0	105600	19	0.074	0.373	0.109	0.122	0.127	0.201
2/0	133100	19	0.084	0.419	0.138	0.967	0.101	0.159
3/0	167800	19	0.094	0.470	0.173	0.0766	0.0797	0.126
4/0	211600	19	0.106	0.528	0.219	0.0608	0.0626	0.100
250	—	37	0.082	0.575	0.260	0.0515	0.0535	0.0847
300	—	37	0.090	0.630	0.312	0.0429	0.0446	0.0707
350	—	37	0.097	0.681	0.364	0.0367	0.0382	0.0605
400	—	37	0.104	0.728	0.416	0.0321	0.0331	0.0529
500	—	37	0.116	0.813	0.519	0.0258	0.0265	0.0424
600	—	61	0.992	0.893	0.626	0.0214	0.0223	0.0353
700	—	61	0.107	0.964	0.730	0.0184	0.0189	0.0303
750	—	61	0.111	0.998	0.782	0.0171	0.0176	0.0282
800	—	61	0.114	1.03	0.834	0.0161	0.0166	0.0265
900	—	61	0.122	1.09	0.940	0.0143	0.0147	0.0235
1000	—	61	0.128	1.15	1.04	0.0129	0.0132	0.0212
1250	—	91	0.117	1.29	1.30	0.0103	0.0106	0.0169
1500	—	91	0.128	1.41	1.57	0.00858	0.00883	0.0141
1750	—	127	0.117	1.52	1.83	0.00735	0.00756	0.0121
2000	—	127	0.126	1.63	2.09	0.00643	0.00662	0.0106

Note: DC Resistance at 75°C, 167°F

3-10 NATIONAL ELECTRICAL CODE®

Several factors are involved in the proper selection of wire sizes and types for electric circuits. The primary aim of the circuit is to supply electric power in an efficient and safe manner. To do this, the wire selected to do the job must be able to carry the current without overheating or producing an excessive voltage drop, the insulation must be able to withstand the voltage and have adequate physical strength, it must be able to operate safely in the temperature ranges likely to be met, and it must be economical to use. Since most commercial wiring is protected by being run in tubing or pipes (called *conduits*), the overall diameter (conductor and insulation) is an important consideration.

Many of the guidelines for the safe and effective selection of wiring and wiring methods are contained in the *National Electrical Code*[1] (often called simply "the Code"® or "the NEC"®). The Code is a set of rules, regulations, recommendations, and guidelines considered essential for the safe installation and operation of electrical systems. In many cases the Code sets forth minimums or maximums. It is not, however, a design or engineering manual. Strict compliance with the Code will not ensure an efficient, convenient, or adequate electrical system. That is the responsibility of the engineer and designer. Further, the Code is not a legal document. It becomes enforceable only when local authorities adopt the provisions of the Code as their own local electrical code. At the federal level the Occupational Safety and Health Act (OSHA) has adopted the Code to cover electrical installations governed by the act.

National Electrical Code® and NEC® are registered trademarks of the National Fire Protection Association, Inc.

The National Electrical Code is issued by the National Fire Protection Association of Quincy, Massachusetts. It is an American National Standard (also called an *ANSI Standard*). The Code is regularly reviewed and periodically revised. The most recent Codes have been issued in 1975, 1978, 1981, and 1984. The current provisions of the Code are the results of decades of study and experience by hundreds of experts in the electrical field. The Code warrants the attention of everyone interested in good electrical service.

Portions of the 1984 National Electrical Code have been reproduced or referred to throughout this text. Copies of the Code, copyright by the National Fire Protection Association, may be purchased in technical bookstores or ordered directly from the association at Batterymarch Park, Quincy, Massachusetts 02269.

3-11 AMPACITY

The National Electrical Code lists what are considered the safe continuous-current-carrying capacities (called *ampacities* by the Code) of standard building wire sizes. These ampacities are related to:

1. The conductor material (copper or aluminum)
2. Conductor sizes from No. 18 through 2000 MCM (2 million circular mils)
3. Approved insulation types (a wide variety of rubber, plastic, asbestos, paper, glass, etc.)
4. Number of conductors being run together
5. Temperature of surroundings (30°C or 86°F)

Tables 3-2 (NEC Table 310-16) and 3-3 (NEC Table 310-18) give the ampacities of building conductors

TABLE 3-2
Ampacities of Insulated Conductors Rated 60° to 90°C

Size AWG MCM	60°C (140°F) TYPES †RUW, †T, †TW, †UF	75°C (167°F) TYPES †FEPW, †RH, †RHW, †RUH, †THW, †THWN, †XHHW, †USE, †ZW	85°C (185°F) TYPES V, MI	90°C (194°F) TYPES TA, TBS, SA, AVB, SIS, †FEP, †FEPB, †RHH †THHN, †XHHW*	60°C (140°F) TYPES †RUW, †T, †TW, †UF	75°C (167°F) TYPES †RH, †RHW, †RUH, †THW, †THWN, †XHHW, †USE	85°C (185°F) TYPES V, MI	90°C (194°F) TYPES TA, TBS, SA, AVB, SIS, †RHH, †THHN, †XHHW*	Size AWG MCM
	COPPER				ALUMINUM OR COPPER-CLAD ALUMINUM				
18	....	....	....	14	....	....	....	....	....
16	....	....	18	18	....	....	....	....	....
14	20†	20†	25	25†	....	....	....	....	....
12	25†	25†	30	30†	20†	20†	25	25†	12
10	30	35†	40	40†	25	30†	30	35†	10
8	40	50	55	55	30	40	40	45	8
6	55	65	70	75	40	50	55	60	6
4	70	85	95	95	55	65	75	75	4
3	85	100	110	110	65	75	85	85	3
2	95	115	125	130	75	90	100	100	2
1	110	130	145	150	85	100	110	115	1
0	125	150	165	170	100	120	130	135	0
00	145	175	190	195	115	135	145	150	00
000	165	200	215	225	130	155	170	175	000
0000	195	230	250	260	150	180	195	205	0000
250	215	255	275	290	170	205	220	230	250
300	240	285	310	320	190	230	250	255	300
350	260	310	340	350	210	250	270	280	350
400	280	335	365	380	225	270	295	305	400
500	320	380	415	430	260	310	335	350	500
600	355	420	460	475	285	340	370	385	600
700	385	460	500	520	310	375	405	420	700
750	400	475	515	535	320	385	420	435	750
800	410	490	535	555	330	395	430	450	800
900	435	520	565	585	355	425	465	480	900
1000	455	545	590	615	375	445	485	500	1000
1250	495	590	640	665	405	485	525	545	1250
1500	520	625	680	705	435	520	565	585	1500
1750	545	650	705	735	455	545	595	615	1750
2000	560	665	725	750	470	560	610	630	2000

AMPACITY CORRECTION FACTORS									
Ambient Temp. °C	For ambient temperatures other than 30°C, multiply the ampacities shown above by the appropriate factor shown below.								Ambient Temp. °F
31-40	.82	.88	.90	.91	.82	.88	.90	.91	87-104
41-45	.71	.82	.85	.87	.71	.82	.85	.87	105-113
46-50	.58	.75	.80	.82	.58	.75	.80	.82	114-122
51-60	....	.58	.67	.71	....	.58	.67	.71	123-141
61-70	....	.35	.52	.58	....	.35	.52	.58	142-158
71-80	....	....	.30	.41	....	....	.30	.41	159-176

TABLE 3-3
Ampacities of Insulated Conductors Rated 110° to 250°C

Size	Temperature Rating of Conductor								Size
	110°C (230°F)	125°C (257°F)	150°C (302°F)	200°C (392°F)	250°C (482°F)	110°C (230°F)	125°C (257°F)	200°C (392°F)	
AWG MCM	TYPES AVA, AVL	TYPES AI, AIA	TYPE Z	TYPES A, AA, FEP, FEPB, PFA	TYPES PFAH, TFE	TYPES AVA, AVL	TYPES AI, AIA	TYPES A, AA	AWG MCM
	COPPER				NICKEL OR NICKEL-COATED COPPER	ALUMINUM OR COPPER-CLAD ALUMINUM			
14	30	30	30	30	40	....	....	....	
12	35	40	40	40	55	25	30	30	12
10	45	50	50	55	75	35	40	45	10
8	60	65	65	70	95	45	50	55	8
6	80	85	90	95	120	60	65	75	6
4	105	115	115	120	145	80	90	95	4
3	120	130	135	145	170	95	100	115	3
2	135	145	150	165	195	105	115	130	2
1	160	170	180	190	220	125	135	150	1
0	190	200	210	225	250	150	160	180	0
00	215	230	240	250	280	170	180	200	00
000	245	265	275	285	315	195	210	225	000
0000	275	310	325	340	370	215	245	270	0000
250	315	335	....	....	....	250	270	....	250
300	345	380	....	....	....	275	305	....	300
350	390	420	....	....	....	310	335	....	350
400	420	450	....	....	....	335	360	....	400
500	470	500	....	....	....	380	405	....	500
600	525	545	....	....	....	425	440	....	600
700	560	600	....	....	....	455	485	....	700
750	580	620	....	....	....	470	500	....	750
800	600	640	....	....	....	485	520	....	800
1000	680	730	....	....	....	560	600	....	1000
1500	785	....	....	....	....	650	....	....	1500
2000	840	....	....	....	....	705	....	....	2000
AMPACITY CORRECTION FACTORS									
Ambient Temp. °C	For ambient temperatures other than 30°C, multiply the ampacities shown above by the appropriate factor shown below.								Ambient Temp. °F
31-40	.94	.95	.96	....	....	.94	.95	....	87-104
41-45	.90	.92	.94	....	....	.90	.92	....	105-113
46-50	.87	.89	.91	....	....	.87	.89	....	114-122
51-55	.83	.86	.89	....	....	.83	.86	....	123-131
56-60	.79	.83	.87	.91	.95	.79	.83	.91	132-141
61-70	.71	.76	.82	.87	.91	.71	.76	.87	142-158
71-75	.66	.72	.79	.86	.89	.66	.72	.86	159-167
76-80	.61	.68	.76	.84	.87	.61	.69	.84	168-176
81-90	.50	.61	.71	.80	.83	.50	.61	.80	177-194
91-100	....	.51	.65	.77	.80	....	.51	.77	195-212
101-120	....	....	.50	.69	.72	....	....	.69	213-248
121-140	....	....	.29	.59	.59	....	....	.59	249-284
141-160	....	....	....	....	.54	....	....	....	285-320
161-180	....	....	....	....	.50	....	....	....	321-356
181-200	....	....	....	....	.43	....	....	....	357-392
201-225	....	....	....	....	.30	....	....	....	393-437

rated for up to 2000 V and insulations with temperature ratings up to 200°C (392°F).

The letter symbols at the top of the tables indicate insulation types. Some common insulations are RH and RHW, which are heat-resistant rubber; TW, THW, and THWN, which are flame-retardant thermoplastic; AVA and AVL, which are asbestos. The NEC Table 310-13 lists the various insulation types as well as approved locations for their use.

Tables 3-2 and 3-3 are the ampacities for conductors run in a raceway (enclosure) or cable or buried directly in earth where not more than three conductors are run together. In the 1984 Code the insulation types marked with a dagger (†) have ampacities that are actually 5 A higher than the highest current they are permitted to carry under the conditions noted above.

3-12 VOLTAGE-DROP CALCULATIONS

The voltage lost or dropped in the conductors feeding a piece of electrical equipment is not available for use by that equipment. It is wasted. In addition, the voltage drop produces heat which can cause damage to insulation and the area around the conductors. The heat also represents power loss which means money is being wasted. For all these reasons voltage drop should be kept as low as economically possible. Of

course it is possible to select a wire size so large that the voltage drop will be negligible. But the saving in power loss would probably not pay for the extra expense of the larger wire size. The larger wire size would probably not significantly affect the efficiency of the equipment being served. Some happy medium must be chosen.

Though the Code allows as much as 5 percent voltage drop, it is good design practice to keep the drop much lower.

Table 3-2 shows that a No. 10 THW (moisture- and heat-resistant thermoplastic) copper wire has a current carrying capacity of 30 A. Table 3-1 shows that No. 10 wire has about 1 Ω of resistance per 1000 ft. The voltage drop on a 500-ft run carrying 30 A in No. 10 wire would be $I \times R$ or $30 \times 1 = 30$-V drop. Remember that there are two conductors in a 500-ft run or 1000 ft of wire. The actual power loss would be

$$P = I^2R$$
$$= 30 \times 30 \times 1$$
$$P = 900 \text{ W loss or } 0.9 \text{ kW}$$

If this circuit operated 8 h a day for 1 year, 365 days, at an 8 cents per kWh rate, the power loss would be 2628 kWh at a cost of $210.24.

Tables 3-2 and 3-3 do not consider voltage drop. They are based only on protection of insulation. It will be noted that No. 10 wires with other than thermoplastic insulation are given higher ampacities, up to 55 A for copper wires with asbestos or high-temperature plastic insulation. The voltage and power loss will, of course, increase correspondingly since the resistance of the copper will not change.

Voltage drop can be calculated beforehand if the size wire and the length of the run are known.

The formula is

$$Vd = \frac{22 \times D \times I}{CM}$$

where Vd = voltage drop, V
D = distance from voltage supply to equipment, ft
I = current, A
CM = area of conductor, CM
22 = constant factor related to resistance of copper

The factor for aluminum wire is 35. The formula assumes a two-wire circuit.

Problem 1

What would be the voltage drop of No. 12 THW copper wiring from a panel to an outlet 250 ft away if the wiring is carrying the full allowable current as per NEC Table 310-16?

Solution

No. 12 THW cannot carry more than 20 A. Using the formula

$$Vd = \frac{22 \times 250 \times 20}{6530}$$
$$= 16.8 \text{ V}$$

If the voltage at the panel was 120 V, the percent drop would be

$$Vd\% = \frac{16.8}{120} = 0.14 \text{ or } 14\%$$

This would be too great for most applications.

If the voltage drop is known, the minimum wire size that would be needed can be found using the following formula:

$$CM = \frac{22 \times D \times I}{Vd}$$

where CM, D, I, and Vd are defined as before.

Problem 2

What is the smallest-size THW copper wire that would produce a maximum voltage drop of 2 percent in the run of Problem 1 if the panel voltage is 120 V?

Solution

A 2 percent voltage drop is

$$Vd = 120 \times 0.02 = 2.4 \text{ V}$$

Using the formula for finding CM

$$CM = \frac{22 \times 250 \times 20}{2.4} = 45,833 \text{ CM}$$

From Table 3-1, No. 3 wire has an area of about 53,000 CM; No. 4 wire has an area of about 42,000 CM. Thus No. 3 would have to be used to make sure the voltage drop does not go above 2.4 V. The ampacity of No. 3 THW copper wire is 100 A. This would seem to be an uneconomical use of copper.

Very often at this point the choice of a proper wire size is a matter of compromise. Is the need for the

2.4-V drop so critical that the extra expense of the larger wire is essential? Can a new design of the installation significantly reduce the long run to the equipment? Would it be more economical to install a small transformer at the equipment to provide the correct voltage? These and many other alternatives must be considered by the designer.

Problem 3

What is the maximum length of run for a two-wire 208-V circuit of No. 10 THW wire if the voltage drop is to be limited to 3 percent?

Solution

The maximum voltage drop is calculated first.

$$Vd = 208 \times 0.03 = 6.24 \text{ V}$$

The formula for conductor area

$$CM = \frac{22 \times D \times I}{Vd}$$

can be rearranged to solve for the distance of the run D.

$$D = \frac{CM \times Vd}{22 \times I}$$

where D = distance from supply to load, ft
   I = load current, A
   CM = area of conductor, CM
   Vd = voltage drop, V
   22 = constant related to resistance of copper wire

In this problem we will use the ampacity of No. 10 THW wire as the load current since it is the maximum value allowed by the code.

   $I = 30 \text{ A}$
   $CM = 10,000$ (the rounded-off figure you memorized will give an accurate-enough answer here)
   $Vd = 6.24 \text{ V}$

Therefore

$$D = \frac{10,000 \times 6.24}{22 \times 30}$$
$$D = 94.5 \text{ ft}$$

Note that if the table value of 10,380 CM had been used, the answer would have been about 98 ft. The difference is negligible. It means only that the voltage drop would be lower with the slightly shorter run.

One thing is obvious from the above. Tables and formulas produce the data on which electricians and designers must base their work. But simply plugging numbers into formulas and reading data from tables will not ensure a good, efficient, economical installation.

3-13 TEMPERATURE CORRECTION FACTORS

Other factors that concern wire-size selection are the number of wires in a raceway or cable, and the ambient or room temperature. For room temperatures exceeding 30°C (86°F), correction factors are applied to the allowable ampacities of conductors given in Tables 3-2 and 3-3. Wires are also temperature rated according to insulation type. Insulation that can withstand high temperatures carries the letter designation "H" or "HH." For example, THW wire is rated for 75°C (167°F). Although the wire can be installed so that it will withstand 75°C when its full ampacity is being carried, it must be derated because it exceeds the 30°C ambient temperature. In this particular case the derating (correction) factor from Table 3-2 would be found in the 75°C column. With an ambient temperature of 65°C (149°F) the correction factor for THW wire is 0.35. Thus No. 12 wire would be derated to 20 A × 0.35 = 7 A.

3-14 DERATING AMPACITY

The ability of conductors to get rid of the heat they generate affects the ampacity of the conductor. The ampacity ratings given in Tables 3-2 and 3-3 apply only if not more than three current-carrying wires are run in the raceway. Obviously more wire would create more heat. Less air space in the conduit would reduce the system's ability to get rid of the heat generated.

The code requires that when more than three current-carrying wires are run in the same raceway (conduit or tubing) the ampacities given in Tables 3-2 and 3-3 must be reduced. The reductions, applying to *each conductor* in the raceway, are as follows:

If four, five, or six current-carrying wires are run together, the ampacity of each must be *reduced to 80 percent of the value given in Tables 3-2 and 3-3*.

For example, No. 12 TW wire is rated at 20 A according to Table 3-2. If four No. 12 TW wires are run in the same conduit, and each is designed to carry

current, they are *derated*, or their ampacities are reduced to 20 × 80 percent = 16 A.

In certain electrical systems, conductors are run in conduit although they are not designed to carry current. For example, grounding wires are run to reduce shock and short-circuit hazards but not as circuit-current-carrying conductors. Such wires are not counted in derating calculations.

If the conduit contains 7 through 24 current-carrying conductors, the ampacity of each must be *reduced to 70 percent of the value given in Tables 3-2 and 3-3.*

For 25 through 42 current-carrying conductors the derating factor is 60 percent and for installations containing 43 or more current-carrying conductors the ampacity is reduced to one-half of the table value. Thus 45 No. 12 TW wires run in the same raceway cannot carry more than 10 A each.

3-15 CONDUIT FILL

To allow for proper dissipation of heat, the Code limits the amount of cross-sectional area of conduit and tubing that can be occupied (or filled) with conductors. The area of the conductor includes not only the cross-sectional area of the wire but also the insulation covering the wire.

Table 3-4 (NEC Table 1 of Chap. 9) gives the allowable percent fill of conduit and tubing. This figure is based upon the total cross-sectional area of conductors and insulation as a percentage of the total cross-sectional area of the conduit or tubing.

The actual number of conductors with various types of insulation that is allowed in conduit in trade sizes from $\frac{1}{2}$ through 6 in. is given in NEC Tables 3A, 3B, and 3C.

The NEC Table 3A is reprinted here as Table 3-5 to illustrate the use of these tables. The first column contains various types of insulation, the second contains conductor sizes, and the remaining columns contain the number of conductors allowed in the various sizes of conduit. For example, according to the table the number of No. 10 conductors with TW, T, RUH, RUW, or XHHW insulation that can be placed in $\frac{1}{2}$-in. conduit is 60.

In combining conductors of various sizes and insulations in conduit, NEC Tables 4 and 5 are used to find the conduit size. Table 4 gives cross-sectional areas in square inches and percent of allowable fill for conduit and tubing. Table 5 gives cross-sectional area in square inches of various sizes of conductors and insulations.

To show how these tables are used, they are reproduced here, in part. NEC Table 4 is Table 3-6, and NEC Table 5 is Table 3-7.

Problem 4

Find the smallest conduit size for a job in which three No. 10 and three No. 8 THW conductors are to be run in the same conduit.

Solution

Table 3-7 (see page 28) shows that No. 10 THW conductors have a cross-sectional area of 0.0311 in.^{2}. Three No. 10 THW conductors have a total cross-sectional area of 0.0933 in.^{2}. No. 8 THW conductors have a cross-sectional area of 0.0598 in.^{2}. Three No. 8 THW conductors have a total cross-sectional area of 0.1794 in.^{2}. The total cross-sectional area of the six conductors is 0.2727 in.^{2}.

To find the minimum conduit size that can be used to carry these six conductors, refer to Table 3-6. Six non-lead-covered conductors are being used. Therefore the column headed "Over 2 cond. 40%" under the main heading "Not Lead Covered" should be used. The 40 percent figure means that no more than 40 percent of the internal cross-sectional area of the conduit can be occupied by the conductors. In this column we find that $\frac{3}{4}$-in conduit can contain only 0.21 in.^{2} of conductor. The six conductors occupy 0.2727 in.^{2}. Thus $\frac{3}{4}$-in conduit is too small. The next trade size

TABLE 3-4
Allowable Conduit or Tubing Fill in Percent of Conduit Cross Section

Number of Conductors	1	2	3	4	Over 4
All conductor types except lead-covered (new or rewiring)	53	31	40	40	40
Lead-covered conductors	55	30	40	38	35

TABLE 3-5
Maximum Number of Conductors in Trade Sizes of Conduit or Tubing

Type Letters	Conductor Size AWG, MCM	½	¾	1	1¼	1½	2	2½	3	3½	4	5	6
TW, T, RUH, RUW, XHHW (14 thru 8)	14	9	15	25	44	60	99	142					
	12	7	12	19	35	47	78	111	171				
	10	5	9	15	26	36	60	85	131	176			
	8	2	4	7	12	17	28	40	62	84	108		
RHW and RHH (without outer covering), THW	14	6	10	16	29	40	65	93	143	192			
	12	4	8	13	24	32	53	76	117	157			
	10	4	6	11	19	26	43	61	95	127	163		
	8	1	3	5	10	13	22	32	49	66	85	133	
TW, T, THW, RUH (6 thru 2), RUW (6 thru 2), FEPB (6 thru 2), RHW and RHH (without outer covering)	6	1	2	4	7	10	16	23	36	48	62	97	141
	4	1	1	3	5	7	12	17	27	36	47	73	106
	3	1	1	2	4	6	10	15	23	31	40	63	91
	2	1	1	2	4	5	9	13	20	27	34	54	78
	1		1	1	3	4	6	9	14	19	25	39	57
	0		1	1	2	3	5	8	12	16	21	33	49
	00		1	1	1	3	5	7	10	14	18	29	41
	000		1	1	1	2	4	6	9	12	15	24	35
	0000			1	1	1	3	5	7	10	13	20	29
	250			1	1	1	2	4	6	8	10	16	23
	300			1	1	1	2	3	5	7	9	14	20
	350				1	1	1	3	4	6	8	12	18
	400				1	1	1	2	4	5	7	11	16
	500					1	1	1	3	4	6	9	14
	600					1	1	1	3	4	5	7	11
	700					1	1	1	2	3	4	7	10
	750					1	1	1	2	3	4	6	9

TABLE 3-6
Cross-Sectional Areas of Conduit and Tubing

Trade Size	Internal Diameter Inches	Area — Square Inches								
		Not Lead Covered				Lead Covered				
		Total 100%	2 Cond. 31%	Over 2 Cond. 40%	1 Cond. 53%	1 Cond. 55%	2 Cond. 30%	3 Cond. 40%	4 Cond. 38%	Over 4 Cond. 35%
½	.622	.30	.09	.12	.16	.17	.09	.12	.11	.11
¾	.824	.53	.16	.21	.28	.29	.16	.21	.20	.19
1	1.049	.86	.27	.34	.46	.47	.26	.34	.33	.30
1¼	1.380	1.50	.47	.60	.80	.83	.45	.60	.57	.53
1½	1.610	2.04	.63	.82	1.08	1.12	.61	.82	.78	.71
2	2.067	3.36	1.04	1.34	1.78	1.85	1.01	1.34	1.28	1.18
2½	2.469	4.79	1.48	1.92	2.54	2.63	1.44	1.92	1.82	1.68
3	3.068	7.38	2.29	2.95	3.91	4.06	2.21	2.95	2.80	2.58
3½	3.548	9.90	3.07	3.96	5.25	5.44	2.97	3.96	3.76	3.47
4	4.026	12.72	3.94	5.09	6.74	7.00	3.82	5.09	4.83	4.45
5	5.047	20.00	6.20	8.00	10.60	11.00	6.00	8.00	7.60	7.00
6	6.065	28.89	8.96	11.56	15.31	15.89	8.67	11.56	10.98	10.11

TABLE 3-7
Dimensions of Building Conductors
A. Sizes No. 18 Through 4/0

Size AWG MCM	Types RFH-2, RH, RHH,*** RHW,*** SF-2		Types TF, T, THW,† TW, RUH,** RUW**		Types TFN, THHN, THWN		Types**** FEP, FEPB, FEPW, TFE, PF, PFA, PFAH, PGF, PTF, Z, ZF, ZFF		Type XHHW, ZW††		Types KF-1, KF-2, KFF-1, KFF-2	
	Approx. Diam. Inches	Approx. Area Sq. In.	Approx. Diam. Inches	Approx. Area Sq. In.	Approx. Diam. Inches	Approx. Area Sq. In.	Approx. Diam. Inches	Approx. Area Sq. Inches	Approx. Diam. Inches	Approx. Area Sq. In.	Approx. Diam. Inches	Approx. Area Sq. In.
Col. 1	Col. 2	Col. 3	Col. 4	Col. 5	Col. 6	Col. 7	Col. 8	Col. 9	Col. 10	Col. 11	Col. 12	Col. 13
18	.146	.0167	.106	.0088	.089	.0062	.081	.0052	...	....	.065	.0033
16	.158	.0196	.118	.0109	.100	.0079	.092	.0066	...	....	.070	.0038
14	30 mils .171	.0230	.131	.0135	.105	.0087	105 .105	.0087 .0087	...	....	.083	.0054
14	45 mils .204*	.0327*	....	....	...	....	... ...	....	...	....	...	....
14	......		.162†	.0206†	...	....			.129	.0131	...	....
12	30 mils .188	.0278	.148	.0172	.122	.0117	.121 .121	.0115 .0115	...		.102	.0082
12	45 mils .221*	.0384*	...	....							...	....
12	......		.179†	.0252†	...	....			.146	.0167	...	....
10	...... .242	.0460	.168	.0222	.153	.0184	.142 .142	.0158 .0158			.124	.0121
10	......		.199†	.0311†	...	....			.166	.0216		
8	...... .328	.0845	.245	.0471	.218	.0373	.206 .186	.0333 .0272	...	....	...	....
8	......		.276†	.0598†	...	....			.241	.0456		
6	.397	.1238	.323	.0819	.257	.0519	.244 .302	.0468 .0716	.282	.0625	...	....
4	.452	.1605	.372	.1087	.328	.0845	.292 .350	.0670 .0962	.328	.0845	...	....
3	.481	.1817	.401	.1263	.356	.0995	.320 .378	.0804 .1122	.356	.0995	...	....
2	.513	.2067	.433	.1473	.388	.1182	.352 .410	.0973 .1320	.388	.1182	...	....
1	.588	.2715	.508	.2027	.450	.1590	.420	.1385	.450	.1590		
0	.629	.3107	.549	.2367	.491	.1893	.462	.1676	.491	.1893	...	....
00	.675	.3578	.595	.2781	.537	.2265	.498	.1948	.537	.2265		
000	.727	.4151	.647	.3288	.588	.2715	.560	.2463	.588	.2715		
0000	.785	.4840	.705	.3904	.646	.3278	.618	.3000	.646	.3278		

B. Sizes 250 MCM Through 2000 MCM

Size AWG MCM	Types RFH-2, RH, RHH,*** RHW,*** SF-2		Types TF, T, THW,† TW, RUH,** RUW**		Types TFN, THHN, THWN		Types**** FEP, FEPB, FEPW, TFE, PF, PFA, PFAH, PGF, PTF, Z, ZF, ZFF		Type XHHW, ZW††	
	Approx. Diam. Inches	Approx. Area Sq. In.	Approx. Diam. Inches	Approx. Area Sq. In.	Approx. Diam. Inches	Approx. Area Sq. In.	Approx. Diam. Inches	Approx. Area Sq. Inches	Approx. Diam. Inches	Approx. Area Sq. In.
Col. 1	Col. 2	Col. 3	Col. 4	Col. 5	Col. 6	Col. 7	Col. 8	Col. 9	Col. 10	Col. 11
250	.868	.5917	.788	.4877	.716	.4026	...	....	.716	.4026
300	.933	.6837	.843	.5581	.771	.4669	...	....	.771	.4669
350	.985	.7620	.895	.6291	.822	.5307	...	....	.822	.5307
400	1.032	.8365	.942	.6969	.869	.5931	...	....	.869	.5931
500	1.119	.9834	1.029	.8316	.955	.7163	...	....	.955	.7163
600	1.233	1.1940	1.143	1.0261	1.058	.8791	...	....	1.073	.9043
700	1.304	1.3355	1.214	1.1575	1.129	1.0011	...	....	1.145	1.0297
750	1.339	1.4082	1.249	1.2252	1.163	1.0623	...	....	1.180	1.0936
800	1.372	1.4784	1.282	1.2908	1.196	1.1234	...	....	1.210	1.1499
900	1.435	1.6173	1.345	1.4208	1.259	1.2449	...	....	1.270	1.2668
1000	1.494	1.7530	1.404	1.5482	1.317	1.3623	...	....	1.330	1.3893
1250	1.676	2.2062	1.577	1.9532	...	....	...	....	1.500	1.7671
1500	1.801	2.5475	1.702	2.2751	...	....	...	....	1.620	2.0612
1750	1.916	2.8832	1.817	2.5930	...	....	...	....	1.740	2.3779
2000	2.021	3.2079	1.922	2.9013	...	....	...	....	1.840	2.6590

* The dimensions of Types RHH and RHW.
** No. 14 to No. 2.
† Dimensions of THW in sizes No. 14 to No. 8. No. 6 THW and larger is the same dimension as T.
*** Dimensions of RHH and RHW without outer covering are the same as THW No. 18 to No. 10, solid; No. 8 and larger, stranded.
**** In Columns 8 and 9 the values shown for sizes No. 1 thru 0000 are for TFE and Z only. The right-hand values in Columns 8 and 9 are for FEPB, Z, ZF, and ZFF only.
†† No. 14 to No. 2.

larger, 1-in conduit, can be filled to 0.34 in.^{2} with conductors. Thus the six conductors in the problem can be safely carried in 1 in. conduit. In fact, the conduit will have spare capacity of 0.34 - 0.2727 in.^{2} or 0.0673 in.^{2}.

3-16 MAGNET WIRE

Wire used in winding electric equipment such as motors, generators, transformers, and coils of various kinds is known as *magnet wire*. It is usually soft-drawn or annealed to permit easy handling. Magnet wire is made with many types of insulations, including synthetic enamel (under trade names such as Formex, Formvar, and Nylclad), glass, silicone, cotton, etc., or combinations of these, applied in thin layers to conserve winding space and flexibility. Magnet wire should be stored in a cool, clean, dry place and protected at all times from mechanical injury. After being wound, coils or coil assemblies are sometimes impregnated with insulating varnishes, epoxies, or plastics for added protection. Care should be used in selecting these materials for the various types of insulation, since not all impregnating materials are compatible with all types of insulations.

3-17 SOLDERING

Soldering is a simple, inexpensive, and practical method of making a permanent and effective mechanical and electrical connection between metals. A good solder connection is one of the most dependable trouble-free connections that can be made.

Soldering is a process for connecting two or more metal parts using another metal, usually an alloy of tin and lead, and heat. The soldering alloy is brought to a molten stage by heat, and it combines with the parts by "wetting" and penetrating the pores of the parts to be joined. It then solidifies in cooling to join the parts permanently. There are four requirements for a good solder job: clean surfaces to be soldered, good flux, good solder, and proper heat. The surfaces to be soldered must be free of all foreign matter such as ox-

ides, oil, protective films, etc. For the desired reaction between surfaces and solder, the surfaces must be completely clean and bare.

Oxides form on all bare metals in contact with air. This formation is faster when metal is heated. Rosin soldering flux in solid or paste form is used (1) to clean surfaces to be soldered and prevent rapid oxidation when heat is applied in the soldering operation, and (2) to lower the effect of capillary attraction, or lessen surface tension, of the solder to permit it to flow freely. These effects of soldering paste allow the molten solder to "wet" the bare work surfaces and firmly adhere to them.

Acid flux should never be used in electrical work unless it can be thoroughly washed away after soldering—which is seldom practicable. Acid, in the presence of moisture, attacks copper conductors and in time will destroy them.

Good electrical solder is a 50-50 tin-lead alloy. It becomes plastic at about 375°F (191°C) and molten at about 425°F (218°C). A 60-40 tin-lead solder alloy has slightly lower melting and flow points. Soldered work should not be moved or otherwise disturbed during the time it is going through the stages of solidifying. A soldered connection must be thoroughly "set" and hard before it is strained, or a weak, high-resistance connection will result.

A soldering iron should have sufficient heat capacity for the job at hand. Three sizes of electric soldering irons are recommended for average jobs as follows: 100-W, 300-W, and 600-W. Heat cannot transfer sufficiently through a layer of corrosion or oxide. The tips of irons must be well tinned to prevent corrosion and permit heat transfer to work. A *tinned* surface is one coated with a thin layer of solder.

Although the NEC prohibits soldered connections in many types of work, modern electronic equipment used in industrial applications does have soldered wires. In older electrical work involving knob-and-tube wiring soldered joints are still common (though knob-and-tube wiring itself is not very common). Motor and transformer windings are typically terminated in solder lugs or contain soldered splices. In most electronic applications soldering irons rated between 25 and 50 W are used. Motor repairs requiring soldering often use gas-fired torches.

SUMMARY

1. An electrician should know the nature and properties of copper, how wire is made and measured, and how it is used in electric circuits.

2. Copper hardens when it is "worked," and resistance increases about 2.5 percent. Hard copper is best for construction purposes because it has less stretch and higher tensile strength. Annealed copper is best for electrical windings because it bends more easily.

3. Copper wire is measured by the American Wire

Gage (AWG), which is the same as the Brown and Sharpe (B & S) gage.

4. Gage numbers run from 4/0, which is the largest, through 50, the smallest. Wire larger than 4/0 is sized by its cross-sectional area measured in circular mils.

5. Wire calculations involving cross-sectional area are usually made in circular mils or square mils. A circular mil is the area enclosed by a circle 1 mil in diameter. A mil is one-thousandths of an inch. A square mil is the area enclosed by a square 1 mil on each side.

6. Copper wire sizes are allowed a plus or minus tolerance and therefore do not always exactly fit a fixed gage. A micrometer caliper is convenient in such cases to find wire sizes, especially for smaller sizes where variation in wire sizes is only a few mils.

7. Two or more wires can be used for their flexibility in rewinding instead of a large wire if they contain the same cross-sectional area as the large wire. If two wires are used in parallel, the winding is said to be two-in-hand.

8. Circular mils can be converted to their equivalent square mils by multiplying by 0.7854. Square mils can be converted to their equivalent circular mils by dividing by 0.7854.

9. A No. 10 wire is about 100 mils in diameter, 10,000 CM in cross-sectional area, and has 1 Ω of resistance per 1000 ft.

10. A wire three gage sizes larger than another wire has twice the cross-sectional area and one-half the resistance of the smaller wire. A wire three gage sizes smaller than another wire has one-half the cross-sectional area and twice the resistance of the larger wire.

11. Three important considerations in designing an efficient electrical wiring system are voltage drop, protection of insulation, and safety.

12. The National Electrical Code is an invaluable aid to an electrician in safeguarding persons and property from possible hazards in the use of electricity.

13. Voltage drop, known as *IR* drop, is voltage reduced by current flowing through a circuit or part of a circuit. Voltage drop in line conductors or feeders is known as line drop. Voltage drop in a line can be found by multiplying the amperes flowing in the line by the resistance of the line in ohms.

14. In selecting the proper wire size for a circuit, it is necessary to know the allowable voltage drop. A voltage-drop formula can then be used to calculate the minimum wire size needed. The code must then be checked to see that the wire chosen has the ampacity to serve the load.

15. If conductors operate in a temperature of over 86°F (30°C), a correction factor is supplied by the Code to find ampacity.

16. The number of wires permitted in a given conduit is determined by the wire size and insulation and the inside cross-sectional area of the conduit.

QUESTIONS

3-1 What two metals are most commonly used as electrical conductors?

3-2 What is the standard gage used to measure round conductors?

3-3 What is the range in wire sizes that can be measured by the standard gage?

3-4 What unit of measurement is used to specify conductor sizes larger than 0000?

3-5 What is meant by "wires parallel"?

3-6 Convert 400 square mils to circular mils.

3-7 Convert 250 MCM to SM.

3-8 What is the derating factor for four to six conductors in conduit?

3-9 Above what ambient temperature must ampacities of conductors be derated? Give your answer in both Fahrenheit and Celsius.

3-10 What alloy is commonly used for soldering copper wires?

3-11 A two-in-hand winding is used to replace a single No. 13 conductor. What size wire should be used?

3-12 What are the rounded-off values for the diameter, cross-sectional area, and resistance per 1000 ft of No. 10 wire?

3-13 Based on your answer to question 12, what is the resistance per 1000 ft of No. 4 wire?

3-14 Based on your answer to question 12, what is the cross-sectional area of No. 16 wire?

3-15 What organization issues the NEC?

3-16 Five No. 12 TW wires are used to feed a piece of equipment 1000 ft from the voltage supply. Assume the wires are run in conduit and are carrying their maximum allowable ampacity. The ambient temperature is 70°F. What is the voltage drop in each wire?

3-17 What is the total power lost in all five wires in question 16? (Total power loss is the sum of all individual losses.)

3-18 In a two-wire circuit, 10 A is taken by a piece of electrical equipment located 250 ft from a 120-V supply. The voltage drop cannot exceed 2 percent. What is the smallest-size wire that should be used?

3-19 From Table 3-2 name two thermoplastic (T) insulation types for copper conductors rated for temperatures up to 75°C (167°F).

3-20 How many No. 2 TW wires can be run in 1¼-in. conduit?

3-21 Three No. 12 TW copper wires are run in a ¾-in. conduit. How many No. 10 TW can be added to the same conduit?

3-22 What type of material is used to remove oxides from copper before soldering?

3-23 Name three types of insulation used on magnet wire.

3-24 How are coils protected after being wound?

3-25 What instrument is used to make accurate measurements of wire diameters?

4

ELECTRIC CIRCUITS

There are three types of electric circuits—*series, parallel,* and *series-parallel.* The physical difference between these circuits is the way the various current-consuming devices, or loads, are connected in relationship to each other. Each type of circuit has different voltage, current, and resistance characteristics. The electrician should know these characteristics in order to understand and work with these circuits.

4-1 SERIES CIRCUITS

In the series circuit all the loads are connected in sequence, as in Fig. 4-1. All the current that flows in a series circuit *flows equally through every part of the entire circuit.* A series circuit has only one path from one terminal of the power source to the other. When parts are connected in series, they are connected one after the other in succession to form one circuit. If any part of a series circuit opens, current stops flowing in the entire circuit.

Series circuits have the following characteristics:

1. There is only one current path in a series circuit.
2. The same current flows in all parts of a series circuit.
3. The sum of the voltage drops around a series circuit is equal to the supply voltage.
4. The voltage drop in a part of a series circuit is in the same proportion to the total applied voltage as the resistance of that part of the circuit is to the total resistance of the circuit.
5. The total voltage of power sources connected in series is the sum of the voltages of the individual sources.
6. The total resistance of a series circuit is the sum of the resistances of the various parts.

The circuit of Fig. 4-1 contains four heating resistors (loads 1 through 4) connected in series across a 120-V generator. An examination of this circuit shows that there is only one path, and so all the current must flow through each part of the circuit. The current flow is the same in all parts of the circuit. The current flowing throughout the circuit is 6 A.

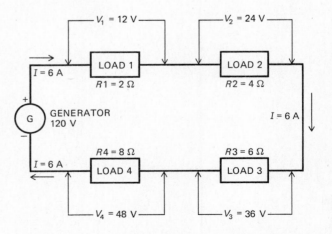

Fig. 4-1 A series circuit showing the relationship of voltage, current, and resistance.

The total resistance of the circuit is the sum of the resistances of the parts of the circuit. In this case the resistance of the lines connecting the loads will not be included. In practice this is not serious since the resistance of the lines is usually much lower than that of the loads. There are four resistors in series in the circuit, and the total resistance is the sum of the resistance of each resistor. The resistors have resistances of 2, 4, 6, and 8 Ω. The total resistance of the series circuit is 20 Ω (neglecting line resistance).

$$2 + 4 + 6 + 8 = 20\ \Omega$$

The voltage drop across the entire circuit is the full line voltage of 120 V. The voltage drop in any part of the circuit is proportional to the resistance of that part of the circuit. The voltage drops are shown as V_1, V_2, V_3, and V_4 across each resistor; $R1$, with 2-Ω resistance, will cause a voltage drop of 12 V, according to Ohm's law, and $R2$ will drop 24 V, $R3$ will drop 36 V, and $R4$ will drop 48 V.

Ohm's law: $V = I \times R$

$$V_1 = 6 \times 2 = 12\ V$$
$$V_2 = 6 \times 4 = 24\ V$$
$$V_3 = 6 \times 6 = 36\ V$$
$$V_4 = 6 \times 8 = 48\ V$$

The sum of 12 V, 24 V, 36 V, and 48 V is 120 V, which equals the line voltage.

$$
\begin{aligned}
V\ (\text{total}) &= V_1 + V_2 + V_3 + V_4 \\
&= 12 + 24 + 36 + 48 \\
V\ (\text{total}) &= 120\ V
\end{aligned}
$$

Since the total resistance is 20 Ω and the voltage is 120 V, according to Ohm's law, 6 A will flow:

$$
\begin{aligned}
I &= \frac{V}{R} \\
&= \frac{120\ V}{20\ \Omega} \\
I &= 6\ A
\end{aligned}
$$

This verifies the value given in the circuit of Fig. 4-1.

Series circuits are rarely used in light and power systems but they do have special applications. A string of series Christmas-tree lights is an example of a series circuit. If one of the lights burns out in the series string, the remaining lights will go out, because the circuit is broken by the defective lamp.

Some incandescent street-lighting systems use a series circuit. The advantage in this case is the low-load current in the lines; this means smaller wires can be used to feed the circuit. The lamp sockets contain a device that automatically short-circuits the socket when the lamp filament burns out. This restores the circuit. Constant-current devices feed such circuits, and when a lamp burns out, the device automatically lowers the voltage to maintain the proper voltage across each lamp. Otherwise, the remaining lamps would be operated at a higher voltage. Operating lamps at more than their rated voltage greatly reduces their life.

Figure 4-2 shows voltage sources connected in series. The total voltage is equal to the sum of the individual voltages. In the case of the cells (dc-voltage sources) in Fig. 4-2 the positive terminal of one cell is connected to the negative terminal of the next cell and so on. The voltages of all cells are thus connected in the same direction.

Flashlight cells, when placed in a flashlight case, are automatically connected in series, the voltage of each cell adding to that of the other cells. Cells should be placed in the flashlight so that they all face in the

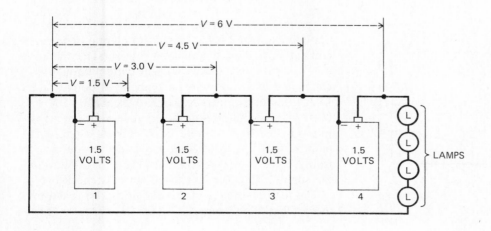

Fig. 4-2 Four 1.5-V cells connected in series. The total voltage across any combination of cells is the sum of the individual voltages in that combination.

same direction, that is, with the top or center post resting against and making contact with the bottom of the cell in front of it. This makes the voltage of each cell add to that of the other cells. Flashlight cells, for example, AA, C, and D cells, furnish about 1.5 V each. Two cells in a flashlight furnish 3 V, three furnish 4.5 V, four furnish 6 V, etc. When bulbs are purchased for replacement, the number of cells in the flashlight must be known so that the bulb voltage rating will match the voltage produced by the cells. Using a bulb with more cells than it is rated for will quickly burn out the bulb.

Direct-current generators can be connected in the same manner as cells if their plus and minus terminals are connected as shown for the cells.

4-2 PARALLEL CIRCUITS

A parallel circuit has two or more paths or *branches,* from positive to negative, or *across the line.* If line resistances are neglected, practically all wiring for power and lighting can be considered of the parallel type, since practically all loads are connected across the line. This permits individual use of each piece of equipment, each of which must be rated for full line voltage.

Parallel circuits have the following characteristics:

1. A parallel circuit has two or more paths or branches across the line.
2. The voltage across each branch is equal.
3. The current in each branch is determined by the resistance of each individual branch.
4. The total line current is the sum of currents in the individual parallel branches.
5. Total resistance of a parallel circuit is *not the sum of the individual resistances or loads of the circuit.* The total resistance is always *less* than the lowest branch resistance in the circuit.

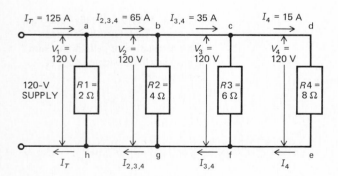

Fig. 4-3 A parallel circuit with four branches.

6. The total resistance of a parallel circuit can be found by dividing the line voltage by the total current supplied by the line.

Figure 4-3 shows a parallel circuit. Four heating elements ($R1$ through $R4$) are connected in parallel across a 120-V supply. The voltage drop across each heater is therefore 120 V (the line resistance is neglected).

The current through each heating element can be found using Ohm's law. The elements can be considered resistances with the values given in Fig. 4-3.

$$\text{For } R1 \qquad I_1 = \frac{120 \text{ V}}{2 \text{ }\Omega} = 60 \text{ A}$$

$$\text{For } R2 \qquad I_2 = \frac{120 \text{ V}}{4 \text{ }\Omega} = 30 \text{ A}$$

$$\text{For } R3 \qquad I_3 = \frac{120 \text{ V}}{6 \text{ }\Omega} = 20 \text{ A}$$

$$\text{For } R4 \qquad I_4 = \frac{120 \text{ V}}{8 \text{ }\Omega} = 15 \text{ A}$$

The total current delivered by the supply is

$$I_T = I_1 + I_2 + I_3 + I_4$$
$$= 60 \text{ A} + 30 \text{ A} + 20 \text{ A} + 15 \text{ A} = 125 \text{ A}$$

Currents entering any point in the parallel circuit must equal the currents leaving that point. This is called *Kirchhoff's current law* (KCL). We can therefore check our answers very easily. For point a, 125 A is entering from the supply. This divides into two currents. Resistor $R1$ uses 60 A, and 65 A continues to point b.

Therefore at point a

$$125 \text{ A} = 60 \text{ A} + 65 \text{ A}$$

At point b the current divides again. Resistor $R2$ uses 30 A, and 35 A continues to point c.

$$65 \text{ A} = 30 \text{ A} + 35 \text{ A}$$

At point c, the current divides so that 20 A goes through $R3$, and 15 A continues to point d. The current at point d is used completely by $R4$ (since there is only one path for it to follow). The current through $R4$ enters and leaves point e for point f. At point f, the 15 A from $R4$ is joined by the 20 A from $R3$: 15 A + 20 A = 35 A. At point g the 35 A leaving point f is joined by the 30 A from $R2$: 35 A + 30 A = 65 A. At point h the 65 A from point g is joined by the 60 A from $R1$: 65 A + 60 A = 125 A. The 125 A returns to the supply. Thus the calculated currents check out.

The total resistance of the circuit can be found by using Ohm's law again.

$$R \text{ (total)} = \frac{V \text{ (supply)}}{I \text{ (line)}}$$

$$RT = \frac{120 \text{ V}}{125 \text{ A}} = 0.96 \ \Omega$$

Notice that this value is less than the lowest branch resistance (2 Ω in this problem). This verifies rule 5 for parallel circuits.

The total resistance of parallel circuits is not the sum of the individual resistances. The more parallel paths there are, the lower the total resistance of the circuit.

If all the parallel branches have the same value of resistance in ohms, the total resistance can be found by dividing the resistance of one branch by the number of branches.

Problem 1

What is the total resistance of three resistors of 9-Ω resistance each, connected in parallel?

Solution

$$R = \frac{9}{3} = 3 \ \Omega \text{ (total)}$$

When the parallel resistances are not the same, two methods can be used for finding the total resistance. One is the product-over-the-sum method. Only two resistances can be combined at a time by this method. The name of the method describes the operations involved.

To find the total resistance of two resistances in parallel, the product of the two resistances in ohms is divided by the sum of the two resistances. The answer is the total resistance of the two parallel resistances.

Problem 2

What is the total resistance of 3 Ω and 6 Ω in parallel?

Solution

The rule given above can be written as a formula:

$$RT = \frac{R1 \times R2}{R1 + R2}$$

Substituting the given values in this formula

$$RT = \frac{3 \times 6}{3 + 6} = \frac{18}{9}$$

$$= 2 \ \Omega$$

If it is desired to find the total resistance of more than two resistances by this method, they can be combined only two at a time. Therefore two can be combined, and the total of these combined with the value of the next, and so on. This can be a cumbersome operation, and so another method is often used.

The other method of finding the total resistance of a parallel circuit uses reciprocals. A reciprocal of a number is written as 1 divided by that number. Thus the reciprocal of 7 is 1/7; the reciprocal of 3 is 1/3; the reciprocal of 1/2 is $1\frac{1}{2}$ or 2; the reciprocal of R is $1/R$.

To find the total resistance of a parallel circuit, first convert all branch resistances to their reciprocals. Then add all the reciprocals. Finally, find the reciprocal of the sum. This is the total resistance. The above method can be written as a formula:

$$\frac{1}{RT} = \frac{1}{R1} + \frac{1}{R2} + \frac{1}{R3} + \cdots$$

(The dots after the last plus sign indicate that the formula applies to as many parallel branches as are in the circuit.)

Problem 3

Using reciprocals, find the total resistance of a 3-Ω and a 6-Ω resistor in parallel.

Solution

$$\frac{1}{RT} = \frac{1}{R1} + \frac{1}{R2}$$

Add the reciprocals

$$\frac{1}{RT} = \frac{1}{3} + \frac{1}{6} = \frac{2}{6} + \frac{1}{6} = \frac{3}{6}$$

The sum of the reciprocals is

$$\frac{1}{RT} = \frac{3}{6}$$

Find the reciprocal of the sum (remember $1/\frac{3}{6} = \frac{6}{3}$).

$$RT = \frac{6}{3} = 2 \ \Omega$$

Problem 4

Find the total resistance of the parallel circuit of Fig. 4-3.

Solution

$$\frac{1}{RT} = \frac{1}{R1} + \frac{1}{R2} + \frac{1}{R3} + \frac{1}{R4}$$

$$= \frac{1}{2} + \frac{1}{4} + \frac{1}{6} + \frac{1}{8}$$

Find the lowest common denominator (LCD). In this case it is 24.

$$\frac{1}{RT} = \frac{12}{24} + \frac{6}{24} + \frac{4}{24} + \frac{3}{24} = \frac{25}{24}$$

$$RT = \frac{24}{25} = 0.96 \ \Omega$$

This checks with the value obtained earlier using Ohm's law.

4-3 SERIES-PARALLEL CIRCUITS

If we were to consider the resistance of the feeders and the other conductors that make up a complete electrical system, then series-parallel circuits would be the most common circuits found in industrial electricity. However, except for voltage drop problems, the resistance of conductors is often neglected. Thus the electrician can consider most circuits as strictly parallel.

Sometimes it is necessary to solve series-parallel circuits for current and voltage. To do this, the circuit is broken down to its basic series and parallel parts, the parts are combined—that is, the series resistances are added together and the parallel resistances are combined according to the rules previously discussed. This combining process is continued until the simplest circuit is found.

Problem 5

Figure 4-4 shows a feeder supplying two welders. The welders can be considered resistance loads. If the resistance of the feeders cannot be neglected, find the voltage across welder 2.

Solution

Notice that branch 2 containing welder 2 is a series circuit consisting of two feeders of 1-Ω resistance each and the 10-Ω welder. The total resistance of this branch is therefore

$$RT \ (\text{branch 2}) = 1 \ \Omega + 1 \ \Omega + 10 \ \Omega = 12 \ \Omega$$

This 12-Ω resistance is actually across branch 1. In other words, it is *in parallel* with welder 1. This

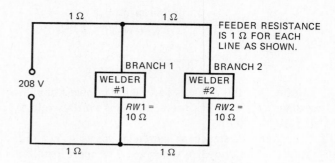

Fig. 4-4 A series-parallel circuit. In this circuit the resistance of the feeders must be considered.

means the 10 Ω of welder 1 and the resistance of branch 2 (the 12 Ω) can be combined as parallel branches.

$$RT \ (\text{branches 1 and 2}) = \frac{10 \times 12}{10 + 12} = \frac{120}{22}$$

$$= 5.45 \ \Omega$$

This is the total effect of welder 2 and its feeders plus welder 1 across the feeders from the supply. Figure 4-5 shows how this circuit would look if the welders and feeders were replaced with an equivalent resistance.

Now we have a simple series circuit. The total resistance of this circuit is

$$RT \ (\text{combined}) = 1 \ \Omega + 1 \ \Omega + 5.45 \ \Omega$$

$$= 7.45 \ \Omega$$

This is the *actual* total resistance across the 208-V supply shown in Fig. 4-4. It includes the resistance of both welders and *all the feeders*. Ohm's law can be used to find the current drawn from the supply.

$$I_T = \frac{208 \ \text{V}}{7.45 \ \Omega}$$

$$= 27.9 \ \text{A}$$

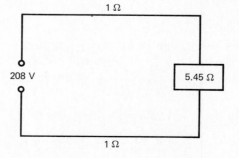

Fig. 4-5 The circuit of Fig. 4-4 simplified by combining the resistances of the two welders and the feeder for welder 2.

This is the current that will flow in the feeder going to welder 1. Since the feeder has a total resistance of 2 Ω, the voltage drop in the feeder (that is, the line drop) will be

$$V \text{ (feeder)} = I \times R$$
$$= 27.9 \times 2 \text{ } \Omega$$
$$= 55.8 \text{ V}$$

Thus the voltage at welder 1 is

$$V \text{ (welder 1)} = 208 \text{ V} - 55.8 \text{ V}$$
$$= 152.2 \text{ V}$$

The current through welder 1 is also found using Ohm's law

$$I \text{ (welder 1)} = \frac{V}{R}$$
$$= \frac{152.2 \text{ V}}{10 \text{ } \Omega}$$
$$= 15.2 \text{ A}$$

The balance of the supply current will go to the branch circuit supplying welder 2.

$$I \text{ (welder 2 circuit)} = 27.9 - 15.2$$
$$= 12.7 \text{ A}$$

The feeders supplying welder 2 will also produce a line drop.

$$V \text{ (feeders for welder 2)} = 12.7 \text{ A} \times 2 \text{ } \Omega$$
$$= 25.4 \text{ V}$$

Since the voltage across this branch was 152.2 V, the voltage across welder 2 will be

$$V \text{ (welder 2)} = 152.2 \text{ V} - 25.4 \text{ V}$$
$$= 126.8 \text{ V}$$

The voltage drops in this problem are extreme. In practice the conditions present in this circuit *would not be acceptable*. The problem might be solved by increasing the size of the feeders or reducing the length of the run. Other solutions include boosting the voltage at the loads by adding a transformer. In any case the problem shows how important it is to consider feeder resistance and how this may require the solution of a series-parallel circuit.

4-4 SHORT CIRCUITS

A short circuit occurs when current takes an accidental path short of its intended circuit. It thus does not flow in its normal circuit and serve the connected load. Short circuits often involve the regular circuit conductors and are caused by the creation of a path of lower resistance than that of the normal circuit.

If two conductors of different polarity make direct contact with each other (i.e., without a "load" between them), a short circuit will result. A short circuit can be damaging to an electrical system because of excessive currents that flow under short-circuit conditions. Such currents can cause damage that does not always show up immediately; it can be several months before the extent of damage is evident in a circuit that has been subjected to a short circuit. If short circuits are not cleared immediately by circuit protective devices, excessive currents can damage insulation, excessively heat the conductors including all connections, destroy the temper in fuse holders, switches, and other parts of the circuit, melt soldered connections such as splices and soldered lugs, and burn contact points.

The only insurance against short-circuit damage is properly located and sized overload protective equipment. All electrical wiring should be protected at, or less than, its rated current-carrying capacity, except motor circuits that sometimes must allow for starting currents in excess of normal load currents.

4-5 GROUND FAULTS

A ground fault is an accidental connection of a nongrounded conductor with conducting materials other than the regular current-carrying parts of a system such as equipment frames or enclosures. A ground fault is a form of short circuit caused by current leaving the circuit conductors and flowing through a path of other conducting materials. It is usually the result of a fault in insulation.

An electric motor or other equipment can be accidentally grounded when the insulation breaks down and some part of the wiring or winding comes in contact with the frame of the motor or equipment. A wire in conduit is grounded if an uninsulated portion of it accidentally comes in contact with the conduit.

A large majority of deaths and injuries by electrical equipment is caused by accidental grounds. In such cases the human body serves as the conductor between the exposed circuit and the ground. Coming in contact with grounded equipment can, under certain conditions, result in death. Persons operating or touching accidentally grounded electrical equipment such as an appliance or power tool in a house not properly wired may be electrocuted if they come in contact at the same time with other grounded objects, such as water pipes, gas pipes, or heating and air-conditioning ducts.

4-6 OPEN CIRCUITS

An open circuit is a break in an electric circuit, either intentional or unintentional. Usually an unintentional or accidental open circuit is caused by a loose connection, a broken conductor due to vibration or constant bending, or a burned out piece of equipment.

SUMMARY

1. There are three types of electric circuits—series, parallel, and series-parallel.

2. The manner in which power-consuming devices are connected determines the type of circuit.

3. Most commercial and domestic circuits may be considered parallel circuits.

4. A series circuit has only one path from positive to negative, or across the line, and is seldom used in light and power circuits. The voltage across any part of a series circuit is in proportion to the resistance of that part of the circuit. The current is the same in all parts of the circuit. The total resistance is the sum of the resistances of all the parts of the circuit.

5. A parallel circuit has two or more paths from positive to negative, or across the line. The voltage is equal in all parts of the circuit, and the total current is the sum of the currents of the various paths. Its total resistance is the reciprocal of the sum of the reciprocals of the resistance values of the various parts of the circuit.

6. The series-parallel circuit is a combination of series and parallel circuits. Its total resistance is found by first combining series loads and then combining the result with parallel loads. Although most circuits found in electrical systems are series-parallel, they can usually be considered as parallel circuits by neglecting the resistance of feeders.

7. A short circuit exists when current flows in a very low resistance path other than its normal circuit. An accidental path allows the current to bypass its normal load.

8. A ground fault occurs when an accidental connection is made through equipment frames or enclosures by circuit conductors.

QUESTIONS

4-1 What is a series circuit?

4-2 What is characteristic of current in a series circuit?

4-3 In a series circuit which load would have the greater voltage drop across it, 10 Ω or 40 Ω?

4-4 How many times greater would be the higher voltage drop than the lower voltage drop in question 4-3?

4-5 How is the total resistance of a series circuit found if the value of all the individual resistances is known?

4-6 When two 1.5-V batteries are connected so that the plus terminal of one battery is connected to the minus terminal of the other battery, what is the voltage across the two remaining terminals?

4-7 Find the current in the circuit of Fig. 4-6.

4-8 Find the voltage across each load in Fig. 4-6.

4-9 If the current needs to be reduced to one-half of the value in question 4-7 but the voltage across each load had to remain the same as in question 4-8, find the new resistance of each load.

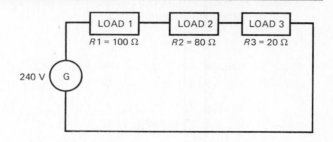

Fig. 4-6 Circuit for questions 4-7 and 4-8.

4-10 What is a parallel circuit?

4-11 What is characteristic about the voltage across each load in a parallel circuit?

4-12 What law covers the characteristics of current at any point in a parallel circuit?

4-13 What happens to the total current in a parallel circuit as loads are added?

4-14 What assumption is made about feeders in a parallel circuit?

4-15 How does a series-parallel circuit differ from a parallel circuit?

4-16 In Fig. 4-7, find the total resistance of the two loads.

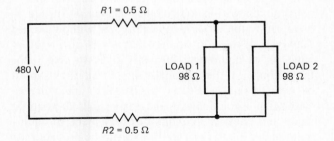

Fig. 4-7 Circuit for questions 4-16, 4-17, and 4-18.

4-17 In Fig. 4-7, $R1$ and $R2$ represent the resistance of the feeders serving loads 1 and 2. Find the total voltage drop in the feeders.

4-18 What is the value of the voltage across load 2 in Fig. 4-7?

4-19 What is a short circuit?

4-20 How do short circuits damage equipment?

4-21 What is a ground fault?

4-22 How can equipment or conductors become grounded?

4-23 How is equipment protected against excessive current?

4-24 What presents more of a shock hazard, a ground or a short circuit?

4-25 What is an open circuit?

5
ELECTRICAL DRAWINGS

An electrician is constantly working with drawings of electrical systems and equipment. The drawing is a graphic description of the electrical system using symbols in a sort of shorthand or trade language. To read electrical drawings, it is necessary to know the meaning of the symbols and to have a knowledge of electricity and electrical equipment. Efficiency in reading drawings can be greatly increased by learning to organize and make drawings.

5-1 NECESSITY FOR DRAWINGS

Drawings are to an electrician what road maps are to a traveler. One would be foolish to start on a long trip without first planning the route with the aid of a road map. It is equally foolish for electricians to start a construction wiring job, or the installation of equipment, or troubleshooting on equipment without proper drawings. If they do not have drawings of the job at hand, or of equipment they are likely to deal with, they should get them or make them. And they should be able to read and interpret them. A maintenance electrician should have diagrams and thoroughly study and know the circuits of every piece of equipment under his or her care in order to be prepared for efficient action in emergencies. Proper diagrams can make this possible.

5-2 TYPES OF DRAWINGS

The electrician usually deals with three types of electrical drawings. These are *electrical plans, schematic diagrams,* and *pictorial diagrams.* On occasion one or more of them is also referred to as a *wiring diagram.* Occasionally the drawing types are combined or used as details for one or another type of drawing. A special type of schematic diagram called a *one-line,* or *single-line, diagram* is sometimes used to describe entire systems in a very simplified fashion. Figure 5-1 gives examples of each of these types of drawings. Note that with the exception of the pictorial diagram, symbols are used to represent the actual wiring and equipment installed or to be installed.

Each electrical drawing has a primary purpose. The electrical floor plan tells the electrician where each electrical device and piece of equipment is to be located relative to the architectural features of the area. It shows the electrician where and how the conductors are to run. It helps locate the incoming service. With the aid of other mechanical-trade drawings, for example, plumbing and heating drawings, the electrician can coordinate his or her work with that of the other building services. Electrical plans are also used for estimating installation and material costs. By examining a plan, the electrician can take off, that is, count, the number of switches, outlets, fixtures, feet of wire, etc., that may be required for a particular job. Although the architectural parts of an electrical floor plan are drawn to some exact proportion, or *scale,* the electrical symbols used on the plans are not.

Electrical plans show the location of equipment and wiring runs. They do not show exactly how the equipment or system is wired from point to point and from terminal to terminal. Details of this type are given in

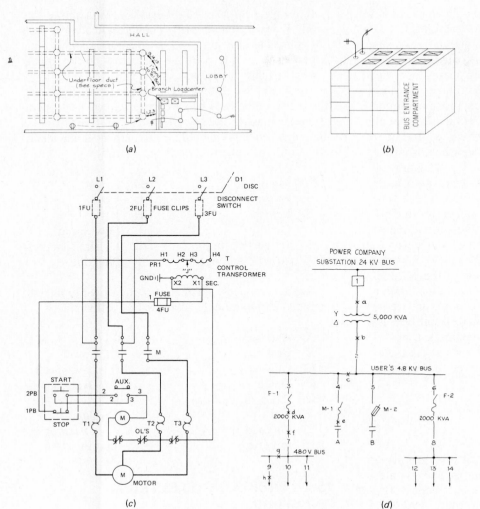

Fig. 5-1 Types of electrical drawings. (*a*) **Electrical floor plan.** (*b*) **Pictorial diagram of a main switchboard.** (*c*) **Schematic diagram of motor circuit.** (*d*) **One-line diagram of electric service to a building.** (*Baer-Ottaway*, Electrical and Electronics Drawing, Fourth Edition, *McGraw Hill Book Company, 1980.*)

schematic diagrams. The schematic diagram shows how each wire in the installation or equipment is connected electrically to the system. The operation of a system or piece of equipment can usually be followed by tracing the flow of current in a schematic diagram. In order to troubleshoot a circuit problem, it is often necessary to have a schematic diagram of the circuit, since the diagram will show the various interlocks and connections to other circuits and equipment. The diagrams found inside the covers or doors of panel boards, motor controls, switchgear, and other equipment are basically schematic diagrams, though they may be modified with pictorial or one-line diagrams. Schematic diagrams are also referred to as *wiring diagrams*.

The pictorial diagram, or drawing, is a realistic pic-

ture of the equipment or components of a piece of equipment. In fact, a pictorial diagram may be an actual photograph of the equipment. The purpose of such diagrams is to help the electrician locate each part exactly as it would appear when it comes from the manufacturer. Pictorial diagrams are therefore usually found in the manufacturer's instruction booklets or literature. They are rarely prepared in the field by the electrician or drafter.

Very often the electrician is required to troubleshoot entire electrical systems. For example, short circuits or overloads may cause circuit breakers to open or fuses to blow out in areas very remote from the actual fault. In order to trace the possible current paths, the electrician must be able to study a diagram of the entire electrical system that is involved. To simplify this investigation and to narrow down the possible areas of fault, the electrician uses a one-line, or single-line, diagram. The one-line diagram uses a single line to represent all the wires and circuit breakers or fuses in a single run. Thus a three-phase, four-wire run, which would normally be represented with four lines for the conductors and three fuses or circuit breakers for the line protection, would be shown with a single line and a single circuit breaker or fuse. The load might be represented by a circle or box, drawn at the end of the line. Although a one-line diagram can show the operations and interconnections of an entire system, it does not show the exact wiring of each part. One-line diagrams are therefore frequently used together with schematic diagrams and electrical plans.

In multistory buildings it is often convenient to show the vertical runs of cable and conduit. This can be done by drawing a simple section of the building showing the various floors. On this section are drawn the vertical lines representing the conduit and cable runs. Often the horizontal branches to panel boards, motor control centers, and heavy equipment are included. Drawings of this type are known as *riser diagrams* since a vertical run of conduit is often called a *riser*.

5-3 STANDARD SYMBOLS

Over the years symbols for electrical drawings have been designed by many organizations and people. Some of these symbols gained wide acceptance. The American National Standards Institute (ANSI), through the cooperation of interested groups, has become the central agency for the standardization of symbols for electrical drawings. The three standards most useful to industrial electrical workers are Graphic Symbols for Electrical and Electronics Diagrams, ANSI Y32.2; Graphic Symbols for Electrical Wiring and Layout Diagrams Used in Architecture and Building Construction, ANSI Y32.9; and Reference Designations for Electrical and Electronics Parts and Equipments, ANSI Y32.16.

The complete list, covering practically the entire electrical field, is too extensive to reproduce here. However, many of the frequently used symbols in industrial electricity are shown in Fig. 5-2. Although they are commonly used, not all these symbols are considered standard.

It will be noted that several symbols have an asterisk (*) that should be replaced with identifying numbers or letters. Operating coils and relays are shown in the circle by numbers or letters. For example, if an operating coil closes a main line contactor, the coil should have an M placed in it, and the contactor

should have an M placed beside it. This shows that coil M closes contactor M. A coil should always carry the same identification as the contactor or contactors it operates when the coil and contactors do not appear close together on a diagram.

Resistors are shown by a rectangular figure and should be identified according to sequence of operation, such as $R1$, $R2$, etc. The value of fixed resistors in ohms and watts should also be shown. Some identifications for operating coils are as follows:

M	Main line	D	Down
F	Forward	U	Up
R	Reverse	H	High speed
OL	Overload	L	Low speed
1A	1st accelerating	DB	Dynamic braking
2A	2d accelerating	P	Plugging

Relays, devices which control their own or another circuit, are commonly identified as follows:

A	Accelerating	JR	Jog
B	Brake	L	Lower
CC	Closing coil	LS	Low speed
D	Down	OL	Overload
DB	Dynamic braking	P	Plugging
F	Field	TR	Time
FR	Forward	UV	Undervoltage
RR	Reverse	U	Up

5-4 PREPARING ELECTRICAL DRAWINGS

While the professional drafter and designer can employ drafting machines and CAD (computer-aided drafting) systems, the industrial electrician and technician must make do with less sophisticated drafting aids.

It is not hard to draw electrical diagrams. You need only a few items of drafting equipment to draw some of the most complicated diagrams. Although rough drawings and sketches may be drawn freehand on the job, a record copy that is to be saved for the future should be drawn with the aid of drafting tools.

Straight lines and curves are drawn with the aid of straightedges, templates, compasses, French curves, and the like. Lettering can be done freehand or with the aid of templates, guides, transfer type, or paste-on labels. Attractive drawings do not require extraordinary artistic abilities or specialized skills. Anyone can develop the sense of balance and proportion needed to produce attractive and pleasing work. Drafting affords a means of getting closely involved in every phase of an electrical system. It allows you to visualize the

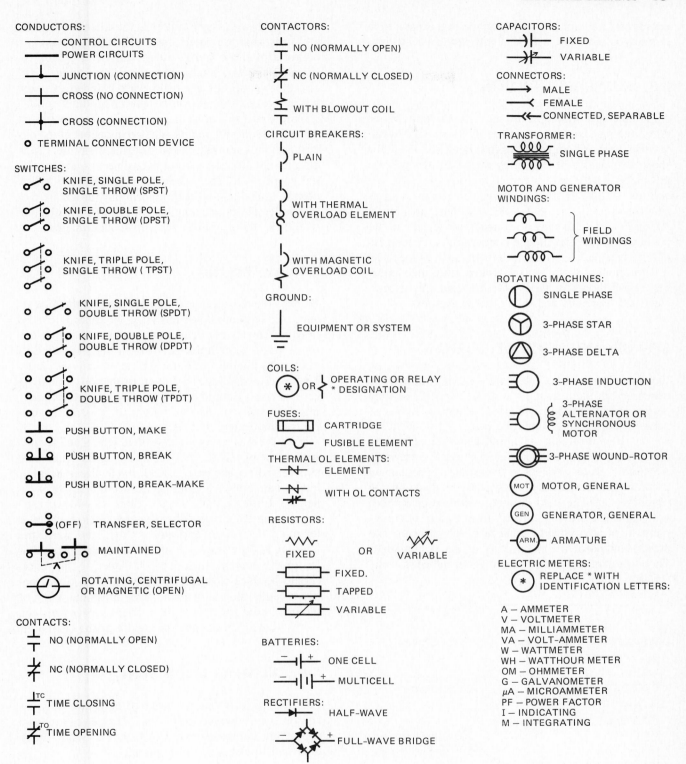

Fig. 5-2 Standard symbols used in electrical drawings.

operation of a complex system while at the same time letting you focus on just one small part of the system.

5-5 RULES FOR DRAWING

All circuit lines should be drawn vertically or horizontally, and parallel when possible. No lines should run at other than 90° angles to each other except 45° for short distances. Heavy lines are used to indicate power or load circuits, and light lines are used for control circuits. If lines cross and do not make connection, simply cross the lines without making a "loop" or "U." If they do connect, place a dot at the connection. The presence of noncurrent-carrying parts that operate the electrical equipment, such as mechanical interlocks, is shown by broken or dashed lines. Figure 5-3 shows the types of lines commonly used in electrical drawings and their relative weights.

5-6 ORGANIZATION OF DRAWING

In organizing a typical electrical drawing, all factors concerned in the drawing should be carefully studied before actual drawing is started. Rough sketches of parts of the drawing should be made. These sketches are considered for size, position, and spacing and are then arranged in a rough sketch of the entire drawing. Any additions or corrections necessary to produce an accurate, balanced drawing should be made in the rough sketch. The final drawing can then be started.

In starting a drawing directly from a piece of equipment, it is advisable to make a rough drawing and draw symbols of the main parts of the system first. Then trace and draw the circuits. For instance, assume

a control system for a motor. Draw the main contactors first, then the power circuit. Then draw the main contacts of the control system, followed by the remaining circuits. The system should be studied first, and the location of the main parts planned to allow proper space for circuit lines and remaining symbols.

Solid-state electronic devices usually are wired on printed circuit (pc) boards. The components of the circuit are often transistors and integrated circuits. Such circuits are difficult if not impossible to trace by visual inspection. They should be treated as "black boxes" and shown in outline form only as a square or rectangle. The connecting lines into and out of the box would be the only lines shown on such diagrams.

5-7 LETTERING

Perhaps the most important single thing in the appearance of a diagram is the lettering. A properly lettered drawing creates confidence and respect. Good lettering is easily accomplished with an understanding of the basic principles of letter formation and with practice in drawing letters. Letter formation should be carefully studied and practiced in drawing. All lettering should be confined to straight single-stroke, vertical, capital letters and figures.

Figure 5-4 is a sample of vertical letters and figures with pleasing shape and balance. These letters are simple to draw and easy to read. In electrical drafting, letters and figures should be drawn about 1/8 in. in height. Extremely light guidelines should first be drawn with 1/8 in. space between the lines, each letter touching the line at top and bottom, and when the lettering is completed, the lines can be easily erased or left on the drawing as desired. All lettering, however, should be done with the aid of guidelines. Space between letters should be about three times the width of the lines used for drawing the letters. Space between words should about equal the width of the capital H.

5-8 DRAWING EQUIPMENT

Learning to draw can be easier if the use of drawing equipment is understood first. Figure 5-5 shows equipment commonly used in electrical work. Each item and its use will be discussed.

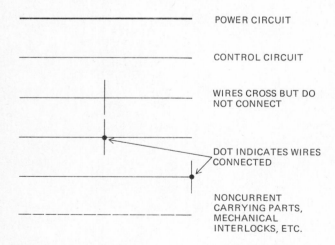

POWER CIRCUIT

CONTROL CIRCUIT

WIRES CROSS BUT DO NOT CONNECT

DOT INDICATES WIRES CONNECTED

NONCURRENT CARRYING PARTS, MECHANICAL INTERLOCKS, ETC.

Fig. 5-3 Examples of the types and weights of lines used in electrical drawings.

Fig. 5-4 Simple, single-stroke capital letters are most suitable for electrical drawings.

Fig. 5-5 Drafting equipment used for making electrical drawings.

1. *Drawing board.* Good drawing boards are made of white pine, basswood, or yellow poplar. No wood with a grain or with hard and soft spots in it should be used. The board should be at least 18 × 24 in.

2. *T square.* A T square should be durable and tough enough to maintain a straight edge and not dent easily. T squares are used to draw horizontal lines and align and support triangles and other equipment for drawing other than horizontal lines. Figure 5-6(*a*) illustrates the correct position of the hands in the use of a T square and pencil (for a right-handed drawer). The left hand should maintain a steady push or pressure to the right on the blade at all times to keep the T square tight against the left edge of the board.

3. *45° triangle.* This triangle is used with the T square for drawing vertical lines or lines at 45°

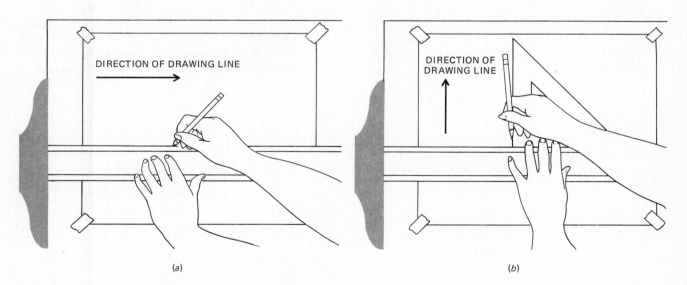

(a)

(b)

Fig. 5-6 (*a*) Drawing a horizontal line using the T square. (*b*) Drawing a vertical line using a T square and triangle.

to the horizontal. Figure 5-6(b) illustrates the proper position of the hands, T square, and pencil in using triangles.

4. *30°–60° triangle*. This triangle is used to draw lines vertical or 30° or 60° to the horizontal. Angles in multiples of 15° can be drawn by using the 30°–60° or 45° triangles separately or in combination. Figure 5-7 illustrates how to use both triangles and a T square to draw a line 15° or 75° to the horizontal. The other angles, 30°, 45°, 60°, and 90°, can be drawn by using the proper triangle.

5. *Template*. Templates aid in rapid and accurate drawing of a variety of figures and designs. They are usually made of plastics with figures and designs of various shapes and sizes cut out. The cutouts are used as pencil or pen guides in drawing.

   Templates are made for many types of work in many trades. Commercial templates are available with many of the standard electrical symbols shown in Fig. 5-2.

6. *French curve*. This is used to draw irregular curves. It is available commercially in several sizes and shapes. Electrical drawings rarely require the use of French curves, though they can be useful for drawing accurate curves on graphs.

7. *Scale*. A 12-in. scale (or rule) is the most practical type to use for general measurement work. A ruler can be used conveniently, as illustrated in Fig. 5-8(a), for dividing a line into equal parts. For example, assume that a line 9 in. long must be divided into 11 parts. In Fig. 5-8(a) the line is shown horizontal, but it could be at any angle from the horizontal. First, draw a second line, as shown, at an angle away from the given line. Divide the second line into 11 equal parts by simply measuring off 11-in. marks using a scale.

Connect the end "tick" mark (the 11th mark) on the line with the end of the 9-in. line. Then, using the T square and triangle as shown, draw a very light parallel line from each of the other 10 tick marks to the 9-in. line. The intersection of the parallel line with the given 9-in. line divides the line into 11 equal parts.

8. *Drafting tape*. Drafting tape is used to fasten the drawing paper to the board. Pieces about 1 in. long and $\frac{3}{4}$ in. wide are allowed to lap about $\frac{1}{2}$ in. each on the board and paper.

9. *Drawing pencil*. Any good black pencil of medium hardness (H or 2H) can be used for drawing. A 4H pencil should be used for the light guidelines in lettering. A good rubber eraser should be used for corrections or removal of guidelines. A ball-point pen can be used for permanent drawings.

10. *Protractor*. A protractor is used for laying out angles that cannot be laid out with the regular triangles. Sheet-metal or plastic protractors are suitable for electrical work.

11. *Divider*. A divider has two sharp steel points at the tips of the legs and is used for transferring measurements and for dividing lines and circles or arcs into a number of equal parts. In dividing a line into equal parts, the divider points are set at the approximate distance of one part, and the divider is "stepped" along the line for a trial. If it is set too wide, it will exceed the line. The amount of excess is divided by the number of parts, and the divider is closed by that amount.

   A circle can be divided into equal parts with a divider by stepping around the circle to establish the division marks, as illustrated in Fig. 5-8(b). First dividing the circle into major parts will make the process easier, if there are to be a large number of parts. If the number of divisions is

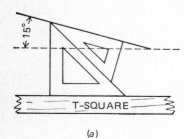

(a)

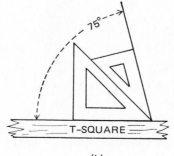

(b)

Fig. 5-7 Use of 30°–60° and 45° triangles to draw lines 15° (*a*) and 75° (*b*) to the horizontal.

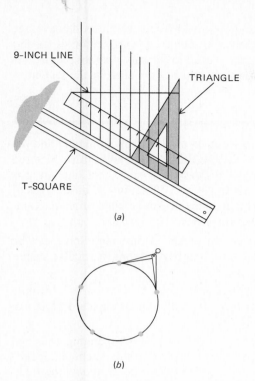

9-INCH LINE

TRIANGLE

T-SQUARE

(a)

(b)

Fig. 5-8 (a) **Using a scale, T square, and triangle to divide a line into equal parts.** (b) **Using a compass to divide the circumference of a circle into equal parts.**

divisible by 2, the circle can be first divided into two equal major parts and then stepped off; if it is divisible by 4, it can be divided into four equal major parts with a 90° triangle and then stepped off.

12. *Compass.* A compass is used to draw circles or arcs. Some compasses have interchangeable legs permitting them to be used with ruling pens for ink drawings, lead holders for pencil drawings, or needle points so that the instrument can be used as a divider.

For durable permanent drawings, the paper should be the equivalent of 20-lb weight or heavier, white or cream, and have at least 50 percent rag content. It should have a good "bite" to take pencil lead easily. For permanent drawings ink drawings can be made on plastic or tracing cloth (starched linen).

5-9 MAKING SIMPLE TEMPLATES

Occasionally some drawings require a number of duplicate designs for which a template cannot be pur-

chased. Templates for such applications can be easily made from stiff paper, cardboard, or plastic. Thickness should be from 0.010 to 0.050 in.

5-10 SCALE DRAWING

A scale drawing is one that is in exact proportion to the object drawn, whether it is larger, smaller, or the same size. The actual size of the drawn object can be determined by measuring the drawing and applying the ratio. Scale drawings are seldom used in schematic diagrams. They are typically used in floor plans of buildings to be wired, though the electrical symbols are rarely drawn to scale. Drawings of equipment are usually drawn to scale.

Assume that floor plans for a building 40×60 ft are to be drawn. The drawing paper is 18×24 in. It will be necessary to select a convenient unit of measurement for the drawings, such as a fraction of an inch, to equal a foot of the house measurement. One-fourth in. is convenient in this case. Since the house is 40 ft on one side, drawing this side of the house on the paper will require 40 one-fourths of an inch, or 10 in. Drawing the other side of the house will require 60 one-fourths of an inch, or 15 in. on the paper. The whole drawing will occupy 10×15 in. on the paper, and it would be termed "drawn to a scale of 1/4 inch equals 1 foot," written as: $1/4'' = 1'$.

If it is desired to have the drawing larger, 3/8 in. will result in a $15 \times 22 \ 1/2$ in. drawing, which will allow about 1-in. margin on the paper. A scale, when used, should be shown in the title block.

5-11 TITLE BLOCK

Certain identification or explanatory information of a drawing should appear on the drawing. This information, such as the name of the drawing, the drafter, date, scale (if used), or any other pertinent information, is recorded in the *title block*. The title block can be a full-length strip across the bottom of the drawing, but usually it is placed in an outlined area at the lower right-hand corner of the drawing.

5-12 LEGEND

Occasionally a supplementary explanation of symbols, devices, or parts of a drawing is necessary to clarify the drawing. Such information is recorded in a table on the drawing called a *legend*. The legend is usually located in a convenient space at the right. Some drawings contain a legend that completely identifies the drawing, while others contain only information necessary to make them clear to the average reader. Information such as an explanation of the letters and numbers used to identify operating coils, relays, and resistors is usually placed in the legend.

SUMMARY

1. An electrical drawing or blueprint is a brief, concise method of recording and conveying information on wiring and equipment.

2. Most electrical work requires the use of some type of electrical drawings.

3. Industrial electricians should be able to draw and interpret blueprints.

4. Drawings are composed of lines and symbols usually made with the aid of drawing instruments, with the exception of lettering, which can be done freehand.

5. Drawing and studying diagrams of equipment is a good method of familiarizing oneself with equipment.

6. Appearance has considerable bearing on the appeal and effectiveness of a diagram. No special skill or knowledge is required to draw neat and appealing diagrams. They can be drawn with understanding of some basic rules, practice, and inexpensive equipment. Copying good diagrams affords excellent practice.

7. An electrician should have diagrams of all equipment under his or her care.

8. There are three common types of electrical drawings: schematic diagrams, plans, and pictorial drawings.

9. A schematic design shows the actual wiring of a piece of equipment or system. A plan shows the physical location of electrical equipment and devices as well as cable and conduit runs. Pictorial drawings show the actual equipment and wiring as it appears to the eye. Variations of these types of drawings include one-line or single-line diagrams and riser diagrams.

10. Templates are handy in drawing. Some can be handmade to simplify drawing particular shapes or objects.

11. The American National Standards Institute (ANSI) has adopted a set of standard electrical symbols.

12. The symbol used to represent operating coils and relays is a circle; this circle should contain identification letters or numbers to designate the function of the part.

13. Resistors are shown by a rectangular figure, and they should be identified according to sequence of operation, such as $R1$, $R2$, etc. The resistance value in ohms, or the wattage in watts, may be given inside the rectangle.

14. Supplementary information for a drawing should be placed in the legend in any convenient location on the drawing. Identification information should be placed in the title block in the lower right-hand corner.

15. Anyone with practice and study can produce neat drawings, and neat drawings produce pride and confidence.

QUESTIONS

5-1 What is an electrical diagram?

5-2 Why should electricians have electrical plans and diagrams of the areas and equipment under their care?

5-3 How are two crossing wires that are not connected shown on a diagram?

5-4 How are two crossing wires that are connected shown on a diagram?

5-5 How can solid-state electronic devices be shown on an electrical wiring diagram?

5-6 What type of drawing is used to indicate vertical conduit runs in multiple-story buildings?

5-7 What is a T square used for?

5-8 What are the two common triangles used for in electrical drawings?

5-9 What is a drawing template?

5-10 Draw the electrical symbol for each of the following: an ammeter; a circuit breaker; a system ground.

5-11 What is a title block?

5-12 Where are title blocks usually located?

5-13 What is a legend table?

5-14 Explain the use of a protractor.

5-15 A scale of 1/4″ = 1′-0″ is used to draw the outline of a building 100 ft. × 150 ft. What is the drawing size of the outline?

6
MAGNETISM AND ELECTRO-MAGNETISM

Most electrical equipment uses magnetism in some part of its operation or control. Magnetism furnishes the force that causes motors to rotate, magnetic switches to close, and other mechanical devices to operate. It also produces the field which can generate a voltage and current in a set of conductors. In addition, magnetism is used to record information on magnetic tape and disks. Information is stored and transferred magnetically to operate and control machines and equipment.

Thus magnetism can do mechanical work. It can generate electrical power and it can store and use data.

Since magnetism plays an important part in the operation of electrical equipment, a good understanding of it is necessary in order to understand electrical equipment. The important things to know are the properties and behavior of magnetism. Like electricity, very little is known about the exact nature of magnetism; however, its effects are well known, and those are what are important to the electrician.

6-1 NATURE OF MAGNETISM

All magnetism is the same regardless of how it is produced. The magnetism in a motor, transformer, or generator is the same as that composing the earth's magnetic field, or that of the familiar horseshoe magnet. Nearly everyone has played with a horseshoe magnet and knows it has the property of attracting and holding certain metals. Magnetism also occurs naturally in certain ores and rocks such as hematite (a reddish oxide of iron) and lodestone (a natural magnetic stone).

A horseshoe magnet holding a large number of nails is shown in Fig. 6-1(*a*). It is evident that some sort of attractive powers are present between the legs of the magnet. Figure 6-1(*b*) shows what happens when a piece of steel (in this case a chisel) is placed across the ends of the magnet. Here the magnetism has been "short-circuited" and most of the nails have dropped away. It is evident, therefore, that the magnetic force is now going through and acting on the chisel instead of the nails. The medium that gives a magnet the powers of attraction and other characteristics is known as *magnetic lines of force,* or simply *lines of force.* These lines make up what is known as the *magnetic field,* or *magnetic flux,* of the magnet.

Figure 6-2 is a photograph of iron filings that were sprinkled on a flat sheet of cardboard resting on the ends of the horseshoe magnet shown in Fig. 6-1. The filings have arranged themselves naturally in a pattern corresponding to the magnetic lines of force produced by the magnet. A study of the pattern indicates that the lines of force are strong at the ends of the magnet as evidenced by the large number of filings attracted to those points. The filings also have been arranged by the magnetism of the lines of force traveling from one magnet end to the other. It can be seen that lines of force do not travel straight and directly from one magnet end to the other.

Figure 6-3 shows a magnet with lines of force sketched to emphasize the nature of their travel. When lines of force travel from the same source, they are said to be of the same *polarity,* and lines of same or like polarity *repel* each other. Hence, the first line in the illustration is straight from tip to tip, the second line is slightly curved owing to repulsion by the first line, and the remaining lines are curved to a greater degree owing to repulsion between all the lines. A study of these lines will explain the curved lines formed by repulsion and shown in the pattern of the iron filings in Fig. 6-2. Lines of force actually radiate from all surfaces of a magnet, not just the ends. However, most of the lines are densely bunched at the ends of the magnets before traveling into open space.

49

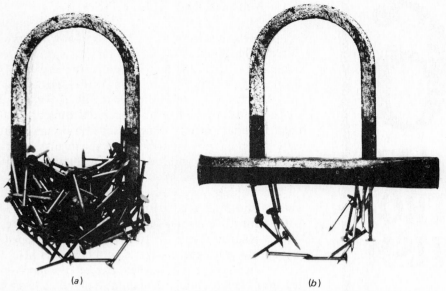

Fig. 6-1 (*a*) Horseshoe magnet attracts nails. (*b*) The chisel "short circuits" the magnetic force and reduces the ability of the magnet to hold the nails.

6-2 THE COMPASS

The earth is a huge magnet. Magnetic lines of force emerge from the earth at the *north* pole and travel through space southwardly and enter the earth at the *south* pole. A compass is used to indicate geographical directions by placing it in the earth's magnetic field and noting the action of the compass needle. The needle is a small magnet mounted on a sensitive pivot which allows it to rotate freely. Lines of force in the earth's magnetic field enter the south pole of the compass needle and emerge from the north pole and swing the needle into alignment with the earth's field. The pole of the needle usually finished black is a south

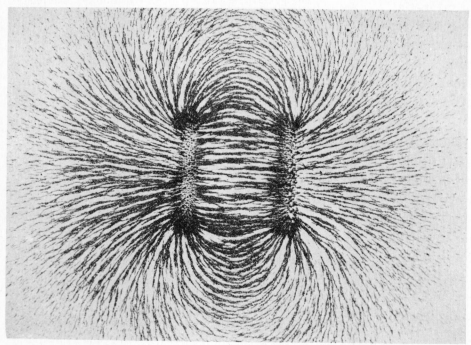

Fig. 6-2 Iron filings arrange themselves naturally along the magnetic lines of force. Iron filings have been sprinkled on a flat cardboard resting on the ends of the horseshoe magnet shown in Fig. 6-1.

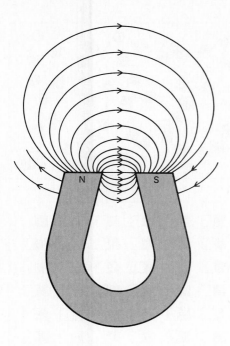

Fig. 6-3 The direction of the magnetic lines of force.

pole but since it points to the north it is often referred to as a *north-seeking* pole.

Since the earth's *north magnetic* pole is at the geographic *north* pole, lines of force emerging at the

north magnetic pole travel southwardly. They enter the north-seeking pole of a compass needle. They then continue on to enter the *south magnetic* pole at the earth's geographic south pole. Hence, the north-seeking pole of the needle points to the earth's geographic north pole. The important thing to remember is that the geographic north pole behaves like the north pole of a magnet and the geographic south pole behaves like the south pole of a magnet. These conditions are illustrated in Fig. 6-4.

6-3 POLARITY

Since magnetism emerges at one pole of the earth and enters at the other pole, the area where it emerges is called the *north magnetic pole*, and the area where it enters is called the *south magnetic pole*. The earth's geographic poles are the same as its magnetic poles. The area where magnetism leaves a body is called the *north pole*, and where it enters is called the *south pole*.

6-4 MAGNETIC MATERIALS

Only a few metals show magnetic properties or are affected by magnetism. These materials are known as

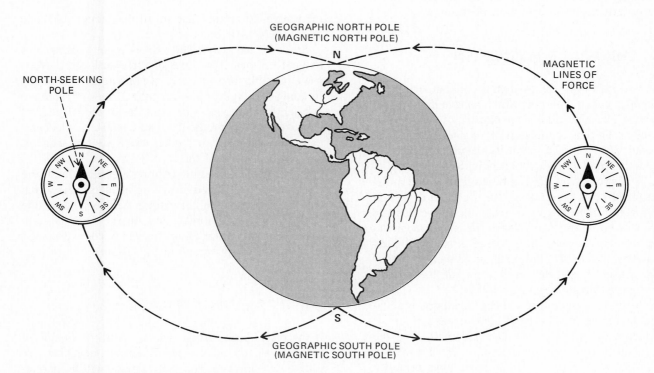

Fig. 6-4 The earth as a magnet. Note that the geographic poles coincide with the magnetic poles.

magnetic materials. Iron, cobalt, nickel, manganese, and chromium are magnetic to some degree. Iron and iron derivatives (such as steel and iron alloys) are the best magnetic materials. Magnetism is conducted by iron about 2000 times better than by air or any other nonmagnetic material.

Magnetism can interfere with the proper operation of certain pieces of equipment, such as electrical meter and watch movements, recording tapes, and television sets. Unlike electricity, there is no magnetic insulator. Rather than repel magnetism, nonmagnetic materials are readily penetrated by the magnetic field. Such materials as wood, glass, rock, plastics, concrete, paper, lead, and air cannot stop the lines of force. Their only advantage is in being able to physically separate the source of the field from the object that may be attracted or influenced by the field. To cancel or reduce the effects of a magnetic field, it is necessary to deflect or divert the field. This is done by surrounding the object to be protected with a soft iron shield or case. This shield provides an easy path for the magnetic field and deflects the field away from the object being protected. Thus a ''nonmagnetic'' watch is actually one in which the movement is mounted in a soft-iron case. Naturally, for the sake of beauty and physical protection, an additional outer case of stainless steel or gold-plated base metal is usually used. An obvious method of protecting against the effects of magnetic fields is to make the device or equipment out of nonmagnetic materials. In this way the field would have no effect on the operation of the object.

6-5 THEORY OF MAGNETISM

Magnetic materials are thought to contain *magnetic molecules* (called *domains*) which behave like tiny bar magnets. These molecules are capable of swinging or turning around in the material. A magnet is a piece of material with some of or all its magnetic molecules lined up so that the same poles point in a given direction. The strength of the combined total of the molecules lined up is concentrated in the same direction. The material then possesses magnetic properties and is said to be *magnetized*. When the magnetic molecules are disarranged, the material is said to be *demagnetized*. Actually, the material contains as many of the magnetic molecules as before, but they are just not aligned to form magnetic poles.

An iron bar containing magnetic domains is illustrated in Fig. 6-5. The black end of each figure is considered the *north pole* of each domain, and the white end is considered the *south pole* of each domain. In Fig. 6-5(*a*) all the domains have arranged

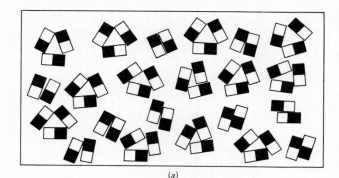

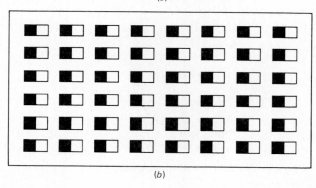

Fig. 6-5 Magnetic materials contain domains that behave much like tiny bar magnets. (*a*) The poles of the magnetic domains are not aligned. The material does not have the properties of a magnet. (*b*) With the poles of the magnetic domains aligned in the same direction, the material becomes a magnet.

themselves in groups that afford the shortest distance of travel of their lines of force.

Since air has about 2000 times more resistance to magnetism than iron, lines of force seek the shortest route possible through space of nonmagnetic materials in going from north to south poles. Lines of force are constantly under tension and have the nature of stretched rubber bands; they are capable of mechanically drawing magnetic bodies into a position that will shorten their path of travel.

If this bar is placed in a strong magnetic field, the domains will be forced to line up in such manner that the lines of force of the strong field will enter the south pole of each domain and emerge from its north pole. This will cause all of them to become arranged in a pattern as illustrated in Fig. 6-5(*b*).

6-6 PERMANENT MAGNET

If the bar is high-carbon steel or an alloy capable of being hardened, it can be heated and quenched, or otherwise hardened, and the domains will be permanently locked in magnetic alignment. When the influencing strong magnetic field is removed, the domains will hold their organized positions as shown in Fig. 6-5(b). Thus a permanent magnet has been produced. All the magnetism possible in the bar from each domain is now concentrated in the same direction. The end of the bar from which the lines of force leave is the *north pole* of the magnet, and the end they enter is the *south pole*.

Since the domains are in a strain because of the distance through space from pole to pole, they are at all times tending to swing out of line and form into groups. A sharp blow on the bar will possibly allow several to swing out of the present order and form small groups. This will weaken the magnet because it destroys part of the concentrated magnetic force.

Magnetism in a permanent magnet can be weakened or destroyed by *striking* it, *heating* it, or otherwise aiding the domains in becoming disarranged. In the manufacture of permanent magnets for use where a constant strength is desired, such as in electrical meters, the magnets are subjected to such demagnetizing conditions as they may encounter in normal use. This causes domains that are likely to become disarranged to swing out of their organized position before the magnet is installed for use. After installation, the remaining domains can be reasonably depended upon to remain in their organized positions, and the strength of the magnet maintained for the life of the equipment.

If a magnet is produced to contain maximum magnetism, it is carefully handled at all times, and a *keeper,* such as a piece of soft iron, is temporarily placed across its poles to provide a low-resistance path for its lines of force. This low-resistance path relieves the strain on the domains and minimizes the likelihood of their becoming disarranged.

When a piece of magnetic material is magnetized by the influence of external magnetism and loses its magnetism when the influence is removed, it is known as a *temporary magnet.*

Magnetic poles can be created in any part of the magnetic material. The position of the poles will depend on the way the magnetic domains are arranged. Figure 6-6(a) illustrates a Y magnet with two north poles and one south pole. The bar magnet in Fig. 6-6(b) has been magnetized with the north pole in the center and a south pole at each end. This type of magnet is used in polarized relays, ac signal bells, and other applications.

6-7 ATTRACTION

A basic law of magnetism is that *unlike magnetic poles attract each other,* or stated simply, *unlike poles attract*. This principle is most easily understood by studying the effects of unlike poles placed close enough together to be influenced by each other. Figure 6-7(a) illustrates conditions that produce the effect of attraction between unlike poles. The lines of force from the north pole of magnet A enter the south pole of magnet B and emerge from the north pole of that magnet and travel through space to reenter the south pole of magnet A. The lines of force in the air gap between the magnets and along the outsides of the magnets, having the nature of stretched rubber bands, *draw the two magnets together*. Thus, if the lines of force, in shortening their travel distance, can pass through a magnetic material, they will magnetize it and draw it, if possible, to form a shorter route. This action is known as *attraction*. The iron-filing pattern formed by the two bar magnets in Fig. 6-7(a) is shown in Fig. 6-7(b).

The force of attraction between two magnets increases as the distance between them decreases. However the force *increases* by the *square of the inverse of the decrease* in distance. Similarly the force *decreases* by the *square of the inverse of the increase* in distance.

For example, if the two magnets are 1-in. apart and they are moved so that they are only $\frac{1}{2}$-in. apart, the increase in the force of attraction would be $(2)^2$ or 4 times greater than before. If the magnets were moved

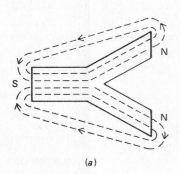

(a)

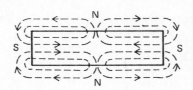

(b)

Fig. 6-6 Magnetic poles can be created anywhere in a material. (a) A Y-shaped magnet with two north poles and one south pole. (b) A bar magnet with a south pole at each end and a north pole at the center.

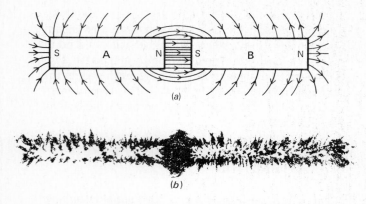

Fig. 6-7 (*a*) Unlike poles attract. In this case a force of attraction exists between the adjacent north and south poles tending to pull the two magnets together. (*b*) The iron filings show the existence of a strong field between the poles of the two magnets.

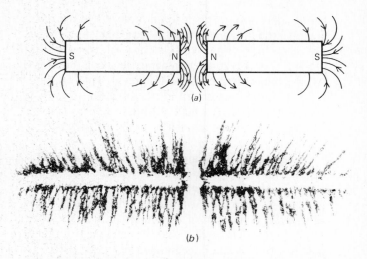

Fig. 6-8 (*a*) Like poles repel. In this case a force of repulsion exists between the adjacent north poles tending to push them apart. (*b*) The iron filings show that the two fields repel one another, forming areas in which there are no lines of force.

so that they were only $\frac{1}{4}$-in. apart, the force would be $(4)^2$ or 16 times greater than it was when they were 1-in. apart. On the other hand, if the magnets were moved 3-in. apart, the force would be $(1/3)^2$ or 1/9 the force that existed when they were 1-in. apart.

6-8 REPULSION

Another basic law of magnetism is that *like magnetic poles repel each other;* stated simply, *like poles repel.*

The conditions that exist when like magnetic poles are close enough to be under the influence of each other are illustrated in Fig. 6-8. The lines of force emerging from the north poles of the two magnets in Fig. 6-8(*a*) are shown crowding in the air gap owing to repulsion between lines of like polarity. This produces a *force of repulsion* between the two like poles. The two poles tend to push one another away. The iron-filing pattern formed by this actual condition is shown in Fig. 6-8(*b*).

6-9 TESTING FOR POLARITY

A compass can be used to determine the polarity of a magnetic pole. In using a compass, it is important that the principle of repulsion is used. That is, it is more accurate to observe the compass needle being repelled by the magnetic pole than being attracted to a pole. A compass needle that becomes attracted to a pole may have had its own polarity changed by the magnet's field. In any case, good compasses should not be kept long in strong magnetic fields. When testing the polarity of strong magnets with a compass, a good

method is to rub the magnet with a piece of hardened steel, such as a chisel or screwdriver, and then test the polarity of the pieces of steel with a compass. Rubbing the steel against the magnet magnetizes the steel with the same polarity as the magnet. If the end of the steel rubbed against the magnet results in a north pole, it will attract the south pole of a compass needle, and this indicates that the pole of the magnet under test is a south pole.

A compass should always be tested before and after a polarity test to be certain of results. It can be tested by checking to see that it still indicates the geographic north.

6-10 PERMEABILITY

The degree to which a magnetic material conducts magnetism is called the *permeability* of the material. It is determined by the number of lines of force that can be contained in a given area of the material. Soft iron is more permeable than hard steel. Permeability in magnetism has a meaning comparable with *conductivity* in electricity. Permeability is generally used to describe the ease with which magnetism can travel through a given material.

6-11 RELUCTANCE

Materials that are poor conductors of magnetism are said to have a high reluctance. *Reluctance* in magnetism has a meaning comparable with resistance in electricity. Reluctance and permeability pertain to opposite conditions. Hard steel has higher reluctance to magnetism than soft iron.

6-12 RETENTIVITY

When the magnetizing influence has been removed from a magnetic material, it loses some of its magnetism. Its ability to retain magnetism is called its *retentivity*. The retentivity of hard steel is considerably higher than that of soft iron. A high degree of retentivity is desirable in materials used for making permanent magnets, while materials with a low degree of retentivity are desirable in the manufacture of many electrical devices, such as magnetic contactors. Retained magnetism will tend to hold contactors closed even after the magnetic circuit has been deenergized. Antimagnetic shields are usually made of soft iron so that they will lose their magnetism once they are out of the influence of a magnetic field.

6-13 ELECTROMAGNETISM

Whenever electrons move, *a magnetic field develops and surrounds the path*. This magnetism is known as *electromagnetism*. It is the same as magnetism produced by any other means—the same as that of a horseshoe magnet or the earth's magnetic field—and it therefore has the same characteristics.

When an electric current flows through a conductor, *a magnetic field surrounds the conductor*. Magnetism develops in the center of the conductor and expands outward into space around the conductor. The strength of the magnetism is in direct proportion to the amount of current flowing.

If the amount of current is doubled, the resultant magnetism will be doubled. Lines of force appear in the form of circles around, and at right angles to, the conductor. They do not remain stationary but travel in a circle around the conductor. The *direction of flow of current determines the direction of flow of lines of force* in the circle.

6-14 RIGHT-HAND RULE FOR CONDUCTORS

The *right-hand rule* for conductors can be used to determine the direction of flow of lines of force around a conductor if the direction of current flow is known. Figure 6-9 illustrates the use of the right-hand rule. If

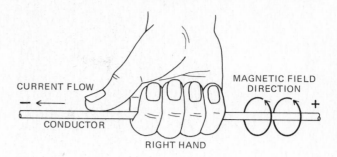

Fig. 6-9 Current flow in a conductor produces magnetic lines of force around the conductor. In using the right-hand rule, the thumb is extended in the direction of current flow (+ to −). The fingers will then grasp the conductor in the direction of flow of the magnetic lines of force.

a wire is grasped with the right hand with the thumb pointed in the direction of current flow, the fingers will point in the direction of flow of the magnetic lines of force around the wire. In this illustration, current flow is from right to left, and lines of force are traveling in the direction the fingers are pointing.

If two wires carrying current in opposite directions are close enough together to be influenced by their magnetic fields, they will be repelled by their fields. Figure 6-10(*a*) illustrates how repulsion results from this condition. A cross is shown in the wire at the right, indicating that current is flowing away from the reader; a dot is shown in the wire at the left, indicating that current is flowing toward the reader. This method of indicating direction of current in the cross section of a conductor is commonly used in textbooks. (A good way to remember the symbol is to think of the dot as the point of an arrow coming *toward* you while the cross indicates the tail of the arrow going *away* from you.)

With the current flow toward you in the left wire (as indicated by the dot) the lines of force will travel counterclockwise. This can be verified by the use of the right-hand rule. By pointing the thumb of the right hand in the direction of current flow, the fingers will grasp the conductor in a counterclockwise direction.

The current in the right wire is flowing away (as indicated by the cross), and the lines of force are traveling clockwise. Between the two conductors the lines of force are flowing in the same direction; they are of like polarity and therefore repel each other.

The force of attraction between two parallel wires carrying current in the same direction is illustrated in Fig. 6-10(*b*). The lines of force around each wire travel clockwise. In the space between the wires, the direction of the circular lines of the left wire is down, and the circular lines of the right wire are up. Since they are traveling in opposite directions, they are of

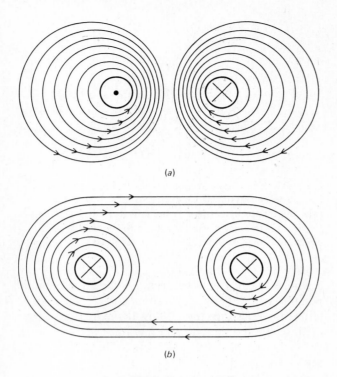

(a)

(b)

Fig. 6-10 (*a*) Current flowing in opposite directions in two parallel wires will produce a magnetic force tending to push the two wires away from each other. (*b*) Current flowing in the same direction in two parallel wires will produce a magnetic force tending to pull the wires toward each other.

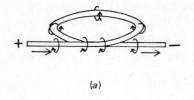

(a)

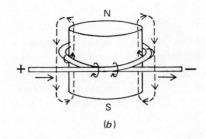

(b)

Fig. 6-11 (*a*) Current in a conductor produces magnetic lines of force. (*b*) When a conductor is formed into a coil, magnetic poles are created at each end of the coil.

unlike polarity and attract each other, so they join and encircle both wires, which tends to draw the two wires together.

6-15 ELECTROMAGNETS

An electromagnet is a magnet *produced by electricity*. If a wire carrying current is wrapped around an iron core, the magnetic lines of force around the wire will align the magnetic molecule-like (domain) structure of the core. These domains will concentrate their forces in the path of the lines of force and produce an electromagnet.

The directions of circular lines of force around a conductor containing a loop and carrying current from left to right are indicated by the arrows in Fig. 6-11(*a*). It will be noted that the circular lines around the loop turn from the top of the loop toward the outside, down and around under the loop, and up on the inside of the loop. If an iron core is passed inside the loop, as shown in Fig. 6-11(*b*), the circular lines of force around the wire will enter the iron core, follow magnetic domains to the top of the core, and emerge, creating a north pole, and continue down through

space and enter the core at the bottom, creating a south pole at the bottom of the core. Thus an electromagnet is created.

A simple electromagnet is shown in Fig. 6-12. Bell wire is shown crudely wound around a large nail. Current is being supplied by a large dry cell. A knife switch controls the current. When the switch is open, the nail will be nonmagnetic. With the switch closed, current will flow and the nail will become magnetized. The strength of the electromagnet will depend on the current, the number of turns of wire around the nail, and the material from which the nail is made.

An electromagnet has the same characteristics and nature as a permanent, or horseshoe, magnet. It has the same properties of attraction and repulsion. It can magnetize bodies of magnetic materials brought into its influence, and it is governed by all rules of magnetism. The advantage of electromagnetism is that it can

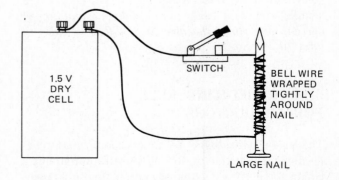

Fig. 6-12 A simple electromagnet.

be increased, decreased, or reversed merely by controlling the electricity producing it.

6-16 POLARITY OF ELECTROMAGNETISM

When electromagnets are used for mechanical functions, such as lifting iron and steel and closing or opening contactors, valves, brakes, and relays, polarity is of no consequence. An electromagnet has equal pull or attractive force at either of the poles. Polarity is the direction of flow of lines of force in a core producing magnetism. In producing electromagnetism for the operation of motors and generators, polarity is of prime importance. A dc motor requires magnetism for its armature to push against, to produce rotation, and a generator requires a magnetic field for its armature conductors to cut in generating electricity.

Magnetic circuits for these pieces of equipment must be orderly and efficient. The *polarity* of the magnetism of a *motor* determines the *direction of rotation of the armature,* and the *polarity* of a *generator* determines the *direction of generated voltage.* Accordingly, polarity of electromagnetism must be understood for connecting field coils in motors and generators and other pieces of electrical equipment.

The polarity of a coil is determined by the *direction of flow of current around the coil.* If the current flows in a certain direction around the core, a certain polarity will result in the core. If the current is reversed, the polarity of the core will reverse.

6-17 RIGHT-HAND RULE FOR COILS

The right-hand rule for coils is frequently used in connecting motor and generator field coils. The rule is: If a coil is grasped with the *right hand* with the *fingers pointing in the direction of current flow in the coil,* the *thumb* will be pointing toward the *north pole.* That is, it will be pointing in the direction of flow of the lines of force through the center of the coil. Figure 6-13 shows arrows on the turns of the coil indicating the direction of flow of current; the fingers of the right hand are pointing in the same direction as the arrows. The thumb is pointing to the north pole. When current enters the coil at the bottom and flows around the core as indicated by the arrows, the north pole will form at the top of the core, and the south pole will form at the bottom, as indicated by the letters.

The two bars A and B in Fig. 6-14 demonstrate how coils are wound to produce a north and south pole in the air gap between the bars. Current entering the coil of bar A from the positive side, and flowing as indi-

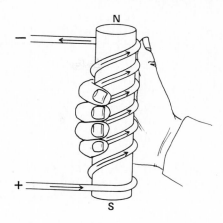

Fig. 6-13 The right-hand rule for coils. If the fingers grasp the coil in the direction of current flow through the coil (+ to −), the thumb will point to the north pole. The thumb points in the direction of flow of the lines of force through the center of the coil.

cated by the arrowheads, produces a north pole at the right end of bar A. The right-hand rule proves this. Current continues into the coil of bar B to magnetize it like bar A, which produces unlike poles in the air gap between the bars.

6-18 STRENGTH OF ELECTROMAGNETS

The strength of the pulling force of an electromagnet is in inverse proportion to the square of the distance at which it acts. If the distance is halved, the pulling force is increased 4 times. If the distance is doubled, the pulling force is reduced to one-fourth. If an electromagnet has a pulling force of 1 lb through a distance of 1 in., it will have a pulling force of 4 lb at a distance of 1/2 in.

Some control systems employing electromagnets to close contactors automatically place resistance in series with the coil of the electromagnet after it has closed its armature and reduced the air space to zero. This is because there would be a large surplus of pulling power with no air gap. (The armature of an electromagnet is the part of the magnetic circuit that moves in operation.)

The relationships governing a magnetic circuit are similar to the relationships governing an electric circuit as expressed by Ohm's law. Current flow in an electric circuit is determined by the values of voltage (pressure) and resistance in the circuit. The amount of

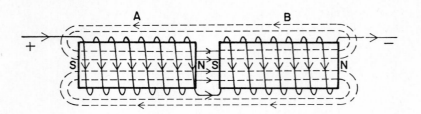

Fig. 6-14 Winding coils to produce opposite polarity in adjacent magnets.

magnetic flux (that is, lines of force) is determined by the values of *magnetomotive force* (pressure) and *reluctance* (resistance) in the magnetic circuit. The relationship is

$$\text{Magnetic flux} = \frac{\text{magnetomotive force}}{\text{magnetic reluctance}}$$

Thus the strength of an electromagnet is determined by the strength of the magnetizing influence and the reluctance of the magnetic path.

6-19 MAGNETOMOTIVE FORCE

The magnetizing influence is the *magnetomotive force* (abbreviated *mmf*). The value of mmf is determined by the *number of turns in a coil* and the *amperes flowing in the coil*. The product of the turns and amperes is *ampere-turns* (abbreviated At). If a coil has 1000 turns of wire and 5 A flowing through it, it is supplying a magnetomotive force of 5000 At.

The ampere-turns of a coil operating at a constant voltage can be changed only by *changing the wire size*. This is true because a change in turns of the same size wire makes a proportionate change in the total resistance of the coil, and this results in an equal but opposite change in current (Ohm's law). Hence, ampere-turns will be the same. For example, a coil at a constant voltage having 1000 turns is carrying 5 A. The mmf will be 5000 At.

$$\text{mmf} = 1000 \times 5 = 5000 \text{ At}$$

If the number of turns were doubled to 2000, the length of wire would about double and the resistance of the wire would be approximately doubled. According to Ohm's law the current will be one-half as much, or 2 1/2 A. Thus the new ampere-turns will be

$$\text{mmf} = 2\ 1/2 \times 2000 = 5000 \text{ At}$$

which is the same as before the change. In a practical case, of course, as we add turns to a coil, the turns become longer so that an equal number of turns wound over another coil would require more wire than the turns wound closer to the core. If 500 turns were wound over a coil consisting of 500 turns, the resistance would *more* than double and the current would *decrease more* than one-half. Thus the ampere-turns of both coils would not be exactly equal; the coil with the fewer turns would produce a strong magnet.

Increasing the number of turns of a coil maintains about the same magnetic strength but reduces the operating temperature by reducing the amperes. This is because the lower current produces less I^2R heating in the wire, thus reducing the overall operating temperature.

6-20 MAGNETIZATION AND SATURATION

There is a limit to the amount of magnetization a core can take. The reluctance of magnetic material increases as the material becomes magnetized. In the early stages of magnetization, a core magnetizes to a high degree for each unit of mmf. As the units are increased beyond a certain point of magnetization, each unit of mmf becomes less effective in producing magnetism. This point is known as the *saturation point*.

Figure 6-15 illustrates the magnetization characteristics of soft annealed sheet steel. The units of magnetic pressure (mmf) are shown in ampere-turns along the horizontal line (the *x* axis), and the degree of resulting magnetic flux density is shown in units called *teslas* (abbreviated T) along the vertical line (the *y* axis). It will be noticed that the first 10 ampere-turns produce about 1.4 teslas, while the first 20 ampere-turns produce 1.6 T for a gain of only about 0.2 T for the second 10 ampere-turns. So it can be seen that annealed sheet steel magnetizes easily in the beginning but reaches a point where additional magnetizing force produces little magnetization. The point on the magnetization curve marked *X* is considered the saturation point of this piece of steel. Beyond this point magnetization can be induced in the steel, but the degree of magnetization becomes less with each additional unit of mmf, as the graph shows. All magnetic materials have a saturation point. Keeping these facts in mind will aid in understanding why the speed and horsepower of motors do not increase proportionately

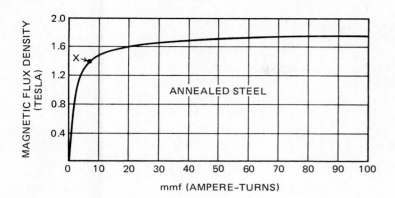

Fig. 6-15 Magnetizing curve of annealed sheet steel.

with increases of current or ampere-turns. This fact is of particular importance in the operation of self-excited generators.

6-21 RESIDUAL MAGNETISM

When a magnet produced by electricity is deenergized by stopping the flow of electricity, some magnetism remains in the core of the magnet. This remaining magnetism is known as *residual magnetism*. The amount is partly determined by the retentivity of the core. Residual magnetism in its pole pieces is necessary for a dc generator to start generating each time it is started. If for any reason it loses its residual magnetism, it cannot start generating. Residual magnetism in some magnetic contactors causes erratic operation since it will not permit positive operation of the contactor arm.

6-22 USES OF ELECTROMAGNETS

Electromagnets are used in numerous and varied ways. One of their chief advantages is remote control. They are used for such mechanical duties as opening and closing valves, braking, clutching, holding magnetic material for machining, operating signaling devices and relays, and lifting scrap iron. A *relay* is an

electrical device for controlling its own or other circuits. An electromagnetic relay is operated by an electromagnet. There are a large number of such relays, usually named to indicate their function and nature, but they are electromagnets in construction. The magnetic poles of most large motors and generators are electromagnets.

6-23 MOTOR AND GENERATOR POLES

Polarities of adjacent poles of a motor or generator are magnetized alternately north and south by the method shown in Fig. 6-16. Figure 6-16(a) illustrates a dc field frame with pole pieces to be magnetized north and south as indicated, and Fig. 6-16(b) shows the winding and direction of current flow to produce the polarities indicated for the poles. In Fig. 6-16(b), current enters the positive line at the left and flows in the direction indicated by the arrows. This magnetizes the poles with the required polarity.

6-24 CONSEQUENT POLES

If one coil were left off one of the poles in Fig. 6-16(b), it would not affect the polarity or the path of

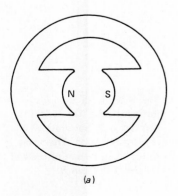

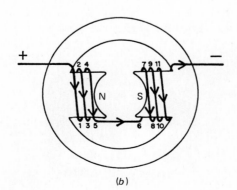

(a)   (b)

Fig. 6-16 (a) A dc motor frame showing the specified polarities for the two poles. (b) The direction of current in the field coils, and the direction of winding around the pole pieces determine the polarity of the poles.

the magnetic circuit. For example, if the coil on the pole piece at the right is omitted, the coil on the pole piece at the left would produce a north pole as usual, and the magnetic flux leaving this pole would enter the pole piece at the right and form a south pole there. The south pole thus formed would be created as a *consequence of the fact that a north pole exists*. When a pole is created in this manner, it is known as a *consequent pole*. In the case of consequent-pole motors, the pole pieces that do contain windings are known as *salient* poles.

Consequent poles are occasionally used in dc motors and are frequently used in multispeed ac motors and in a few single-speed ac motors.

6-25 COIL CONSTRUCTION

Coils used in motors and generators are frequently wound and taped, and leads are brought out for connection. Figure 6-17 illustrates the way coils are wound and taped and the leads brought out. To produce a north pole on the face of the pole piece in Fig. 6-17(c), current will have to enter in lead 2 and leave from lead 1. Flexible lead wire is usually connected to the magnet wire of the coil and brought out, but in cases of large coils sometimes screw terminals are taped to the coil.

Coils made like the ones illustrated in Fig. 6-17 are called *open-type* coils and are sometimes designated by the letter O. Coils are also made with the leads

crossed inside the tape. These coils are known as *crossed* type and are designated by the letter X, or they have an arrow taped near one lead showing how that lead enters the coil.

An open coil is shown in Fig. 6-18(a). To produce a north pole on top of this coil, the current would have to enter the coil in lead B. Figure 6-18(b) shows a coil in which the leads are crossed under the tape and the arrow taped under the right lead. To form a north pole at the top of this coil, current would have to enter in the A lead. Leads are crossed in the coil before taping and brought out for convenience in connecting coils in sequence. This is illustrated in Fig. 6-19. Figure 6-19(a) shows the external crossing of leads to coils A and C to produce proper polarity of the coils when leads are not crossed inside under the tape. Figure 6-19(b) shows the external connections for proper polarity with the leads crossed inside coils A and C before taping. This makes a neater connection and minimizes mechanical strain on the leads but causes confusion in connecting them, and so special care is necessary to make the connections properly. This crossed condition can usually be detected by examining the coil. Coil leads are extra-flexible stranded wire with durable insulation and can usually be detected under the tape in determining if they are crossed.

In some cases terminals are placed one directly over the other on coils. This is illustrated in Fig. 6-20. An arrow is taped near a lead to indicate the direction it is wound around the coil.

In cases of doubt, a coil should be tested with a

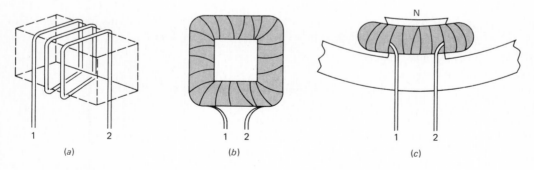

(a)   (b)   (c)

Fig. 6-17 Winding and taping coils. (*a*) A coil wound on a square form. (*b*) Taped coil. (*c*) Coil installed on pole piece.

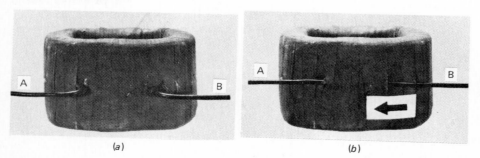

(a)   (b)

Fig. 6-18 Lead identification for taped coils. (*a*) An open coil (leads do not cross). (*b*) Crossed leads. The arrow shows the direction of the coil winding from lead *B*.

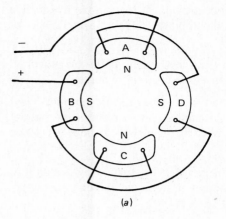

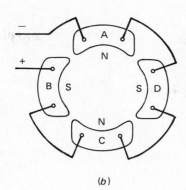

Fig. 6-19 (*a*) Open coils, with leads brought out and taped according to general practice, require external crossing of leads to produce the needed polarities. (*b*) The leads of coils A and C are crossed internally and taped to eliminate external crossing.

(*a*)   (*b*)

compass to determine polarity. Figure 6-21 shows a method of testing. According to the right-hand rule, if the lead A at the bottom is connected to the positive terminal of a battery, and lead B to the negative terminal, the coil will produce a north pole at the right of the coil if the leads are not crossed. The compass is shown with the south pole being attracted by the coil, indicating that a north pole is being formed here, which proves that the leads are not crossed. The north pole lines of force are attracting and entering the south pole of the compass needle and going through the needle to emerge at the north pole of the needle and continue around to enter the left of the coil. Care must be used in making this test to avoid reversing the magnetism of the needle. The compass should be started at a distance of 2 to 3 ft and brought slowly into the influence of the magnetism of the coil.

A compass should never be placed against a pole piece, or in a coil, before the coil is energized, because its magnetism might be reversed before it could swing into proper alignment.

When coils are tested, they should be marked to avoid confusion during installation. If the leads are open, the coil should be marked O. If the leads are crossed, the coil should be marked X. If a coil is of the type shown in Fig. 6-20, an arrow should be placed on one lead to show which way that lead enters the coil.

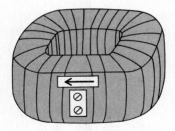

Fig. 6-20 Coil windings are sometimes terminated at screw terminals mounted side-by-side on the taped coil. The arrow indicates the direction of winding of the top lead.

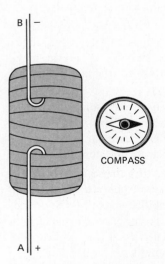

COMPASS

Fig. 6-21 Using a compass to determine the direction of winding of a coil.

SUMMARY

1. Most electrical equipment uses magnetism in some part of its operation.

2. Magnetism is used to produce electricity, and electricity is used to produce magnetism.

3. Electric motors and generators operate by the use of magnetism.

4. Magnetism consists of invisible lines of force produced by magnetic materials or electricity.

These lines of force are also known as magnetic field, magnetic flux, field flux, or flux.

5. Magnetic lines of force leave a material from the north pole and enter a material at the south pole. Stated simply, magnetic lines of force travel from north to south poles.

6. The north and south poles denote the polarities of the magnet.

7. The earth is a large magnet with its magnetic north pole at its geographic south pole and its south magnetic pole at its geographic north pole. Its magnetic poles are opposite its geographic poles.

8. The north pole of a compass needle points to the earth's geographic north pole.

9. Unlike magnetic poles attract; like magnetic poles repel. Lines of force traveling in the same direction are of like polarity and repel each other.

10. Magnetic lines of force in space exert a pulling force at all times.

11. The force of attraction or pulling power of a magnet is in inverse proportion to the square of the change in distance from the magnet.

12. Magnetic materials contain magnetic domains. Iron and steel are the best magnetic materials and therefore the best conductors of magnetism.

13. When the magnetic molecules or domains of a magnetic material are aligned in the same direction, the material is magnetized. When the magnetic domains are disarranged, the material is not magnetized.

14. When a compass is used for polarity tests, it should be held several feet from a magnet and brought slowly into the field under test, to avoid reversing the magnetism of the compass needle.

15. To test for the polarity of a magnet, observe the effects of repulsion rather than attraction.

16. In magnetism, permeability is similar to conductivity and reluctance is similar to resistance in electric circuits. Retentivity is the ability of a material to retain magnetism.

17. Residual magnetism is magnetism that is retained in a magnetic material after the magnetizing influence has been removed. Its comparative intensity is determined by the degree of retentivity of the material.

18. Saturation is the maximum point of practical magnetization of a magnetic material.

19. Electromagnetism is magnetism produced by electricity. All magnetism has the same nature and characteristics, regardless of how it is produced.

20. An electromagnet can be created by winding a conductor around an iron core and energizing the winding with electricity.

21. The strength of intensity of a magnetic field or flux produced by an electromagnet is determined by the magnetizing force and the reluctance of the magnetic path. The magnetizing force is the ampere-turns and is called magnetomotive force, abbreviated mmf.

22. The right-hand rule for conductors is used to determine the direction magnetic lines of force travel around a wire carrying current. If a wire is grasped by the right hand with the thumb pointing in the direction of current flow, the fingers will then point in the direction of travel of the lines of force around the wire.

23. The polarity of an electromagnet is determined by the direction of current flow in the coil around the core.

24. The right-hand rule for coils frequently used in connecting motor and generator fields is employed to determine the polarity of an electromagnet. If a coil is grasped with the right hand with the fingers pointing in the direction of current flow, the thumb will point in the direction of the north pole.

25. Motor or generator field coil leads should be carefully checked or tested to determine the direction the wire is wound in the coil before it is installed and connected.

26. After a coil is tested, it should be carefully marked to indicate its polarity before it is installed.

QUESTIONS

6-1 What two laws of magnetism are applied in the operation of motors and generators?

6-2 Explain the internal structure of a material that becomes a magnet.

6-3 What terms are used to denote magnetic lines of force?

6-4 How is the polarity of a magnet determined?

6-5   Describe the difference between the geographic and the magnetic north poles.

6-6   What types of materials are affected by magnetism?

6-7   How is a permanent magnet created?

6-8   How can a steel watch movement be protected against magnetism?

6-9   What is permeability?

6-10   Is the permeability of soft iron high or low?

6-11   What is reluctance?

6-12   Is the reluctance of tempered steel high or low?

6-13   What is retentivity?

6-14   Is the retentivity of soft iron high or low?

6-15   What is residual magnetism?

6-16   What is the relationship between retentivity and residual magnetism?

6-17   What is meant by saturation?

6-18   Explain the right-hand rule for a magnetic field around a current-carrying wire.

6-19   Explain the right-hand rule for the magnetic field produced by a current-carrying coil.

6-20   What is a consequent pole?

6-21   The north pole of a bar magnet is facing the south pole of another bar magnet as in Fig. 6-7(a). As the magnets are pulled further apart, does the force of attraction between them increase or decrease?

6-22   What are open-type coils?

6-23   What are crossed-type coils?

6-24   Find the magnetomotive force of a coil having 200 turns if its resistance is 10 Ω and it is connected across 120 V.

6-25   If the coil in question 6-24 were connected across 6 V, how many times weaker would the magnetomotive force be?

7

CELLS AND BATTERIES

The largest quantities of alternating current and direct current energy are produced by rotating equipment, generators and alternators. However, vast amounts of dc electricity are produced chemically. Portable equipment uses stored energy in the form of batteries; backup systems often rely on such means also. Noninterruptible electrical systems use large battery installations on a standby basis. Although these systems provide direct current, the direct current can be converted efficiently and easily to alternating current as required.

7-1 TYPES OF CELLS

Chemical energy is converted into electrical energy in devices called cells. Cells fall into two general types—*primary cells* and *secondary cells*. In the primary cell the materials that make up the cell are involved in a nonreversible process that uses up the materials as energy is taken from the cell. In the secondary cell the composition of the materials involved in the chemical reaction is slowly changed as energy is taken from the cell. However, the process can be reversed by delivering energy to the cell. This cycle can be repeated many times, and unless the cell is damaged the capacity of the cell to deliver energy will be restored fully each time.

7-2 THE VOLTAIC CELL

In its simplest form a cell consists of two dissimilar materials that are capable of conducting electricity immersed in a conducting solution. The conductors are called *electrodes*. In the typical cell there is a positive electrode and a negative electrode. The solution is called an *electrolyte*. The chemical nature of the materials involved determines the voltage produced.

A basic cell is shown in Fig. 7-1. This type of cell is also called a *voltaic cell,* after the Italian physicist Alessandro Volta. The original voltaic cell used zinc and copper and the solution was vinegar. It produced a voltage of approximately 1.05 V.

The electrochemical action producing the voltage is as follows. The electrolyte, because of its chemical nature, breaks down into negative and positive ions. These are atoms with extra electrons or a deficiency in electrons. At the same time the electrodes in the electrolyte become ionized also. In Fig. 7-2(*a*) the electro-

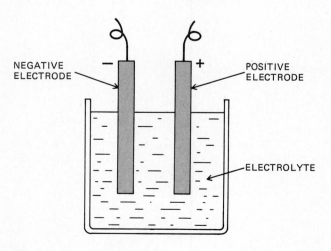

Fig. 7-1 A basic electrochemical cell.

NEGATIVE ELECTRODE

POSITIVE ELECTRODE

ELECTROLYTE

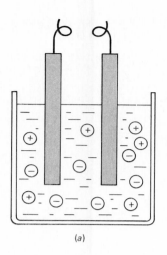 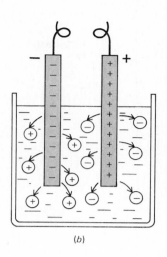 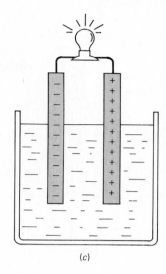

(a)   (b)   (c)

Fig. 7-2 The electrochemical action that produces a voltage. (*a*) The electrolyte breaks down into positive and negative ions. (*b*) The electrodes become ionized. (*c*) A potential difference occurs between the electrodes and current flows.

lyte is shown ionized. The electrode on the left, because of its chemical composition, is ionized and goes into the solution as a positive ion. That leaves that electrode negative. The positive ions in solution move to the other electrode and combine with electrons ionized from that electrode. This makes the electrode on the right positive [Fig. 7-2(*b*)]. If the two electrodes are connected to an external load, as in Fig. 7-2(*c*), current will flow. The movement of charge from the electrodes will keep the process going until all the materials from the electrode are dissipated.

7-3 PRIMARY CELLS

The most common primary cell is the *dry cell* used in portable radios, flashlights, toys, and other electrical and electronic equipment. These cells are available commercially in a number of different combinations of electrode materials and electrolytes.

Carbon-Zinc Cell The most widely used dry cell is the *carbon-zinc cell* (Fig. 7-3). This cell consists of a zinc can which serves as the negative electrode and a black powder compound called manganese dioxide, which is the positive electrode. A solid carbon rod makes contact with the manganese dioxide and serves as the positive terminal. The carbon, however, does not enter into the electrochemical process that takes place in the cell. The electrolyte is a damp paste of zinc and ammonium chlorides and a binder. In certain cells only a paste solution of zinc chloride is used as the electrolyte.

Carbon-zinc cells produce a voltage of approximately 1.5 V. They are relatively inexpensive and are available commercially in many shapes and sizes. Many of the popular sizes of dry cells are denoted by letters such as AA, C, and D. They do have disadvantages in that their operating characteristics are poor at low temperatures; their output voltage drops sharply as energy is withdrawn; their weight and volume are relatively high as compared with the energy available; and they have high internal resistance. By using the zinc chloride electrolyte, the drop in voltage due to the load is decreased and the internal resistance is

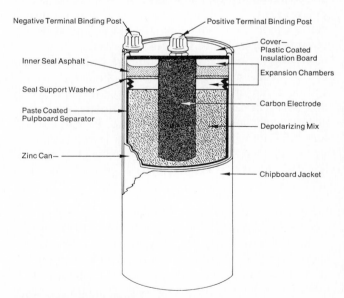

Fig. 7-3 Cross-section view of a carbon-zinc cell. (*Union Carbide Corp.*)

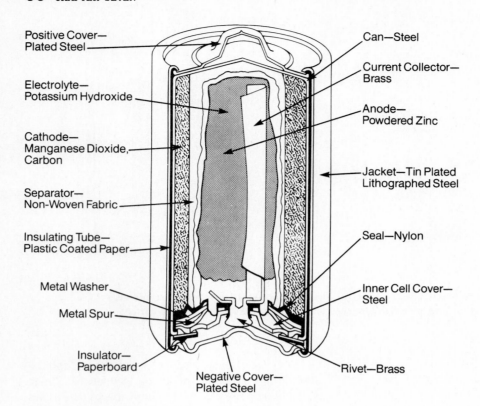

Positive Cover—
Plated Steel

Electrolyte—
Potassium Hydroxide

Cathode—
Manganese Dioxide,
Carbon

Separator—
Non-Woven Fabric

Insulating Tube—
Plastic Coated Paper

Metal Washer

Metal Spur

Insulator—
Paperboard

Negative Cover—
Plated Steel

Can—Steel

Current Collector—
Brass

Anode—
Powdered Zinc

Jacket—Tin Plated
Lithographed Steel

Seal—Nylon

Inner Cell Cover—
Steel

Rivet—Brass

Fig. 7-4 Cutaway view of an alkaline cell. (*Union Carbide Corp.*)

improved somewhat. Zinc chloride cells are sometimes sold as "heavy-duty" cells. The carbon-zinc cell is also known as the *Leclanche cell*.

Alkaline Cell Another popular type of primary cell is the *alkaline cell*. The electrodes in this type of cell are the same as used in the carbon-zinc cell. The difference in the two cells is the electrolyte. Alkaline cells use potassium hydroxide in place of the chloride electrolytes (Fig. 7-4). The output voltage of an alkaline cell is about 1.5 V. This type of cell has a number of advantages over the carbon-zinc cell. While they are more expensive than the carbon-zinc cell, they can operate at much lower temperatures; the output voltage does not drop sharply as energy is withdrawn; they have a better weight- and volume-to-energy ratio; and they have lower internal resistance. Primary alkaline cells are available commercially in most of the same sizes and shapes as the carbon-zinc cell. Alkaline cells are also available in a rechargeable (secondary) type.

Mercury Cell The *mercury cell* consists of a negative electrode made of zinc and a positive electrode made of mercuric oxide. The electrolyte is sodium or potassium hydroxide. Figure 7-5 shows a cross-sectional view of such a cell. Mercury cells have an output voltage of 1.35 V which remains relatively constant as the cell is used. For this reason mercury cells are often used as standards against which other cells and meas-

uring devices are checked. Although mercury cells are expensive, they do have advantages over both the carbon-zinc and alkaline cells. While low-temperature performance is poorer, their high-temperature performance is considerably better than that of the other two cells. Mercury cells have a lower internal resistance and a better weight- and volume-to-energy ratio. Mechanically they are extremely rugged. Mercury cells are available in many sizes and shapes including small button-shaped cells.

Silver Cell A cell similar to the mercury cell is the *silver oxide cell*. In construction, the two cells are very much alike. Their electrical characteristics are also similar, but the output voltage of the silver oxide cell is 1.5 V. Both have good volume- and weight-to-energy ratios; both have good high-temperature

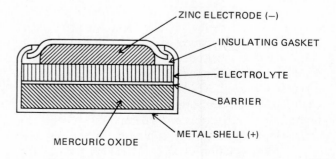

ZINC ELECTRODE (−)

INSULATING GASKET

ELECTROLYTE

BARRIER

METAL SHELL (+)

MERCURIC OXIDE

Fig. 7-5 The mercury cell.

performance characteristics but poor low-temperature performance. The positive electrode of this cell is made of silver oxide and the negative cell is made of zinc. The electrolyte is sodium or potassium hydroxide. Silver cells can tolerate a constant current drain but only light loads. For that reason they are used for continuous service in watches, calculators, cameras, and hearing aids.

7-4 SECONDARY CELLS

When current is being drawn from a cell, we say the cell is being *discharged*. In primary cells this discharging process cannot be reversed. Secondary cells, also called *storage cells,* however, can be recharged, or simply *charged* by sending energy into the cell. Secondary cells include the lead-acid type used in automobiles; a rechargeable alkaline type; nickel-cadmium cells; and the nickel-iron cell, also known as the Edison cell.

Lead-Acid Cell The lead-acid cell consists of a positive electrode of lead peroxide, a negative electrode of metallic lead in a spongy form, and an electrolyte of dilute sulfuric acid. The lead peroxide is formed into a positive electrode by applying a paste of the peroxide onto a grid made of a lead-antimony or lead-calcium alloy. The purpose of antimony and calcium is to facilitate casting of the grids and harden them. To prevent the plates from touching one another, thin plates of wood or hard rubber are generally placed between the positive and negative electrodes. Figure 7-6 shows a single cell of an automobile battery. The voltage produced by this cell is about 2.1 V. As current is drawn from the cell, the cell becomes discharged; the

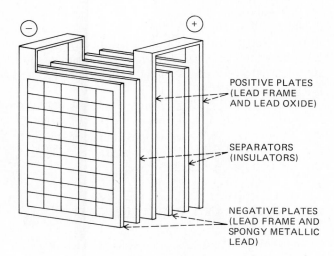

Fig. 7-6 A single cell of an automobile lead-acid battery.

POSITIVE PLATES (LEAD FRAME AND LEAD OXIDE)

SEPARATORS (INSULATORS)

NEGATIVE PLATES (LEAD FRAME AND SPONGY METALLIC LEAD)

lead and lead peroxide gradually change to lead sulfate. The electrolyte becomes more dilute. If the cell is not completely discharged, energy can be supplied to the cell to reverse the process of discharge. Charging the cell restores the lead peroxide and metallic lead electrodes and increases the strength of the acid. To charge the cell, the positive terminal of the charging supply is connected to the positive terminal of the cell and the negative terminal of the supply is connected to the negative terminal of the cell. Current is therefore forced to flow in the opposite direction through the cell during the charging process.

The strength of the acid is determined by a unit called *specific gravity*. This unit is a ratio of the weight of a volume of a liquid to the weight of the same volume of water. For example, if a certain volume of acid weighs 1.5 lb and the same volume of water weighs 1.2 lb, the specific gravity of the acid would be

$$\text{Specific gravity} = \frac{\text{weight of given volume of liquid}}{\text{weight of same volume of water}}$$

$$\text{SG} = \frac{1.5}{1.2} = 1.25$$

This figure can be interpreted as meaning that the acid is 1.25 times heavier than water.

The sulfuric acid in a fully charged lead-acid cell as used in an automobile has a specific gravity of about 1.26 to 1.30. Large lead-acid cells used for standby or stationary dc power supplies have specific gravities somewhat lower, ranging between 1.21 and 1.23. The acid of a discharged cell will have a specific gravity of about 1.12. Thus the drop in specific gravity between the fully charged state to the discharged state is about 0.14 to 0.15.

Specific gravity is sensitive to temperature. As temperature increases, the specific gravity decreases. The specific gravity of a cell may be varied intentionally to cope with different climates. In cold climates a higher specific gravity is necessary to prevent the electrolyte from freezing; similarly in hot climates the specific gravity should be reduced to prevent boiling of the electrolyte.

The specific gravity of the acid in a cell is measured by an instrument called a *hydrometer* as shown in Fig. 7-7. This type of instrument is used by dipping its nozzle into the electrolyte and withdrawing enough electrolyte to allow the small weighted vial to float within the larger tube. The vial is calibrated in specific gravity, and the level at which it floats produces the reading of the specific gravity of the electrolyte.

If the specific gravity is too high or the level of the water too low, water can be added, but only distilled water should be used. The minerals and impurities in

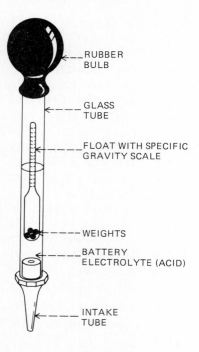

RUBBER BULB

GLASS TUBE

FLOAT WITH SPECIFIC GRAVITY SCALE

WEIGHTS

BATTERY ELECTROLYTE (ACID)

INTAKE TUBE

Fig. 7-7 A hydrometer used to measure the specific gravity of sulfuric acid.

plain tap water can react chemically with the materials in the cell to produce sediment and particles that settle to the bottom of the cell and eventually short the electrodes, preventing the cell from keeping a charge.

As a cell is charged, the water in the electrolyte is broken down into its components of gaseous oxygen and hydrogen. These compose a highly explosive combination. Stationary installation of lead-acid cells must be properly vented, and flames and sparks carefully avoided. In some cases automobile batteries present the same explosive mixture; therefore flames should never be allowed near charging cells, especially in enclosed areas. Goggles and gloves are also recommended when handling cells to prevent injury from the sulfuric acid.

The usual maintenance for lead-acid cells involves checking the specific gravity and level of the cell acid and seeing that all electrical connections are tight and free of corrosion. The white powdery residue around cell terminals is lead sulfate, a product of the discharging process of a cell. It should be removed with a wire brush.

Cells that are labeled "maintenance free" are sealed so that water (or acid) need not and cannot be added. In addition, the specific gravity of the acid cannot be checked, though some maintenance-free cells contain built-in indicators for specific gravity. These cells are constructed so that spacing and materials reduce gassing conditions.

Lead-acid cells, despite their wide usage, are sensitive to extremes in temperature and highly susceptible to damage from overcharging and full discharge.

Nickel-Cadmium Cell The nickel-cadmium cell consists of a positive electrode of nickel hydroxide and a negative electrode of metallic cadmium. The electrolyte is potassium hydroxide. Unlike the electrolyte in the lead-acid cell, the potassium hydroxide does not enter into the chemical process of charge and discharge. The output voltage of this cell is about 1.2 to 1.3 V.

The full name of the cell is *nickel-cadmium-alkaline cell*. It is available in both vented and sealed types. In its sealed form it is available in shapes and sizes similar to the primary dry cells. In this type the gases usually produced during the charging cycle are either eliminated or designed to combine with the metal of the electrodes.

Nickel-cadmium cells have extremely rugged mechanical and electrical characteristics. They are much less affected by complete discharges or overcharging. They are capable of hundreds of charging cycles in the sealed form and thousands of cycles in the vented form. They may be left in their discharged state for long periods and then recharged. Because they are not easily damaged by overcharging, it is common to leave them on a permanent trickle charge. Other advantages of the nickel-cadmium cell are good low-temperature operating performance; low operating cost (though they have a relatively high initial cost); low internal resistance; very gradual drop in output voltage during discharge; and very long shelf life, which means the cell will not deteriorate significantly in storage.

A popular use of nickel-cadmium cells is in cordless electrical appliances such as knives, toothbrushes, shavers, lamps, and the like. Although the appliances themselves are cordless, they are usually stored in a stand which contains a continuous charger. Each time the appliance is returned to its holder, the cell receives a continuous charge until removed again for use.

A version of the nickel-cadmium cell is known as the *Edison cell*. This cell uses iron oxide in place of the cadmium electrode. The output voltage of this cell is about 1.4 V. Edison cells find use in heavy industrial applications as standby power.

7-5 SERIES AND PARALLEL CELLS

Up until this point, the discussion has been concerned with cells ranging in output voltage between 1 and 2 V. In order to achieve higher voltages and greater

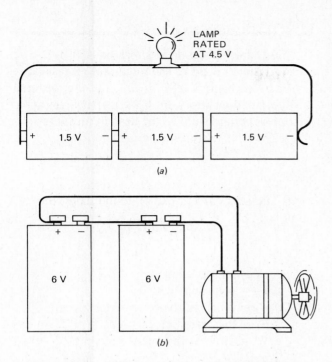

Fig. 7-8 Cells connected in series. (*a*) Three D cells are connected in series to light a 4.5-V lamp. (*b*) Two 6-V batteries connected in series to operate a 12-V motor. Note that 1.5 V cells are connected internally in series to produce the 6 V batteries.

capacity, cells are normally combined in series- or parallel-circuit arrangements or both. A combination of cells is usually called a *battery*. Although the terms *battery* and *cell* are often used interchangeably, in the technical sense they are not the same. Thus the 1.5 V flashlight source is properly called a cell since it contains a single unit of carbon and zinc producing an output voltage of 1.5 V. On the other hand, the 9-V source should be referred to as a battery since it contains a number of individual 1.5-V cells. In order to obtain a higher voltage than that produced by a single cell, multiple cells are connected in series as in Fig. 7-8.

Problem 1

The modern automobile uses a 12-V system. How many lead-acid cells must be connected in series to obtain this voltage?

Solution

Each cell of the 12-V automobile battery produces about 2.1 V. Thus six cells in series will produce

$$6 \times 2.1 = 12.6 \text{ V}$$

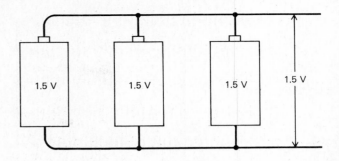

Fig. 7-9 Cells are connected in parallel to increase their capacity.

The capacity of a cell to deliver current is governed by the area and quantity of its electrodes. For example, to increase the capacity of a lead-acid battery, the size and number of the electrodes (called *plates*) are increased. This increases the area of exposure between the plates and thereby increases the capacity of the battery. In smaller batteries the effective area of the battery is increased by connecting several cells in parallel as in Fig. 7-9. If each cell was capable of delivering 2 A over a period of time, the parallel arrangement would increase the amount of time the 2 A could be drawn, or it would allow 6 A to be drawn over a little less time.

Problem 2

Construct a battery with an output voltage of about 24 V capable of delivering 10 A. The basic cell from which this battery is to be constructed has an output voltage of 1.2 V and a capacity of 2 A per given time.

Solution

To obtain 24 V, it is necessary to connect 20 cells in series. However, this string of cells would still be limited to 2 A. To obtain 10 A, five series strings would need to be connected in parallel. Figure 7-10 shows how this battery might be constructed.

7-6 INTERNAL RESISTANCE

The total resistance between the terminals of an unloaded battery or cell is called the *internal resistance* of that battery or cell. The resistance is made up of a

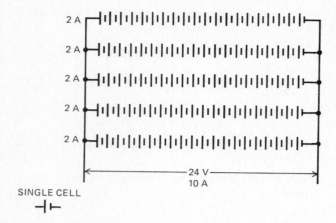

SINGLE CELL

Fig. 7-10 Series-parallel combination of 100 1.2-V cells produces a 24-V, 10-A battery.

number of factors: the materials (metals, electrolytes, etc.), the temperature, the length of the current path, and the resistance of the connections along the path. If the circuit showing a battery serving a load is drawn as in Fig. 7-11, the internal resistance becomes part of the series circuit and must be added to the load resistance to obtain the total resistance of the circuit.

The terminal voltage of the battery must reflect the internal voltage drop due to the internal resistance. But the internal drop is dependent upon the load current delivered by the battery. A formula for finding the terminal voltage is as follows:

$$V_t = V_B - IR_i$$

where V_t = terminal voltage of battery
V_B = nominal voltage produced by battery
I = load current
R_i = internal resistance

The nominal voltage V_B is the voltage produced by the electrochemical action of the electrodes and electrolyte.

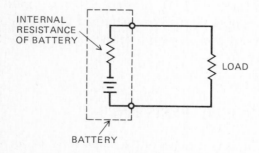

INTERNAL
RESISTANCE
OF BATTERY

LOAD

BATTERY

Fig. 7-11 The internal resistance of the battery causes the terminal voltage of the battery to drop when load current flows.

Problem 3

A 50-Ω load is connected to a bank of lead-acid cells producing 90 V. The total internal resistance of the cell bank is 0.18 Ω. (*a*) Find the terminal voltage of the battery. (*b*) If the internal resistance doubles, what would be the drop in terminal voltage from the value in part (*a*)?

Solution

(*a*) The load current can be found using the formula

$$I = \frac{V_B}{R_T} = \frac{V_B}{R_L + R_i}$$

where I = load current
V_B = voltage of battery
R_T = total resistance of series circuit
R_L = load resistance
R_i = internal resistance

Substituting the known values

$$I = \frac{90}{50 + 0.18} = \frac{90}{50.18}$$
$$= 1.79 \text{ A}$$

Thus

$$V_T = V_B - IR_i$$
$$= 90 - 1.79 \times 0.18$$
$$= 90 - 0.322$$
$$= 89.68 \text{ V}$$

(*b*) $R_i = 2 \times 0.18 = 0.36 \ \Omega$
$$I = \frac{90}{50 + 0.36} = \frac{90}{50.36}$$
$$= 1.79 \text{ A}$$

which to two decimal places is the same as the current found in part (*a*). But the internal resistance is twice that of part (*a*). Thus

$$V_t = V_B - IR_i$$
$$= 90 - 1.79 \times 0.36$$
$$= 89.36 \text{ V}$$

The drop in terminal voltage due to the increase in internal resistance is therefore

$$89.68 - 89.36 = 0.32 \text{ V}$$

7-7 CAPACITY RATINGS

The capacity of cells and batteries is related to the current that can be drawn and the time over which it can be drawn. A common unit for rating cell or battery capacity is the ampere-hour (abbreviated Ah). This unit is simply the product of the discharge current and the period over which the discharge takes place. However, the point at which discharge is considered over must be specified, usually in terms of voltage. For lead-acid cells this voltage ranges between 1.5 and 1.8 with 1.75 V a common value. Another standard is the discharge duration which for certain types of battery installations is taken as 8 h. For example, a battery that delivers a constant 40 A for a period of 8 h would have a rating of 40 A × 8 h = 320 Ah.

In addition to the ampere-hour rating, lead-acid batteries used in automobiles are often rated by the maximum cranking current that can be drawn for short periods (usually taken as 30 s) and the amount of reserve capacity they can deliver. Reserve capacity is specified in time and denotes the period over which 25 A can be drawn before the battery voltage falls below about 10 V.

Although the ampere-hour rating of a cell or battery is the product of current and time, it is not a constant except under the specified conditions of voltage, time, and temperature. As the temperature decreases, the capacity of the cell decreases. Automobile batteries can command less energy in cold weather than in warm. Low discharge rates and longer discharge periods increase the capacity of a cell. For example, a battery with a capacity of 240 Ah could deliver 30 A over the standard 8-h discharge period. However, it could not deliver 60 A for 4 h or 80 A for 3 h. On the other hand it could most likely deliver 10 A for more than 24 h.

SUMMARY

1. Direct current can be produced chemically.
2. Chemical energy is converted into electrical energy in cells.
3. Primary cells consume the materials that constitute the cells as energy is taken from the cell.
4. Secondary cells contain materials that change composition as energy is taken from the cell. The process can be reversed and the materials restored to their original state by delivering energy to the cell.
5. The simplest cell consists of two dissimilar conducting materials immersed in a conducting solution.
6. The two conductors in a cell are called electrodes. The solution is called an electrolyte.
7. The most widely used dry cell is the carbon-zinc cell.
8. The electrodes of a carbon-zinc cell are zinc and manganese dioxide, the electrolyte is zinc and ammonium chlorides.
9. The negative electrode of a carbon-zinc cell is the metallic zinc case; the positive electrode is manganese dioxide. A solid carbon rod makes contact with the manganese dioxide and serves as the positive terminal.
10. The voltage produced by a carbon-zinc cell is about 1.5 V.
11. The carbon-zinc cell is also known as the Leclanche cell.
12. Carbon-zinc cells are available in such popular sizes as AA, C, and D.
13. The alkaline (primary) cell consists of a positive electrode of manganese dioxide and a negative electrode of zinc. The electrolyte is potassium hydroxide.
14. The alkaline cell produces about 1.5 V.
15. The mercury cell consists of a positive electrode made of mercuric oxide and a negative electrode of zinc. The electrolyte is sodium or potassium hydroxide.
16. The mercury cell produces about 1.35 V.
17. The silver cell consists of a positive electrode made of silver oxide and a negative electrode of zinc. The electrolyte is sodium or potassium hydroxide.
18. The silver cell produces 1.5 V.
19. Secondary cells are also called storage cells.
20. The lead-acid cell is the most common secondary cell. It is used in automobiles and trucks.
21. The positive electrode of a lead-acid cell is metallic spongy lead. The negative electrode is lead peroxide packed in a grid made of a lead-antimony or lead-calcium alloy. The electrolyte is sulfuric acid.
22. The voltage produced by a lead-acid cell is about 2.1 V.
23. Specific gravity is the ratio of the weight of a volume of liquid to the weight of the same volume of water.
24. The sulfuric acid in a fully charged lead-acid cell used in an automobile has a specific gravity of

about 1.26 to 1.30. The specific gravity of the acid in a discharged cell is about 1.12.

25. The white powdery residue around the terminals of a lead-acid cell is lead sulfate.

26. The nickel-cadmium cell is a secondary alkaline cell.

27. The positive electrode of a nickel-cadmium cell is nickel hydroxide. The negative electrode is metallic cadmium. The electrolyte is potassium hydroxide.

28. The nickel-cadmium cell produces 1.2 to 1.3 V.

29. Nickel-cadmium cells are used for cordless electrical appliances.

30. Cells may be connected in series, parallel, or series-parallel.

31. Combinations of cells are called batteries.

32. An automobile battery consists of six lead-acid cells connected in series to produce 12.6 V.

33. The terminal voltage of a cell or battery is the voltage produced by the chemical reaction less the internal voltage drop due to the cell or battery internal resistance.

34. Batteries and cells are rated according to ampere-hours (Ah).

QUESTIONS

7-1 Explain the difference between primary cells and secondary cells.

7-2 What are the three elements of a voltaic cell?

7-3 What is the most common dry cell?

7-4 How does a heavy-duty dry cell differ from a regular dry cell?

7-5 Name three popular sizes for dry cells.

7-6 List four dry cells in the order of their voltage with the highest-voltage cell first.

7-7 What does the addition of antimony and calcium do in a lead-acid battery?

7-8 A certain volume of acid weighs 5 lb. An equal volume of water weighs $4\frac{1}{2}$ lb. What is the specific gravity of the acid?

7-9 Does specific gravity increase, decrease, or remain the same when the temperature goes down?

7-10 Why should the specific gravity of the lead-acid electrolyte be intentionally increased above normal in extremely cold climates?

7-11 What is the range of specific gravity for the acid of a normal fully charged lead-acid battery?

7-12 Describe a good maintenance routine for lead-acid batteries.

7-13 List three advantages of nickel-cadmium batteries over lead-acid batteries.

7-14 Indicate the output voltage of each of the cell arrangements in Fig. 7-12(a) through Fig. 7-12(e). Each cell is rated 1.5 V.

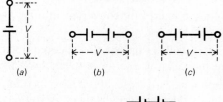

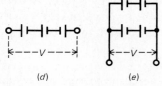

Fig. 7-12 Series-parallel combinations of cells for question 7-14.

7-15 What advantage is there in connecting three cells in parallel?

7-16 Construct a battery with an output voltage of 95 V using nickel-cadmium cells if each cell produces 1.3 V.

7-17 A 9-V battery with an internal resistance of 0.35 Ω is connected to a 100-Ω load. What is the voltage across the load?

7-18 A battery is rated at 400 Ah. What current can the battery deliver over the standard discharge period?

8

DIRECT-CURRENT MOTORS

A dc motor is a machine that converts electric energy to mechanical energy. It uses the principles of electromagnetism to turn its armature and produce mechanical energy at its shaft. This twisting or turning effect on the armature shaft is known as *torque.*

Direct-current motors can be controlled precisely and are therefore used for jobs requiring exacting control characteristics. Because of their relatively simple control system, dc motors are easily adapted to automatic operation of a variety of machines especially where precise control is needed. Cranes, hoists, elevators, draglines,electric shovels, and derricks are usually operated by dc motors. Direct-current motors are also extensively used for operation of battery-powered mobile equipment in materials handling.

Railway systems are another important user of direct-current motors. In the case of diesel-electric railway operations, direct current is generated by diesel-powered dc generators to operate dc motors for traction. Overhead cables, below-ground trolley wires, and batteries furnish direct current to other types of portable work.

8-1 CONSTRUCTION OF DC MOTORS

Because dc motors are constructed almost identically with dc generators, the study of motors is a distinct aid in understanding generators. In fact, a dc motor is a generator and depends on generation as a part of its feature of operation as a motor.

There are three types of dc motors—*series, shunt,* and *compound.* They differ in the way the fields are connected in relation to the armature. Different operating characteristics result from the different field connections. Certain principles apply to all motors, and these will be discussed first; then the three types will be discussed individually. Figure 8-1 is a cutaway view of a dc motor showing the frame, field pole pieces and field coils, and the armature with its winding, brushes, and commutator. Brushes ride on the commutator and supply current from the line to the rotating armature winding.

8-2 DIRECT-CURRENT MOTOR TORQUE

Direct-current motors get their ability to rotate through the *interaction of two magnetic fields.* One of these fields is produced at the face of stationary pole pieces by current flowing in the field coils. The other magnetic field is produced by current flowing in the armature winding. The interaction of these two fields causes the armature to rotate and furnish torque or mechanical power. In very small dc motors the pole pieces are permanent magnets, and so field windings are not used.

The magnetism and magnetic circuit produced by the field coils in the field poles and frame of a dc motor are shown in Fig. 8-2(*a*). This magnetism and the magnetism produced by current flowing in the armature coils repel one another and cause the armature to rotate.

The direction of travel of magnetic lines of force around one turn of an armature coil is illustrated in Fig. 8-2(*b*). In the wire at the left, current is flowing toward the back, away from the reader, as indicated by the cross in the conductor. The current in the conductor at the right is flowing forward, toward the reader, as indicated by the dot in the conductor.

The circular lines of force, according to the right-hand wire rule, are traveling clockwise around the left conductor and counterclockwise around the right conductor. If these conductors are placed in the frame shown in Fig. 8-2(*a*), the interaction of the magnetism in the frame and the circular lines around the conductors will force the conductor on the left to move down and the conductor on the right to move up.

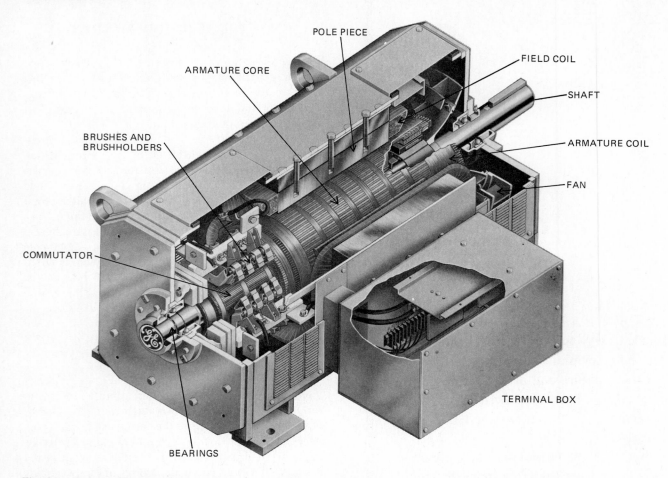

POLE PIECE

ARMATURE CORE

FIELD COIL

SHAFT

BRUSHES AND
BRUSHHOLDERS

ARMATURE COIL

FAN

COMMUTATOR

TERMINAL BOX

BEARINGS

Fig. 8-1 Cutaway view of a direct-current motor. (*General Electric Co.*)

The cause and effect of this interaction is illustrated in Fig. 8-3(*a*). When magnetic lines of force leaving the north field pole meet the circular lines of force of the left conductor, they are forced *up* and *over* this conductor as they travel by it and are forced *down* and *under* the right conductor as they pass it going to the south pole. Being under tension like stretched rubber bands, they are under force to straighten and therefore exert a force against the armature conductors. This force is *downward* against the left conductor and *up-*

ward against the right conductor, which would produce *counterclockwise force* on the armature.

If the current flow is reversed in the two conductors, the force would be reversed, or in a clockwise direction, as illustrated in Fig. 8-3(*b*). Reversing either the *polarity of the main fields* or the *direction of current in the armature* will *reverse the direction of rotation* of a dc motor. It is customary to reverse the armature current to reverse rotation.

If the polarity of the lines feeding a dc motor are

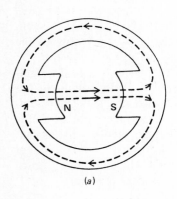

(*a*)

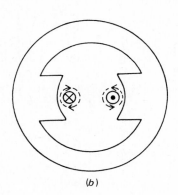

(*b*)

Fig. 8-2 The magnetic fields of a dc motor. (*a*) Field coils produce electromagnetism at the field poles. (*b*) Current in the armature winding produces electromagnetism in the armature.

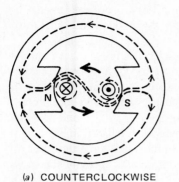 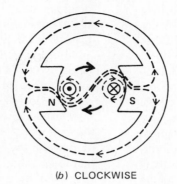

(a) COUNTERCLOCKWISE   (b) CLOCKWISE

Fig. 8-3 Interaction of the field and armature magnetic fields. (*a*) Counterclockwise rotation. (*b*) Clockwise rotation.

reversed, the effect would be to reverse the current in both the field coils *and* the armature. This *would not* reverse the direction of rotation.

In small dc motors with permanent magnet pole pieces, reversing the polarity of the line *will* reverse the direction of rotation of the motor. This is because the magnetic field produced by the permanent magnet pole pieces will remain the same while the polarity of the armature will change. In other words, it is the *relative* change in magnetic fields that produces the change in the direction of rotation of dc motors.

8-3 LEFT-HAND MOTOR RULE

To determine the direction of movement of a conductor carrying current in a magnetic field, the *left-hand*

motor rule is used. The thumb and first two fingers of the left hand are arranged to be at right angles to each other with the forefinger pointing in the direction of the magnetic lines of force of the field, the middle finger pointing in the direction of current flow in the conductor. The thumb will be pointing in the direction of movement of the conductor as shown in Fig. 8-4.

8-4 ARMATURE CURRENT FLOW PATTERN

To understand how all the coils in a motor or generator armature carry current and produce torque in a motor (and generate current in a generator), we will examine the four-pole lap winding of a dc machine as

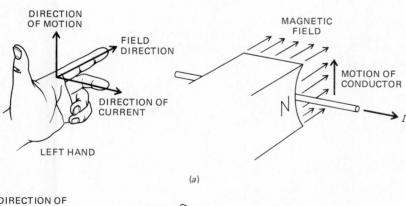

(a)

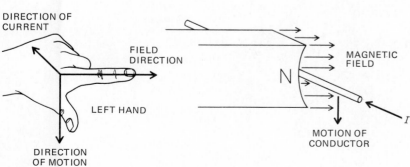

(b)

Fig. 8-4 The left-hand rule for determining direction of rotation. (*a*) Upward movement produces clockwise rotation. (*b*) Downward movement produces counterclockwise rotation.

shown in Fig. 8-5. Current enters from the positive brush and flows through the lap winding and leaves the armature at the negative brush. There are two types of armature windings, *lap* and *wave*. A four-pole lap winding differs from a wave winding only in the way the leads connect to the commutator and in the number of current paths from positive to negative brushes. In a lap winding, the leads of a coil lap toward each other and connect to adjacent commutator bars (or segments). It provides as many paths for current to flow from positive to negative brushes as there are brushes or poles. The four-pole lap winding shown in Fig. 8-5 has four brushes and is a four-pole armature; therefore it has four paths from the positive brushes through the winding to the negative brushes.

When current flow is traced in Fig. 8-5, it will be found that there are four circuits from positive to negative brushes. The current flows according to a very definite plan and forms a definite pattern. This *pattern of flow* is a clue to how a motor actually develops torque and is a great aid in further studies of dc motors and generators. Before current flow and pattern of flow are traced, Fig. 8-5 should be thoroughly studied because it will be referred to several times in further discussions of dc motors and generators.

Figure 8-5 shows 12 numbered commutator bars in the center. Positive brushes are making contact with commutator bars 1 and 7, and negative brushes are making contact with bars 4 and 10. For clarity in the diagram the brushes are drawn on the inside of the commutator; in reality they touch the outside of the commutator.

An armature coil is usually wound so that one of its sides, the top side, is in the top of a slot in the core, and its other side, the bottom side, is in the bottom of another slot. There are 12 numbered slots in the core, with each slot containing the top half of one coil and the bottom half of another coil. Each coil in Fig. 8-5 is numbered; the numbers in the small circle are the coil numbers. The circles in the slots represent the cross section of the coil. Current direction is indicated by X (current into the page) and by • (current out of the page). In this case a dot indicates current flowing toward you and into the commutator.

We will now trace the current through the armature coils to show that there is a definite pattern of flow. Notice that in slots 12, 1, and 2 at the top, and slots 6, 7, and 8 at the bottom, current in both windings is flowing away from you (or into the page). In slots 9, 10, and 11 on the left, and 3, 4, and 5 on the right, current is flowing toward you (or out of the page). The effect of this pattern is to provide maximum torque in the armature when it is placed in the magnetic field produced by the field coils.

There are four complete circuits in the armature winding of Fig. 8-5. The first circuit begins at commutator segment 1. This segment is touching the top brush, which is connected to the positive side of the

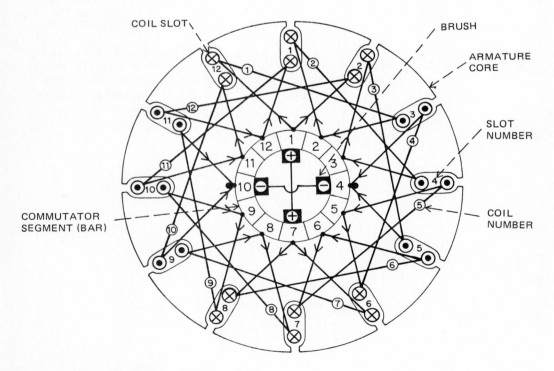

Fig. 8-5 A four-pole lap winding of an armature.

line. Current flows from segment 1 to slot 12, through coil 1 to slot 3, to segment 2, to slot 1, through coil 2, to slot 4, to segment 3, to slot 2, through coil 3, to slot 5, to segment 4. Segment 4 is touching the brush connected to the negative side of the line, thus completing the circuit that started at segment 1.

Segment 1 also provided a second path for current. From segment 1 current flowed to slot 2, through coil 12 to slot 11, to segment 12, to slot 1, through coil 11, to slot 10, to segment 11, to slot 12, through coil 10, to slot 9, to segment 10. Segment 10 is touching the brush connected to the negative side of the line, thus completing the circuit that started at segment 1.

The third circuit begins at the lower brush, which, like the upper brush, is connected to the positive side of the line. (Actually the two brushes are connected in parallel and only a single connection is made to the line.) From commutator segment 7, current flows to slot 6, through coil 7 to slot 9, to segment 8, to slot 7, through coil 8, to slot 10, to segment 9, to slot 8, through coil 9, to slot 11, to segment 10. As noted previously, commutator segment 10 is connected to the negative side of the line through the brush.

Finally, the fourth circuit traces a pattern of flow similar to the other three circuits. Starting from segment 7, current flows to slot 8, through coil 6 to slot 5, to segment 6, to slot 7, through coil 5, to slot 4, to segment 5, to slot 6, through coil 4, to slot 3, to segment 4. Again commutator segment 4 is connected to the negative side of the line through the brush, thus completing the fourth circuit.

The patterns of flow can be shown more graphically through a table such as Fig. 8-6.

To summarize: When the current flows in the four circuits from positive to negative brushes, all the current in the top and bottom coil sides in the top one-fourth of the armature is flowing back, all the current flowing in the one-fourth of the right of the armature is flowing forward, the current in the bottom one-fourth is flowing back, and all the current flowing in the left one-fourth is flowing forward. This produces a definite, orderly pattern of current flow.

If this pattern of flow is set in a field frame, as in Fig. 8-7, one can readily see why the pattern is necessary to produce maximum torque in a motor.

All the coil sides under the north poles contain crosses, indicating that the current is flowing backward, and all the coil sides under south poles contain dots, indicating that the current is flowing forward. Therefore all the coils under north poles are being forced counterclockwise, and all the coils under south poles are being forced in the same direction, so torque produced by all the coils is counterclockwise. Arrows around the coils indicate the direction of circular magnetic lines of force around each coil, according to the right-hand wire rule.

The circular lines force the main lines around them by stretching the main lines, which sets up a strain or force between the circular and main lines. This force is counterclockwise against all the coil sides in this instance, and if the coils producing the circular lines are free to move, they will be moved counterclock-

Fig. 8-6 Patterns of flow in a four-pole lap-wound armature.

START (+) ... FINISH (−)

CIRCUIT		SEG	SLOT	COIL	SEG	SLOT	COIL	SEG	SLOT	COIL	SEG	SLOT	COIL	SEG
1	SEG	1				2				3				4
	SLOT		12		3		1		4		2		5	
	COIL			1				2				3		
2	SEG	1				12				11				10
	SLOT		2		11		1		10		12		9	
	COIL			12				11				10		
3	SEG	7				8				9				10
	SLOT		6		9		7		10		8		11	
	COIL			7				8				9		
4	SEG	7				6				5				4
	SLOT		8		5		7		4		6		3	
	COIL			6				5				4		

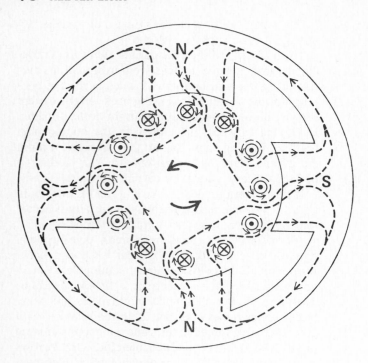

Fig. 8-7 Interaction of the field magnetism and armature magnetism produces a torque which turns the armature in a counterclockwise direction.

wise. Since the coils are on an armature in a motor and can move, the armature will move or rotate counterclockwise under these conditions.

The brushes are stationary and maintain the pattern of flow as shown in Fig. 8-5. With the pattern interacting continually with the field, the armature will continue to rotate as the result of this continued reaction.

It can be seen in Fig. 8-5 that the position of the brushes and the lead connections to the commutator determine the location of the pattern on the armature. The brushes are in a *mechanically neutral position* when they produce a pattern of crosses and dots *directly under main poles* as shown.

If the brushes are shifted away from this position, the pattern will be moved correspondingly. If the brushes are shifted clockwise so that the top positive brush is contacting bar 2 instead of bar 1, the pattern will be moved clockwise one slot distance. Referring to Fig. 8-5, it will be seen that this condition would place slot 2 containing two coil sides with crosses under the south main pole at the right. The circular lines of force around the coils in this slot are opposite in direction to the circular lines of the coil in slots 3 and 4, and so the torque developed in slot 2 by the south main pole would be clockwise, while the torque developed in slots 3 and 4 would be counterclockwise. These opposite forces would considerably reduce the total torque of the armature and result in severe sparking between the brushes and commutator. So for maximum torque the pattern must be located properly in the main field flux.

8-5 HARD NEUTRAL

When the pattern is positioned directly under pole pieces, the brushes are on *hard neutral*. Under certain conditions the main field flux is distorted and does not appear exactly at the pole pieces, so that the brushes must be shifted to a new position to place the pattern in the main field flux. This position is known as the *working neutral*.

8-6 ARMATURE REACTION

When a motor is running under load, the magnetism produced by the armature acting against the magnetism of the main fields *pushes* the main fields out of normal position. This is known as *field distortion* and is produced by what is known as *armature reaction*.

Field distortion and armature reaction reduce the horsepower of a motor and cause sparking at the commutator and must be minimized. Two means of correcting these conditions are used in motors.

Interpoles are used for correcting field distortion to some degree, but their main function is to *minimize sparking* at the commutator. Interpoles are small poles located midway between main poles. Their operation will be discussed in detail later. The best means of correcting severe cases of field distortion is the use of *compensating winding*. This winding is inserted in slots in the face of the main poles and is considered an extension of the interpoles.

8-7 WORKING NEUTRAL

In Fig. 8-7 the magnetism from the top north pole is being pushed to the *right of the pole* so that the effective magnetic main field is not at right angles to the pole piece. This causes the brushes to be off neutral because the pattern is not directly in the main field but is somewhat to the left of it. On a noninterpole motor, under conditions such as are represented in Fig. 8-7, it is necessary to shift the brushes to the right, or clockwise, for the working neutral position. This brings the pattern at right angles to the armature coil.

On a noninterpole motor, neutral position shifts, because of field distortion, as the load on the motor varies, and so the best setting of the brushes is for average loads. This position of the brushes produces less sparking and maximum torque on average loads. Working neutral is the proper position of the brushes for nonreversing noninterpole motors.

The working neutral position for interpole motors and noninterpole motors requiring reversing is the *mechanical* or *hard* neutral. The hard neutral position of the brushes places the armature current flow patterns directly under the pole pieces. Most motors equipped with adjustable brushes contain a reference line on the brush holder and the frame of the motor for guidance in setting brushes on hard neutral. This setting varies in a direction *opposite the rotation* of the armature on noninterpole motors according to average load conditions.

If hard neutral is not marked on a motor, there are several methods by which it can be approximately found. But before studying these methods, an examination of Fig. 8-8 will give a better understanding of conditions and objectives in establishing neutral position to produce maximum torque and minimum sparking.

Coil 1 of Fig. 8-5 is considered to be the coil in Fig. 8-8 when the armature turns counterclockwise the distance of one commutator bar. In Fig. 8-8(a), the positive brush is on bar 1 and is supplying current into the left side of the coil, as indicated by the X. The current in the right side of the coil is forward, as indicated by the dot. In position Fig. 8-8(b) the positive brush is across bars 1 and 2, and the coil is carrying no current, since it is bypassed and short-circuited by the brush. It is bypassed because current from the positive brush can flow directly from bars 1 and 2 into adjacent coils, and it is short-circuited because there is a short-circuit path around the coil, through the bars and the brush.

When this coil entered this short-circuit condition, it was carrying current in the direction shown in Fig. 8-8(a). When it enters the condition shown in Fig. 8-8(c), the current flow will be opposite to conditions in Fig. 8-8(a). So between Fig. 8-8(a) and (c), the direction of current flow is *reversed*. It takes time for this reversal to be completely accomplished.

8-8 COMMUTATION

When a coil is carrying current in one direction, the coil is surrounded by circular magnetic lines of force. When the current flow is suddenly stopped, the lines of force collapse and in doing so cut across the coil and *induce a current* in it in the same direction of flow as the original current. This current creates a second magnetic field opposite to but weaker than the original field, which is collapsing. This delays its collapse. Hence, it takes time for the original field to collapse in opposition to the second field and completely clear the coil of magnetism.

The coil must be clear of magnetism and circulating currents before it moves from the position shown in Fig. 8-8(b) to the position in (c) to avoid sparking when bar 1 moves away from the positive brush. If some magnetism still exists during this time, it will collapse and induce a current in the coil which will jump across the gap between the brush and the bar and cause a spark.

The clearing of a coil of its current and magnetism

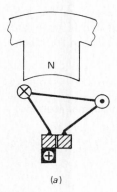

(a)

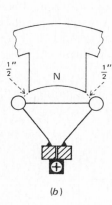

(b)

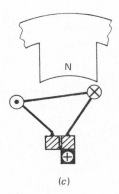

(c)

Fig. 8-8 The effect of commutation on current flow in an armature coil. The black square represents the stationary brush and the shaded square represents the rotating commutator. In practice the brush rides on the outside of the commutator. (a) Current flow as the armature rotates counterclockwise. (b) The commutation period. (c) Current flow as the next commutator segment is contacted.

in preparation for the reversal of the current is called *commutation,* and the period of time required in the process is known as the *commutation period.* Commutation is aided by the use of interpoles.

8-9 INTERPOLES

A quick collapse of the magnetism surrounding a coil during the period of commutation can be accomplished by the use of *interpoles* (occasionally called *commutating poles*). Interpoles are small poles located between main poles. Interpole coils are connected *in series with the armature.*

A study of Fig. 8-5 will aid in understanding the principles of operation of interpoles. It will be noticed that the current in slots 12, 1, and 2 is flowing back and the current in slots 3, 4, and 5 is flowing forward. This pattern of current flow causes magnetic poles to form in an armature.

The formation of armature poles is further illustrated in Fig. 8-9. With current flowing in the direction shown in conductors around a straight core as shown in Fig. 8-9(*a*), a south pole would develop at the top of the core. Similarly, a south pole would develop in a conical core as shown in Fig. 8-9(*b*). A south pole would also form in a core of the shape shown in Fig. 8-9(*c*), which is the same shape as the section of the armature discussed in Fig. 8-5. It can be seen, then, that a south armature pole forms between slots 2 and 3 in Fig. 8-5. A north pole forms between slots 5 and 6, a south pole between slots 8 and 9, and a north pole forms between slots 11 and 12.

It is the function of an interpole to repel armature poles and magnetism as much as possible because this is magnetism in the area of commutation. So the polarity of an interpole should be the same as the armature pole under it. In Fig. 8-5 an interpole should be located between slots 2 and 3, and it should be a south pole so it can repel the armature south pole.

8-10 INTERPOLE POLARITY

The polarity of interpoles in motors or generators should always be the same polarity as the armature

poles. But the armature polarity, in practical work, is seldom known.

Armature polarity in a motor is opposite to that in a generator. It reverses when rotation is reversed in either a motor or a generator. So *polarity of the interpoles* depends on the *polarity of the main poles,* which determines direction of rotation and whether the machine is a motor or a generator.

8-11 M-G INTERPOLE RULE

A convenient rule for determining interpole polarity is the *motor-generator (M-G) interpole rule.* This rule applies to motors or generators rotating in either direction and can be equally applied looking at either end of the machine.

The M-G interpole rule is simple. Figure 8-10 is a pictorial representation of the rule, showing an interpole at the top bearing a question mark. This can be called the *problem pole* since its polarity must be determined. An arrow under the interpole is drawn for clockwise rotation. Two main poles are shown with one on each side of the interpole. The one on the left is marked ''M'' for motor, and the one on the right is marked ''G'' for generator.

The rule is as follows:

> For a *motor* rotating in the direction of the arrow (clockwise), the interpole should have the same polarity as the M pole. For a *generator* rotating in the direction of the arrow (clockwise), the interpole should have the same polarity as the G pole.

The rule should always be applied on a clockwise basis. If the machine is actually rotating counterclockwise, the polarity found by the rule should be reversed. Any interpole in the machine can be selected in applying the rule, but one should be selected that is next to a main pole whose polarity can be easily established, such as a main pole connected directly to the positive line.

This M-G rule should be studied and committed to memory for aid in mental calculations on the job. When it is sketched for use, the M and G poles are lettered clockwise, and the arrow is drawn for clock-

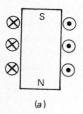
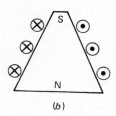
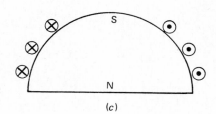

(*a*)   (*b*)   (*c*)

Fig. 8-9 The formation of magnetic poles by windings.

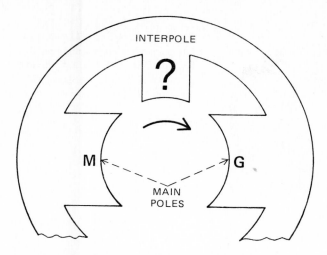

Fig. 8-10 The M-G interpole rule is used for determining the polarity of the interpole.

wise rotation. This rule also applies to the polarity of compensating windings discussed in Sec. 8-12.

Fig. 8-11 shows the correct polarity for interpoles in motors and generators for both clockwise and counterclockwise rotation. In each case the polarity of the main poles is the same.

In applying the M-G rule to the motor in Fig. 8-11(*a*), the interpole at the top is chosen as the problem pole and the main pole at the left is the M pole. With the motor running clockwise, the interpole should be of the same polarity as the M pole, which in

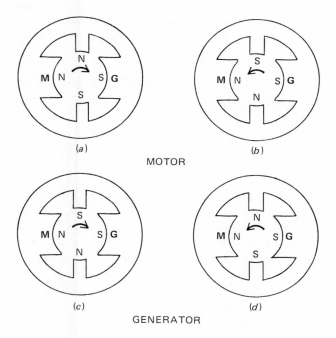

Fig. 8-11 Application of the M-G interpole rule.
(*a, c*) Clockwise rotation. (*b, d*) Counterclockwise rotation.

this case is north. So the correct polarity of the interpole is north. The same check can be made on the generators, except that the polarity of the G pole is used for the interpoles in generators.

Interpoles, like main poles, are connected for *alternate polarity* when there are as many interpoles as main poles. But in some cases there are fewer interpoles than main poles. A motor or generator can have one for each main pole, or one for each pair of poles, or in some cases only one interpole. Where there is less than one for each main pole, the correct polarity of each interpole should be found individually with the M-G rule.

If a machine is operated with wrong polarity of the interpoles, *heating, reduced horsepower,* and *severe sparking* at the brushes will result. Wrong interpole polarity greatly intensifies the trouble interpoles are designed to correct. It is better to leave the interpoles disconnected and out of operation than to have them of wrong polarity.

The armature and interpole windings are connected in series; therefore they carry the same current. Hence, the relative strengths of the interpoles and the armature magnetism are automatically balanced for proper operation at all times.

8-12 COMPENSATING WINDINGS

Field distortion in large motors or generators subject to widely varying loads presents an unusual problem. An extremely heavy load causes an extreme distortion of the field. To compensate for this, coils are placed in the face of the main poles and around the interpoles. These coils are connected in series with the armature so that armature current passes through them. The resulting field is the same polarity as the interpoles it surrounds and can be considered as an extension of the interpoles since usually a machine with a compensating winding has fewer turns on the interpoles. A motor with interpoles and slots in the main poles for a compensating winding is shown in Fig. 8-12(*a*), and it is shown with the compensating winding in the slots in Fig. 8-12(*b*).

8-13 NEUTRAL BRUSH SETTING

Referring to Fig. 8-8(*b*), it will be seen that the coil undergoing commutation has both its sides equidistant outside the main field flux in a nonflux or neutral area. The brushes are now set at hard neutral, and this position is essential for good commutation. Brushes on interpole motors should be set on hard neutral. Brushes on noninterpole motors will have to be

(a)

(b)

Fig. 8-12 Direct-current motor with interpoles and compensating winding. (a) Slots in the face of the main poles accept the compensating winding shown in (b). (*Siemens-Allis.*)

shifted a few degrees opposite the direction of rotation from hard neutral to working neutral.

8-14 LOCATING HARD NEUTRAL

There are several methods of approximately locating hard neutral on a motor with no load on it. The best method is to set the brushes for maximum torque and minimum sparking with the motor under its normal load on the job.

1. *Specific-coil method.* One coil of an armature winding is selected, and the core slots containing the coil, as well as the commutator bars containing the leads from that coil, are marked. The armature is positioned in the motor where the two coil sides are equidistant from the edges of a main pole. In this position a brush should be located on the marked commutator bars containing the coil leads of the marked coil. This condition is shown in Fig. 8-8(*b*). In case of a wave winding, either brush can be located over the bar containing either lead.

2. *Interpole method.* Connect the armature and interpoles to less than rated voltage to allow less than full-load current to flow in this circuit, leaving the main fields out. Shift the brushes to where the armature will not turn. This is neutral position. On some motors a slight shift either way from this position will occur before the armature turns in either direction. The center of this space of shift before rotation is then to be considered neutral. If the residual magnetism in the main poles is strong, it may cause this test to be slightly inaccurate.

3. *Field kick method.* Connect a low-reading voltmeter across a positive and negative brush. Make and break the circuit through the main field only and note the kick on the voltmeter. When the brushes

are in neutral position, no kick will occur on the meter.

4. *Forward-reverse method.* A dc motor will run at the same speed in both directions under the same load when the brushes are on neutral. A tachometer is used to test the speed.

8-15 SPEED OF DC MOTORS

The speed of dc motors is determined by the counterelectromotive force the motor produces and the load on the motor. When a conductor cuts magnetic lines of force, a voltage is generated or induced in the conductor. The strength of voltage thus induced in a conductor is in proportion to the speed of movement of the conductor and the strength of the magnetic field.

8-16 COUNTERELECTROMOTIVE FORCE (CEMF)

A motor armature, rotating, and therefore cutting the magnetic field of the motor, generates a voltage which is opposite in direction to the original voltage driving the motor. Since the generated voltage is opposite the original voltage, it is known as *counterelectromotive force,* abbreviated *counter-emf* or *cemf*. The strength of cemf in a motor is determined by the strength of the field, the number of armature conductors in series between the brushes, and the speed of the armature.

At the moment a motor starts, the armature is not rotating and thus there is no cemf; consequently full line voltage is applied across the motor, and it draws current through the armature circuit in accordance with Ohm's law. The only factor limiting current is the resistance of the motor windings. But as the armature increases speed or accelerates, it *generates cemf,* and this limits the flow of current into the motor. As the *speed increases, cemf increases,* and the current drawn by the motor *decreases.*

When a motor reaches its full no-load speed, it is designed to be generating a cemf *nearly equal to line voltage.* Only enough current can flow to maintain this speed. If a load is applied to the motor, its speed will be decreased, which will reduce the cemf, and more current will be drawn to drive the load. Thus the load of a motor regulates the speed by affecting the cemf and current flow.

When the load increases, the speed and cemf decrease and the current increases. Likewise, as the load decreases, the speed and cemf increase and the current decreases.

If the main magnetic field is weakened, less cemf will be generated, more current will flow, and the motor speed will increase. A motor is designed to produce its rated horsepower at full-load speed. Its normal (full load) speed is known as the *base speed* of the motor. Speeds other than base speed can be obtained by regulation of the current flow in either or both the armature or field windings by certain types of control devices.

8-17   FIELD POLE SHIMS

Some motors have shims made of magnetic material placed between the pole pieces and frame. These shims regulate the air gap between the armature core and the pole pieces. Since air has about 2000 times more reluctance than the core and pole pieces, a slight change in the air gap results in a considerable change in the strength of the magnetic field. An increase in the air gap will decrease the strength of the field, and this results in decreased cemf and increased speed. Likewise, a decrease in the air gap results in increased cemf and decreased speed. Thus the speed of a dc motor is greatly influenced by the strength of its magnetic field which, in turn, is related to the air gap between the main pole and the armature.

Since the air gap bears such an important relationship to the strength of the magnetic field, great care should be taken in handling the shims in the disassembly and assembly of a motor. Because shims are not always of the same thickness, each pole and its shims should be marked to aid in assembly. The air gap should be as nearly the same all around the armature as possible. Substitution of steel shims with nonmagnetic material is the equivalent of increasing the air gap, since the reluctance of all nonmagnetic material is the same as air.

An armature core should never be machined, because this reduces its diameter and increases the air gap. When the air gap is increased by any means, the armature will draw more current to produce its rated horsepower at its rated speed. The increased heating effect of current is in proportion to the square of the increase in amperes.

It is estimated that a *10°C rise in operating temperature halves the expected life of average motor insulation,* and so machining an armature core can lead to inefficient operation of a motor and early burnout of its insulation.

8-18   PRONY BRAKE TEST

A prony brake test is used for calculating the horsepower, starting torque, and other characteristics of electric motors (Fig. 8-13). A basic prony brake may be constructed of two pieces of 2-in. wooden blocks about 8 in. long, hinged at one end and connected at the other end with a threaded rod welded to a hinge. A wing nut on the rod is used for clamping the blocks together. A semicircle is cut out of each block so that when the blocks are put together there is a 4-in. circle in the center. The inside of the circle is lined with automobile brake lining material firmly attached to each block. The arm can be made of ¾-in. "thin wall" conduit (EMT) with hooks spaced at 1-ft intervals beginning 1 ft from a line at right angles to the center of the circle.

In use, the blocks are clamped around a flat motor pulley not less than 4 in. in diameter. A scale of sufficient capacity is connected to one of the hooks on the arm. The prony brake and scale must be attached so that the rotation of the motor will tend to pull the prony brake arm *down* from the scale. *Great caution should be observed in being certain of the direction of rotation of the motor.* Severe damage and personal

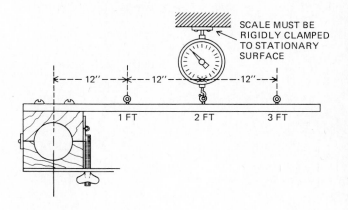

Fig. 8-13   Basic construction of a prony brake.

injury can result if the actual motor rotation is opposite to intended rotation.

For measurement of starting torque, the blocks are securely clamped to the pulley, and the motor is energized at normal voltage. The reading of the scale in pounds is multiplied by the length of connection to the lever arm to determine pound-feet of starting torque. Thus, if a motor connected to the brake shown in Fig. 8-13 pulls 10 lb on the scale, and the scale is connected in the second hook from the blocks, the pound-feet of starting torque will be $10 \times 2 = 20$ lb-ft.

The *horsepower* of a motor is in proportion to its torque in *pound-feet and its speed*. If a motor produces twice the pound-feet torque of another motor at the same speed, its horsepower is twice that of the other motor.

To determine the horsepower of a motor with a prony brake, a formula is used in the calculations. The formula is

$$hp = \frac{6.28 \times rpm \times L \times P}{33,000}$$

where 6.28 = constant
    rpm = revolutions per minute of motor shaft at full load
    L = distance from center of pulley to hook in arm, ft
    P = pull as measured on scale, lb
    33,000 = constant equal to number of pound-feet per minute in 1 hp

In testing for horsepower of a motor, adjust the friction blocks to full load of the motor. This can be done by connecting an ammeter in the motor circuit and tightening the wing nut on the blocks until the ammeter reads full-load nameplate current. A tachometer can be used to read rpm. When the motor is under full load, its pull in pounds on the scale and speed should be recorded and applied to the formula.

As an example, if a motor pulls 16 lb on the hook 1 ft from the pulley while rotating at 1750 rpm, its horsepower would be

$$hp = \frac{6.28 \times 1750 \times 1 \times 16}{33,000}$$
$$= 5.32 \text{ hp}$$

In making the prony brake test, the motor temperature should be checked to make sure the motor is being operated within its safe temperature range.

8-19 TYPES OF DC MOTORS

There are three distinct types of dc motors—*shunt, series,* and *compound.* Each type can be equipped with interpoles and compensating windings. The construction of each type is practically the same, but the fields in each type are connected differently in relation to the armature, which results in different operating characteristics.

In the discussion of the individual motors that follows, full-voltage starting torques are given. Direct-current motors above 2 hp should not be started on full voltage unless designed for it. Starting currents should be limited by controls to about 150 to 200 percent of the full-load current given on the motor nameplate.

8-20 SHUNT MOTOR

A shunt motor has its field connected *in parallel* with its armature. Figure 8-14 shows a shunt motor and a shunt interpole motor. The parallel connection of fields and armature can be easily traced. There are two circuits from positive to negative line. One is through the shunt field, and the other is through the armature.

Shunt motors have starting torques of about 275 percent of full-load torque and a speed variation of about 5 percent from no load to full load. This variation, called *speed regulation,* is very low so that shunt motors are used where a fairly constant speed is required even with varying loads.

With the shunt field across the line, this field receives constant line voltage at all times and is not affected by cemf from the armature. Therefore the magnetic flux from the shunt fields is *constant* at all times. The torque varies *nearly directly with armature current* within saturation limits of the magnetic paths.

When a shunt motor starts, it draws a high starting

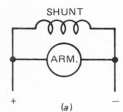

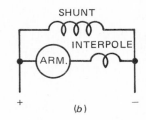

Fig. 8-14 Shunt motor diagram. (*a*) With shunt field only. (*b*) With shunt field and interpole winding.

current through the armature circuit. This current is reduced as the armature accelerates because of the generation of cemf by the armature. When it reaches full no-load speed, the armature generates cemf which is nearly equal to line voltage. With motor cemf nearly equal to line emf, or voltage, only enough line current flows to provide sufficient torque to maintain the speed. Hence *cemf determines maximum no-load speed of a shunt motor.*

When load is applied to the motor, the speed is reduced and cemf is reduced, which allows sufficient current to flow in the armature circuit to provide more torque to drive the load. If more load is added and additional torque is required, the speed is further decreased, which decreases the cemf, allowing more current for the additional torque requirements of load. If the load is reduced, the motor will increase speed and cemf, which will reduce armature current. If the motor is relieved of all load, it will increase speed to the point where cemf nearly equals line emf, and this is full no-load speed. Thus the load of a shunt motor regulates the speed, which determines the amount of cemf, which regulates current flow to provide required torque to drive the load.

A shunt motor with a constant current in its field coils and a constant magnetic field does not vary much over 5 percent in speed from no load to full load. It is classed as a constant-speed motor.

8-21 SPEED-LOAD CURRENT RELATIONSHIPS

Typical speed-load current relationships of a shunt motor are shown in Fig. 8-15. The no-load speed for this motor is about 1150 rpm with about 3 A. At full load the speed has been reduced to about 1100 rpm and the current increased to about 40 A. This is about a 4 percent variation in speed from no load to full load. It can be seen that the speed of a shunt motor varies little with load from no load to full load, while load current increases significantly (more than 13 times in this case).

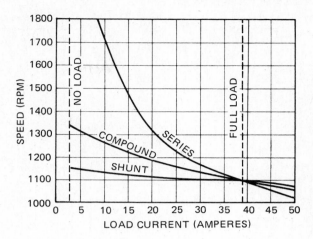

Fig. 8-15 Speed vs. load current characteristics for typical dc motors.

8-22 BASE SPEED OF MOTORS

Normal speed is the *base speed* of a motor. A shunt motor can be operated at below base speed by insertion of resistance in the armature circuit or above base speed by insertion of resistance in the field circuit. When the field is weakened, a shunt motor will increase speed if the load permits, since cemf is reduced and more current flows in the armature circuit.

The polarity and circuit connections of a shunt motor and shunt interpole motor are shown in Fig. 8-16. In the shunt interpole motor main fields and interpole fields are connected for clockwise rotation. For counterclockwise rotation, the armature and interpole circuit should be reversed.

8-23 STANDARD TERMINAL IDENTIFICATION

A standard system of marking terminal leads on shunt reversing and nonreversing, and shunt interpole reversing and nonreversing, motors is shown in Fig. 8-17. This system is followed by practically all manu-

Fig. 8-16 Field and armature connections of a shunt motor. (*a*) Without interpoles. (*b*) With interpoles.

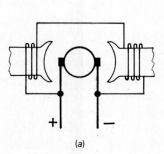

(a)

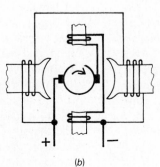

(b)

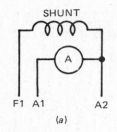

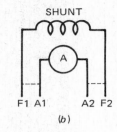

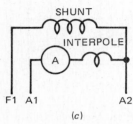

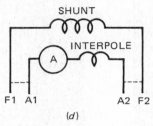

Fig. 8-17 Standard motor terminal identification for shunt motors. (*a*) Nonreversing. (*b*) Reversing. (*c*) Nonreversing with interpoles. (*d*) Reversing with interpoles.

facturers. Letters are used to identify the circuits, and numbers are used to indicate direction of current flow, or relative polarity of the circuits. For example, on the reversing shunt motor, if A1 and F1 are connected to the same side of the line, the motor will rotate counterclockwise.

8-24 STANDARD ROTATION

Counterclockwise rotation as viewed from the end opposite the shaft or pulley end is considered standard rotation. The standard terminal marking is referred to this rotation. Because the end of a dc motor opposite the shaft end is almost always the commutator end, rotation is generally viewed from the comutator end of the motor. If the terminals with like numbers shown in Fig. 8-17(*b*) and (*d*) with the dashed lines between them are connected to the same side of the line, the motor will rotate in the standard direction, which is counterclockwise. That is, if F1 and A1 are connected to one line, positive or negative, and A2 and F2 are connected to the other line, rotation will be standard, or counterclockwise.

To reverse these motors, the connections of A1 and A2 are reversed; that is, F1 and A2 will be connected together and to one line, and F2 and A1 will be connected together and to the other line. Line polarity does not affect rotation of dc motors.

It will be noticed that the interpoles are connected to the armature on the A2 side, which is standard practice, and are included as part of the A1-A2 connection. This automatically reverses the interpoles with rotation, which is necessary. The M-G rule will prove that the polarity of interpoles should be reversed with rotation.

8-25 NONREVERSING MOTORS

In a nonreversing motor some of the leads are connected inside the motor and are not available externally for reconnection. The only leads brought out from the motor are those necessary for operation and connection through a controller. It will be noticed in the nonreversing diagrams [Fig. 8-17(*a*) and (*c*)] that the field and armature circuits are separate on one side for individual control purposes. This three-lead nonreversible system is usually applied to equipment motors, and the lead identification system does not indicate direction of rotation.

General-purpose motors are usually externally reversible by having all the leads brought out for external wiring.

8-26 TESTING FOR TERMINAL IDENTIFICATION

For efficiency in the management of motors, all motors should have nameplates and properly identified leads. If the terminal markings are missing or wrong on a shunt-reversing interpole motor, lift the positive or negative brushes to open the armature circuit, and test the leads with a test light or ohmmeter. The leads that afford a circuit or a spark on contact are the shunt field leads and should be marked F. (Shunt field leads are usually a smaller-size wire than armature leads.) The two remaining leads are armature leads and should be marked A. Lower the brushes. Connect an A and F lead together and to one power supply line, and the other A and F lead to the other power line. If the armature rotates counterclockwise as viewed from the end opposite to the shaft (usually the commutator

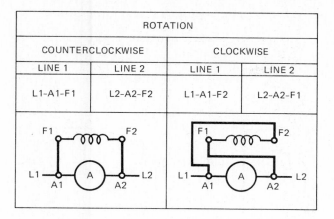

ROTATION			
COUNTERCLOCKWISE		CLOCKWISE	
LINE 1	LINE 2	LINE 1	LINE 2
L1-A1-F1	L2-A2-F2	L1-A1-F2	L2-A2-F1

Fig. 8-18 Shunt motor connections.

end), number the A lead from the interpole winding A2 and the F lead connected to it F2. The other A and F leads should be numbered A1 and F1. If the armature rotates clockwise, number the A lead from the interpole winding A2 and the F lead connected to it F1. The remaining A lead is A1, and the remaining F lead is F2. (In some cases it is more convenient to find armature leads by touching the commutator with a test prod and finding two leads that give a circuit with the other prod.)

Before starting a noninterpole shunt motor, mark one of the F leads F1. Start the motor, and if the armature rotates counterclockwise, the A lead connected to the F1 lead is A1; if the armature rotates clockwise, this lead is A2. The remaining A and F leads can be numbered accordingly. If an interpole motor has been overhauled, the polarity of the interpoles should be checked by use of the M-G rule (Sec. 8-11).

To test nonreversible motor leads, locate the two leads that give the brightest light or lowest resistance reading. These are the A leads and should be marked A. Raise or insulate the positive or negative brushes from the commutator. One A lead should give no reading between it and the other leads, and this is A1. A dim light or spark should be obtained between the other A lead, which is A2, and the remaining lead, which is F1.

A chart with schematic diagrams showing standard connections for shunt motors, either interpole or noninterpole, for counterclockwise and clockwise rotation is shown in Fig. 8-18. If the motor has inter-

poles, they will be included in the A1-A2 circuit on the A2 side of the armature.

8-27 SERIES MOTORS

In the series motor the fields are connected in *series* with the armature. The same current that flows through the fields flows through the armature. The fields, therefore, are wound with wire of sufficient size to carry total motor current.

Figure 8-19 shows the armature and field connections of series and series interpole motors. As shown, interpoles are also connected in series with the armature.

A series motor differs considerably from a shunt motor in operating characteristics. A series motor has an *extremely high* starting torque, about 450 percent of full-load torque. *Its speed varies with its load.* As load decreases, speed increases. In large sizes, a series motor with no load will tear itself apart unless no-load governors or brakes are used. These characteristics are due to the way the magnetism of the fields and armature interact.

When a series motor is operating under a heavy load, its speed and cemf are low. Low cemf allows a high value of current to flow through both the field and armature. The resultant torque increase is nearly in proportion to the *square of the increase in current*. If the current is *doubled* because of increased load and decreased speed, magnetic fields of both the fields and armature are *nearly doubled;* so the resultant torque is nearly *4 times* as great. This is true for conditions below the saturation point of the magnetic circuits.

As load is decreased, speed increases and cemf increases, but cemf does not increase in proportion to speed since less current flow weakens the field. By weakening its field as speed increases, a series motor does not have a natural upper limit to its speed. Because of the lack of a no-load speed ceiling, it is dangerous to use a series motor unless it is directly connected to the load. Series motors cannot be safely used on belt drives.

The speed-load current relationship of a typical series motor operating in a range of its full-load speed is illustrated in Fig. 8-15. It will be noticed that the curve for the series motor shows that it is drawing about 40 A at its rated full load. As the load current

SERIES
(a)

SERIES   INTERPOLE
(b)

Fig. 8-19 Field and armature connections of a series motor. (*a*) Without interpoles. (*b*) With interpoles.

decreases, the curve turns sharply up toward higher speeds. This curve will continue upward as the load decreases.

Because of their high torque and low speeds at heavy loads, series motors are ideal for traction drives such as railway, streetcar, and mine trolley service, as well as for crane, hoist, and elevator work. Series motors also are extensively used in portable hand tools because of the high horsepower they deliver per pound of weight. Since horsepower is the product of speed and torque, small high-speed motors can deliver relatively high horsepower at rated loads. It will be noticed that in the above applications, series motors are usually under the direct control of the operator, as would be the case with varying loads.

Series motors are also designed for use on alternating current. The type of current a series motor is designed for is usually given on the nameplate of the motor. Small series motors designed to operate on either ac or dc and, at the same voltage, with equal efficiency are called *universal motors*. Many electrical appliances used in the home have universal motors.

Standard terminal identification for series and series interpole motors is given in Fig. 8-20. It will be noticed that the interpole winding in the series interpole motor is connected in the armature circuit as part of the A circuit and is connected on the A2 side of the armature.

If the two leads with the dashed line between them in either diagram are connected together and the remaining two leads are connected to the lines, rotation will be standard rotation, which is counterclockwise. For clockwise rotation, A1 and A2 must be reversed.

Standard series motor connections for counterclockwise and clockwise rotation are given in Fig. 8-21. Connections for interpole or noninterpole motors are the same. The leads listed in the "Tie" column are to be connected together. According to Fig. 8-21 for counterclockwise rotation, A1 connects to line 1 (which can be positive or negative), A2 and S1 connect together, and S2 connects to line 2. Line polarity does not affect rotation of a dc motor, except motors with permanent magnet or separately excited fields.

If terminal identification numbers are wrong or lost, they can be established by external testing. The

ROTATION			
COUNTERCLOCKWISE		CLOCKWISE	
LINE 1	LINE 2	LINE 1	LINE 2
L1-A1	L2-S2	L1-A2	L2-S2
TIE A2-S1		TIE A1-S1	

Fig. 8-21 Series motor connections.

series field leads and armature leads of a series motor are usually the same-size wire since they carry the same current.

To test for identification with a test lamp, touch the commutator with a test prod. The leads that cause the light to go on when touched by the other test prod are A leads. If the motor has interpoles, the A lead coming directly from an interpole winding should be marked A2. The other *A* lead should be marked A1. On a noninterpole motor the A leads can be arbitrarily marked A1 and A2. Connect the A2 lead to one of the S leads.

If the motor rotates counterclockwise, the S lead connected to A2 is S1 and should be so marked. The other S lead should be marked S2.

If the motor rotates clockwise, the S lead connected to A2 is S2. The S lead should be marked S1.

If the motor has been overhauled, the polarity of the interpoles should be checked by use of the M-G rule (Sec. 8-11).

8-28 COMPOUND MOTORS

A compound motor is a *combination series and shunt* motor. It contains a *series* winding and a *shunt* winding on its main poles. The series winding is connected in *series* with the armature as in a series motor, and the shunt winding is connected in *parallel*, or across the line, as in a shunt motor. This arrangement gives a

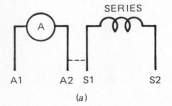

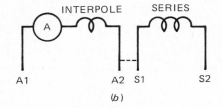

Fig. 8-20 Standard motor terminal identification for series motors. (*a*) Without interpoles. (*b*) With interpoles.

compound motor some of the operating characteristics of a series motor and some characteristics of a shunt motor.

A compound motor has a starting torque of from 300 to 400 percent of full-load torque at rated voltage, depending on the relative strength of the series and shunt fields. Its speed varies with load up to 25 percent of no-load speed.

In construction and installation, the fields, in some cases, are made with the shunt winding wound first on a form and insulated, and the series winding wound over the shunt winding. Two leads are brought out from each winding, and the two windings are taped into one coil and installed on a pole piece. In other cases, two separate complete coils are made and taped together and installed as one coil. Another method used is to make and install two coils separately.

Figure 8-22 shows the circuit connections for compound interpole and noninterpole motors. A compound motor has high starting torque similar to series motors and a no-load speed ceiling similar to shunt motors. It does not have the constant-speed characteristics of a shunt motor.

When starting or overcoming heavy loads, the armature draws comparatively *heavy current* through the series field. This produces a torque in proportion to the *square of the increase in current*. To this is added the torque supplied by the shunt field. As load decreases, the armature speed increases. The increased speed produces a greater cemf so that the armature and series field circuit draw less current. This weakens the series fields; but the shunt field, remaining constant at all times, provides a ceiling to the no-load speed. Thus a compound motor has a high starting torque and the heavy load torque of a series motor but does not have the runaway characteristic of a series motor. Speed of a compound motor from no load to full load will vary up to 25 percent of its no-load speed. The degree of series motor and shunt motor characteristics depends on the relative strengths of the two windings. If the shunt field of a compound motor becomes disconnected or open-circuited, the motor will, in effect, become a *series motor* and race to destruction if it is not restrained by a load.

The speed-load current curve for a compound motor (Fig. 8-15) shows that the compound motor has a higher no-load speed and a wider variation between no load and full load than a shunt motor. Depending on the comparative strength of the series and shunt fields, there are three general types of compound motors: *overcompound, flat compound,* and *undercompound*. The overcompound motor has the strongest series field, and the undercompound the weakest field. Overcompound motors have more series motor characteristics, while undercompound motors have more shunt motor characteristics.

A motor with fewer turns in the series field than undercompound motors is known as a *stabilized shunt* motor. This motor is similar to a shunt motor except that it has slightly more starting torque.

8-29 CUMULATIVE AND DIFFERENTIAL COMPOUND

There are two ways the series fields of a compound motor can be connected in relation to the shunt field. The series fields can be connected for the *same polarity* of the shunt fields so that the magnetic effect is the same or *cumulative,* or they can be connected in *opposite polarity* so that the magnetic effect of one *opposes* the other. If current enters the series and shunt fields at the same terminal number, the series and shunt poles are of the same polarity and the motor is *cumulative*. Thus, if S1 and F1 are connected to the same side of the line, the connection is cumulative. For a *differential* connection, S2 and F1 will be connected to the same side of the line.

When the polarities of the two windings are the same, the motor is known as a *cumulative compound* motor, and the operating characteristics are as previously described for compound motors. If the fields are different or oppose each other in polarity, the motor is known as a *differential compound* motor.

The operating characteristics of a differential connection are quite different from those of a cumulative connection. A differential connection affords a more constant speed through the no-load to full-load range of the motor, but as the load increases beyond full load, torque decreases rapidly, and the motor stalls quickly. In some cases this is a desirable safety fea-

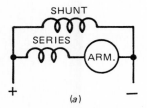
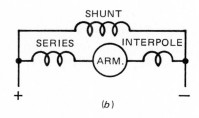

Fig. 8-22 Series and shunt field, and armature connections of a compound motor. (*a*) Without interpoles. (*b*) With interpoles.

ture that prevents breakage or damage to overloaded equipment.

If the load is too great, the current in the series and armature circuit can become so high on an *over-compound differential* motor that the series field overcomes the magnetic effect of the shunt field. When this happens, the motor will *stall and start in reverse.* Under some conditions an inadvertent connection of this kind can be dangerous. Therefore great care should be used in connecting the series fields of a compound motor.

If the shunt field of a differential compound motor is disconnected or open-circuited while running, the motor will, in effect, become a series motor and stop, reverse, and race to destruction if it is not restrained by a load.

The differential connection is seldom used, but where it is desired, the controller can be made to automatically disconnect the series field on starting or on sudden heavy load and let the motor operate as a shunt motor until the load is adjusted to normalcy.

Standard terminal markings for compound and compound interpole motors are shown in Fig. 8-23. In the reversing motors, if the leads of each group connected by dashed lines are connected together, rotation will be counterclockwise. Terminal connections for a compound motor are shown in Fig. 8-24.

If terminal identification numbers on a compound motor are wrong or missing, they can be established by testing with a test lamp or ohmmeter.

To test, touch one prod of a test lamp or ohmmeter on the commutator. A low-resistance reading will be obtained from the two armature leads. The lead connected to an interpole should be marked A2 while the other lead should be A1. Without interpoles, the armature leads can arbitrarily be marked A1 and A2.

A bright light or low-resistance reading will be obtained between the series leads, which should be marked S. A dim light or spark or high-resistance reading will

be obtained on the two shunt leads, which should be marked F.

Connect an S lead to A2 and connect the other S lead and A1 lead to the lines. Start and stop the motor briefly without the shunt winding to determine direction of rotation. If rotation is counterclockwise, the S lead connected to A2 is S1 and should be marked S1, and the other S lead should be marked S2. If rotation is clockwise, the S lead connected to A2 is S2 and should be so marked.

If the motor has been overhauled, the polarity of the interpoles should be checked by use of the M-G rule (Sec. 8-11).

Actual connections for a four-pole compound interpole motor for cumulative counterclockwise rotation are shown in Fig. 8-25. This motor is equipped with open-type coils.

8-30 LONG- AND SHORT-SHUNT CONNECTIONS

Compound motors and generators are further distinguished by the way the shunt windings are connected across the armature. Figure 8-26(*a*) shows the shunt winding connected across both the series field and the armature. This configuration is called a *long-shunt* connection. In Fig. 8-26(*b*) the shunt winding is connected across the armature only. This configuration is known as a *short-shunt* connection. Notice that in the short shunt the current through the series field is not the same as the current through the armature. In addition, the shunt field is not connected across the line so that it can be rated for a lower voltage.

There is so little difference in performance between these two connections that the difference may usually be ignored. In most cases compound motors are connected as long-shunt machines.

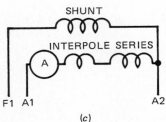

(a)

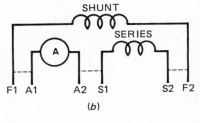

(b)

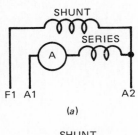

(c)

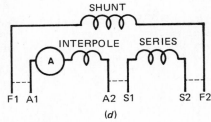

(d)

Fig. 8-23 Standard motor terminal identification for compound motors.
(*a*) Nonreversing without interpoles.
(*b*) Reversing without interpoles.
(*c*) Nonreversing with interpoles.
(*d*) Reversing with interpoles.

ROTATION			
COUNTERCLOCKWISE		CLOCKWISE	
LINE 1	LINE 2	LINE 1	LINE 2
L1–A1–F1	L2–F2–S2	L1–A2–F1	L2–F2–S2
TIE A2–S1		TIE A1–S1	

Fig. 8-24 Compound motor connections.

8-31 THE MOTOR NAMEPLATE

It is always important to protect equipment nameplates. Too often motor nameplates are carelessly handled and become lost at great expense to all concerned. The nameplate on a motor contains valuable information for the electrician when installing the motor; when selecting proper wire size, controller, and overload equipment; and when ordering parts, blueprints, or winding materials or data from the manufacturer. When a motor is selected for a job, the horsepower, speed, and heat rating must be known. When it is installed, the voltage and full-load current of dc motors must be known. For ac motors the voltage, full-load current, frequency, phases, and locked rotor current must be known.

In ordering parts or corresponding with the manufacturer regarding a motor, the complete information on the nameplate should be supplied. All the information needed by a manufacturer to identify a motor is contained on the nameplate. Each manufacturer has its own method of recording information pertinent to

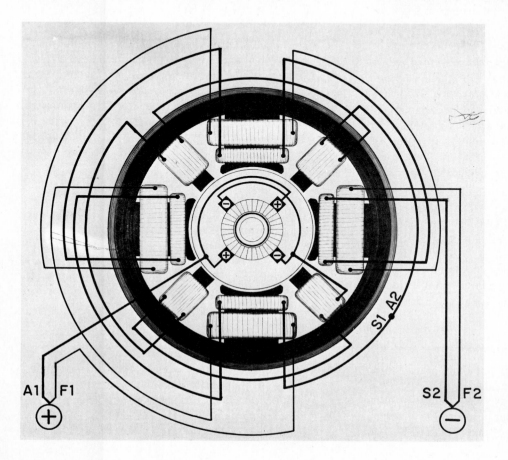

Fig. 8-25 Winding connections for a four-pole cumulative compound motor with interpoles.

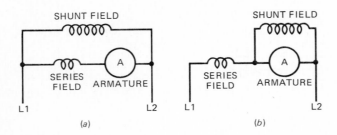

Fig. 8-26 **Long and short shunt connections of compound motors.**

its motors, and some of this information is coded individually, in combinations, or in some of the headings of type, style, frame, or serial number.

Information of interest to the electrician usually appears on the nameplate as follows:

Horsepower. The horsepower the motor can deliver under full load for the time period given on the nameplate under the heading *Temperature rise* or *Degree C rise.*

Rpm. Revolutions per minute of the armature or rotor shaft.

Volts. The normal operating voltage at the motor terminals required by the motor. Most motors will operate satisfactorily on voltages ranging from 10 percent below to 10 percent above the rated voltage.

Full-load current. The number of amperes the motor will draw per line under full load.

Phase and frequency. The number of phases of alternating current required by a motor which always appears on polyphase motors and occasionally is given on single-phase motors. The frequency in hertz (or cycles per second) is always given on an ac motor. (See Chaps. 14 and 15.)

Degrees C rise. The final determining factor in the horsepower rating of a motor is heat. Nearly all general-purpose motors carry a heat rating in the form of heat rise in degrees Celsius above ambient or room temperature. This rating means, for example, that if a motor is rated at 40°C rise, it will deliver its rated horsepower for the time given and will not heat more than 40°C above room tempera-

ture, other conditions being favorable. The time is usually stated as "Cont." which means continuous operation at rated load, or "Int." which means intermittent operation, or the time may be indicated in minutes of operation. A simplified method of converting Celsius to Fahrenheit, for the specific purpose of interpreting motor nameplate data, is to multiply degrees Celsius by 1.8 and add 32. Thus 40°C equals 1.8 × 40°C plus 32 or 104°F. A motor rated at 40°C Cont. should drive its full rated load continuously and not heat over 104°F above room temperature. If room temperature is 80°F, normal operating temperature will be 184°F for this motor. Although this motor will be too hot to touch, it will be at its safe operating temperature.

Service factor. Some motors have reserve capacity for use for short periods of time. The amount of this reserve is expressed by a multiplier that is used with the motor's rating to determine the reserve. Thus, if a 10-hp motor has a service factor of 1.2, it can drive a 12-hp load for a short period of time without heating to the extent of injury to the motor.

Code. A code letter is used here to indicate the kilovolt-amperes an ac motor will draw under locked-rotor conditions. The values of these code letters are contained in NEC Table 430-7(*b*). The code letter is used in determining motor branch-circuit overcurrent protection, as given in NEC Table 430-152.

Miscellaneous information. If a series motor is labeled "Universal," it will operate on either alternating or direct current. "Fan Duty" means that the motor must operate in a current of air and is provided with proper thrust capacity to support a fan load. In some cases part numbers for a capacitor, overload device, or controller for a motor appear on the nameplate.

If a nameplate is lost or damaged, blank plates can be obtained and necessary information can be stamped on the blank with steel dies. In case of lost nameplates, information data may be found in an inventory list, in preventive-maintenance schedule files, in a stock file, in purchase invoices, or in repair records. In every case, a motor should have its own nameplate.

SUMMARY

1. Direct-current motors are popular because they are easily adapted to control and can operate from portable batteries. They are well suited for jobs requiring frequent starting, accelerating, decelerating, reversing, plugging, and jogging.

2. Direct-current motors and direct-current generators are practically identical in construction.

3. Interaction between the field and armature magnetism gives a dc motor armature the ability to rotate.

4. The force of rotation is known as torque.

5. Horsepower is determined by torque and speed in revolutions per minute.

6. A dc motor can be reversed by reversing current flow through either the field or the armature, but not both. Rotation will remain unchanged if both the field and armature are reversed. It is common practice to reverse current in the armature to reverse rotation.

7. Brush position determines where the pattern of current flow appears in the armature. The pattern of current flow remains stationary with the brushes regardless of the speed of the armature.

8. Each current flow pattern should be positioned under a pole of main field flux.

9. A dc motor armature distorts the main field flux away from the direction of rotation. This is known as armature reaction. For the flow pattern to be positioned in the main field flux on noninterpole motors, it is necessary to slightly shift the brushes opposite to the direction of rotation, to working neutral, where the flow pattern is positioned in the main flux.

10. Interpoles are small poles located between main poles. Interpoles help reverse current flow in the armature coils during commutation. Their effect is to reduce sparking at the brushes.

11. Interpole polarity can be found by use of the M-G interpole rule. Interpoles with wrong polarity increase the distortion of the magnetic field in the air gap. This results in severe sparking, charring and damage to the commutator, overheating of the motor, and loss of horsepower.

12. The M-G rule can be applied to a motor or generator, viewed from either end for either direction of rotation.

13. Hard neutral is the proper position of the brushes on interpole reversing motors. Working neutral is the proper position of the brushes on noninterpole nonreversing motors.

14. Several methods are used for locating hard or working neutral, depending on the motor and conditions of the job.

15. Compensating windings are used to minimize field distortion. They are wound in slots on the main poles and surround the interpoles. They are connected in series in the armature circuit. The polarity of a compensating pole is the same as that of an interpole it surrounds and can be found by the M-G rule.

16. The speed of dc motors is determined by the load and the strength of the countervoltage, called cemf, generated when running.

17. At a constant applied voltage, cemf regulates the current flow in the armature circuit.

18. Weakening of the field of a dc motor reduces cemf, which allows more armature current to flow and increases the speed if the load permits.

19. The main factor in rating a motor's horsepower is the amount of heat it can safely dissipate.

20. The three main types of dc motors are shunt, series, and compound. They differ in the way their fields and armature are interconnected and in their speed vs. load current characteristics.

21. A shunt motor has its fields connected in parallel with the armature. It has medium starting torque (about 275 percent of full-load torque at rated voltage) and a practically constant no-load to full-load speed, varying about 5 percent of full-load speed.

22. A series motor has its fields connected in series with the armature. It has high starting torque (up to about 450 percent of full-load torque at rated voltage). Speed varies with the load. Unless controlled, the motor will run away at no load.

23. A compound motor has both shunt and series fields. It has more starting torque than a shunt motor, or about 350 percent of full-load torque at rated voltage, and a limited but higher no-load speed than a shunt motor. Its speed varies up to 25 percent between no load and full load, depending on the comparative strength of the series fields.

24. If the series fields of a compound motor are of the polarity of the main fields, the motor is cumulative compound. If the series fields are of opposite polarity to the main fields, it is differential compound. Cumulative compound motors are used in practically all cases.

QUESTIONS

8-1 For what types of applications are dc motors best suited?

8-2 Name the principal parts of a dc motor.

8-3 How is current conducted to the rotating armature winding of a dc motor?

8-4 What is the torque of a motor and how is it produced?

8-5 List the three basic types of dc motors, draw a schematic diagram of each, and give their starting torque characteristics.

8-6 What causes field distortion?

8-7 What methods are used to minimize field distortion?

8-8 What is the hard neutral position of the brushes?

8-9 How is commutation aided by interpoles?

8-10 What is the rule for determining the proper polarity of interpoles?

8-11 Name four methods used for finding the hard neutral position of the brushes.

8-12 What factors determine the speed of a dc motor?

8-13 How does weakening the field of a shunt motor affect its speed?

8-14 How does weakening the field of a series motor affect its speed?

8-15 How can the air gap between the field poles and the armature be decreased?

8-16 How would decreasing the air gap in a motor affect its speed?

8-17 What is speed regulation of a motor?

8-18 Would a constant-speed motor have high- or low-speed regulation?

8-19 What percentage of full-load current is considered acceptable for starting dc motors?

8-20 What is the base speed of a motor?

8-21 For the purpose of terminal identification, what is considered standard rotation of a motor?

8-22 What is the direction of rotation of a shunt motor when F1 and A2 are connected to the same line?

8-23 Why should a series motor always be directly connected to its load?

8-24 What can cause a compound motor to race excessively?

8-25 How does the shunt field connection of a short-shunt compound motor differ from that of a long-shunt motor? Draw diagrams to show the difference.

8-26 What end of a motor is used to specify the direction of rotation of its armature?

8-27 How can the direction of rotation of a dc motor be changed?

8-28 What part of the usual dc motor is eliminated through the use of permanent magnets?

8-29 How is the prony brake test used?

8-30 What is the horsepower of a motor rotating at 1150 rpm, pulling 22 lb on a hook 18 in. from the pulley of a prony brake?

8-31 How does a cumulative compound motor differ from a differential compound motor?

8-32 What is the strongest field in an undercompounded motor?

8-33 Why is it important to retain the nameplate of a motor?

8-34 What is meant by the motor specification ''Degrees C rise''?

8-35 A temperature is given as 20°C; what is the Fahrenheit temperature?

8-36 What is a universal motor?

9
ARMATURE WINDING

In order to best understand the principles of a dc motor or generator and be able to test, diagnose, and correct troubles, a thorough knowledge of armature windings is essential. The armature is the heart of a motor or generator. *It is the rotating member* of these machines. When rotating, the armature is subjected to unbelievably high strains and pressures resulting from centrifugal forces and vibration, which are sources of trouble.

The armature is also subject to more abuse in performing its function than any other part of these machines. The fields seldom give trouble and are not severely affected by overloads, but the armature is damaged to some extent every time a motor or generator is overloaded.

Main shunt fields are connected directly to the line, and their current is *constant,* irrespective of the load. All the current an armature receives is supplied through brushes and the commutator, which creates a possible source of trouble.

Armature current is *in proportion to the load.* Every time a motor or generator is overloaded, an overloaded armature results, leading to early damage or burnout of the armature. Knowledge of the nature and duty of armatures will lead to better protection and care of them.

9-1 THE ARMATURE

Figure 9-1 shows an armature with the names of the principal parts. The entire armature assembly consists of a shaft with a core having slots which contain the winding or coils. The winding is connected to a commutator from which it *receives current* from the line *through brushes,* in the case of a motor, or *delivers current* to the line, in the case of a generator. Direct-current motors and generators use the same kind of armatures.

The core of an armature is assembled from thin sheets of steel known as *laminations* (see Sec. 14-20). Winding slots and the hole through which the shaft will pass are punched out of the laminations. The laminations are stacked to make up the length of the armature. The shaft is then press fit through the core.

The commutator is made by assembling copper bars securely around a rigid insulated holder (see Sec. 10-1). Mica or other thin rigid insulation is placed between adjacent segments to provide electrical isolation. The bars in the commutator, when connected to the winding, can be considered as *merely extensions of the coil ends,* or leads, to present a suitable surface from which the rotating coils can deliver or receive line current through the brushes.

9-2 FUNCTION OF THE WINDING

An armature of a motor or generator contains a winding that conducts current through it in an orderly manner to produce a desired pattern of current flow. A review of Fig. 8-5 will illustrate what is meant by a "desired pattern of current flow." In the case of a motor, the current flow from positive to negative brushes in all the coil sides under north poles is in one direction, forward or backward from the commutator, and the direction of current flow in all coil sides under south poles is in the opposite direction.

The pattern of current flow of a motor running counterclockwise is shown in Fig. 8-6. The direction of current flow of a generator running counterclockwise is opposite to that of a motor, but the general pattern is the same. Motors and generators generally have the same armature windings.

9-3 LAP AND WAVE WINDINGS

There are two classifications of armature windings—*lap* and *wave.* Armatures in ac or dc motors or generators regardless of size are either lap or wave wound. Some armatures are *wound by hand,* and some are wound by the use of *form-wound* coils.

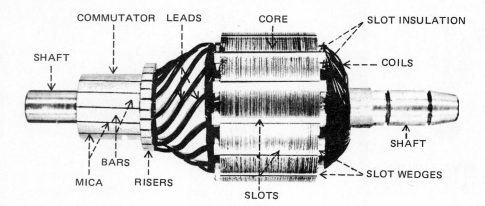

Fig. 9-1 A typical dc armature.

The physical difference between a lap and a wave winding is in the *lead connections to the commutator*. In a lap winding, the leads of a coil connect to *adjacent commutator bars*. In a wave winding, the leads of a coil connect to commutator bars *several bars apart*. These differences are shown in Fig. 9-2. There is no difference between the actual coils of a lap winding and a wave winding of an armature up to the point of connecting the leads to the commutator.

Two identical winding conditions are shown in Fig. 9-3. Both windings have their lead 1 connected to commutator bars. Figure 9-3(*a*) is to be a lap, and Fig. 9-3(*b*) is to be a wave. Both windings are the same up to this point. Either winding can now be connected lap or wave. Where lead 2 connects now determines whether each winding is lap or wave.

The final connections of the right leads of these two windings are shown in Fig. 9-4. In Fig. 9-4(*a*) lead 2 is connected to produce a lap winding. Figure 9-4(*b*) shows lead 2 connected to produce a wave winding. In the lap winding, the leads of coil 1 connect to adjacent bars 1 and 2, and all other coils connect in the same order. In the wave winding, one lead of coil 1 connects to bar 1 and the other lead connects to bar 3. One lead of coil 2 connects to bar 3 while its other lead connects to bar 5, which is adjacent to bar 1, the beginning of this circuit. A circuit once around the armature of a single wave winding always ends on a

bar before or after the beginning bar. A continuous circuit can be traced by beginning with bar 1 and coil 1 and ending on bar 1.

The electrical difference in a lap and a wave connection is that a *lap winding has two or more poles and as many paths or circuits through it from positive to negative brushes as it has poles or brushes,* while a *single wave winding has four or more poles and only two circuits through it from positive to negative brushes, regardless of the number of poles.* A single wave winding requires only two brushes, although more are generally used.

Since a single wave winding has only two circuits through it regardless of the number of poles, while a lap winding has several circuits depending on the number of poles, a *wave winding requires higher voltage* for operation when used in a motor, or it will *generate a higher voltage* when used in a generator. So wave windings are commonly used for *higher-voltage operations*. The circuits through the two windings are different, but the pattern of current flow is the same.

9-4 REWINDING AN ARMATURE

Rewinding of armatures is not a complicated or difficult job, as it is generally believed to be. With the

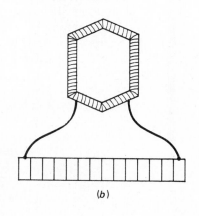

(*a*)   (*b*)

Fig. 9-2 Methods of connecting coil ends to a commutator. (*a*) Lap winding. (*b*) Wave winding.

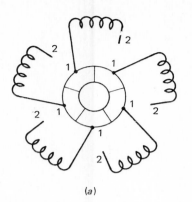

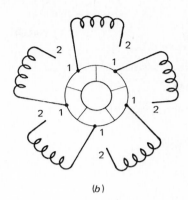

Fig. 9-3 The start of a lap and a wave winding are identical. (*a*) The start of a lap winding. (*b*) The start of a wave winding.

(a)

(b)

understanding of a few basic principles and practices, an armature can be wound or rewound accurately and efficiently.

To rewind an armature, it is necessary to get and record certain information from the armature before it is stripped of the old winding. In many ways the new winding must be an exact duplicate of the old winding. Recording information from the old winding is known as *taking data,* and the individual coil from which data are taken is known as the *data coil.* Since all coils are connected in the same manner, any coil can be selected as the data coil. It is advisable to select the *last coil* in the winding, since it is on top of the other coils and its sides and ends are visible and accessible.

The most important data items are the *coil span,* the *number of turns of wire* in a coil, the *wire size,* and the *connection of the coil leads to the commutator.*

9-5 THE DATA FORM

Data forms vary according to the needs and desires of various rewinding shops. The data form shown in Fig. 9-5 is a typical general form that covers most situations. Data forms should be made in duplicate so that one copy can be filed for safekeeping immediately after data are taken and one copy can be used on the

job. Data in some cases are *literally priceless* and should be stored in permanent equipment files.

The top of the form shown is filled in chiefly with data from the nameplate of the motor or generator. The sketch shown is based on the armature in Fig. 9-6, with one coil, shown in heavy lines, connected to the commutator and the commutator bars properly numbered for this armature. Guidelines in the sketch are used to draw coils as needed.

Vertical lines are drawn in between the two horizontal lines below the sketch to show the relative position of the commutator bars to the coil that connects to them. The core span, or coil span, is shown in the rectangular figures in the slots. For a 1–7 span, 1 should be placed in the left figure, and 7 should be placed in the right figure, as illustrated.

Items on this form are arranged to be filled in typical order as the armature is stripped. Each item is defined as follows:

Number of core slots. The total number of slots in the core.

Coil span. The number of slots spanned by a coil from left side to right side, beginning with 1 for the left side. The recording for a coil occupying slots 1 and 10 would be 1–10.

Number of poles. The number of poles of an armature, usually determined by dividing the number of slots

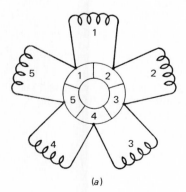

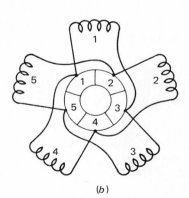

Fig. 9-4 The connection of lead 2 in Fig. 9-3 determines the type of winding, (*a*) lap or (*b*) wave.

(a)

(b)

ARMATURE DATA

Fig. 9-5 Armature data form.

Customer_____ Job No._____ Date_____

Make_____ Frame_____ Model_____ Type_____

HP_____ RPM_____ Volts_____ Amperes_____ AC or DC_____

Miscellaneous_____

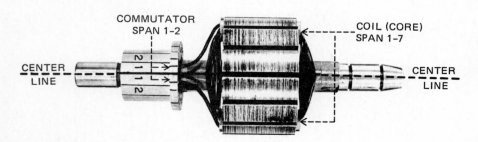

(Commutator Connection)

Number Core Slots __14__ Coil Span __1-7__ Number Poles __2__

Number Commutator Bars __14__ Commutator Span __1-2__ Wires Per Bar __2__

Lap or Wave __Lap__ Do Leads Cross? __no__ Equalizer Connection __none__

Left Leads to __1 L__ Center Line. Right Leads to __1 R__ Center Line

Coils Extend From Core: Front __3/4"__ Back __3/4"__ Overall Length Coils __4"__

Length of Core __2½"__ Diameter of Core __3½"__ Semi- or Open Slots __Semi-__

Distance From End of Shaft to End of Commutator Bars __1⅛"__

Number Turns Per Coil __10__ Wire Size __16__ Wire Insulation __Formex__

Coils Per Slot __1__ Total Turns Per Bundle __10__ Number Wires Parallel __1__

Pounds of Scrap Wire __2__ Fan (On-Off) __Off__ Ball Bearings (On-Off) __On__

Direction of Rotation (PE) __C W__ Data by_____

Miscellaneous_____

by the coil span and taking the nearest whole even number.

Number of commutator bars. The total number of commutator bars, sometimes called *segments,* in the commutator.

Commutator span. The number of bars occupied and spanned by the leads from the data coil. The span of

a single lap winding is always two consecutive numbers such as 1–2, 2–3, etc.

Wires per bar. The total number of wires contained in the riser or slot of a bar. In a single lap winding, a bar would contain the right lead of one coil and the left lead of another coil for a total of two wires. If the coils were wound with wires parallel, for exam-

COMMUTATOR SPAN 1-2

COIL (CORE) SPAN 1-7

CENTER LINE

CENTER LINE

Fig. 9-6 Hand wound two-pole lap armature used in filling out the data form in Fig. 9-5.

ple, two No. 13 wires instead of one No. 10, for convenience in winding, the total would be four wires per bar.

Lap or wave. State here the type of winding in the armature—lap or wave.

Do leads cross? State "yes" or "no," as the case may be.

Equalizer connection. The commutator span of the equalizers, if used, is shown here.

Left leads to ——— centerline. Record here the number of the bar left or right of the centerline to which the left lead is connected. For example, the left lead in Fig. 9-5 connects to bar 1 left of the centerline, so it connects to 1L.

Right leads to ——— centerline. Record here the number of the bar left or right of the centerline to which the right lead is connected. For example, the right lead in Fig. 9-5 connects to bar 1 right of the centerline, or 1R.

Coils extend from core. Front: Record the distance from the extreme front ends of the coils to the iron part of the core (commutator end). Back: Record the distance from the extreme back ends of the coils to the iron core.

Overall length of coils. Record the distance between the extreme ends of the coils. A winding should be measured and compared with these data during the process of winding to be certain it is within its winding space limits. A new set of form-wound coils should be measured and compared with these data to be certain they fit properly before winding is started.

Length of core. This information, with the diameter of the core and type of slots and other data, is commonly used to identify an armature.

Distance from end of shaft to end of commutator bars. This is important information needed in replacing a commutator, and it should be taken with careful measurement. The measurement is made from the end of the shaft to the front ends of the commutator bars.

Number of turns per coil. This is the number of turns in one coil from left lead to right lead. All coils in a set should have the same number of turns. In replacing a hand winding, it is advisable to count the turns of more than one coil to verify the correct number of turns.

Wire size. The size of wire, measured with an AWG gage, is recorded here. The diameter of a wire can be measured with a micrometer, and its size can be determined by reference to a wire table, such as Table 3-1.

Wire insulation. Magnet wire, used for winding armatures, is insulated with many types and combinations of insulation, such as glass, cotton, silk, asbestos, enamel, synthetic enamels, silicones, and plastics such as Teflon fluorocarbon resins. The synthetic-enamel insulated wires, such as Formex, etc., are usually made in single heavy, triple, and quadruple thickness of enamel.

Coils per slot. The number of coils per slot is determined by the ratio of commutator bars to slots. Only the data coil or bundle is counted. A coil or bundle occupying the other half of the slot is not counted. Most armatures have ratios of bars to slots of 2 to 1, 3 to 1, or 4 or more to 1. Coils per slot can be determined by dividing total slots into total commutator bars or segments. Total coils in a slot compose a bundle. A bundle is usually referred to as a *coil,* although it may contain more than one coil.

Total turns per bundle. If coils are wound with one wire, the total turns in a bundle can be determined by cutting the bundle at the back end and counting the turns. If smaller wires are used parallel instead of one larger wire, the total wires in a bundle divided by the number of wires parallel will give total turns per bundle. Total turns per bundle divided by the number of coils per bundle will give total turns per coil.

Number of wires parallel. Occasionally, commutator slots or risers are not wide enough for a needed wire size, or the size is too large for convenience in winding. In these cases, wires are paralleled to equal the cross-sectional area of the size needed. This condition must be considered in determining wires per bar, number of turns per coil, and total turns per bundle.

Pounds of scrap wire. This is the total weight of the copper in a winding. This information is helpful in determining the cost of a job or returning scrap wire to a customer if it is requested.

Fan (on-off) and ball bearings (on-off). These conditions, including pulley or gear on-off, should be recorded when an armature is received for rewinding.

Direction of rotation (PE). In some cases it is necessary to know the required direction of rotation of a motor or generator after it is assembled. The direction given is as viewed from the pulley end.

Data by. This space is signed by the person responsible for the accuracy of the data recorded from an armature.

Miscellaneous. Any information not previously recorded in the data form but necessary or pertinent to the armature or winding should be recorded in this space.

9-6 TAKING DATA

A hand-wound two-pole lap armature with one connected coil is shown in Fig. 9-6. The coil span is the number of slots occupied and spanned by the coil sides. Full coil span of an armature is determined by dividing the number of poles into the number of slots and adding 1. Few armatures, however, are wound full span. The coil span is usually shortened, or *chorded* up, to about 20 percent less than full span.

The coil span in the armature shown is 1–7. The left coil side is in slot 1, and the right coil side is in slot 7. Sometimes this span, or *pitch*, is said to be 6, since one coil side is in one slot and the other coil side is six slots from it. (In this book when the span is referred to in this manner, with a single number, it will be called *pitch*.)

This armature was chosen for simplicity in illustrating rewinding. It has 14 commutator bars and 14 core slots; therefore it has one coil per slot. In most armatures the number of bars is two or more times the number of slots; therefore each slot will contain two or more coils.

The connections of the coils to commutator bars and the circuits through the winding are illustrated in Fig. 9-7(*a*). In the illustration, there are 14 coils connected to 14 commutator bars. The left-hand lead of coil 8 connects to bar 8, and the right-hand lead of coil 7 also connects to bar 8 at the right, which is to be considered the same bar as the bar 8 at the left.

The circuits through the coils are shown from the positive brush on bar 1 through the coils in the directions of the arrows to bar 8 at each end, which contains the negative brush. This illustrates the relationship of one coil to another and the circuits.

These same bars and coils are shown connected in circular fashion in Fig. 9-7(*b*), which is more like the actual condition of a lap winding in an armature. It will be seen that the coil relationship, circuits, and direction of current flow from positive to negative brushes are the same as in the preceding straight-line representation.

9-7 THE CENTERLINE

In taking data on the connection of a coil to the commutator, it is necessary to be *extremely cautious* in determining and recording this information. There is a definite, required relationship between the proper lo-

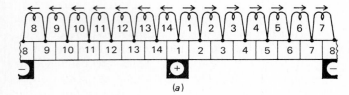

(a)

Fig. 9-7 Lap winding showing the current path from positive to negative brush. (*a*) Straight-line representation. (*b*) Circular representation.

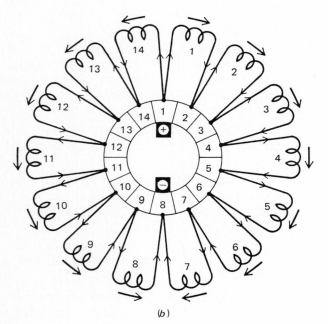

(b)

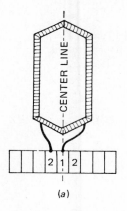 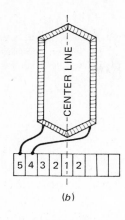 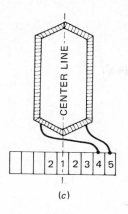

(a) (b) (c)

Fig. 9-8 **Examples of lead swing.**

cation of the *main motor poles*, the *brushes*, the *coil in the armature*, and its *connection to the commutator*. To record this required relationship as it exists on the armature, a string is used to form a centerline in the center of the data coil, parallel to the shaft and extending over the commutator. The lead connection of the data coil to the commutator is found and recorded from the centerline at the commutator.

A centerline is shown established in Fig. 9-6. It runs parallel with the shaft and through a slot which is the center of the data coil in slots 1 and 7 and over the mica between commutator bars. The bars on each side of this line are numbered 1, and bars to the right or left beginning with bar 1 are numbered as far as necessary to include bars containing leads from the data coil.

In the illustration, the data coil leads are connected to bars 1 each side of the centerline. A sketch showing the centerline, numbered bars from the line, and coil lead connections is made and recorded from these data on a data sheet (Fig. 9-5).

9-8 LEAD SWING

Coil leads sometimes *swing* or *throw* to the left or right of the centerline. The *relative positions of the motor poles*, the *brushes*, and the *data coil* in the ar-

mature determine the swing. Three conditions of swing where the centerline runs over the center of a bar are illustrated in Fig. 9-8. In Fig. 9-8(a) the leads would be recorded as connecting as follows: Left lead, 2L, meaning it connects to bar 2 left of the centerline, and the right lead 1CL, meaning it connects to bar 1 on the centerline.

In Fig. 9-8(b), the left lead connects to bar 5L, and the right lead connects to 4L. In Fig. 9-8(c) the left lead connects to 4R, and the right lead connects to 5R.

If the centerline runs between two commutator bars, the bars on each side of the centerline are numbered 1. This method of numbering bars and various lead swings is illustrated in Fig. 9-9. In Fig. 9-9(a) this condition would be recorded as follows: Left lead, 1L, and right lead 1R, since the left lead connects to bar 1 left of the centerline, and the right lead connects to bar 1 right of the centerline.

The condition illustrated in Fig. 9-9(b) would be recorded: Left lead, 4L, and right lead, 3L. The condition in Fig. 9-9(c) would be recorded: Left lead, 3R, and right lead, 4R.

If the coil span were an odd number, such as 1–5, the centerline would run in the center of a slot. In some cases, slots are skewed; that is, they are not parallel with the shaft, and so the centerline must be established parallel with the shaft and over a point in

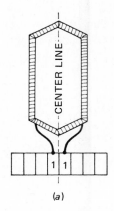 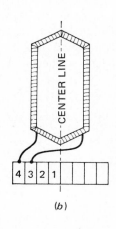 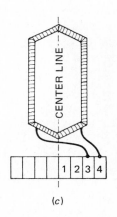

(a) (b) (c)

Fig. 9-9 **Examples of lead swing when the centerline falls between commutator bars.**

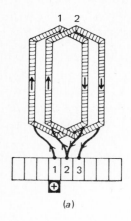

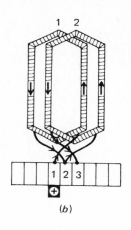

Fig. 9-10 Coil connections and current flow.
(*a*) Progressive connection; leads not crossed.
(*b*) Retrogressive connection; leads cross.

the center of the coil sides midway between the ends of the core.

9-9 PROGRESSIVE AND RETROGRESSIVE CONNECTIONS

If the leads of coils of a lap winding cross at the commutator, the winding is said to be a *retrogressive winding*. If the leads *do not cross*, it is said to be a *progressive winding*. There is no standard way of connecting leads in this respect.

A lap winding with a progressive connection (leads not crossed) is shown in Fig. 9-10(*a*). A retrogressive connection is shown in Fig. 9-10(*b*).

If a winding is inadvertently *changed* from progressive to retrogressive, or vice versa, in rewinding the armature, the effect will be to *reverse* direction of current flow in the armature, and therefore to *reverse* direction of rotation of a motor, or to *reverse* polarity of a generator. The reason for this is shown in Fig. 9-10. In the progressive connection, current flow (as shown by the arrows) is from the positive brush clockwise in the coils, while in the retrogressive connection, current flow is counterclockwise.

Proper direction of rotation of a motor or polarity of a generator can be restored by reversing the positive and negative line leads at the brushes. Usually, a mistake of this nature in the armature lead connections also results in a *wrong lead swing*, which causes sparking at the brushes. Sparking can be corrected by

shifting the brushes. If the brushes are in a fixed position, the lead connections to the commutator bars will have to be changed to the proper connection.

9-10 SLOT INSULATION

Methods of insulating armature core slots are shown in Fig. 9-11. In (1), the slot insulation is cut to reach fully to the opening at the top of the slot. When a slot is insulated in this manner, care must be taken while winding to avoid scraping and damaging the wire against the teeth. After the winding is in the slot, a wedge is driven in to close the slot and hold the wire.

Slot insulation should extend at least 1/8 in. from each end of the slot. In (2) the slot insulation is cut to extend out of the top of the slot to serve as a guide and protection to the wire when winding. After the winding is in, the tops of the insulation are folded into the slot over the winding as shown in (3), and a wedge is driven in over the insulation as shown in (4).

9-11 SLOT WEDGES

Various materials are used for slot wedges. The most commonly used wedge is the wooden wedge made of maple. Examples of typical wooden wedges are shown in Fig. 9-12. These wedges are made in sizes to meet nearly all needs.

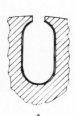

Fig. 9-11 Insulating armature coil slots.

Fig. 9-12 Typical shapes of armature slot wedges.

Formed wedges of vulcanized fibre are used in slots where space is lacking. These wedges are formed in rectangular and U shapes, usually 30 in. long, and are cut to the necessary size. Laminated silicone-bonded glass cloth is used in making formed wedges for high-temperature windings. Wedges are also cut from flat sheets of vulcanized fibre or other insulating materials to fit specific requirements.

9-12 STRIPPING ARMATURES

Methods of stripping and cleaning armatures preparatory to rewinding vary widely in accordance with available equipment, volume of work, and type of armature. The suggested methods given here for small armatures are commonly used in average conditions.

After data have been recorded for an armature, the armature is placed in a lathe and the front ends of the coils are cut with a cutoff tool as near the core as possible without damaging the core.

While the armature is in the lathe, the core and commutator can be cleaned by taking a fine cut on the commutator to the depth of clear mica and using a coarse file or wire brush on the core to remove varnish or foreign matter. Sand blasting is another good method of cleaning.

Following operations in the lathe, the armature is placed in a holder and the wedges are removed. Various tools are available for removal of wedges. A hacksaw blade can be set along the length of a wedge and tapped on the top to set it in the wedge. It can then be driven on the end to drive the wedge out of the slot.

The winding is removed by forcing the point of a large punch or drift in between the back ends of the coils in the core and driving the coils outward with a hammer. Everything in the slot, including slot insulation, will usually be removed with the coils if the core

is heated in an oven or with a torch before the coils are driven out.

The commutator should be given a bar-to-bar test for shorts with a test light or ohmmeter, and a bar-to-core test for grounds. (See Sec. 10-2 on Commutator Repair.)

9-13 HAND WINDING

Hand winding is used chiefly on *small armatures.* In this process, wire is wound directly into the slots by hand. In *form winding,* coils are made on a form and inserted into the slots as a unit. Large armatures are form-wound.

The step-by-step processes of hand winding a two-pole lap armature such as in Fig. 9-6 is shown beginning with Fig. 9-13. In Fig. 9-6, the first coil is wound in the core slots, and the leads from the coil are lap-connected to bars of the commutator. Coil leads are usually not connected to the commutator *until the winding of coils is finished,* but for illustrative purposes the coil in the illustration is shown wound and the leads are connected to the proper bars. All other coils in the armature are wound and connected in the same order as the first coil.

In the method of hand winding shown, the wire between coils is cut. In some cases, especially where small wire is used, the wire between coils is not cut. This is known as a *loop winding.*

In a loop winding, the wire is looped and twisted at the end of each coil, and winding is continued for the next coil. Thus the twisted loop contains the left, or finish, lead of one coil and the right, or start, lead of the next coil. Since the finish lead of one coil and the start lead of an adjacent coil are connected in the same commutator bar, the twisted loops can be soldered to the commutator without cutting the wire.

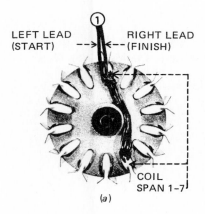
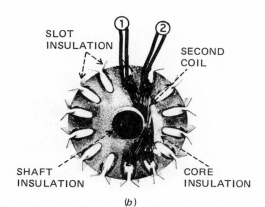

LEFT LEAD (START) → ← RIGHT LEAD (FINISH)

COIL SPAN 1-7

(a)

SLOT INSULATION

SECOND COIL

SHAFT INSULATION

CORE INSULATION

(b)

Fig. 9-13 Hand winding a small armature. (*a*) Coil 1 is wound with a 1-7 coil span. (*b*) Coil 2 is wound clockwise from the first coil.

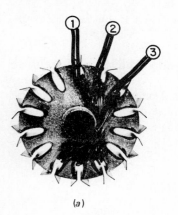

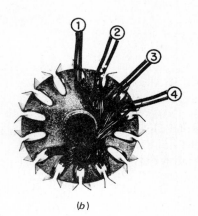

Fig. 9-14 Wire is pulled right in hand winding a small armature. Coil ends are pressed firmly against the core to conserve space. (*a*) Coil 3 is wound on the core. (*b*) Coil 4 is wound on the core.

In some cases the leads of a coil are held in place by folding them back through the slot containing one side of the coil; when the winding is completed, a string or cord is wound around the winding at the back of the core. For connection to the commutator the leads are bent over the string band and forward through their slot, or an adjacent slot, to the commutator. In this method, all leads come from the top of the slot to the commutator. When taking data, *one must be careful in all cases to determine the proper commutator connection.*

Places to be insulated on an armature are shown in Fig. 9-13(*b*). Insulation is shown in the slots, at the ends of the core, and on the shaft. Rag-content insulation paper, or combinations of paper and varnished cambric, or paper and other types of insulation media are generally used for slot insulation. The *ends of the core,* the *shaft,* and *any other part of the armature* likely to come in contact with the winding are thoroughly insulated to avoid grounds.

In Fig. 9-13(*a*) the commutator has been removed to facilitate winding, and the first coil has been wound. The leads from the left side of the coil, when the armature is viewed from the commutator end, are known as the *left* (start) leads, and the leads from the right side of the coil are known as the *right* (finish) leads. Both left and right leads of the first coil are identified by the number 1 in the circle, since they are

from the first coil. The right (finish) lead is brought over and tied to the left lead to hold the last turn of the coil in place. Sometimes leads are twisted together to hold them.

When the first coil is wound, the ends of the coil should be flattened and pressed tightly against the insulation at the end of the core at both ends to conserve winding space. Each time a coil is wound, it should be *shaped to fit the preceding coils and be pressed toward the core at the ends.*

In Fig. 9-13(*b*) the second coil has been wound and the leads marked 2. The lead to the left is the starting lead, or left-hand lead, and the lead to the right is the finishing lead, or right-hand lead, of coil 2.

The third and fourth coils have been wound in Fig. 9-14(*a*) and (*b*). All coils of an armature are wound the same and should have the same number of turns. A piece of vulcanized fibre as long as the slot, and about 3 in. wide, can be used to press or tamp the wires in place in the slots. The fibre should be as thick as possible but still allow easy manipulation in the slots.

In Fig. 9-15(*a*) and (*b*) coils 5 and 6 have been wound. All slots with winding in them contain only one coil side, and no wedges have been installed. Coils can be wound clockwise or counterclockwise. In these illustrations the coils are wound clockwise.

In Fig. 9-16(*a*) the seventh coil has been wound, and the right side of the coil is in the slot with the left

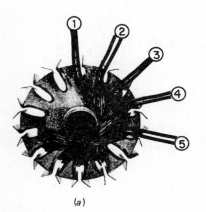

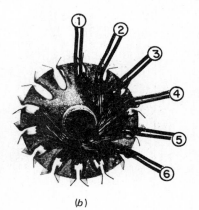

Fig. 9-15 To this point, only one coil side occupies each slot used; (*a*) and (*b*) show the addition of coils 5 and 6, respectively.

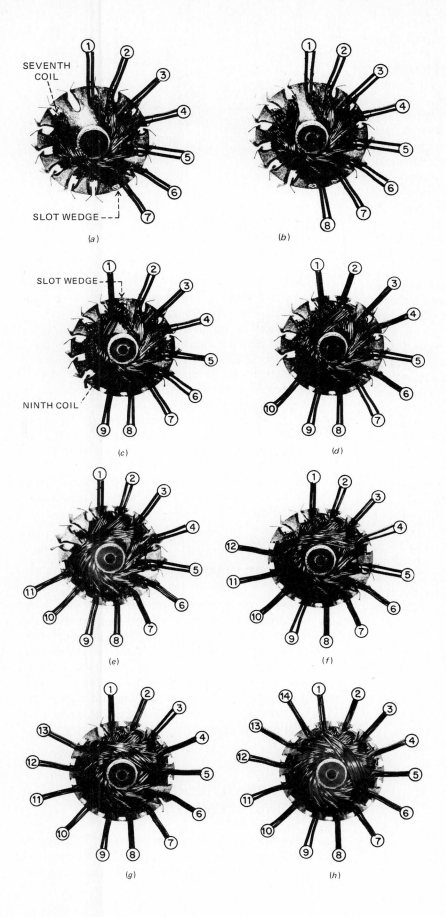

Fig. 9-16 The final coils are wound in the slots and slot wedges are added; (*a*) through (*h*) show the sequence of winding coils 7 through 14, respectively.

side of coil 1. This fills the slot, and a wedge is inserted in this slot, closing it.

In Fig. 9-16(*b*) the eighth coil has been wound. Its right side is in a slot with the left side of coil 2, and a wedge is installed to close the slot.

In Fig. 9-16(*c*) the ninth coil is wound in the slots, and its sides, with other coil sides, fill its slots, and both slots are wedged. Two slots will be closed and wedged every time a coil is wound for the remainder of the winding. Leads of the finished winding will come out more evenly if they are brought out over the coil ends. To arrange for this the leads should be treated as shown in Fig. 9-16(*c*). Leads from coil 1 are shown bent back, and the ninth coil is wound under them. The remaining leads should be treated similarly, as shown in Fig. 9-16(*d*).

The eleventh and twelfth coils are shown wound in Fig. 9-16(*e*) and (*f*). The coil ends are shown flattened and pressed firmly toward the end of the core to preserve winding space.

In Fig. 9-16(*g*) the thirteenth coil is wound. One more coil is needed to complete the winding. In Fig. 9-16(*h*) the fourteenth and last coil has been wound, and the winding is complete with all slots closed. All coil ends have been carefully formed to fit closely and conform to other coil ends. The winding is now ready for connection to the commutator. The back end of the winding is shown in Fig. 9-17.

In Fig. 9-18 the commutator centerline is established, and the left-hand (start) lead of coil 1 is shown connected to commutator bar 1 left of the centerline. The right-hand (finish) lead of the same coil is shown connected to bar 1 to the right of the centerline.

The leads of coil 2 will connect to bars 1 and 2 right of the centerline. Bar 1 right will contain the finish lead of coil 1 and the start lead of coil 2. This order will prevail throughout the remainder of the bar connections. However, when the leads are actually connected, all the start (left) leads are placed in their proper bars *first,* and insulation is placed over them *before* the *finish* (right) leads are connected.

The slots in commutator risers must be clean, and the leads must be scraped well to permit a good soldering job when the leads are soldered to the risers.

The risers can be cleaned by sawing the slots in them with a hacksaw. Wide slots in risers can be sawed by placing two or three hacksaw blades in the saw, or placing a sheet-metal spacer of the proper thickness between the blades to fit the slot. (For soldering, turning, and undercutting commutators, see Chap. 10.)

The completed winding, except for a cord band over the leads, is shown in Fig. 9-19. (Cord banding is discussed in Sec. 9-16.)

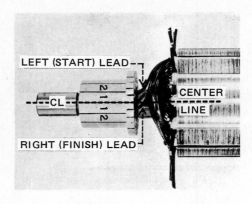

Fig. 9-18 Connecting the coil ends to the commutator bars.

9-14 COIL TAMPERS

Wires of coil sides of armatures and stators are tamped in slots by means of coil tampers shown in Fig. 9-20. Great care must be taken in using coil tampers to *avoid damaging wire insulation* and causing shorts. The thin section of the tamper fits between the slot teeth with the lower end on the coil, and the upper end is pressed by hand or *lightly* tapped with a rawhide mallet to tamp the coil in the slot, or to tamp insulation separators in a slot.

9-15 ARMATURE HOLDERS AND WINDING HORNS

An armature holder is known as a *buck*. Commercial holders are available, but a perfectly adequate buck can be made in the shop. Figure 9-21 is an example of

Fig. 9-17 The back end of the armature winding shown in Fig. 9-16 (*h*).

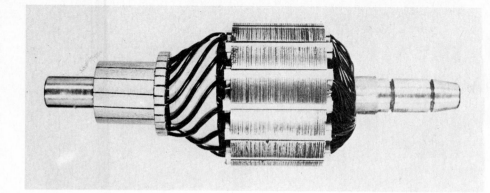

Fig. 9-19 The complete armature winding with leads soldered to the commutator. A cord band over the leads is not shown.

Fig. 9-20 Coil tampers used in tamping coils in slots. (*Crown Industrial Products Co.*)

a shop-made buck. Overall size and proportions can vary according to the armatures used.

In Fig. 9-22 a *winding horn* is shown being used to shape armature coils. These horns are sawed about 8 to 12 in. long from hard sheet fiber, 1/2 to 1 in. thick, and are used to shape or form windings of all kinds in places where a rawhide mallet cannot be used. Rawhide mallets are used when it is necessary to drive, tap, or shape coils. A crutch or chair leg tip slipped over the head of a ball-peen hammer can be used in an emergency.

9-16 CORD BANDING

A temporary cord band is placed over the leads before soldering to hold them in place. After soldering, a permanent band is installed. The hot flux in soldering and heat from the soldering iron damages the temporary band.

Fig. 9-21 A shop-made armature holder, or buck.

Fig. 9-22 A winding horn in use. Note the position of the armature in the buck.

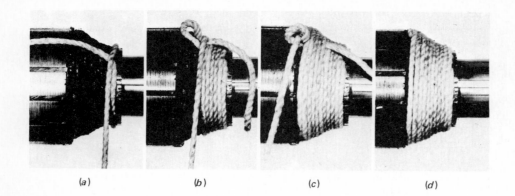

(a)   (b)   (c)   (d)

Fig. 9-23 The procedure for cord banding.

Steps in installing a cord band are shown in Fig. 9-23. In Fig. 9-23(*a*), the starting end of a cord is laid over the core and one turn of the cord holds it against the commutator risers. Several more turns are added, and the starting end is brought under the desired turn to form a loop as shown in Fig. 9-23(*b*). After more turns are added, the finishing end of the cord is brought through the loop as in Fig. 9-23(*c*).

The starting end of the cord is pulled, which pulls the finishing end of the cord under the band. The two ends of the cord extending from under the band are pulled as they are cut as close as possible to the band, and the tension draws them into concealment under the band. The finished band is shown in Fig. 9-23(*d*) with no free ends exposed.

9-17 STEEL BANDING

Steel bands are wound on large armatures in a banding lathe. This lathe makes it possible to wind steel wire under proper tension. Tinned metal strips are spaced under the wire to form a clip to be bent over the band and soldered. After banding is finished, the entire band is soldered in a solid mass.

In taking data before stripping an armature, one should carefully note and record the size of banding wire, methods of insulating it, and number, location, and construction of bands.

A nonmagnetic and insulating banding material, made with a glass base and uncured resins, is used extensively on some armatures. It is in the form of tape and is applied in a banding lathe with a suitable tension device. It cures to a high tensile strength when the armature is baked. This material eliminates possibilities of short-circuiting of main field magnetism and hysteresis loss and heating in the bands, and short-circuiting or grounding of the winding, which is possible with steel bands.

9-18 SIMPLEX AND DUPLEX LAP WINDINGS

The lap windings previously discussed in this chapter have been *single,* or *simplex,* windings; that is, *one winding* fills all slots and bars. A lap winding known as a *duplex* lap provides *two windings, separate* and *insulated* from each other, on *one armature.* This winding is seldom used.

The coils with odd numbers connect to commutator bars with odd numbers, and coils with even numbers connect to commutator bars with even numbers. There is no electrical connection between coils with odd numbers and coils with even numbers. The brushes, covering at least two bars, connect the two windings parallel. Since the brushes parallel and supply two windings, this is known as a *doubly reentrant duplex* winding. This winding requires an even number of bars. It can be connected either progressive or retrogressive.

Another type of duplex lap winding is called *singly reentrant.* In this type of winding a circuit can be traced twice around the armature. A singly reentrant duplex lap winding can be connected either progressive or retrogressive, and it requires an odd number of bars in the commutator.

9-19 ARMATURE EQUALIZERS

Uneven air gaps between the armature and pole pieces of a motor or generator with four or more poles cause unequal cemf in various parts of an armature lap winding. These unequal voltages cause circulating currents between the winding and brushes. As a result there will be sparking at the brushes.

This trouble is minimized by connecting commutator bars or coils, that should be at the same potential, with a conductor known as an *equalizer.*

the bottom is stronger than the field magnetism at the top.

Counterelectromotive force generated in the armature as it rotates is higher in the section marked *B* than in the area marked *A*. This unequal cemf causes current to flow from the *B* area of the winding through the commutator and positive brush at the bottom and through the brush jumper to the positive brush at the top, and into the commutator and winding. This current, in flowing between the commutator and brushes, causes sparking. If a circuit other than through the brushes is provided for this current, sparking from this cause will be minimized. The bars making contact with the positive brushes 1 and 7 are shown connected by a heavy dashed line representing the equalizer circuit. Current can now flow from the lower section of the winding to the upper section without passing through the brushes and commutator.

Some windings are equalized by equalizers connected behind the commutator as illustrated in Fig. 9-25, or by connections to the coils at the back of the armature.

In Fig. 9-24 the two positive brushes are connected together by a jumper. If either positive brush is removed, current can flow through the equalizer to the bar the removed brush contacted, and the armature will continue to operate. Some small multiple lap-

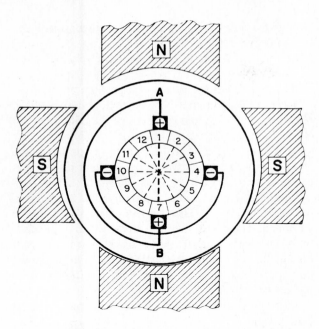

Fig. 9-24 Use of equalizers to correct an uneven air gap.

This is shown in Fig. 9-24. A four-pole motor with worn bearings permits the armature to run too low in the air gap. The air gap at the top is greater than the air gap at the bottom; therefore, the field magnetism at

Fig. 9-25 Commutator with equalizers.

wound armatures are equipped with equalizers and only two brushes.

Equalizers normally carry little, if any, current but should have sufficient current-carrying capacity, usually the same as the winding, in case a brush sticks, breaks, or for any cause fails to carry its load, and current flows through the equalizers instead.

The commutator span for equalizers is determined by dividing the total number of commutator bars by the number of pairs of poles of the armature and adding 1.

Problem

Find the span of the equalizers for a four-pole armature with a 48-bar commutator.

Solution

There are four-poles or two pairs of poles. Thus

$$\text{Span} = \frac{48}{2} + 1 = 24 + 1$$
$$= 25$$

Thus the span would be from bar 1 to bar 25 or 1–25.

Equalizers are used only on lap windings of four or more poles. Two-pole lap and all wave windings are self-equalizing and do not need equalizers.

There is a combination lap-wave winding, known as a *frog-leg* winding, which is a form of equalized lap winding. It has advantages but it is seldom used because of its complexity.

9-20 WAVE WINDINGS

Wave and lap windings differ physically only in the lead connections to the commutator and the resultant circuits formed. A wave winding is shown in straight-line schematic form in Fig. 9-26. This is the lead connection of 11 coils connected to an 11-bar commutator.

All wave windings are for four or more poles, since a wave winding cannot be connected for two poles. Notice in Fig. 9-26 that the leads of the coil at the left "wave out" in connecting to bar 1 and bar 6. The span is 1–6. The coil at the right starts on bar 6 and ends on bar 11, which would be adjacent to bar 1. A coil that starts on bar 11 ends on bar 5.

Figure 9-27 is a circular diagram of the same armature as in Fig. 9-26. The coil lead span is 1–6, or the pitch is 5. The coil at the right starts on bar 1 and ends on bar 6. The coil at the left starts on bar 6 and ends on bar 11, one bar before starting bar 1. This is known as a *retrogressive* wave connection since the leads do not cross.

With a span of 1–7, the lead would cross between bars 11 and 1. This would be a *progressive* winding, and the current flow in the winding would be *reversed*. This would produce motor rotation in a reverse direction or a generator voltage of reverse polarity.

The current flow from the positive brush in Fig. 9-27 is from bar 1 into the coil at the right and to bar 6 to the coil at the left. A 1–7 span connection is shown in Fig. 9-28. The leads are crossed between bars 1 and 11, and current flow from the positive brush is from bar 1 into the coil at the left, and to bar 6 to the coil at the right. This direction of current flow is in opposite direction to the flow in Fig. 9-27.

Extreme care must be used in determining the lead span of a wave winding. The counting of commutator bars must be done on the side of the armature containing the data coil and between the leads of that coil. It is very easy to make the mistake of counting bars on the opposite and wrong side of the armature. In Fig. 9-27 the lead span of the coil at the right, counting on the right side of the commutator, is 1–6. If the counting were done on the left side of the commutator, the span would be mistaken as 1–7.

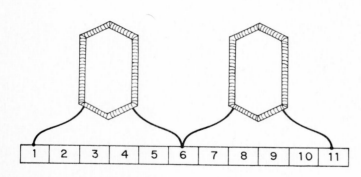

Fig. 9-26 Straight-line diagram of a four-pole wave winding.

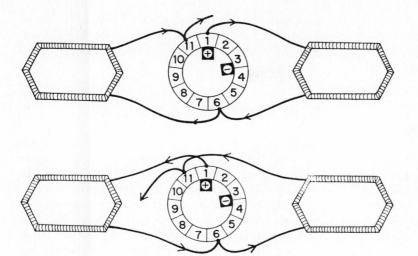

Fig. 9-27 Circular diagram of the armature in Fig. 9-26.

Fig. 9-28 A wave winding with progressive connection.

9-21 WAVE CURRENT PATTERN

The pattern of current flow set up in a four-pole wave winding is shown in Fig. 9-29. If the path is traced through the winding as it flows from positive to negative brushes, it will be noted that a pattern through the slots is formed similar to the pattern produced by the four-pole lap winding in Fig. 8-5.

In Fig. 9-29, beginning with the positive brush, there are two circuits into the winding. This is the case in all single-wave windings regardless of the number of poles. One circuit leads to the bottom side of a coil in slot 2, and the second circuit leads to a coil side in

the top of slot 11. The Xs indicate that current is flowing away from the observer. Dots in the circles indicate that current flow is toward the observer.

In tracing the path of current beginning with bar 1 to slot 2, the current flows toward the back of the coil in this side, and to the right side of the coil in the top of slot 5 and forward through the coil to bar 6. From there it flows through coils in slots 7 and 10, bar 11, coils in slots 1 and 4, bar 5, coils in slots 6 and 9, bar 10, coils in slots 11 and 3 and bar 4. The negative side of the line is at brush 4.

The second circuit begins at the positive brush on bar 1 and flows to the top side of a coil in slot 11, and

Fig. 9-29 Pattern of current flow in a wave winding.

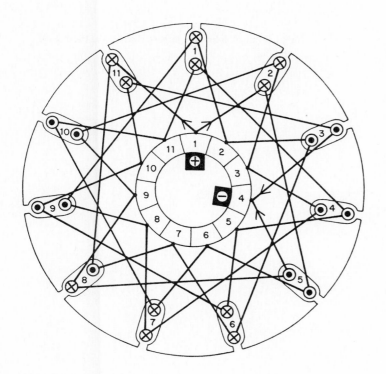

backward, as indicated by the X, and downward in the back of the coil to the coil side in slot 8. In this coil side it flows forward, as indicated by the dot, and to bar 7. From here it continues through coils and bars to the negative brush and leaves the armature.

9-22 WAVE LEAD PITCH

The lead pitch of a wave winding is determined by *dividing* the total number of commutator bars *plus or minus 1* by the number of *pairs of poles* of the armature. If the total number of bars minus 1 is used, the winding will be retrogressive. If the total number of bars plus 1 is used, the winding will be progressive.

For example, consider the winding in Fig. 9-27. This is a four-pole (two pairs of poles) armature with 11 bars; 11 minus 1 equals 10, divided by 2 equals 5. The *lead pitch* is 5, the *lead span* is 1–6 for a *retrogressive* connection (leads not crossed), as shown in the illustration.

For a progressive connection, 11 plus 1 equals 12, divided by 2 equals 6. The *lead pitch* is 6, and the *lead span* is 1–7 for a *progressive* connection (leads crossed), as shown in Fig. 9-28.

Four-pole and eight-pole wave windings require an odd number of commutator bars. Other windings may or may not require an odd number of bars. In any case, when a circuit is traced from a beginning bar around the armature once and back to the commutator, it should lead to one bar before or one bar after the beginning bar. It must not end at the same bar at which it began. If it does, the circuit would be closed on itself, with no connection to the remainder of the windings.

9-23 DUPLEX WAVE WINDINGS

Occasionally, two wave windings are wound in one armature. This type of winding is known as a *duplex wave* winding. There are two types of duplex wave windings, *doubly reentrant* and *singly reentrant*.

A doubly reentrant duplex winding contains two separate windings not connected together. Alternate coils, leads, and commutator bars compose one winding, and the remaining coils, leads, and commutator bars compose the other winding. This winding requires an even number for the lead pitch.

A four-pole duplex doubly reentrant wave winding is shown in Fig. 9-30. The lead span is 1–9, with a pitch of 8, an even number. The brushes for this armature must cover at least two commutator bars to parallel the two windings. In tracing this winding, it will be noted that one circuit, starting on bar 1 through coil 1, contains all the odd-numbered bars and coils in one winding and continues to the beginning bar 1.

The second circuit, beginning on bar 2 and through coil 2, contains all the even-numbered bars and coils in the second winding and continues to the beginning bar 2.

This winding can be either progressive or retrogres-

Fig. 9-30 A four-pole duplex doubly reentrant wave winding.

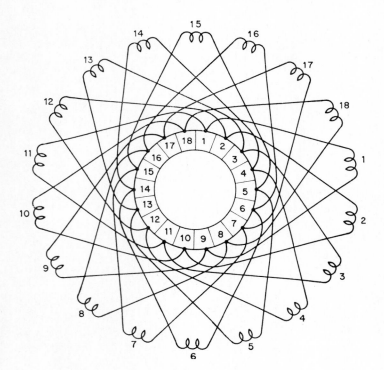

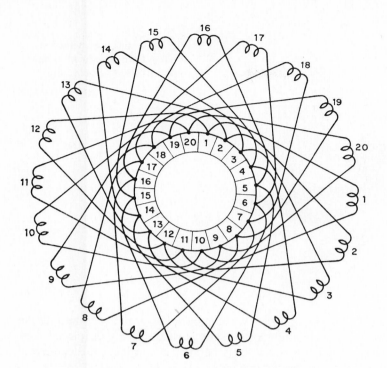

Fig. 9-31 A four-pole duplex singly reentrant wave winding.

sive. By use of test light or ohmmeter, continuity should be obtained only between alternate commutator bars.

A four-pole duplex singly reentrant wave winding is shown in Fig. 9-31. This winding requires an uneven lead pitch. The lead span is 1–10, the pitch 9, an uneven number. The two windings are connected together.

In tracing a circuit, beginning on bar 1 through coil 1, it will be noted that the circuit goes through alternate odd- and even-numbered bars and coils to include the entire winding before it ends at bar 1 from which it began.

9-24 DEAD COILS AND JUMPERS

Occasionally, the number of commutator bars is not a multiple of the number of coils in an armature winding. In a case of this kind, *dead coils* or *jumpers* on the commutator are used. A dead coil is a coil in the winding that is not connected to the commutator. The leads of a dead coil are cut off and taped to avoid damage to the adjacent coils.

As an example of the use of a dead coil, an armature with a 28-slot core and a 55-bar commutator would have two coils per slot for a total of 56 coils, which would be one coil more than commutator bars. Fifty-six coils would be wound to balance the armature and fill the slots, but one coil would *not* be connected to the commutator. In taking data on a winding

with a dead coil, data should be taken on the bundle containing the dead coil, and this bundle should be included in the sketch on the data sheet.

A jumper on the commutator is used when there are more bars in the commutator than there are coils. Two bars are connected together with a jumper; these two bars are considered as one bar, thus subtracting one bar from the total number of bars. Occasionally, it is more convenient to omit one turn from a coil and include an extra turn to be used as a jumper. This extra turn is connected in the same manner as a coil is connected.

9-25 FORM WINDING

Coils for some small and all large armatures are wound on a form, shaped, taped, varnished, and baked before they are placed in the armature. In stripping a form-wound armature, a sample coil should be carefully preserved to be used in setting the form for making new coils.

In winding with form coils, one side of a coil occupies the *bottom* of a slot and the other side occupies the *top* part of a slot. The *bottom* sides of coils for a *number equal to the coil span* are installed in the slots, and the top sides of these coils are left out of the slots. After this, both sides of each coil are placed in slots. For example, if the span is 1–10, both sides of the tenth coil are placed in slots. This is known as the *span-up* method of winding and is shown in Fig. 9-32.

Fig. 9-32 Span-up method of winding form coils in an armature. (*Whitfield Electric Co.*)

Winding is continued around the armature, and the last coils installed are placed in under the raised span. Then top sides of the coils in the span are placed in their proper slots.

Form-wound coils usually have slot insulation taped in as part of the coil; therefore additional slot insulation is not used. This type of coil is known as a *slot-size* coil.

The bottom leads in a form winding are placed in the proper commutator bars as the coils are installed, but the top leads must be left out until sufficient coils and bottom leads are in place to allow the proper connection of top leads. This is earlier in lap windings than in wave windings. The space between top and bottom leads is carefully insulated when the top leads are placed in the bars.

When leads are sufficiently flexible, all the top leads are left out and bent back out of the way until all the coils are in the slots. Insulation is then placed over the bottom leads before the top leads are connected to the commutator.

A hand-wound lap winding is shown in Fig. 9-33 in an adjustable holder. The bottom or right-hand leads have been connected to the commutator, and insulation has been placed over them. The left-hand leads are now being connected, and a piece of glass tape is being woven under one lead and over the next as they are connected to provide insulation between them.

9-26 TESTING ARMATURES

In rewinding an armature, testing should begin when the armature is stripped and cleaned. The commutator

Fig. 9-33 Leads being connected to the commutators. (*Whitfield Electric Co.*)

should be given a *bar-to-bar* test for short circuits between bars and a *bar-to-shaft* test for grounds.

During winding, a *ground test* should be made when all the bottom leads are in commutator slots. This can be done by placing a test prod of a high resistance ohmmeter on the *shaft* or *armature core* and then touching each of the commutator bars with the other prod. If a ground exists in either the winding or commutator, the meter will give a reading. Under normal conditions, that is, no ground, there should be almost no movement of the meter's pointer. At this stage a ground can still be easily cleared.

After all leads are in the commutator, and before soldering, the winding should be tested for short circuits. Short circuits can exist in a winding between turns, between coils, between leads, or in the commutator. A growler test or bar-to-bar test with an ohmmeter will indicate any of these shorts. (These methods are not satisfactory on an armature with equalizers.)

9-27 GROWLER TEST

A growler is used to test for short circuits. It consists of a U-shaped electromagnet made of laminations such as used for transformer cores. When it is used, the armature is placed with the core in the opening formed by the upright legs, the coil is energized with alternating current, and a *feeler,* such as a hacksaw blade, is moved over the core slots. The feeler is held parallel with the slots. The armature is turned in the growler by hand as each slot is tested by the feeler. The coil is deenergized while the armature is turned.

Coils with sides forming a plane at right angles to the flux are in position to be tested. Thus the sides of coils of a two-pole armature would be at the top, and in a four-pole armature they would be about 45° from the top.

A growler works as follows. Alternating current produces an alternating magnetic flux through the core of the growler and the armature. This alternating flux induces a voltage in the armature coil. If the coil is good, no path will exist for current flow. However, if a short circuit exists, there will be a path for current flow. When current flows in the armature coil, it produces magnetism in the slots containing the shorted coil, and this magnetism is detected by the feeler. The feeler will complete the magnetic path and vibrate in time with the changing flux.

High-carbon steel feelers with high retentivity should be used; alloy steel is not reliable. When a good feeler capable of vigorous response to a short is found, it should be marked and kept for future use.

A shop-made growler such as in Fig. 9-34(*a*) can be made of 1 1/2 in. transformer laminations stacked

3 in. thick. The legs are 7 in. high. The bolts are brass. The coil contains 200 turns of No. 18 magnet wire. A switch is provided so that the coil can be energized and deenergized as the armature is turned. Power for a test lamp or other instruments is provided by a receptacle. A commercial growler is shown in Fig. 9-34(*b*).

9-28 BAR-TO-BAR TEST

A bar-to-bar test of a winding basically is determining the degree of voltage drop across the circuit formed by a coil beginning at the commutator. Direct current is passed through a section of the winding from the commutator several bars apart. The degree of voltage drop in a coil or group of coils is determined by touching the prods from a millivoltmeter or milliammeter to adjacent commutator bars. An average reading around the armature is obtained. Any variation from the average reading at any point around the commutator indicates trouble.

A shop-made tester, such as the one shown in Fig. 9-35, can be used to supply current to the armature and carry the prods of a meter.

Variation from the average reading can be diagnosed as follows:

Extremely high reading. Open circuit; all current is flowing through the meter.

High reading. High resistance; probably poor solder joint or too many turns of wire in coil. More than average current is shunted through the meter.

Low reading. Partial short; probably due to deteriorated mica, solder between bars, short circuit between a few turns of wire in the coil, or too few turns of wire in the coil.

No reading. Complete short circuit in commutator or coil leads behind commutator, or transposition of leads. The reason for no reading in the case of transposed leads is shown in Fig. 9-36. Coil B, with both leads connected to bar 2, is not connected to the remainder of the winding. The right leads of coils A and B are transposed and should be interchanged.

9-29 AUDIBLE TESTING

An earphone and pulsating direct current can be used instead of a meter for the bar-to-bar test. It is efficient and convenient as a portable tester, and it will indicate trouble not detectable by a meter. The reversed-coil

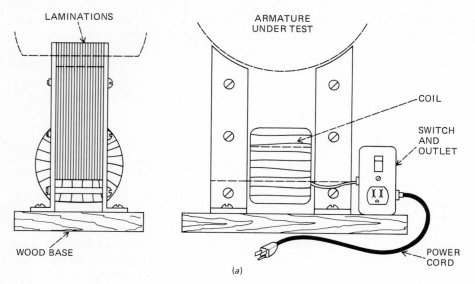

LAMINATIONS

ARMATURE UNDER TEST

COIL

SWITCH AND OUTLET

WOOD BASE

POWER CORD

(a)

Fig. 9-34 A growler used for making short-circuit tests on armatures. (a) Shop-made growler. (b) Commercial growler. (*Crown Industrial Products Co.*)

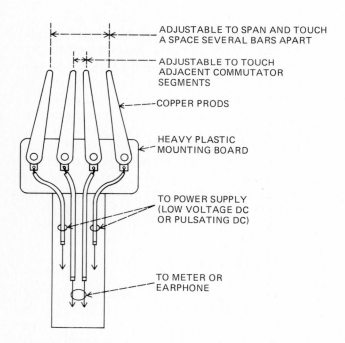

(b)

ADJUSTABLE TO SPAN AND TOUCH A SPACE SEVERAL BARS APART

ADJUSTABLE TO TOUCH ADJACENT COMMUTATOR SEGMENTS

COPPER PRODS

HEAVY PLASTIC MOUNTING BOARD

TO POWER SUPPLY (LOW VOLTAGE DC OR PULSATING DC)

TO METER OR EARPHONE

Fig. 9-35 A bar-to-bar commutator tester.

condition in Fig. 9-37 can be detected with an earphone test but not with a meter test.

The intensity of a tone in the earphone instead of a meter reading is used to locate and diagnose trouble. Pulsating direct current can be produced easily using a buzzer in series with a battery as in Fig. 9-38. This arrangement can be connected to the tester in Fig. 9-35.

In use, the tester is moved around the commutator with the four contact points touching bars determined by the spacing between tester prods. The pulsating current produces a pulsating magnetic field which results in inductive reactance in the coil. This increases the impedance in the coil which shunts part of the current through the earphone. The intensity of the tone from each coil is compared with the average tone for the armature and determines the condition of the coil.

In testing an armature in a motor or generator, the armature should be turned so the test can be made in the same position in relation to the main poles. If testing is done around a stationary armature, a higher tone

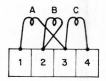

Fig. 9-36 Transposed-lead connections.

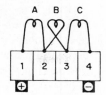

Fig. 9-37 Reversed-lead connections.

will result from coils under pole pieces because of greater inductive reactance. Under pole pieces the coils are practically surrounded by iron, which increases inductive reactance.

Armature windings with equalizers cannot be tested by the methods used for other armatures. Equalizers indicate short circuits in all coils when a growler or earphone test is made. The bar-to-bar test with a millivoltmeter is satisfactory in some cases. To test ac single-phase repulsion-type armatures with equalizers, lift or insulate the brushes from the commutator with the motor completely assembled. Connect the motor to the regular line voltage. If the armature rocks and tends to lock in certain positions, the winding is shorted. If it can be turned freely, it is free of shorts. Bearings must be in fair condition for this test.

9-30 MAGNETIC TEST

The principle of current producing a magnetic field can be used in certain tests. By connecting a voltage across two commutator bars, lead swing and coil connections can be checked without disturbing bands or the windings. For example, in Fig. 9-37, if an ac voltage is connected across bars 1 and 2, a feeler can locate the slots containing the sides of coil A. A centerline can be established, and coil span and lead connection data can be taken to be compared with correct data for the connection.

To determine if leads are crossed at the commutator, direct current is used and a compass is laid over the left side of coil A. The positive lead of the dc test voltage is touched to bar 1, and the negative lead is touched to bar 2. If the leads are not crossed and current flow is as indicated for coil A, the compass will point with its north to the right. If the leads are crossed, the compass will point its north to the left.

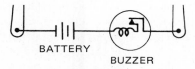

Fig. 9-38 A pulsating dc supply.

Thus complete coil connection data and whether the leads cross can be determined by this method.

If a reversed coil is suspected by an audible test, it can be verified by this dc test, using a compass. If dc current is supplied to adjacent bars, each coil in the winding should cause the compass to point in the same direction. A reversed coil, such as coil B in Fig. 9-37, will reverse the compass since current flow is opposite in it, as compared with the remaining coils.

9-31 BALANCING ARMATURES

After an armature winding job is completed, the armature should be tested for balance. An unbalanced armature causes damaging vibration that is destructive to the entire motor, depending on the severity of the vibration. Vibration causes severe wear and strain on every part of a motor or generator, and it also results in severe brush and commutator trouble.

9-32 STATIC BALANCE

There are two types of balance—static and dynamic. Static balance can be checked while an armature is at rest. It can be placed in balancing ways or on perfectly level knife-edges. In balancing ways, each end of the armature shaft rests on precisely balanced disks containing precision bearings. This provides an almost frictionless surface on which the shaft can turn. The armature is then made to rotate slightly and allowed to come to rest. This procedure is repeated several times. If it is out of balance, the armature will stop rotating with the light side up each time. To balance it, weights are added to the light side in the form of nonmagnetic metal slot wedges, additional solder on steel banding wire, or weights screwed or bradded to fans or any metal part of the core. In some cases it may be necessary to drill the core lightly on the heavy side of the armature, but this is not considered good practice.

9-33 DYNAMIC BALANCE

An armature can be statically balanced but still be severely out of balance when it is running. An armature can be heavy at one end on one side, and equally heavy at the other end on the other side, which will result in a perfect static balance. But when this armature rotates, centrifugal force will cause it to wobble, since the heavy ends on opposite sides tend to pull the armature at right angles to its axis. At high speeds this wobbling results in severe vibration. It is therefore necessary to also test an armature for dynamic balance.

Dynamic balancing equipment is somewhat expensive, and therefore it is rarely found in small to average-size repair shops. In a shop without such equipment, it is common practice to send small high-speed armatures to shops that are equipped for dynamic balancing and that specialize in winding these armatures.

Dynamic imbalance causes small high-speed armatures to spark severely at the commutator and heat and throw solder from the commutator. This results from the brushes not being able to maintain even pressure and contact with the vibrating commutator.

Dynamic balancing of an armature is sometimes attempted by trial and error. The results, at best, are crude and not precise. Weights are added and adjusted at various points about an armature until a degree of minimum vibration is obtained. In critical situations such methods do not achieve the degree of balance required and are rarely used.

SUMMARY

1. The rotating member of a dc motor or generator is the armature. In ac equipment it is called the rotor.

2. The principal parts of an armature are (a) the core, made of laminated sheet steel with slots to contain the winding; (b) the commutator, consisting of copper bars securely held in an insulated holder; (c) the winding, which conducts current through the armature; (d) the shaft on which all parts are mounted for rotation.

3. The function of an armature winding in a motor is to conduct current in an orderly fashion to produce maximum torque by interaction of armature magnetism with field magnetism. In a generator, the armature winding delivers current to the load.

4. The commutator provides a convenient means of connecting the rotating armature coils and stationary brushes connected to the line.

5. There are two types of armature windings—lap and wave. The difference between these windings is in the coil lead connections to the commutator. The leads of a coil in a lap winding connect to adjacent commutator bars, and the leads of a coil in a wave winding connect several bars apart.

6. A lap winding has as many circuits through it from positive to negative brushes as it has brushes or poles. A wave winding has only two circuits.

7. Taking data means determining and recording certain conditions of an armature winding preparatory to rewinding it.

8. The relative position of a coil and its commutator bars is recorded from a centerline running parallel to the shaft and through the center of the coil.

9. The number of poles in an armature is usually determined by dividing the number of core slots by the coil pitch and taking the nearest whole even number.

10. Full coil span of an armature is determined by dividing the number of slots by the number of poles and adding 1. Most windings are up to 20 percent less than full span.

11. Equalizer connections or span is determined by dividing the number of commutator bars by the number of pairs of poles of the armature and adding 1.

12. Coil lead span is the number of commutator bars occupied and spanned by the leads of a coil. Coil lead pitch is the number of bars from, but not including, the first bar to and including the last bar in the lead span.

13. Coil span is the number of core slots occupied and spanned by the sides of a coil.

14. In a progressive lap winding the coil leads do not cross at the commutator. In a retrogressive lap winding the leads cross at the commutator. In a progressive wave winding the leads cross at the commutator. In a retrogressive wave winding the leads do not cross at the commutator.

15. If an error in rewinding results in changing a winding from progressive to retrogressive, or vice versa, it will reverse the rotation of a motor, or reverse the polarity of a generator.

16. Winding can be started at any place on an armature. Coils can be wound clockwise or counterclockwise. Winding can progress clockwise or counterclockwise around the armature in a hand winding.

17. Equalizers are usually connected to commutator bars of equal potential to equalize cemfs of varying values due to unequal air gaps around the armature. Sometimes equalizers are connected to the back ends of coils.

18. A simplex lap or wave winding is one winding on an armature. A duplex lap or wave winding is two windings on an armature. The brushes, covering at least two bars each, parallel the two windings. The two windings of a duplex doubly reentrant winding are not interconnected. The two windings of a duplex singly reentrant winding are interconnected.

19. An armature should be balanced as nearly perfect as possible. An unbalanced armature produces damaging vibration in a motor or generator.

20. An armature should be tested for short circuits and grounds at intervals during winding when possible defects can be more easily corrected.

21. A growler is used to test for short circuits. It induces current in a shorted coil. The current produces magnetism that is detected with a feeler.

22. A meter or an earphone is used to test for shorts, reversed coils, more or fewer turns in coils, and other defects in the coils. A test light, ohmmeter, or voltmeter is used in testing for grounds. Direct current and a compass are used to test for reversed coils.

QUESTIONS

9-1   What are the principal parts of an armature?

9-2   What are the physical differences between a lap and a wave winding?

9-3   What are the electrical differences between a lap and a wave winding?

9-4   List the most important items in taking data on an armature.

9-5   In taking data, how are the relative positions of a coil and its commutator bars expressed?

9-6   What is the function of a commutator?

9-7   How is full coil span determined?

9-8   A four-pole armature contains 32 slots. If the coil span is chorded up to 80 percent of full coil span, what is the coil span of this armature?

9-9   What is the difference between simplex and duplex windings?

9-10   How is the centerline of an armature with skewed slots determined?

9-11   How can the number of poles in an armature be determined?

9-12   Why should an armature be balanced?

9-13   What is the function of armature equalizers?

9-14   How is the span of equalizers determined?

9-15   The span of the equalizers for a four-pole armature is 1–19. How many bars are there in the commutator?

9-16   What tests should a commutator be given before an armature is wound?

9-17   How are armature windings tested for shorts?

9-18   How is a growler used for testing armatures?

9-19   How can a reversed coil be detected?

9-20   How are a short and an open circuit detected in an armature using an earphone?

9-21   What would be the result in a motor if its coil leads were changed from a progressive connection to a retrogressive connection?

9-22   What determines the direction and degree of lead swing?

9-23   What is the purpose of a dead coil?

9-24   What are form coils and where are they used?

9-25   In a bar-to-bar test a lower than average reading is obtained. What are the probable troubles with this armature?

10

COMMUTATORS AND BRUSHES

Commutators and brushes are used to connect the rotating windings of the armature with the lines or field windings. A commutator consists of specially shaped copper bars mechanically secured in a rigid holder in a cylindrical form around the armature shaft. The bars are individually insulated from each other and the rest of the armature.

Commutators range in size from a fraction of an inch to many feet in diameter. A picture of commutators in many sizes and shapes is shown in Fig. 10-1.

Commutators and brushes have an *enormous* and *exacting* job to do. They must transfer current from fast-moving surfaces to stationary conductors with a minimum of sparking, heating, chatter, resistance, and wear. A commutator 12 in. in diameter running at 1750 rpm has a surface speed passing under the brushes of 60 mph. Under these conditions a commutator and its brushes must receive proper care for efficient operation.

10-1 COMMUTATOR CONSTRUCTION

Commutator bars are insulated from each other by *segment mica,* which is mica bonded with a stiff, rigid, bonding agent. Bars are insulated from the holder by *molding mica,* which is mica with a bonding agent that provides softness and pliability when heated. Molding mica can be shaped or formed into mica rings while hot and becomes rigid when cooled. A piece of segment mica and a copper bar are shown in Fig. 10-2.

When arranged in cylindrical form, the bars and segment mica provide a circular groove at the front and back of the assembly for rigid clamping. A front and back view of a bar and mica assembly, showing the grooves, are seen in Fig. 10-3(*a*) and (*b*).

Molded mica rings, for insulating the bars from the holder, hub, or clamping rings, are shown in Fig. 10-4. A completely assembled commutator is seen in Fig. 10-5.

Radial or vertical commutators are used for brush-lifting repulsion-start induction-run single-phase motors. A front view of a radial commutator is shown in Fig. 10-6.

10-2 COMMUTATOR REPAIR

A commutator can be removed from an armature by pressing in an arbor or hydraulic press or by use of pullers similar to those used for pulling gears. Pullers may be screw operated or hydraulically powered.

Small commutators are usually manufactured as a unit and cannot be disassembled for repair. Medium to large commutators can be disassembled for repair. Badly pitted and burned bars can be welded and machined. Segment mica, in the proper shape and thickness, can be purchased or by the use of templates can be cut from sheet mica for replacement. A bar can be used in making two tin or vulcanized fibre templates of the exact shape of the bar. The templates are placed on each side of a stack of segment mica strips of proper size and secured with friction tape or rubber bands. The entire assembly can be secured in a vise, and the segment mica can be sawed to size and shape with a hacksaw. Finishing can be done with a coarse file.

Molding mica can be used to repair mica and rings, but in most cases replacement with new rings is advisable. For obsolete equipment, when new rings are not available, the steel end rings of the commutator can be used in forming molding mica in making replacement rings. A cord band should be wound over the exposed area of mica rings and sealed with a commutator

Fig. 10-1 Commutators come in many sizes and configurations. (*Kirkwood Commutator Co.*)

(a)   (b)

Fig. 10-2 Parts of a commutator. (*a*) Segment mica used between bars. (*b*) A copper commutator bar.

(a)   (b)

Fig. 10-3 Commutator clamping grooves. (*a*) Front view. (*b*) Back view of commutator assembly.

COPPER SEGMENTS
AND SHEET MICA

MICA
COLLAR

Fig. 10-4 Exploded view of commutator showing copper segments and mica insulation. (*General Electric Co.*)

MICA CONES

sealer to protect the mica from mechanical damage and avoid flaking when an armature is running.

In the assembly of a commutator following repair, a torque wrench should be used in accordance with the manufacturer's recommendations to avoid commutator distortion or warped bars. A commutator should be heated, cooled, and tightened several times before final tightening to ensure a solid and "fixed" set.

10-3 SOLDERING AND SURFACING COMMUTATORS

Soldering a commutator requires thorough preparation and care to produce a good job. A poorly soldered commutator connection produces heat, melts the solder, and results in an open circuit and arcing that can ruin the commutator. Excess solder can cause short circuits that are difficult, at best, to clear.

For a thorough solder joint, all parts involved must be clean with bare metal surfaces exposed to the solder. Commutators with risers not over 1/4 in. high are usually soldered in a horizontal position. Larger commutators are usually soldered with the axis of the

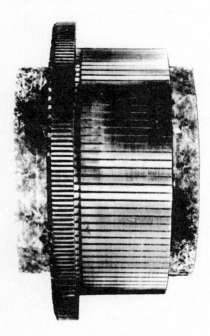

Fig. 10-5 Copper segments and mica insulation in a completely assembled commutator.

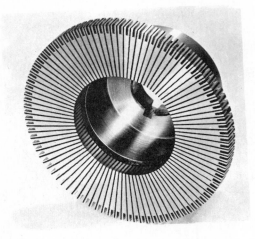

Fig. 10-6 Radial commutator for brush-lifting action. (*Kirkwood Commutator Co.*)

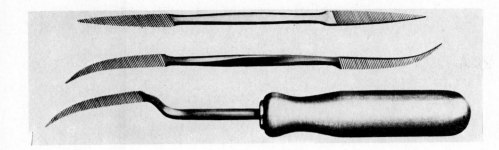

Fig. 10-7 Files for undercutting mica in a commutator. (*Ideal Industries, Inc.*)

commutator about 20° from horizontal to minimize the likelihood of solder running behind the bars and causing short circuits. Risers are treated with a non-acid flux in paste or stick form. An electric soldering iron, held approximately vertical with the chisel edge of the tip in the slot of a riser, is used to melt solder and heat the riser to accept solder. The solder wire or bar can be used to scrape excess solder running out of the front of the riser and to restore it to the top of the riser.

When a large soldering iron cannot furnish enough heat, a gas torch can be used to preheat the commutator and aid the iron. The flame of the torch should not be allowed close enough to ignite the flux in the risers and should be at right angles to the bars and directed ahead of operations, so as to preheat the bars that are to be soldered and allow bars that have been soldered to cool as soon as possible. Care must be used not to overheat the commutator and injure the mica insulation.

Following repair or a rewind job, it is necessary to resurface the commutator in a lathe. A commutator should be turned at a medium speed and cut with an extremely sharp, round-nose cutting tool. The tool should be flat on top and should have 30° or more front and side relief. The cutting edge parallel with the commutator should overlap at least two previous feed cuts to avoid threading cuts. Chatter must be avoided at all times; it can usually be eliminated by changing the lathe speed.

Extreme care must be taken to prevent gouging by the cutting tool. Most gouging is caused by a dull tool which, failing to cut, forms a work-hardened glaze until sufficient pressure is applied to force cutting. When the tool is forced through the glazed area, the pressure causes the tool to gouge. Gouging can destroy a commutator, bend an armature shaft, break the cutting tool, and damage the lathe.

10-4 UNDERCUTTING MICA

The mica between the commutator bars should be undercut below the surface of the bars to a depth equal to about the thickness of the mica. Mica undercutting can be done in several ways.

Mica undercutting files, shown in Fig. 10-7, can be used on small commutators or in difficult places. A three-cornered file, with the end broken off, can be used in an emergency. There are several types of portable electric undercutters that are suitable for undercutting large commutators.

Mica dust, when breathed into the lungs, can cause a disease similar to silicosis. Therefore all work with mica should be done wearing a mask rated for protection against mica dust. *Good ventilation is absolutely necessary* in any case.

10-5 U AND V SLOTS

Two shapes of undercut slots are used—V and U. Each type has advantages, but generally the V slot is most commonly used, especially for high-speed commutators. In either case, no mica should be left at the top edge of the bars in undercutting.

Besides causing sparking and other problems, high mica creates conditions that rapidly worsen with continued operation. High mica prevents proper brush contact with the commutator, causing current to flow across an air gap as shown in Fig. 10-8. This produces arcing that burns and destroys the commutator surface, leaving the mica higher than the bars. The mica

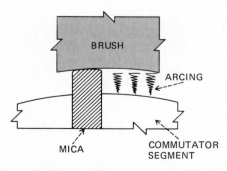

Fig. 10-8 High mica causes arcing between brush and copper segment.

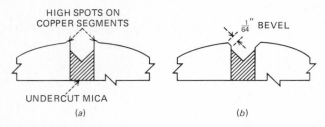

Fig. 10-9 (a) High spots in copper segments due to the undercutting operation. (b) A 1/64-in. bevel on the segment edges eliminates the high spots.

is glazed, and the bars are burned and sooty. Sparking, heating, and brush wear are excessive. The sooty appearance of the commutator is similar to that produced by short brushes. Insufficient brush contact is the cause in both cases.

Undercutting work-hardens and slightly raises the surface of the edges of the bars, as illustrated in Fig. 10-9(a). After undercutting, these edges should be scraped or filed to a level as shown in Fig. 10-9(b).

10-6 COMMUTATOR POLISHING

A commutator brush surface should have a moderately fine polish. A suitable finish can be produced by the use of 4/0 sandpaper. Polishing should be done in a lathe when practicable. If a commutator is polished on the job, great care must be taken to avoid getting copper dust or sanding grit in the machine windings, starting mechanisms, or bearings.

Commutators should be polished just prior to operation, or after baking, to remove oxides and foreign matter that form during idleness or baking.

When a commutator in a machine is to be polished, a sanding block such as that shown in Fig. 10-10(a) can be used. If a commutator surface is slightly out-of-round or uneven, it can be trued with a handstone, such as the one shown in Fig. 10-10(b). The span of contact of the stone should be large enough to extend beyond uneven or low places in the commutator.

10-7 CLEANING A COMMUTATOR

A commutator, in good condition otherwise, will accumulate a deposit of oil, carbon from wear of the brushes, dust from the air, and other foreign matter that should be removed. If the accumulations of these materials are allowed to continue, they will glaze the commutator surface and result in brush chatter, broken brushes, sparking, streaking and etching of the

bars, and destruction of the lubricating film. Foreign matter should be removed before it hardens on the commutator.

A convenient method of cleaning a commutator is the use of a pad of hard woven canvas attached to a long stick. A handle about 3 ft long allows the operator to stay clear of danger. The pad is held on the commutator surface while the machine is running.

10-8 COMMUTATOR LUBRICATING FILM

The only lubrication *suitable* for a commutator is a *copper oxide base film* that is produced by a properly designed brush for specific working conditions. All other forms of lubricants are detrimental to a commutator.

Oil or grease traps foreign matter and causes a hard, high-resistance glaze to form on a commutator. The oil can come from oil vapor that condenses on the commutator, or it can be stray oil or grease from bearings. A glazed surface leads to selective threading of the commutator as current seeks paths through weak spots of the surface and causes brush chatter, which in turn causes the brushes to chip, brush springs to lose their tension, and connecting wires to break owing to metal fatigue.

Many factors contribute to the formation of a suitable lubricating film on a commutator. A new commutator of clean bright copper and its brushes, in the beginning of operation, immediately start the establishment of a suitable copper oxide lubricating film.

The combination of a properly designed brush, a copper commutator, a set of environmental conditions—including oxygen, gases, and moisture—and the load current, armature speed, and brush pressure produces a satisfactory lubricating film. The common notion that any piece of black carbon that fits the brushholder will make a brush is erroneous.

10-9 FILM CONTAMINANTS

Many airborne contaminants are highly injurious or destructive to the average commutator lubricating film. Where these contaminants are constantly present in the air, specially designed brushes are necessary for satisfactory operation.

Especially troublesome or destructive are combustion products of coal, gas, and fuel oils and also such industrial chemicals as fluorine, chlorine, ammonia, and bromine. Nicotine in tobacco smoke, vapors from

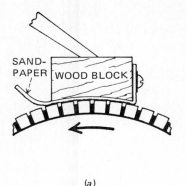
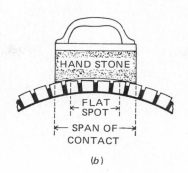

SAND-PAPER · WOOD BLOCK

(a)

HAND STONE / FLAT SPOT / SPAN OF CONTACT

(b)

Fig. 10-10 Polishing the surface of a commutator. (*a*) A wood block shaped to fit the curvature of the commutator. (*b*) Commercial handstones will fit themselves to the commutator after some use. (*Union Carbide Corp., Carbon Products Div.*)

alcohol, turpentine, and other paint solvents and cleaning compounds are sometimes present in concentrations that can cause trouble for commutator films.

10-10 CHARACTERISTICS OF A GOOD FILM

A good lubricating film, depending to an extent on weather conditions, can range in color from a *light tan* to *chocolate brown*. It has a *soft, satiny sheen*. Good commutating conditions, as evidenced by films, are shown in Fig. 10-11. In Fig. 10-11(*a*) the commutator has a light tan film and a smooth, soft, satiny sheen.

In Fig. 10-11(*b*) the commutator is chocolate brown in color with a smooth, satiny film. If the film is uniform and not too thick, good commutation is indicated.

The most commonly found film pattern in good commutations is shown in Fig. 10-11(*c*). This is a mildly mottled color with a random film pattern and a soft, smooth sheen. Occasionally, the condition shown in Fig. 10-11(*d*) is found in good commutation. Alternate bars, or groups of alternate bars, vary in color from light to dark. This is known as *slot bar-marking*. It is related to the number of coils per slot of the armature and does not necessarily mean faulty commutation if the dark bars are smooth and free of

dirt. It usually can be corrected by polishing the commutator and allowing a new film to form.

10-11 MOTOR AND GENERATOR BRUSHES

Brushes for motors and generators are usually made of either *carbon, graphite,* or *metal* or a combination of these. Basic materials are mixed with additives for various grades of brushes and, with a binder, are baked at high temperatures to a hardened stage.

Carbon brushes are shown in Fig. 10-12. These brushes are equipped with shunts to conduct current to or from the brushes. When replacing brushes, always use the brush grade originally used by the motor manufacturer unless operating conditions require that changes be made.

Carbon brushes should have 0.002- to 0.020-in. end clearance and 0.002- to 0.010-in. side clearance in the brushholder. The brushholder should be clean and true to avoid sticking. A stuck brush will cause pitting and etching of bars. Evidences of stuck or short brushes are sparking and a blackened path on the commutator and a sooty deposit in the path. The lubricating film is also destroyed by the arcing produced by improper contact.

Brush shunts should be carefully inspected and pre-

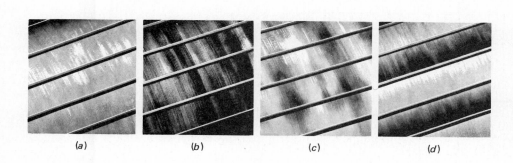

(a) (b) (c) (d)

Fig. 10-11 Examples of good commutator surfaces. (*a*) Light tan, smooth satiny. (*b*) Chocolate brown, smooth satiny. (*c*) Mottled but regular film. (*d*) Slot bar-marking with regular film. (*General Electric Co.*)

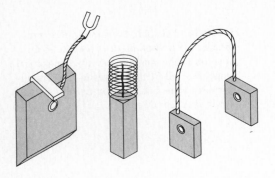

Fig. 10-12 Carbon brushes used for motors.

served. If a brush shunt becomes loose or broken, current will arc and flow between the brush and holder. Arcing will heat, pit, and burn the holder and brush and cause excessive wear or sticking.

10-12 BRUSH PRESSURE

Brush pressure on the commutator is usually expressed in pounds per square inch. Recommended brush pressures in pounds per square inch for various types of commonly used brushes are as follows: Carbon and carbon-graphite, 2 to $2\frac{1}{2}$ lb/in.^{2}; graphite-carbon and electrographitic, 2 to 3 lb/in.^{2}; metal graphite, 2 to $3\frac{1}{2}$ lb/in^2; fractional horsepower motors, 4 to 5 lb/in.^{2}.

To calculate the force (in pounds) that the spring must exert on the brush, multiply the contact area of the brush by the recommended pressure. Although the contact surface is curved, it is accurate to find the area by multiplying the lengths in inches of two adjacent sides. The brush spring pressure can be measured with a small scale.

If brush pressure is less than recommended pressure, arcing between the brush and commutator is increased, and electrical wear of brushes and the commutator is greatly increased. Too much pressure increases friction and mechanical wear.

The rate of brush wear under various pressures for a typical case is shown in Fig. 10-13. The recommended pressure of the brush in this case was 2 to 3 lb/in.^{2}. At $\frac{1}{2}$ lb/in.^{2}, the rate of wear is excessive, with electrical wear being dominant, and wear decreases rapidly as pressure is increased to 2 lb/in.^{2}. Beyond 3 lb/in.^{2} wear slowly increases, with mechanical wear being dominant.

10-13 INSTALLING NEW BRUSHES

A brush face should make full and complete contact with the commutator. Most new brushes need facing to fit or seat the commutator.

Brush facing can be done by placing the brush in its holder and drawing a piece of coarse sandpaper between it and the commutator. The sanded side of the paper must be next to the brush, and the paper should be drawn back and forth in line with the curvature of the commutator. Finishing can be done with about 2/0 sandpaper. The brush and commutator should be thoroughly cleaned following sanding.

All brushholders should be set exactly parallel with the axis of the armature and within $\frac{1}{16}$ to $\frac{1}{8}$ in. of the commutator. In setting an adjustable brushholder, a spacer of the proper thickness can be placed between the brushholder and commutator as a guide while tightening the clamping screws of the brushholder.

Brushholders should be evenly spaced around the commutator. Spacing can be checked by placing a long strip of paper tightly around the commutator under the brushes and marking the paper at the toe of

Fig. 10-13 Effects of brush pressure on the rate of wear.

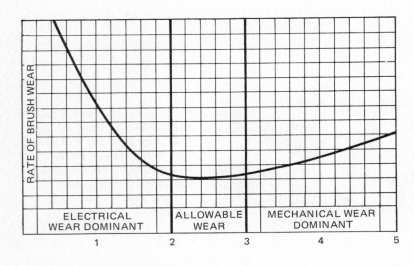

each set of brushes. After marking, the paper is removed, and the distance between marks is measured. The distance between all marks should be the same.

10-14 BRUSH SPARKING

Brush sparking cannot be eliminated completely. Some sparking is expected and acceptable. Excessive sparking in commutation is destructive in many ways. Excessive sparking in commutation is what excessive temperature is in a sick person—a symptom of something wrong. Nearly all troubles characteristic of motors and generators cause excessive sparking in varying degrees. Causes of excessive sparking can be divided into two types—*electrical* and *mechanical*.

Some of the major electrical causes of excessive sparking are:

Wrong interpole polarity

Shorted field or armature coils

Grounded field or armature coils

Open armature circuit

Wrong field polarity

Overload

Severe load conditions

Some of the major mechanical causes of excessive sparking are:

Rough commutator

Out-of-round commutator

High mica

Poor lubricating film

Vibration

Unbalanced commutator

Poor brush fit

Incomplete contact

Incorrect pressure

Improper alignment and position

Excessive copper in brush face

Brush chatter

Unequal air gap

10-15 FLASHOVERS

Under certain conditions a heavy electric arc or flash will occur between brushes of opposite polarity or a brush and ground or other electrical parts. This arc is known as a *flashover*. A *flashover is aided by ionization*. Ionization is the result of severe sparking.

A flashover is usually accompanied by a bright flash and a loud report. It usually clears itself by severe agitation of the air, and no visible damage remains. In some cases, damage is done before a protective device opens the circuit.

A flashover is usually triggered by a heavy transient current, such as a short circuit or sudden overload. Flashovers can be eliminated by reducing sparking and increasing air circulation around the commutator.

10-16 FAULTY COMMUTATION

Trouble in commutation shows itself in many ways. In Fig. 10-14(a) a condition known as *streaking* is shown. Streaking can be caused by a light load, contaminated atmosphere, light brush pressure, low humidity, oily commutator, copper deposits in the brush faces, or improper brush grade. Most of these causes contribute to a poor film. Trouble due to the load being too light can often be eliminated by removing one or more brushes from multiple-brush brushholders. Figure 10-14(b) shows a condition known as *threading*. In most cases this is an advanced or serious case of streaking.

A case of *pitch bar-marking* is shown in Fig. 10-14(c). The number of burned or etched areas in early stages of development usually equals one-half the number of poles in the machine. In more developed stages, they may equal the number of poles. Pitch bar-marking can be caused by dynamic unbalance, soft bars, periodic high mica, light brush pressure, high resistance connection at the commutator, or a flat commutator.

A condition of heavy *slot bar-marking* is shown in Fig. 10-14(d). The frequency of these marks is related to the number of coils per slot in the winding. Heavy slot bar-marking can be caused by misaligned brushes, wrong interpole polarity or strength, too great a load, or wrong brush grade.

A case of *commutator grooving* is shown in Fig. 10-14(e) Grooving is rapid wear to the commutator caused by abrasive brushes, low current density of the brushes, incorrect brush pressure, abrasive dust in the air, too high or too low humidity, wrong brush grade, or excessive copper deposits on the brush face.

A microscopic view of copper in the face of a brush is shown in Fig. 10-14(f). In normal operation the commutator copper is vaporized by minute electric arcs between the commutator and brushes, and copper is deposited on the brush face. Oxidation of the copper with moisture is chiefly responsible for good film formation. Heavy loads and wrong brush grades are

Fig. 10-14 Signs of faulty commutation. (*a*) Streaking. (*b*) Threading. (*c*) Pitch bar-marking. (*d*) Heavy slot bar-marking. (*e*) Grooving. (*f*) Embedded copper in brush face. (*General Electric Co.*)

(a) (b)

(c) (d)

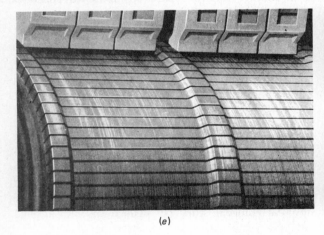

(e) (f)

chiefly the cause of heavy oxidation of copper and excessive copper transfer to brushes. Other causes of excessive copper transfer are high mica, oily commutator, too-low or too-high humidity, low current density, and excessive sparking.

Excessive copper deposit on faces of brushes is detrimental to good commutation. Some results of too much copper transfer to brushes are grooving, sparking, heating of brushes, streaked or raw commutator, brush chatter and pitting, and breaking of brush shunts.

The result of an open circuit in an armature is shown in Fig. 10-15(*a*). In an armature with an open circuit, current will arc across the mica between the two bars connected to the open-circuit coil. The arc will melt the affected area of the bar, and centrifugal force will throw the molten copper from the bar. This process creates a gap between the bars. The arc will travel from end to end of the bars as the gap widens.

The arc, rotating with the commutator, presents the appearance of a circle of sparks around the commutator. This is known as *ring fire*. The rough surface of the commutator will break the brushes.

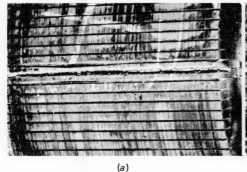 

(a)   (b)

Fig. 10-15 Damaged commutators. (*a*) **Open-circuit damage.** (*b*) **Flashover damage.** (*Kirkwood Commutator Co.*)

A commutator running with broken brushes and receiving current through ionization of the space between the brushholder and the commutator is shown in Fig. 10-15(*b*). This condition continued until the brushholders melted and disintegrated, damaging the commutator to the extent shown. In this case a circuit breaker had been bypassed and provided no protection for the motor.

SUMMARY

1. Commutators and brushes connect stationary and rotating members of an electric circuit.

2. Average commutators rotate with surface speeds of up to 60 mph.

3. The only suitable lubrication for commutators and brushes is a copper oxide film that forms during normal operation.

4. Brushes are chemically designed to form a desirable lubricating film. Stray gases, chemicals, or other materials can seriously interfere with the proper formation of a lubricating film.

5. A good commutator lubricating film will vary in color from light tan to chocolate brown. Variation in color is due to moisture content of the air and operating conditions.

6. A good film has a soft, smooth, satiny sheen free of streaks and dirt smudges.

7. Short or stuck brushes, failing to make proper contact with a commutator, cause arcing between the brush face and commutator. This burns the brushes, pits the commutator, and results in the formation of a sooty film on the commutator.

8. Commutator bars are specially shaped copper segments, clamped in a cylindrical form, and insulated from each other and the rest of the armature assembly.

9. Commutators, following installation or repair, are trued in a lathe. An extremely sharp, round-nose cutting tool is recommended for commutator truing.

10. The segment mica between bars should be undercut below the surface of the bars. The depth of the undercut should be about equal to the thickness of the mica.

11. A commutator should have a round surface with a reasonably high polish. A high polish can be attained with the use of 4/0 sandpaper. A commutator should be polished immediately before installation and operation.

12. A commutator should be kept free of all foreign matter such as oil, grease, oxides, moisture, grit, and dust.

13. Brushes should have free movement in the holders, and they should have the proper pressure on the commutator. Too-light pressure causes greater brush wear than correspondingly too great pressure.

14. Replacement brushes should be of the same grade as the original brush unless operating conditions indicate otherwise.

15. Nearly any trouble in dc motors and generators will result in increased brush sparking.

16. Carbon brushes should have a pressure of about 2 to 2 1/2 lb/in.^{2}.

17. Brushholders should be spaced 1/16 to 1/8 in. from the face of the commutator.

QUESTIONS

10-1 What is the function of a commutator and its brushes?

10-2 What is segment mica?

10-3 What is molding mica?

10-4 Name two ways of removing a commutator from an armature.

10-5 Why should a torque wrench be used to tighten a commutator?

10-6 When resurfacing a commutator on a lathe, at what speed should it be turned and what type of cutting tool should be used?

10-7 How deep should the mica between the commutator bars be undercut?

10-8 Why should a breathing mask be worn when resurfacing a commutator and undercutting segment mica?

10-9 What are the two shapes of undercut slots and which of the two is most commonly used?

10-10 What are some of the problems caused by high mica?

10-11 Why should the edges of the commutator segments be scraped or filed after undercutting?

10-12 Why are commutators polished?

10-13 What precautions must be taken when polishing a commutator on the job?

10-14 What is the only suitable lubricant for a commutator?

10-15 List the characteristics of a good lubricating film for a commutator.

10-16 Name some materials used to make good brushes.

10-17 What can happen if a brush shunt becomes loose or broken?

10-18 What units are used to specify the pressure of a brush on a commutator?

10-19 What brush pressure is generally recommended for carbon brushes?

10-20 How close to the commutator should the brushholder be mounted?

10-21 List some major electrical causes of sparking.

10-22 List some major mechanical causes of sparking.

10-23 What causes flashovers and how can they be eliminated?

10-24 What causes commutator grooving?

10-25 What effect does an open circuit in the armature have on the commutator?

11
DIRECT-CURRENT MOTOR CONTROLS

Electrical controls are to a motor what the accelerator, steering wheel, clutch, and brakes are to an automobile.

There are two distinct types of control systems—*constant-voltage systems* and *adjustable-voltage systems*. On a constant-voltage system, control is provided for each *individual* motor operating on a *constant-voltage line*. On an adjustable-voltage control system, the supply generator voltage and polarity are controlled to control a *single* motor.

The most common functions of controllers for motors are as follows: (1) Start. (2) Stop. (3) Accelerate. (4) Decelerate. (5) Regulate speed. (6) Plug (reverse for quick stopping). (7) Jog (inch, run slightly forward, or reverse). (8) Brake. (9) Reverse. (10) Protect (protect motor, equipment, and the operator).

Direct-current motors over 2 hp in size must have reduced-current starting. Direct-current motors 2 hp and less are usually started and operated across-the-line on full line voltage.

11-1 DRUM START-REVERSE CONTROL

Drum switches are commonly used for starting and reversing dc motors across the line without resistance up to 2 hp in size.

The principles of operation of a reversing drum switch are illustrated in Fig. 11-1. The numbered circles represent stationary contact points, and the black bars represent movable cams mounted on a drum that is turned manually by a handle on the drum shaft.

Most drum switches have three-position switches with *forward, reverse,* and *off* positions. Figure 11-1(*a*) shows the circuits with a drum switch in one running position. The circuit, beginning on the positive line, is from contact 1 through a cam to contact 4, through the circle to the right, in the direction of the arrow, to contact 3. From contact 3 the circuit continues through a cam to contact 6, the jumper to contact 8, and the cam to contact 7 and the negative line.

In Fig. 11-1(*b*) the drum switch is shown in the other running position. Current flow from positive to negative line is now to the left through the circle as indicated by the arrow. In either position, the switch makes circuits directly across from contact 1 to contact 2, from 5 to 6, and from 7 to 8. The polarities of contacts 3 and 4 are reversed when the switch is reversed. Contacts 3 and 4 can be considered the "reversing area" in this switch. A jumper is shown installed between contacts 6 and 8 to complete the circuit. This jumper is removed for series field connections as discussed in the next section.

11-2 DRUM REVERSING OF A SERIES MOTOR

A drum-reversing switch is shown connected to a series motor for across-the-line operation in Fig. 11-2. In Fig. 11-2(*a*) current from the positive line enters the armature on lead A1 from contact 4 and enters the series field on lead S1 from contact 6 for ccw rotation. In Fig. 11-2(*b*) current from the positive line enters the armature on lead A2 but continues to enter the series field on S1. Since the current was reversed in the armature only, the direction of rotation will be reversed to cw.

In nearly all dc motor reversing controls, current in the armature circuit is reversed. The armature circuit usually contains an interpole winding which must be reversed when rotation is reversed. Thus current in the armature and interpoles is reversed simultaneously.

To make the forward and reverse positions of the drum controller correspond with forward and reverse

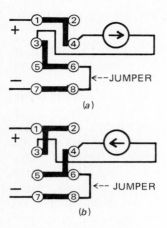

Fig. 11-1 Reversing drum switch. (a) In one direction points 1, 2, and 4 and points 3, 5, and 6 are shorted. (b) In the other direction, points 1, 2, and 3 and points 4, 5, and 6 are shorted.

motions of the driven machine, it may be necessary to interchange the armature lead connections at contacts 3 and 4.

Some type of overload and no-voltage protection, such as a magnetic starter with overload protection, should be provided ahead of a drum switch for motor protection.

11-3 DRUM REVERSING OF A SHUNT MOTOR

A shunt motor can be reversed using a drum switch as shown in Fig. 11-3. In Fig. 11-3(a), current from the positive line enters the armature on A1 from contact 4

and the shunt field on F1 from contact 2 for ccw rotation. Armature lead A2 and field lead F2 are connected to the negative side by the switch.

In Fig. 11-3(b), current enters the shunt field on F1 from contact 2 as before, but it enters the armature on A2 from contact 3 for cw rotation. Terminals F2 and A1 are connected to the negative line by the switch.

11-4 DRUM REVERSING OF A COMPOUND MOTOR

A compound motor can be reversed using a drum switch as shown in Fig. 11-4. In Fig. 11-4(a) current from the positive side of the supply enters F1 from contact 2 and enters the armature on A1 from contact 4 for ccw rotation. The armature current flows from A2 through contact 3 to contact 6 and enters the series field on S1 for *cumulative* compounding. For *differential* compounding, the series field connections are interchanged.

In Fig. 11-4(b) the switch is thrown in the reverse position. This causes current to enter the shunt field on F1 from contact 2 as before, but the armature current enters on A2 from contact 3 for cw rotation.

11-5 DIRECT-CURRENT REVERSING WITH A DPDT KNIFE SWITCH

A shunt motor connected to a double-pole double-throw (DPDT) switch can be reversed as shown in Fig. 11-5. Jumpers are connected between contacts 5 and 4 and between contacts 2 and 6.

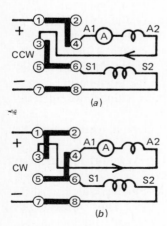

Fig. 11-2 Using a reversing drum switch to control a series motor. (a) Connections for ccw rotation. (b) Connections for cw rotation.

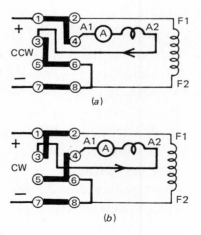

Fig. 11-3 Using a reversing drum switch to control a shunt motor.
(a) Connections for ccw rotation.
(b) Connections for cw rotation.

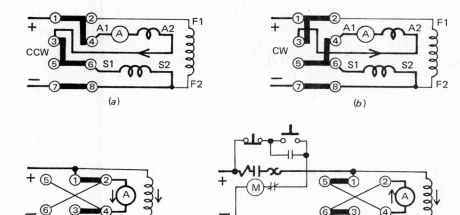

Fig. 11-4 Using a reversing drum switch to control a compound motor. (*a*) Connections for ccw rotation. (*b*) Connections for cw rotation.

Fig. 11-5 Reversing a shunt motor with a double-pole double-throw switch. (*a*) Switch poles 1 and 3 are connected to the line. The switch is shown thrown to contacts 2 and 4. (*b*) A magnetic switch is used to open and close the main line contactor. The DPDT switch has been thrown to contacts 5 and 6.

In Fig. 11-5(*a*), pole 1 is connected to the positive side and pole 3 is connected to the negative side. In the "throw" position shown, the armature is connected in the reversing area.

In the case of a series or compound motor, the series field would be connected between terminal 3 and the negative line. Interpoles are always connected in the armature circuit between the armature and A2.

In Fig. 11-5(*b*) a magnetic switch is shown connected ahead of the reversing switch. The magnetic

switch provides start-stop control and overload protection of the motor.

A standard four-way switch can provide the same reversing action as the DPDT switch described above. These should be used for small dc motors only.

Two types of four-way switches are generally available as shown in Fig. 11-6. The parallel-action four-way switch can be used to reverse both series and shunt motors as shown in Figs. 11-7 and 11-8.

11-6 DIRECT-CURRENT MAGNETIC CONTROLS

Magnetic control of a motor consists of a starting switch operated by an electromagnet and a pilot device to initiate the operation of the control. Magnetic control is used for both across-the-line (full voltage) and reduced-voltage starting. It permits extended remote control of motors or other electrical equipment.

A magnetic contactor used for motor starting is shown in Fig. 11-9. The contacts are normally open. The term *normally* is used to describe the state of the contacts in their *unenergized* condition.

The *closing coil*, also known as *operating coil*, *holding coil*, or *main line coil*, receives its energizing current from an automatic device or a manually operated pushbutton. The operating device may be located at the contactor, at the motor itself, or at some remote location. Drum switches are sometimes used as master switches to control power contactors for controlling large motors. Thus the drum switch carries only control current and not the heavier power current.

There are two general types of magnetic control systems—*two-wire* and *three-wire*. A two-wire control simply makes and breaks the circuit to the closing coil of the control contactors for starting and stopping a motor. It requires only two wires and does not in-

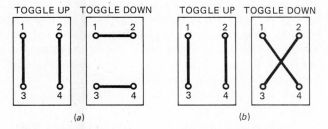

Fig. 11-6 Four-way toggle switches. (*a*) Parallel-action type. (*b*) Cross-action switches type.

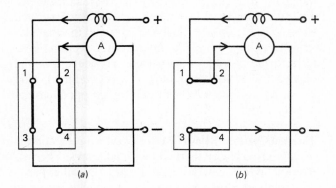

Fig. 11-7 Cross-action four-way switch used for reversing a series motor. (*a*) Counterclockwise rotation. (*b*) Clockwise rotation.

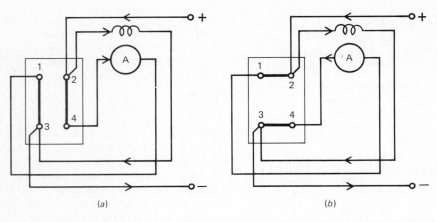

Fig. 11-8 Cross-action four-way switch used for reversing a shunt motor. (*a*) Counterclockwise rotation. (*b*) Clockwise rotation.

clude a provision for no-voltage or low-voltage protection.

A three-wire pushbutton provides no-voltage release, which will not allow a motor to start automatically on restoration of voltage following voltage failure.

A three-wire control is shown in Fig. 11-10. A start-stop pushbutton station [Fig. 11-10(*a*)] contains two momentary contact switches. Three wires must run to this station: one wire to the stop button, one wire to the start button, and one wire from the sealing contact to the connection between the stop and start buttons. The connection between the stop and start buttons is made internally in the pushbutton station; however, the sealing contact is located external to the station (often quite remotely) and so an extra wire must run to the pushbutton station. When the start button is pushed, a connection is made between two contacts. As soon as the button is released, the connection is broken. Similarly, when the stop button is pushed, contact between two points is broken. Upon

release the connection is remade. Figure 11-10(*b*) is a control diagram for a start-stop pushbutton and magnetic contactor. Pushing the start button makes a circuit from the positive line through the start switch, stop switch, overload contacts, and main closing coil to the negative side. The main closing coil, on being energized, closes its main power contacts to start the motor and also closes an *auxiliary contact* known as a *holding contact*, *maintaining contact*, or *sealing contact*.

The holding contact, when closed by the main coil, seals a circuit around the start button, which allows the operator to release the start button as the motor continues to operate. Pressing the stop button breaks the circuit to the main coil, causing it to be deenergized, to open the main contactor, and to stop the motor. This also opens the holding contact, and the system will not operate until the start button is pressed. In case of no voltage or low voltage during operation, the main contactor and holding contact open, and the three-wire system will not restart upon restoration of voltage.

A three-wire control system connected to a series motor is illustrated in Fig. 11-10(*c*). The power circuit of the motor, beginning at the positive line, contains the main line contactor, thermal overload element, and motor armature and series fields. Heat generated by overload current through the overload thermal element causes the normally closed (NC) overload contacts to the closing coil to open and deenergize the closing coil to stop the motor. A short waiting period is usually necessary before resetting a tripped thermal overload unit.

Thermal Overload Relays Motor overload protection is usually provided by thermal overload relays. Thermal overload relays, operating on the principle of expanding metals or alloys, require time to operate and open a control circuit. This time delay makes these relays desirable for motor overload protection since it allows heavy inrush starting currents to pass without causing

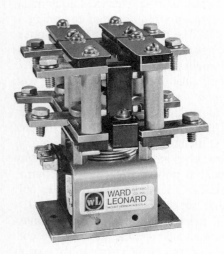

Fig. 11-9 Magnetic contactor. (*Ward-Leonard Electric Co.*)

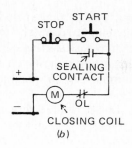

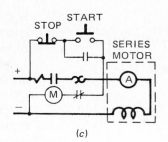

(a)

(b)

(c)

Fig. 11-10 Three-wire magnetic control. (*a*) Start-stop pushbutton. (*Allen-Bradley Co.*) (*b*) Control diagram. (*c*) Magnetic control of a series motor.

the relay to open the control circuit before a motor has accelerated to running speed.

Thermal overload relay heater elements are connected in the motor power circuit, and their normally closed control contacts are connected in the main coil circuit. When they operate, they open the main coil circuit, thus causing the main and holding contacts to open.

There are two basic types of thermal overload relays: *molten alloy* and *bimetallic*. The molten-alloy type is shown in Fig. 11-11. A ratchet is fastened to a

shaft by a low-melting point alloy. The ratchet presses a movable contact against a stationary contact. These contacts are in series with the main coil. A thermal coil carrying the motor current surrounds the ratchet assembly. If the motor is overloaded, the thermal coil becomes hot enough to melt the alloy and the ratchet moves freely on the shaft, releasing the movable contact and thus opening the main coil circuit. This opens the main and holding contacts and the motor stops. Meanwhile the thermal coil cools and the alloy solidifies, thus allowing the thermal relay to be reset.

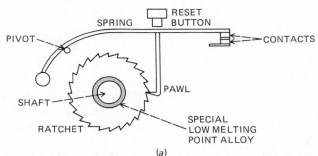

(a)

(b)

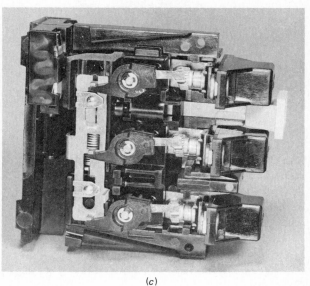

(c)

Fig. 11-11 Molten-alloy thermal relay. (*a*) Simplified sketch of relay. (*b*) Relay in open position. (*c*) Relay in closed position. (*Allen-Bradley Co.*)

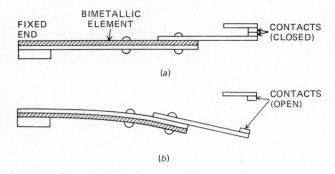

Fig. 11-12 Bimetallic thermal relay. (*a*) Normal position; contacts closed. (*b*) Thermal overload bends bimetallic element breaking contact.

A bimetallic-type relay operates on the principle that metal expands as it is heated. Figure 11-12 is a simplified version of a bimetallic overload relay. Two metals that expand at different rates are welded or riveted together. As they heat, the two metals expand; however, one metal expands at a greater rate than the other. This causes the assembly to bend, breaking contact in the circuit. After the circuit is broken, the bimetallic element cools and returns to its original shape, making contact again.

A three-element molten-alloy overload relay for three-phase motors is pictured in Fig. 11-13. The overload contacts are normally closed.

Fig. 11-13 Three-element overload relay for three-phase motors.
(*Eaton Corp.*)

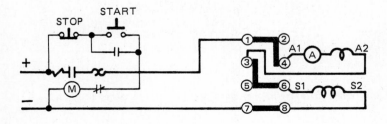

Some overload systems include a thermal unit and a magnetic circuit breaker. The magnetic breaker is set high enough to pass starting currents without tripping but to trip instantaneously on currents in excess of starting currents such as resulting from short circuits or grounds.

In Fig. 11-14, a three-wire start-stop control with thermal, low-voltage, and no-voltage protection is shown connected ahead of a drum-reversing switch for the operation of a series motor.

11-7 REDUCED-VOLTAGE STARTING

A dc motor in starting has comparatively little resistance in its armature circuit. Motors above 2 hp do not have sufficient resistance to properly protect the brushes, commutator, and winding of the motor while starting.

A 5-hp dc motor at 230 V draws about 20 A at full load. The armature circuit, including the interpole winding (if any), the brushes, brush contact with the commutator, and armature winding will have about 0.5-Ω resistance. The inrush, or starting, current can be calculated using Ohm's law.

$$I = \frac{V}{R} = \frac{230}{0.5}$$
$$= 460 \text{ A}$$

This is 2300 percent, or 23 times, the full-load current. Starting current of a dc motor should never exceed 200 percent (twice full-load current) if damage to the commutator, brushes, and windings is to be avoided.

A resistor in series with the armature circuit is required to safely limit the starting current of this motor. The size of the resistor in ohms can be calculated by Ohm's law. Current should be limited to not more than 200 percent of 20 A, which would be 40 A. Using Ohm's law

$$R = \frac{V}{I} = \frac{230}{40}$$
$$= 5.75 \ \Omega$$

This is the total resistance required across the line. The armature circuit is 0.5 Ω. Therefore a resistor

Fig. 11-14 Three-wire control with thermal, low-voltage, and no-voltage protection.

equal to 5.25 Ω (that is, 5.75 − 0.5) is needed in series.

After a dc motor begins rotating, the armature generates counterelectromotive force against the line voltage to protect the motor. The extra series resistor is then shorted out of the circuit.

Problem 1

Calculate the size of a series resistor needed to limit the inrush current of a 480-V dc series motor to 55 A. The armature resistance is 0.35 Ω and the series winding is 0.20 Ω.

Solution

The total armature circuit resistance is 0.35 + 0.20 = 0.55 Ω. Using Ohm's law

$$R = \frac{V}{I} = \frac{480}{55}$$
$$= 8.73 \ \Omega$$

This is the total resistance across the line. Since the armature circuit has a total resistance of 0.55 Ω, an additional resistance of 8.18 Ω (or 8.73 − 0.55) must be used.

A resistor in grid form for starting and speed control of dc motors is shown in Fig. 11-15. Resistors are arranged either in series or in parallel and tapped to afford several steps in starting. Resistors are usually rated in ohms and watts.

11-8 DRUM AND RESISTANCE STARTING

A series motor can be started under reduced voltage and gradually brought up to speed using a drum controller [Fig. 11-16(a)] and a bank of tapped resistors as shown in Fig. 11-16(b). This system uses a magnetic across-the-line starter with pushbuttons, for overload, no-voltage, and low-voltage protection.

Pressing the start pushbutton energizes the main coil to close the line contactor and the holding contactor. The main coil provides no-voltage and low-voltage release, and the thermal overload unit provides running protection to the motor.

A series of cams are mounted on a shaft that is turned by the handle of the drum controller. Although Fig. 11-16 and the following explanation would make it appear the cams move in a straight line, their movement is actually rotary around a shaft. In Fig. 11-16(b) the positions 1 through 4 indicate the positions of the cam relative to resistor taps 1 through 5. For example, in the first position, the dotted line marked 1 aligns with contacts 1 through 5; in position 2, the dotted line marked 2 aligns with the five resistor taps, and so on.

Operation is as follows. With the main contactor closed, the drum controller is turned to position 1. This aligns the dotted line from position 1 over contacts 1 through 5. The circuit to the motor is completed from the + line, to contact 1, through cams A and B, to contact 2, through resistors R1, R2, and R3, through the motor, and returning by way of the − line.

When the handle is turned to position 2, dotted line 2 is aligned with contacts 1 through 5. In this position

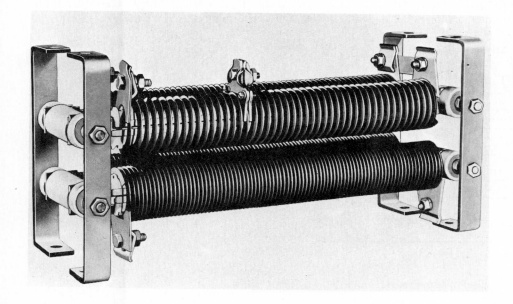

Fig. 11-15 Resistor grid for starting and speed control of dc motors. (*Westinghouse Electric Corp., Control Division.*)

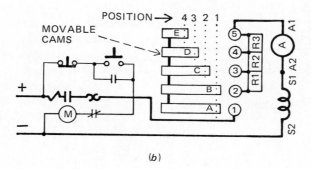

(a) (b)

MOVABLE CAMS

Fig. 11-16 Drum controller. (a) With cover removed, a view inside controller showing contactors and cams. (*General Electric Co.*) (b) Wiring diagram for a series motor connected to a drum controller and magnetic starter.

cams A, B, and C are touching contacts 1, 2, and 3, respectively. When cam C touches contact 3, $R1$ is shorted out and the circuit to the motor is the + line to contact 1, through cam A and C to contact 3, through resistors $R2$ and $R3$, and out to the motor.

In position 3, cam D touches contact 4, shorting out resistor $R2$ so that only $R3$ is in series with the motor.

Finally, with motor coming up to speed, the controller is turned to position 4. In this position cam E touches contact 5, shorting out resistor $R3$. The circuit is thus from contact 1, through cam A to cam E, to contact 5 and then to the motor. In effect, the motor is now across the line with full voltage applied.

11-9 DRUM RESISTANCE STARTING AND REVERSING

A drum controller for resistance starting, speed control, and reversing a compound motor is illustrated in

Fig. 11-17. The controller is energized by a magnetic switch. Parts of the drum controller are identified in Fig. 11-17(b). The lettered cams, mounted on a drum on a shaft, are movable and are moved by an operating handle. Combinations of cams are connected by jumpers. The numbered contacts are stationary, and as the cams are moved in contact with them, they form circuits between the line, resistance, and motor for the various stages of operation.

In operation, the main line contactor is closed by pressing the start button. This energizes the controller and the motor shunt field. Moving the controller handle to position 1 for ccw rotation engages cams A, B, C, and D with contacts 1, 2, 3, and 5, respectively. This completes the armature circuit as follows. From the + line to contact 1, through cam A to cam B, to contact 2 to the armature, entering at A1, through the armature winding, exiting at A2, to contact 3 and cam C, then to cam D, contact 5, through resistors $R1$ and $R2$ to the series winding S1, exiting at S2 to the

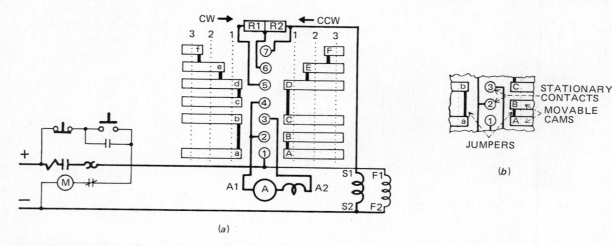

(a)

(b)

Fig. 11-17 Start, reverse, and speed control of a compound motor using a magnetic switch and a drum controller.

minus (−) line. Thus the motor starts with full resistance in the armature circuit.

In ccw position 2, cams A through E are touching contacts 1, 2, 3, 5, and 6, respectively. When cam E touches contact 6, resistor R1 is shorted out and only resistor R2 is in series with the armature circuit.

In ccw position 3, all cams are touching contacts with the exception of contact 4. By touching contact 7, cam F shorts out R1 and R2 so that full voltage is applied to the armature circuit.

Similarly when the handle is turned for clockwise rotation, cams a through f play the same role as cams A through F except that current through the armature circuit is reversed.

For example, in cw position 1 current enters contactor 1 from the + line, through cam a to cam b to contactor 3, and then to the armature circuit entering at A2 and exiting at A1. This is opposite to the direction of current through the armature when the handle was in the ccw positions. As a result the direction of rotation of the motor will be reversed. The rest of the circuit is the same as for the ccw positions with resistors R1 and R2 in series with the armature circuit. Tracing the circuits for cw positions 2 and 3 is left as a student exercise.

While the discussion above relates primarily to resistance starting of motors, if the resistors are capable of carrying the motor currents continuously, this method can also be used as a speed controller. Resistors are usually rated according to the wattage they can dissipate safely. By multiplying the resistance by the square of the motor currents at the various speeds, the required power ratings of the resistors can be calculated.

Problem 2

A reduced-voltage starter uses two resistors R1 and R2 each rated at 10 Ω in series with the armature circuit. With the two series resistors in the circuit, the armature current is 25 A. As the motor comes up to speed R1 is shorted out and the armature current drops to 18 A. When the motor comes up to full speed, both R1 and R2 are

shorted out. The armature current at this point is 10 A. The motor is rated at 230 V, three-phase, 60 Hz. If the starter is also to be used as a speed controller, what is the minimum wattage rating of resistors R1 and R2?

Solution

The only consideration necessary in specifying the wattage rating of the resistors is the maximum current they must carry. In this case, the maximum current is 25 A. Using the power formula

$$
\begin{aligned}
P &= I^2 R \\
&= (25)^2 \,(10) \\
&= 625 \times 10 \\
&= 6250 \text{ W or } 6.25 \text{ kW}
\end{aligned}
$$

In order to be used continuously for speed control, resistors R1 and R2 must be rated for at least 6.25 kW.

It should be understood that the power ratings merely indicate the maximum power the resistors can safely dissipate, not the actual power they will draw. Thus if the same resistors were used in a circuit which actually draws less current, the power dissipated by R1 and R2 would be less. If they were put in a circuit in which more current was drawn, they would have to dissipate more power than their maximum rating. For a short time they might be capable of doing this, but they would soon burn out, creating an open circuit.

11-10 THREE-POINT MANUAL STARTERS

A three-point manual starter employs a faceplate rheostat connected in series with the motor armature circuit to cut out resistance as the motor accelerates.

A three-point starter system is illustrated in Fig. 11-18. A magnetic across-the-line starter with pushbuttons is used to connect and disconnect the main lines to the three-point starter. The starting box is

Fig. 11-18 Three-point face-plate starter.

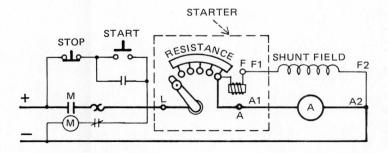

known as a three-point box because there are only *three points* for connection to the system. The three connection points are at L, F, and A in the box.

In operation, the starting box is energized by the magnetic starter. Moving the manually operated lever clockwise connects the motor, with full resistance in the armature circuit, to the line for starting. The lever is gradually moved clockwise, as the motor accelerates, cutting out resistance until it reaches the last contact when the motor is across the line for full line voltage.

An electromagnet, known as a *holding coil,* is shown between the resistance and F terminal in series with the motor shunt field. This holding coil holds the spring-loaded lever in running position when it reaches the last contact. Low voltage or no voltage in the coil will release the lever, and a spring will return it to the "off" position. Also, an open circuit in the shunt field will deenergize the holding coil and stop the motor. This is a safety feature of a three-point starter since an open circuit in the shunt field of an unloaded shunt motor or a compound motor on the line will allow the motor to race to destruction. Some shunt motors can "run away" on residual magnetism.

A three-point starter cannot ordinarily be used for speed control. A four-point starter is usually used for this purpose.

11-11 FOUR-POINT MANUAL SPEED CONTROL STARTERS

A four-point starting box can be used for speed control if the wattage rating of the series resistors is high enough to allow continuous operation. Resistance in the armature circuit is used for speeds below base speed, and resistance in the shunt field circuit is used to obtain speeds above base speed. Base speed is the normal speed of the motor under full load when it is

connected directly across the line without any additional resistances.

A four-point starter with above- and below-base-speed control is illustrated in Fig. 11-19.

The starter in this figure is connected to the line by a magnetic switch. The starting box contains a face-plate rheostat with armature and field resistance. The lever hub is equipped with detents which engage a pawl held by a holding coil. This holds the lever in any desired operating position. The holding coil is connected to the line by way of a series resistor. This connection to the line is the fourth point of a four-point box. However, the holding coil does not provide field-loss protection for the motor, and the motor can run away in the event of an open in its shunt field circuit.

In operation, the motor is started by moving the control lever clockwise. Any desired speed below base speed can be obtained by stopping the lever at the desired speed. The detents and pawl will hold it there. The pawl is held in place by the holding coil.

Further movement of the control lever beyond the armature resistance places resistance in the field circuit to reduce the shunt-field strength. A reduction in shunt-field strength reduces the motor's ability to generate counterelectromotive force against line voltage. This permits armature current to increase and drive the motor above base speed.

Field accelerating and decelerating relays are commonly used with speed-regulating equipment. Usually, these relays, connected in the armature circuit, contain two coils, both used for starting. One coil is used alone for running after starting resistance is bypassed. On starting, the combined strength of both coils is sufficient to close the relay. The relay contacts are connected across a permanently set field speed rheostat which is shorted out when the contacts close. Thus the motor starts with full field voltage. After the last motor starting resistance is bypassed, one coil of the relay is shorted. The second coil then causes the

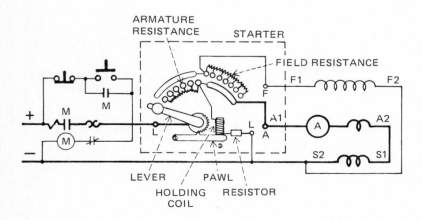

Fig. 11-19 Four-point face-plate starter and speed controller.

relay to flutter and alternately insert and bypass the rheostat until the motor reaches normal speed and armature current.

If the field resistance is suddenly decreased for lower speed, the relay acts during the decelerating period to prevent excessive armature current and braking action from damaging driven equipment.

Field-loss relays are commonly used with shunt and compound motors not otherwise protected against "runaway" from loss of their shunt fields. The field-loss relay coil is connected in series in the shunt-field circuit, and the contacts are connected in the closing or holding coil control circuit. An open circuit in the shunt-field circuit deenergizes the relay, and it opens the holding coil circuit to stop the motor.

11-12 AUTOMATIC DC STARTERS

Automatic starting of dc motors consists in starting a motor with resistance in the armature circuit and automatically reducing it in steps or removing it as the motor accelerates. Several types of mechanical and electrical time delay methods will be discussed in this section. Electronic methods are covered in Chap. 24.

Electrical time-delay systems are *counterelectromotive force, definite magnetic time, current lockout,* and *electrical timers.*

Mechanical time-delay systems are the *dashpot* and numerous forms of *escapement* and *friction* mechanisms.

While the introduction of solid-state electronic starters and speed controllers has made many of these methods obsolete for new installations, maintenance of existing systems is often critical.

Counterelectromotive-Force Starters A cemf starter starts a motor with resistance in the armature circuit and bypasses the resistance in time when the cemf of the armature reaches a predetermined value. The timing is accomplished by a coil connected across the armature and equipped with a bypass contactor which is connected in parallel with the armature resistance. This coil is sometimes called the *accelerating coil.* A cemf starter connected to a shunt motor is illustrated in Fig. 11-20.

In operation, the motor is started through resistance when the start button of the magnetic switch is closed. As the motor accelerates, cemf in the armature increases to a value that causes coil R to close contactor R and bypass the armature resistance for full-voltage operation of the motor. Any number of steps of starting can be provided with additional cemf coils, contactors, and resistors. The additional cemf coils are connected in parallel with the armature, and their operation periods are set by adjusting operating springs or air gaps in their magnetic circuits.

Definite Magnetic Time Starters. Definite magnetic time starters employ a specially constructed coil assembly in which *magnetic flux decay is delayed* to produce time delay action of a contactor.

A definite magnetic time contactor contains a one-turn closed-circuit coil of extremely heavy copper under the conventional wire-wound coil. The one-turn coil somewhat resembles a plain bronze sleeve bearing. The contactor contains one set of normally closed power contacts, which are connected in parallel with the motor starting resistance, and a set of normally open (NO) control contacts. A one-step definite magnetic time starter is illustrated in Fig. 11-21. Pressing the start button makes a circuit from the positive line through the definite magnetic time contactor coil MT to the negative line. This operates the NC contactor MT1 across the resistance in the motor armature circuit and closes its control contacts MT2 in series with the main coil M. Energizing coil M closes the main line contactor M1 to start the motor with resistance in the armature circuit. The main coil also opens its NC control contacts M2 in series with the magnetic time coil MT which deenergizes it. However, MT1 does not close immediately. After a time delay, contact MT1 closes, which shorts out the motor armature resistance and places the motor across the line for full-voltage operation. Contact MT2 is also opened, putting the holding contact M3 in series with the main coil. Overload contact O.L. is also in series with the main coil. Should an overload occur, this contact will automatically open and the coil M will be deenergized.

The time of operation of a magnetic time contactor is adjusted by adjusting air gaps in the magnetic circuit of the contactor or by adjusting the tension of

Fig. 11-20 A cemf starter with one-step resistance starting connected to a shunt motor.

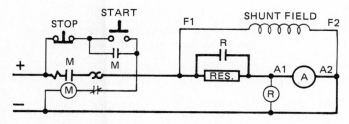

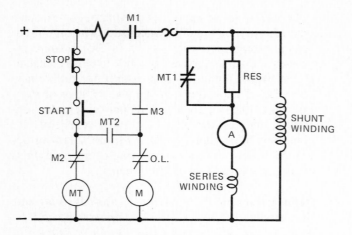

Fig. 11-21 A one-step definite magnetic time starter connected to a compound motor.

operating springs. Some magnetic time contactors contain several heavy rings for the inner coil, and operating time adjustments are made by changing the number of rings, using more rings for more time and fewer rings for less time.

A two-step definite magnetic time starter is also available. This starter contains two definite magnetic time contactors and two steps of resistance.

Current-Lockout Starters. A current-lockout starter contains a normally open contactor with its coil connected in series with the motor armature. This contactor is designed to close on a certain value of armature current. The magnetism created by the coil has two separate paths in which it can flow. One magnetic path contains an air gap, the other a restriction in the magnetic material.

When the initial heavy starting current flows through the coil, magnetism builds up in both paths. Magnetism in the air gap keeps the contactor open, while magnetism in the restricted path tends to close it against the restraining force of the air-gap magnetism. As the motor accelerates, armature cemf is reduced, which reduces current in the coil. The magnetism in the air-gap path weakens faster than that in the re-

stricted path. This process will in time allow the restricted-path magnetism to be stronger and close the contactor, shorting out the motor resistance and providing full-voltage operation of the motor.

A two-step two-coil current-lockout starter is shown in Fig. 11-22. In this type of starter the two magnetic paths are created by two separate coils, a lockout coil and a closing coil. Both coils are in series with the armature and with each other. They are also designed to operate at different current values. In starting, lockout coils 1 and 2, respectively, hold contactors R1 and R2 open, and the motor starts with resistors $R1$ and $R2$ in the armature circuit. As the armature accelerates, the cemf increases and the armature current decreases. Lockout coil 1 weakens and closing coil R1 closes contactor R1. This shorts out resistor $R1$. As armature current continues to decrease, lockout coil 2 weakens, allowing closing coil R2 to close contactor R2. This shorts out resistor $R2$ and the motor is now running at full line voltage.

Dashpot Starters A dashpot starter is a type of mechanical time-delay starter (Fig. 11-23). The time control consists of a solenoid with a plunger connected to a dashpot. The dashpot is filled with a fluid that restrains the upward movement of its plunger. The speed with which the plunger rises can be regulated by an adjusting screw on the dashpot.

In operation, the starter is energized by the magnetic switch. This starts the motor on full resistance and energizes the dashpot solenoid. The solenoid, with a short-circuiting bar across the upper end of its plunger, rises slowly with the bar and short-circuits contacts 1 and 2. This bypasses $R1$. The solenoid continues to rise slowly and short-circuits the remaining contacts until all the resistors have been shorted. The motor is then across the line for full-voltage operation.

Some dashpot starters are equipped with a resistor in series with the solenoid coil. The resistor is bypassed in starting to give the coil more power during starting. Once the motor is up to speed, the resistor is put in the circuit for running to protect the coil.

Ordinary petroleum oils, because of severe changes in viscosity, are not satisfactory for dashpot opera-

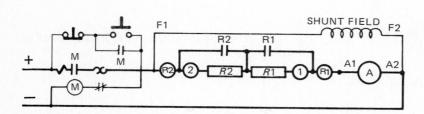

Fig. 11-22 Two-step two-coil current-lockout starter connected to a shunt motor.

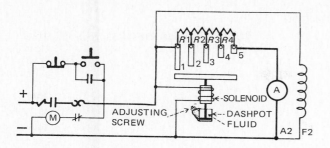

Fig. 11-23 Dashpot starter connected to a shunt motor.

tion. A special oil is required. Some dashpots, known as *pneumatic dashpots,* operate by drawing air into the cylinder; others use a silicone fluid for operation.

11-13 DYNAMIC BRAKING

Dynamic braking and plugging are methods generally used for stopping dc motors quickly. Plugging is reversing a motor while it is running to brake it and bring it to a quick stop.

In dynamic braking of a dc motor, resistance is connected across the armature when it is disconnected from the line and the shunt field is left across the line. This converts the motor to a dc generator. The resistance across the armature serves as a load on the generator. The generated current in the armature is in reverse of the motor current. The action of the field on this new current is to produce a torque opposed to the direction of rotation of the armature. In effect, this puts a brake on the motor and brings the armature to a quick stop.

A dynamic braking system is illustrated in Fig. 11-24. The shunt field of the motor is connected ahead of the main line contactor. The two-pole main line con-

tactor has NO contacts in the line and NC contacts connected in series with a resistor across the armature.

In Fig. 11-24(*a*), the system is shown in running position. The main line contactor is closed, and the braking contactor is open. Conditions for braking are shown in Fig. 11-24(*b*). The main line contactor is open and the braking contactor is closed to connect the resistor across the armature.

The degree of braking can be adjusted by using a variable resistor. Increasing resistance decreases the braking effect. Decreasing resistance increases the braking effect.

Another form of braking is known as *regenerative braking*. This occurs automatically when a load overdrives a motor or the strength of the motor's fields is suddenly increased in order to decrease its speed. In regenerative braking, the motor acts as a generator and tries to deliver current back into the supply line. Usually circuit protective devices will open, thus preventing damage to the line or equipment.

11-14 ADJUSTABLE-VOLTAGE SYSTEM

An adjustable-voltage control system for dc motors consists of a dc generator driven by an ac motor (typically three phase). A separate exciter generator mounted on the generator's armature shaft provides the current for both the generator fields and the motor fields.

The *adjustable-voltage system,* also known as the *Ward Leonard system,* is shown in Fig. 11-25. The dashed line indicates that the shaft of the generators and the ac motor are mechanically coupled. The generator brushes are directly connected to the dc motor brushes. Speed control of the dc motor is obtained by varying the output voltage of the generator. The output voltage of the generator, however, is dependent on the generator field strength. Since the field strength

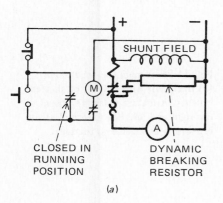

CLOSED IN
RUNNING
POSITION

DYNAMIC
BREAKING
RESISTOR

(*a*)

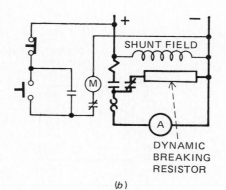

DYNAMIC
BREAKING
RESISTOR

(*b*)

Fig. 11-24 Dynamic braking system. (*a*) Running. (*b*) Braking.

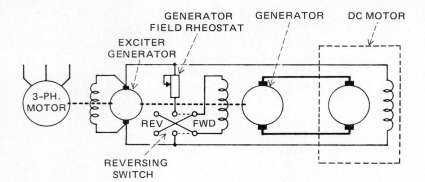

GENERATOR FIELD RHEOSTAT
EXCITER GENERATOR
GENERATOR
DC MOTOR
3-PH. MOTOR
REV   FWD
REVERSING SWITCH

Fig. 11-25 Adjustable-voltage control system.

is controlled by the field rheostat, the rheostat, in effect, is the speed controller of the dc motor. The motor's field is directly connected to the exciter generator so that it receives a constant voltage.

In order to reverse the rotation of the motor, the direction of current in the armature must be reversed. This is accomplished by throwing the reversing switch to the REV position. This will reverse the direction of current through the generator field and thereby reverse the polarity of the generator brushes. Thus the current through the dc motor armature will be reversed.

The advantage of this system is that it provides very good speed control while not requiring starting resistors or relays. In addition speed and rotation are controlled by a relatively small generator field rheostat and simple reversing (DPDT) switch.

11-15 AMPLIDYNE ADJUSTABLE-VOLTAGE CONTROL

The amplidyne generator is a sensitive control generator developed and refined by General Electric Company. Although its basic principle of operation is similar to that of a conventional generator, its main difference is the ability to amplify voltage.

At this point it is worthwhile reviewing the principles of operation of a conventional generator and dc

armature. When current flows in a dc armature, magnetic poles are produced on the armature core at right angles to the main field. Ordinarily these poles are unwanted and interpoles are used to counteract their effect. In the amplidyne generator these poles are used to produce a voltage and current.

The basic arrangement of brushes in an amplidyne generator is shown in Fig. 11-26. The main field of this generator is separately excited, though the wiring for this has been deleted for the sake of simplicity. The main field and the armature field polarities are indicated by N and S. A set of brushes is short-circuited with a heavy jumper as in Fig. 11-26(a). As the armature turns in the separately excited main field, a voltage is generated. With the brushes short-circuited, a large current will flow through the armature. This current will produce the strong armature poles at right angles to the main poles. Although the armature is rotating, the armature poles will appear to be stationary at the positions shown in Fig. 11-26(a). Thus the armature windings will be cutting across this strong armature field, producing a higher voltage. A second set of brushes, at right angles to the original set, is used to connect the amplified voltage to a load. To minimize the effect of armature reaction, a compensating winding is connected in series with the load as shown in Fig. 11-26(b).

When the amplidyne generator is used in a Ward

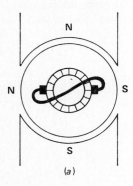

(a)

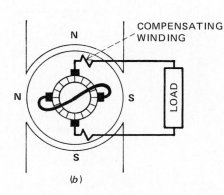

COMPENSATING WINDING
LOAD
(b)

Fig. 11-26 Amplidyne generator. (a) Generator with one set of shorted brushes. Armature poles are shown to the right and left. (b) A second set of brushes is connected to the load.

Leonard system, speed control is effected by varying the relatively low current going to the amplidyne's separately excited field. Speed responses to changes in field current are almost instantaneous, ranging between 50 and 60 *milliseconds (ms)* (about 1/20 s).

While amplidynes can be used to control power output up to about 50 kW, when used in a Ward Leonard system, the motor under control should be less than 10 hp.

11-16 CONTROLS MAINTENANCE

The chief cause of trouble in electrical controls is loose or faulty electrical connections in circuits. A faulty electrical connection presents resistance, and current through resistance generates heat. Heat expands metal, and expansion causes increased looseness. Thus a faulty connection never improves but successively worsens to the point of failure of equipment. This is true whether the controls are electromechanical, solid state, or a combination of the two.

Three main causes of faulty connections are (1) carelessness in the original assembly and later repairs of equipment, (2) heat from normal and abnormal operation, and (3) vibration. To eliminate these, care must be used in assembly, installation, and later repairs of equipment; all connections should be routinely checked for looseness. Vibration should be kept at a minimum.

Connections in the vicinity of thermal overload elements are especially susceptible to looseness because of constant heating and cooling of the thermal elements with each operation of the equipment. These heating and cooling cycles cause further heating and cooling and expansion and contraction of adjacent parts, which results in further loosening of connections. Loose connections in the vicinity of thermal elements cause excessive heating of the elements, destruction of the proper setting, and premature or faulty operation of the overload units (see Sec. 2-15).

A large proportion of trouble in control equipment can be eliminated simply by maintaining good electrical connections. A faulty connection or faulty switch contact can be found by a voltage drop registered on a low-reading voltmeter or by sensing the heat generated in a loose connection.

Other control maintenance tips are as follows: Keep controls dry and clean. Keep mechanical contacts smooth and free from copper oxide. Replace contacts in pairs and keep spring pressure proper and even. Keep all moving parts free in movement and free of oil, gum, and dirt. Generally, no oil should be used in controls.

The resistances of operating coils should be recorded and kept on file for the purpose of comparison with the resistance of a coil suspected to be defective. Only the resistances of coils known to be good should be recorded for this purpose.

SUMMARY

1. The two distinct types of control systems are the constant voltage systems and the adjustable-voltage systems.

2. The most common functions of a motor controller are (*a*) starting, (*b*) stopping, (*c*) accelerating, (*d*) decelerating, (*e*) regulating speed, (*f*) plugging, (*g*) jogging, (*h*) braking, (*i*) reversing, and (*j*) protecting.

3. Direct-current motors over 2 hp should not be started directly across-the-line at full voltage.

4. Drum switches are used to start and reverse dc motors under 2 hp directly across-the-line at full voltage.

5. Most drum switches have forward, reverse, and off positions.

6. Nearly all dc-motor reversing is done by reversing the current in the armature.

7. Overload and no-voltage protection should be provided ahead of a drum switch.

8. A DPDT switch can be used to reverse dc motors. A four-way switch can be used as a reversing switch for small dc motors.

9. Magnetic starters for dc motors consist of contacts operated by an electromagnet.

10. Magnetic starters can be used for across-the-line (full-voltage) starting or reduced-voltage starting.

11. There are two general types of magnetic control systems—two-wire and three-wire.

12. Two-wire controls make and break the circuit going to an electromagnet. There is no low-voltage or no-voltage protection.

13. Three-wire controls use a start-stop pushbutton which is a momentary contact switch. Low-voltage and no-voltage protection is possible with this system.

14. Thermal overload relays protect a motor from damage due to a continuous overload.

15. There are two general types of thermal overload relays—the molten alloy and bimetallic types.

16. Upon overload, the thermal relays break the main coil circuit which in turn opens the main contactors in the motor lines.

17. A drum controller and a bank of resistors can be used for reduced-voltage starting.

18. A three-point faceplate starter can be used to manually start a dc motor with reduced voltage.

19. A four-point faceplate starter can be used to manually start a dc motor with reduced voltage and control its speed from below base speed to above base speed.

20. Direct-current motors can be started automatically with reduced voltage and the resistance automatically dropped as the motor reaches its full speed.

21. Automatic starters use several time-delay methods. Electrical time-delay methods are counterelectromotive force, definite magnetic time, current lockout, and timers. Mechanical methods are the dashpot delay and numerous escapement and friction mechanisms.

22. Dynamic braking is used to stop a motor quickly by reversing the current in its armature.

23. An adjustable-voltage control system (Ward Leonard system) provides starting and speed control by supplying power to the motor's armature directly from a separately excited generator.

24. An amplidyne generator is a voltage or power amplifier. It is sometimes used as the generator in a Ward Leonard system.

25. The chief cause of trouble in electrical controls is loose or faulty connections. Faulty connections can result from carelessness in assembly, heat from operating the equipment, or vibration.

26. Connections should be constantly checked visually, mechanically, and electrically.

27. Always keep controls dry and clean. Keep contacts clean and free of oxides and other insulating films, and keep moving parts free to operate. In general moving parts should not be oiled.

28. The operating coils of starters should be periodically checked for correct resistance.

QUESTIONS

11-1 List the most common functions of motor controllers.

11-2 What is the maximum hp of a dc motor that can generally be started across the line at full voltage?

11-3 What electrical characteristic of dc motors makes it necessary to start them at reduced voltage?

11-4 Explain the operation of a drum switch for full-voltage starting.

11-5 How is the direction of rotation of a dc motor generally reversed?

11-6 Draw a double-pole double-throw switch and show how it is wired for use as a reversing switch.

11-7 What does the term *normally open contact* mean?

11-8 What are other names for a closing or main line coil?

11-9 Explain the function of a sealing coil.

11-10 Why does a two-wire magnetic control system usually lack low-voltage and no-voltage protection?

11-11 How does a three-wire magnetic control system provide low-voltage and no-voltage protection?

11-12 What is the principle of operation of a bimetallic thermal overload relay?

11-13 How are thermal overload relays connected in a control circuit?

11-14 Why shouldn't molten-alloy relay be used for short-circuit protection?

11-15 If the starting current of a dc shunt motor should not exceed 200 percent of its full-load (FL) current, calculate the size of the resistor to be used for reduced voltage starting of the following motor. Neglect field current.

Voltage = 480 V
Load rating = 10 hp
FL current = 18 A
Armature circuit resistance = 0.75 Ω

Specify the wattage of the resistor based on twice the highest wattage the resistor will have to dissipate.

11-16 Explain the operation of a reduced-voltage, reversing-drum controller.

11-17 Explain how a three-point starter provides reduced voltage starting.

11-18 How is low-voltage and no-voltage protection provided by the three-point faceplate starter?

11-19 How does a four-point controller control speed above and below the base speed of a motor?

11-20 What disadvantage does the four-point controller have compared with the three-point controller?

11-21 What is the function of the accelerating coil in an automatic starter?

11-22 How is the time delay adjusted in a magnetic time contactor?

11-23 How does dynamic braking stop a motor quickly?

11-24 What is another name for an adjustable-voltage control system?

11-25 What makes an adjustable-voltage system advantageous in controlling a dc motor?

11-26 Explain the amplifying properties of an amplidyne.

11-27 List some important causes of poor electrical connections.

11-28 List some major maintenance functions designed to prevent or correct faulty connections.

12
DIRECT-CURRENT GENERATORS

Direct-current generators convert mechanical energy to electric energy. In construction, a dc generator is identical to a dc motor. In fact, a dc motor can usually be used as a generator, or a dc generator can usually be used as a motor. For this reason, the discussions of dc motor frames, windings, armatures, commutators, and bearings in this book apply equally to generators with the exception of brush position, lead identification, and standard rotation. Standard rotation of a dc generator is clockwise when one is observing the shaft from opposite the drive end, which is usually the commutator end.

A dc generator produces voltage by passing conductors through a magnetic field. The voltage thus produced is determined by the rate at which magnetic lines are cut and the total number of turns of armature coils connected between brushes. If an external load is connected to the generator, current will flow through the armature winding, commutator, and brushes to the load.

12-1 RIGHT-HAND GENERATOR RULE

When a conductor passes through a magnetic field, electrons in the conductor are displaced toward one end of the conductor. The direction of displacement is determined by the direction of the magnetic lines of force and the direction the conductor is moved through them.

A convenient rule for determining the direction of electron displacement, or induced voltage, is the *right-hand generator rule*. If the thumb, forefinger, and middle finger of the right hand are arranged at right angles with the thumb pointing in the direction of movement of a conductor in a magnetic field and the forefinger pointing in the direction of lines of force from north to south poles, then *the middle finger will point in the direction of induced voltage*.

The right-hand rule is shown in Fig. 12-1. In Fig. 12-1(*a*), the thumb of the right hand indicates downward movement of a conductor, the forefinger indicates lines of force from right to left, and the middle finger indicates that induced voltage will be toward the back of the conductor. The polarity at the terminals of the generator is indicated by the + and −. If an external circuit is connected to the conductor, current will flow in the direction of the arrow shown in the conductor. In Fig. 12-1(*b*), the direction of generated voltage is shown for upward movement of the conductor. It would seem, from the figures, that current is flowing from − to +. However, remember that the plus and minus signs indicate the generator *terminal voltage*. As far as the external load is concerned, current is flowing out of the + side, through the load, and returning to the − side of the generator. *Internally the current will enter the generator at the − terminal, travel through the generator windings, and exit to the load at the + terminal.*

12-2 PRINCIPLES OF A GENERATOR

The principles of operation of generators are shown in Fig. 12-2. A single loop is shown equipped with a ring, cut through the middle, to form two segments. These segments, or bars, together are called a *commutator*. The loop is rotating counterclockwise as viewed from the commutator end. In Fig. 12-2(*a*), the white side of the loop under the north pole in the first half of a revolution is moving to the left. According to the right-hand generator rule, the current in the white side of the loop will flow in the direction of the commutator. Since current is flowing out to the load from the top brush, the top brush is the positive side (+). After making one-half revolution, the loop sides are as shown in Fig. 12-2(*b*). Now the black side is produc-

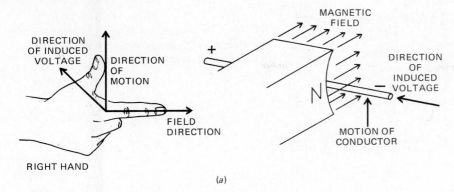

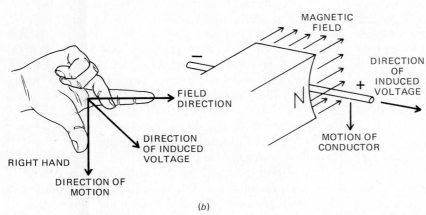

Fig. 12-1 The right-hand generator rule for determining the polarity of induced voltage. (*a*) The conductor moves down through the field. (*b*) The conductor moves up through the field.

ing current in the direction of the commutator. Although the direction of current in the black side reversed in the 180° turn, it is now connected to the top brush. Thus the top brush always is the one through which current leaves the generator and the bottom brush is the one through which current returns to the

generator. This is a basic characteristic of a dc generator. In the single-loop generator, however, the value of voltage will vary since the conductor passes through the field at a constantly changing angle. The result is a pulsating direct current as shown in Fig. 12-2(*c*).

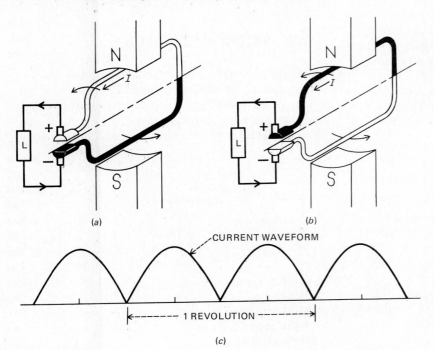

Fig. 12-2 A single-loop dc generator. (*a*) The loop rotates counter-clockwise through the magnetic field. (*b*) After rotating 180°, the position of the loop sides is reversed. (*c*) Pulsating direct current produced by the single-loop generator.

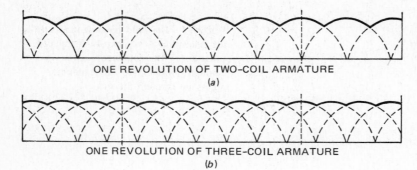

ONE REVOLUTION OF TWO-COIL ARMATURE
(a)

ONE REVOLUTION OF THREE-COIL ARMATURE
(b)

Fig. 12-3 Waveforms produced by simple dc generators. (a) Two-loop armature waveform. (b) Three-loop armature waveform.

If another loop at right angles to the first loop and two more commutator bars were added to the single-loop armature, a voltage waveform, such as shown by the solid black line in Fig. 12-3(a), would be produced.

If three loops with adjacent sides 60° apart and six commutator bars were used on the armature, the resultant voltage waveform would be as illustrated by the solid black line in Fig. 12-3(b).

As more loops, or coils, and commutator bars are added to an armature, the resultant voltage (and current) waveforms become straighter. However the individual waveforms for each loop, as indicated by the broken lines, remain the same as in Fig. 12-2(c).

12-3 CLASSES AND TYPES OF GENERATORS

Direct-current generators are divided into two classes, *self-excited* and *separately excited*. If a dc generator supplies its own field current, it is self-excited. If its field current is supplied from an external source, it is separately excited. Separately excited generators are used chiefly in those applications requiring adjustable output voltages.

Self-excited dc generators are divided into three types, according to the way the fields are connected with the armature. These three types are *series, shunt,* and *compound.*

12-4 SERIES GENERATORS

Series generators, like series motors, have their *fields connected in series with the armature.* All the armature, or load, current flows in the field circuit to energize the fields. If no load is connected to the terminals of the generator, no current will flow within the generator, and the field coils will not be energized. The only voltage generated will be a result of *residual magnetism in the field pole pieces.* This voltage is too low to serve an external load.

A schematic diagram of a series generator is shown in Fig. 12-4. Note that all the load current through the armature must also flow through the fields. As load increases on a series generator, the series fields become stronger, and voltage increases proportionately up to the point of saturation of the pole pieces. Hence the *voltage output* of a series generator *varies with the load.* This makes it unsuitable for applications involving varying loads. Series generators are seldom used because of this voltage characteristic.

A voltage-current curve of a typical series generator is shown in Fig. 12-5. With no load current flowing in the generator, residual magnetism in the poles will produce an output of 6 V. Once a load is connected to the output terminals of the generator, a current will flow, the field coils will be energized, and the output voltage will increase. At a load current of 10 A, the output voltage will be 48 V. Voltage continues to increase with load to about 115 V at 60 A. At this point, saturation of the field core and voltage drop, due to the heavy load, decrease the voltage.

12-5 SHUNT GENERATORS

A shunt generator, like a shunt motor, has its main fields connected across, or *parallel with*, its *armature.* When a shunt generator starts, its armature coils pass through the residual magnetic field in its pole pieces. This produces an initial voltage of about

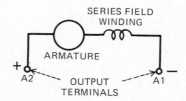

Fig. 12-4 Schematic diagram of a series generator. The symbols *A*1 and *A*2 are standard terminal markings for series generators.

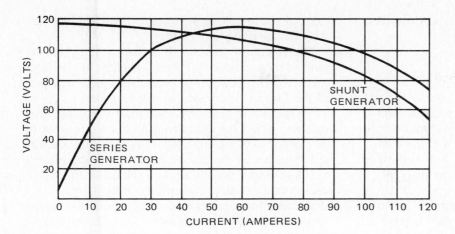

Fig. 12-5 Voltage-current characteristic curves for self-excited generators.

5 percent of full voltage in some generators. Since the fields shunt the armature, the initial voltage produced by residual magnetism causes current flow in the fields. This current flow, when it is in the proper direction, increases the magnetism in the pole pieces. This further increases the output voltage.

The voltage continues to increase as the fields become stronger. The fields become stronger as voltage increases, until the pole pieces reach saturation. When the pole pieces are saturated, a shunt generator has reached its maximum no-load or open-circuit voltage.

A schematic diagram of a shunt generator is shown in Fig. 12-6. Note that the field is connected parallel with the armature and receives full armature voltage. Both the load and the field are connected across the armature. As the *load-current increases,* armature voltage and voltage across the field *decrease* because of *IR* drop in the armature. This weakens the fields, which in turn *decreases the output voltage.* Thus, a shunt generator does not maintain a steady voltage with a varying load.

A voltage-current curve of a typical shunt generator is shown in Fig. 12-5. The *voltage is highest at no load* and drops slightly as load is increased up to about 70 A, where the curve downward is more pronounced as load is increased.

12-6 STARTING SHUNT GENERATORS

A shunt generator is driven, or turned, by its prime mover, such as a diesel engine, steam or water turbine, or electric motor. As the generator speed increases, its output builds up to full rated voltage. Occasionally, some generators require several minutes to build up to rated voltage. A field rheostat is used to vary the strength of the fields. Since voltage depends on field strength, the rheostat serves as a means of varying the output voltage.

Standard terminal markings for shunt generators with interpoles are shown in Fig. 12-7. Noninterpole machines use similar markings. Standard rotation for generators is clockwise. If leads are connected by the dashed lines as shown, the generator is rotated clockwise.

12-7 INTERPOLES AND COMPENSATING WINDINGS

Interpoles and compensating windings perform practically the same function in generators as they do in motors. Polarity of interpoles and compensating

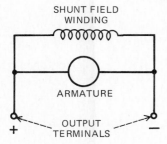

Fig. 12-6 Schematic diagram of a shunt generator.

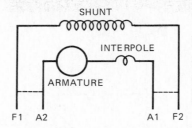

Fig. 12-7 Standard terminal markings for a shunt generator with interpoles. Similar markings are used with noninterpole generators also.

windings is found by the M-G rule, which is discussed in Sec. 8-11.

12-8 BRUSH POSITIONS

Hard neutral position for brushes of generators with interpoles, and working neutral position for brushes of noninterpole generators, are found by the methods discussed for dc motors in Sec. 8-13, but working neutral is slightly in the direction of rotation for generators.

Generators displace the main field magnetism in the *direction of rotation*, while motors displace it *opposite to rotation*.

12-9 SHUNT GENERATORS IN PARALLEL

Shunt generators operate well in parallel since a shunt generator's voltage drops if it attempts to take more than its share of the load. Care must be exercised in starting and stopping shunt generators when operating in parallel.

Two shunt generators connected for parallel operation of bus bars are shown in Fig. 12-8. Each generator is equipped with a field rheostat, a voltmeter, an ammeter, and circuit breakers.

If generator A is supplying the load alone and more energy is needed, generator B will have to be started and connected to the bus. Proper procedure for starting generator B is as follows: Start the prime mover and bring the generator to full speed. Adjust the voltage of generator B with its field rheostat to equal the voltage of generator A. Close the circuit breakers to

the bus. Generator B will now "float" on the line without delivering any current.

The voltage of generator B is raised slightly and the voltage of generator A lowered slightly by the field rheostats until generator B takes its proportionate part of the load as determined by the ammeters.

To stop either generator, the proper procedure is as follows: Adjust the field rheostats of the generator to be stopped until the ammeter reads zero; open the circuit breakers to the bus; and stop the prime mover.

If a generator fails to build up voltage, the probable causes could be that it is rotating in the wrong direction, there is high internal resistance due to the copper oxide film on the commutator, the commutator is dirty, the generator is overloaded, the resistance of the field rheostat is too high, there is an open circuit in the shunt field circuit, the shunt fields have wrong polarity, the fields or armature are grounded or shorted, or there is a loss of residual magnetism.

12-10 PROPER DIRECTION OF ROTATION

A shunt generator starts building up voltage owing to its residual magnetism, which of course is of a set polarity. If a generator is running in the wrong direction, it usually will produce a low voltage of reverse polarity, which will then drop to zero. This is because the initial generated voltage will produce a current flowing in a direction opposite to the normal direction. Thus the magnetism produced in the field will oppose the residual magnetism. As a result the residual magnetism will be canceled and the generator will stop producing voltage.

Depending on the relative strength of residual mag-

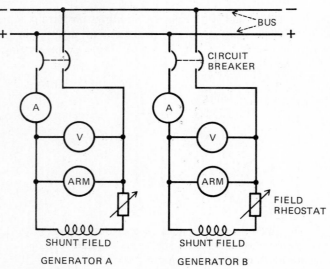

Fig. 12-8 Parallel operation of shunt generators.

netism, field strength, and other factors, a generator running in the wrong direction occasionally will build up voltage of wrong polarity. This will be indicated by the needle of the voltmeter trying to go off-scale to the left or reading negative in the case of digital voltmeters.

Proper connections of a shunt generator for clockwise and counterclockwise rotations are shown in Fig. 12-9. Standard rotation of generators for the purpose of terminal identification is *clockwise* as viewed from the end opposite the prime mover. The commutator end is usually opposite the drive end.

Standard rotation of dc generators is opposite to standard rotation of dc motors. When rotation of a shunt generator is purposely changed, the A1 and A2 leads, which include interpoles, if any, must be interchanged.

12-11 TESTING FOR RESIDUAL MAGNETISM

If rotation is proper and a generator fails to build up voltage, the field rheostat should be adjusted to zero ohms. If this is insufficient, the load should be removed. If the generator still does not build up voltage, the commutator should be cleaned, and a test should be made for copper oxide film resistance on the commutator. It is also important to check for residual magnetism.

A generator sometimes loses its residual magnetism from causes such as idleness, severe vibrations, strong stray magnetic fields, severe electrical and magnetic disturbances just before it was shut down, or reversal of current in its fields.

To test for residual magnetism, the test prods of a voltmeter are touched to the commutator, one near a positive and one near a negative brush with the generator running. If the voltmeter reads zero, it is probable that residual magnetism in the poles has been lost.

If there is a reading during this test, take a reading directly across the brushes. A lower reading (if any) across the brushes means the residual magnetism is too weak or the resistance of the copper oxide film or dirt is too high to permit current to flow in the field windings.

If residual magnetism is lost or too weak, it will have to be restored by "shooting" the fields. To "shoot" the fields of a shunt generator, the fields should be excited from an external source, such as a storage battery, dc arc welder, another generator, or dc bus bars. The positive of the source should be connected to the positive terminal of the generator, with the positive brushes raised or insulated from the commutator so that all current will go through the shunt fields in the proper direction.

12-12 REVERSED RESIDUAL MAGNETISM

Reversed residual magnetism will cause a generator to build up voltage of *wrong polarity*. Proper residual magnetism can be reestablished by shooting the fields in the proper direction.

Reversed residual magnetism can be caused by such severe electrical and magnetic disturbances in a system as short circuits, voltages from other generators, or charging batteries backing up in a system.

12-13 COMPOUND GENERATORS

A compound generator, like a compound motor, contains a *shunt winding* connected *across the armature* and a *series winding* connected in *series with the ar-*

ROTATION			
CLOCKWISE		COUNTERCLOCKWISE	
LINE 1	LINE 2	LINE 1	LINE 2
L1-F1-A2	L2-F2-A1	L1-F1-A1	L2-F2-A2

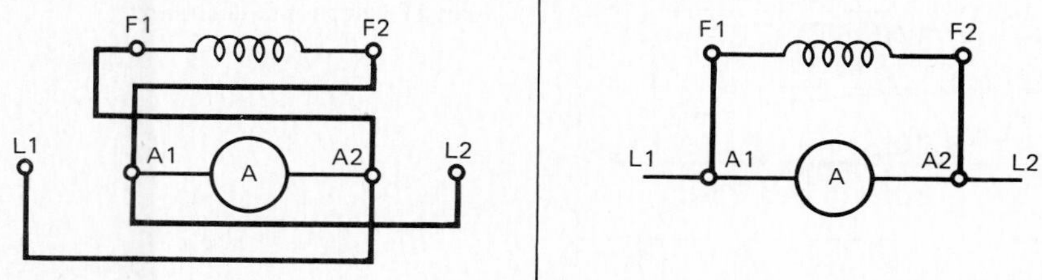

Fig. 12-9 Shunt generator connections for clockwise and counterclockwise rotation.

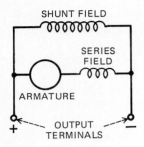

SHUNT FIELD

SERIES FIELD

ARMATURE

OUTPUT TERMINALS

Fig. 12-10 Schematic diagram of a compound generator.

mature. A schematic diagram of these connections is shown in Fig. 12-10. The series winding is *mounted on the same main pole pieces* as the shunt winding.

With this arrangement of windings, a compound generator is capable of delivering practically a constant voltage from no load to full load. The shunt field provides a constant voltage at no load, and the series field, carrying load current, strengthens the main field in proportion to the load to maintain a constant voltage for a varying load.

There are three degrees of compounding commonly available in compound generators: *flat compound, overcompound,* and *undercompound.* Figure 12-11 shows typical voltage-current curves for these generators. The characteristics of these generators from no load to full load are shown.

An overcompound generator has a sufficient number of turns in its series field to produce a voltage increase with additional load. A flat compound generator has sufficient turns in its series field to maintain the same no-load voltage at full load but produces a slightly higher voltage between no load and full load. In an undercompound generator, the voltage decreases slightly as the load increases.

12-14 COMPOUND GENERATORS IN PARALLEL

Compound generators do not operate properly in parallel if both are simply connected to the same lines. If

two compound generators are operating in parallel, and the load in one is reduced, its series field will weaken. As a result, its voltage will drop, further reducing its load. Meantime, the additional load is carried by the other generator, which strengthens its series field. This series of events continues until the loaded generator takes on the entire load of the other generator. It then delivers current to the first generator and drives it as a motor.

Compound generators can be satisfactorily operated in parallel if the *series fields* of the generators are *paralleled.* The conductor used in paralleling series fields is known as an *equalizer.* Two generators, A and B, are shown in Fig. 12-12 connected parallel to a line with an equalizer installed between them. The equalizer is connected between the negative brushes of the generators. This connects the series fields in parallel. If a generator has interpoles, an equalizer is connected between the interpoles and series field, with the interpoles next to the armature.

If for any reason, other than a defect, generator A drops some of its load, generator B will carry the load *equally through both series fields,* since they are in *parallel.* This maintains the strength of the fields of generator A and enables it to resume its share of the load.

Figure 12-12 also shows the typical connections of compound generators with meters, field rheostats, equalizer, and circuit breakers. The procedures in starting compound generators are the same as those for shunt generators. The equalizer circuit breaker is opened when the generators operate singly.

To operate properly in parallel with an equalizer, the degree of compounding or voltage characteristics of the individual compound generators must be practically the same.

12-15 STANDARD TERMINAL MARKINGS

The standard terminal markings for compound generators are given in Fig. 12-13. Leads connected by dashed lines are connected together for clockwise ro-

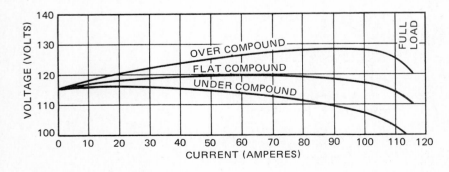

Fig. 12-11 Voltage-current characteristics of compound generators.

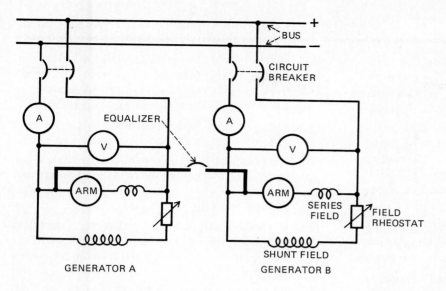

Fig. 12-12 Operating compound generators in parallel using an equalizer connection.

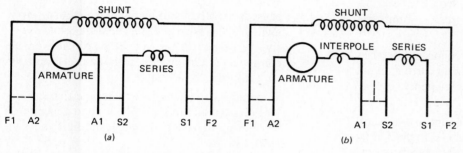

Fig. 12-13 Standard terminal markings for compound generators. (*a*) Without interpoles. (*b*) With interpoles.

tation, which is standard rotation for generators. To reverse rotation of a generator, A1 and A2 are interchanged.

In Fig. 12-13(*b*) the extra dashed line shows where an equalizer is connected to a compound generator with interpoles, that is, between A1 and S2.

The connections for clockwise and counterclockwise rotation of compound generators is shown in Fig. 12-14.

Note that current flow in the armature and series field of a generator is opposite the direction of flow in a motor for the same direction of rotation. A generator delivers current to the positive line through the armature circuit, while a motor draws current from the positive line.

If a compound generator loses its residual magnetism, it is usually restored by shooting the *series* fields in the proper direction.

Fig. 12-14 Connections of a compound generator armature and field windings for clockwise and counterclockwise rotation.

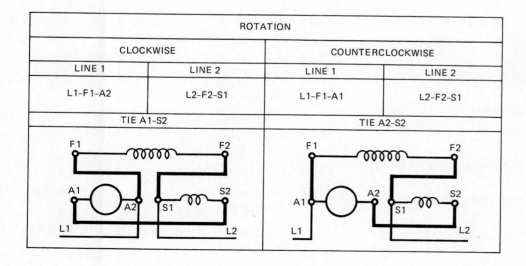

ROTATION			
CLOCKWISE		COUNTERCLOCKWISE	
LINE 1	LINE 2	LINE 1	LINE 2
L1-F1-A2	L2-F2-S1	L1-F1-A1	L2-F2-S1
TIE A1-S2		TIE A2-S2	

SUMMARY

1. Direct-current generators and motors are constructed alike. Standard rotation of generators, for terminal identification purposes, is clockwise. This is the opposite of standard motor rotation.

2. A generator produces current by causing armature conductors to cut magnetic lines of force, produced by the fields. The value of the voltage is determined by the speed, strength of the field, and number of turns in the armature coils between brushes.

3. A single-coil dc generator produces a pulsating direct current. Additional coils and commutator bars in an armature smooth the output waveform.

4. Generators are divided into two classes—self-excited and separately excited. They are further divided into three types—series, shunt, and compound.

5. A series generator has its field connected in series with the armature. The voltage and current output vary directly with the load. Series generators are seldom used.

6. A shunt generator has its field connected in parallel with the armature. Its voltage is highest at no load. The voltage decreases as its load increases.

7. A compound generator contains a shunt field connected in parallel with the armature and a series field connected in series with the armature.

8. There are three common degrees of compounding in a compound generator—overcompound, in which the voltage increases with load; flat compound, in which the voltage is practically steady from no load to full load; and undercompound, in which voltage decreases as the load increases.

9. A dc generator begins generating by cutting lines of force of residual magnetism in the pole pieces.

10. If a generator loses its residual magnetism, it can be restored by exciting the fields from an external source, which is usually a battery, dc arc welder, another generator, or dc power supply.

11. If residual magnetism is reversed, a generator will build up wrong polarity. The remedy for wrong polarity is to shoot the fields in the proper direction.

12. When a generator fails to build up voltage, the cause is usually loss of residual magnetism, overload, high resistance in the field rheostat, excessive resistance in the copper oxide film on the commutator, dirt on the commutator, or an open circuit in the field circuit.

13. To parallel two shunt generators, the first generator is started and loaded. The second generator is brought up to voltage to equal the first and connected to the line. The load is then divided by adjusting the field rheostats of both generators.

14. When compound generators are operated in parallel, they must be connected with an equalizer. An equalizer is connected to the negative brushes. If the generators have interpoles, the interpole is connected at the A1-S2 connections.

QUESTIONS

12-1 In what ways do dc generators differ from dc motors?

12-2 How does a generator produce electricity?

12-3 What is the function of a commutator in a generator?

12-4 What rule is used to find the direction of generated voltage?

12-5 What is a self-excited generator?

12-6 How is the field connected in a series generator?

12-7 How does the load affect voltage output of a series generator?

12-8 How is the field connected in a shunt generator?

12-9 How does the load affect the output voltage of a shunt generator?

12-10 How are the fields connected in a compound generator?

12-11 What are the three degrees of compounding of a compound generator?

12-12 How does the load affect the voltage output of each of the generators in question 12-11?

12-13 Why do shunt generators operate well in parallel?

12-14 Explain how compound generators are connected to operate in parallel.

12-15 How are generators started in parallel?

12-16 What are some common causes of reversed residual magnetism?

12-17 What causes a generator to lose its residual magnetism?

12-18 How is residual magnetism restored in a generator?

12-19 List the common reasons why a generator may fail to build up its voltage.

12-20 How is the voltage output of a generator regulated?

12-21 How does reversed residual magnetism affect output voltage?

12-22 What is the indicated direction of rotation for a generator with F1-A2 connected to one line?

12-23 Explain the difference, in terms of current direction in the armature, between the positive terminal of a generator and the positive terminal of a motor.

12-24 How is the pulsating voltage generated in an armature coil smoothed out in the output voltage of the generator?

12-25 Why doesn't the voltage generated in a series generator continue to build up as more and more current is drawn by the load?

13
ALTERNATING CURRENT

Alternating current is by far the most widely used form of electricity today. Practically all the current generated by power companies is alternating current, and when direct current is needed, it is usually provided by rectifiers, or motor-generator sets.

The main reason for the popularity of alternating current is that it produces an alternating magnetic field. The alternating magnetic field produced in one wire by alternating current can induce a voltage in another wire even though there is no electrical connection between the two wires. This property makes possible the operation of transformers, which can convert an ac voltage to a higher or lower value. It also permits the operation of ac motors which have no commutator or slip rings and no wire windings in their armature. These motors, called squirrel-cage induction motors, are discussed in Chaps. 14 and 15.

Alternating-current generating equipment does not have some of the limitations found in dc generators, since alternating current is induced directly into stationary windings by a constant rotating field, and the load current is conducted through a continuous circuit directly from these windings. Thus load current does not pass through a commutator or brushes.

13-1 GENERATION OF ALTERNATING CURRENT

Alternating current can be generated simply by cutting magnetic lines of force with a conductor. Figure 13-1 demonstrates this principle.

If the ring in Fig. 13-1(a) is moved downward through the magnetic field of a horseshoe magnet as shown, a voltage will be induced clockwise in the ring in the direction of the arrows around the ring.

If the ring is moved upward as shown in Fig. 13-1(b), a voltage will be induced counterclockwise in the ring in the direction of the arrows around the ring.

Movement *downward* induces voltage in one direction, and movement *upward* induces voltage in the reverse direction. Alternate movement down and up will produce a voltage in alternate directions. Since the ring forms a closed circuit, a current will flow each time a voltage is induced. Thus, an alternating current will flow in the ring each time it is moved downward and upward through the magnetic field of the horseshoe magnet.

13-2 DIRECTION OF GENERATED CURRENT

The direction of flow of a generated current is determined by the direction of movement of the conductor and the polarity or direction of flow of magnetic lines of force being cut.

A convenient rule for use in determining direction of flow of generated current is the right-hand generator rule, explained in Sec. 12-1 and illustrated in Fig. 12-1.

The generation of alternating current in a coil containing a single loop connected to two slip rings is illustrated in Fig. 13-2. Brushes are in contact with the slip rings. The loop is connected to an external load through the brushes. One side of the loop is black to identify it. When the coil in Fig. 13-2(a) is rotated counterclockwise about its axis, the white coil side and the black coil side will move through the magnetic field of the poles. A voltage will be induced in the loop. With a circuit through the brushes and load, current will flow in the direction of the arrows through the coil and brushes to the load. Brush A at this point is positive.

After the loop has turned 180°, the white side of the coil will be at the top [Fig. 13-2(b)]. As the coil passes through this position, induced current will flow in the direction indicated again by the arrows. This is opposite to the current flow in the first half turn of the

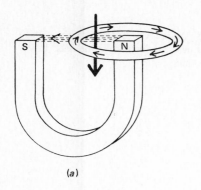 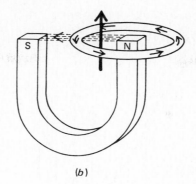

Fig. 13-1 The principle of generating an alternating voltage and current.

coil. Brush B is now positive. The direction of current flow thus alternates in each half turn of the coil, or each time a coil side cuts the magnetism from a pole.

Alternating current and its resultant changing magnetic field produce effects in certain electrical devices and components unlike those in dc circuits. As a result the behavior of ac voltage and current in a circuit differs from that of dc in a circuit.

13-3 SINE WAVE

An alternating voltage or current is graphically represented by a curve known as a *sine wave*. This figure is an aid in the study and calculations of alternating current. The generation of a voltage sine wave is shown in Fig. 13-3.

A conductor is represented in cross section by the dot A, which is rotated clockwise around a center point. The path followed by A is a large circle through the magnetic fields of the poles shown. In order to follow A around the 360° of the circle, its position will be noted at 12 points on the circle. Thus, each point represents 30° of rotation.

The amount of voltage generated in the conductor is

directly related to the number of lines of force it cuts. Notice that as the conductor moves from 0 to 1, it is moving almost parallel to the flux. It cuts across relatively few lines, and, therefore, relatively little voltage is generated in the conductor. As the conductor moves from 2 to 3 and from 3 to 4, it is cutting almost directly across the flux and a much higher voltage is generated. In fact, at 3 the conductor is cutting the flux at a right angle, and at that instant the maximum possible voltage is being generated. As the conductor continues to move around the path, it once again is cutting fewer lines of force and the generated voltage decreases. Thus, the voltage generated in the conductor is directly related to its position around the circular path. No lines of force are cut at positions 0 and 6 (at those points the conductor is moving exactly parallel to the flux lines), and so no voltage is generated at those points. Maximum voltage is generated at 3 since the conductor is cutting the flux at a right angle.

When the conductor moves past point 6 in the clockwise direction, it again cuts flux lines. This time, however, it is cutting the lines from right to left. Using the right-hand generator rule, the voltage is seen to have a polarity opposite to that of the conductor in positions 1 through 5. In fact, if the conductor

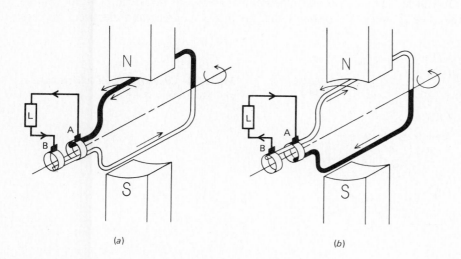

Fig. 13-2 A single-loop generator. (*a*) With the black side of the coil passing under the north pole, brush A is positive. (*b*) With the white side of the coil passing under the north pole, brush A is negative.

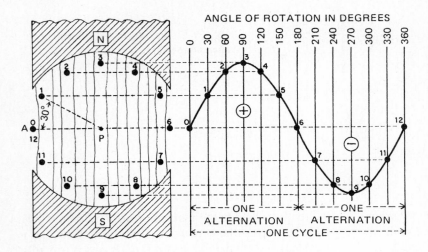

Fig. 13-3 A sine wave represents the voltage generated by a conductor rotating 360° in a magnetic field.

were connected to a complete circuit, current at 5 would appear to be going into the paper while current at 7 would appear to be coming out of the paper.

If the conductor is rotating at a constant speed and the voltage is measured at the instant it passes through each of the points 0 through 11, a series of points on a graph will result. By connecting these points, we will have a sine wave (Fig. 13-3).

13-4 CYCLES AND FREQUENCY

When the conductor in Fig. 13-3 passes through point 12, it has generated two voltage waves called *alternations,* one positive and one negative. These two alternations together are known as *one cycle.* Thus one cycle of alternating voltage is generated when a conductor cuts the flux of a north and a south magnetic pole.

The *frequency* of an ac voltage and current is *the number of cycles that occur per second.* The unit of frequency is the *hertz,* abbreviated Hz, though the old unit, cycles per second (cps), is still widely used. The standard frequency of alternating current in the United

States is 60 Hz. Other frequencies are still in use, though on a very limited basis. For example, 25 Hz is used in electric railways, and some 50-Hz current is used mostly in the western United States, but 60 Hz is considered standard. The 50-Hz frequency is common outside the United States and standard throughout Europe.

13-5 ELECTRICAL DEGREES

When the conductor in Fig. 13-3 made one complete revolution, it traveled through 360°. From geometry we remember that a complete circle contains 360°. The conductor in Fig. 13-3 also passed a north pole and a south pole. In electric motors and generators we say the distance between adjacent north and south poles is 180°. Thus when the conductor passed from north to south and back to north it covered 360 *electrical degrees.* Geometric degrees are not always the same as electrical degrees. In Fig. 13-4(*a*), we see that in a two-pole machine, geometric degrees and electrical degrees are the same. Figure 13-4(*b*) shows that geometric degrees and electrical degrees

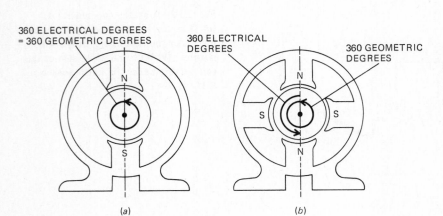

Fig. 13-4 Comparison of electrical degrees and geometric degrees.

are not the same in a four-pole machine. In the case of the four-pole machine, a conductor has to rotate through only half a circle (180°) to pass a complete north pole and an adjacent south pole. The formula for finding the number of electrical degrees in one complete revolution of a conductor is

$$\text{Electrical degrees} = 360 \times PP$$

where *PP* is the number of pairs of north-south poles in the machine. A two-pole machine has one pair of poles, a four-pole machine has two pairs of poles, and so forth.

The number of electrical degrees in one complete revolution is important since one cycle of alternating voltage is generated for each 360 electrical degrees (Fig. 13-3). In making one complete revolution of a four-pole machine, a conductor will generate two complete cycles of voltage. If the conductor is rotating at a speed of 1800 revolutions per minute (rpm), it will produce 1800×2 cycles each minute or 3600 cycles per minute. Since there are 60 s in a minute, we can convert 3600 cycles per minute to a frequency measured in cycles per second (that is, hertz).

$$\text{Frequency (hertz)} = \frac{3600 \text{ cycles per minute}}{60 \text{ seconds per minute}}$$
$$= 60 \text{ Hz}$$

In order to produce a constant 60-Hz frequency, a four-pole ac generator must be turned at a constant speed of 1800 rpm.

The formula for calculating the speed at which an ac generator must be turned in order to produce any given frequency is

$$\text{Speed (rpm)} = \frac{\text{frequency (hertz)} \times 60}{PP}$$

To find the speed of a six-pole machine (three pairs of poles) generating 60 Hz, substitute the given values in the formula

$$\text{Speed} = \frac{60 \times 60}{3} = 1200 \text{ rpm}$$

13-6 AC VOLTAGE

The value of an ac voltage varies constantly during a cycle. It begins at zero, rises to some maximum or peak value, decreases to zero, reverses its polarity, rises to its peak value again but with a reversed polarity, then decreases to zero and begins a new cycle all over again. With this changing voltage, it is necessary to find some single value to describe or specify ac voltage. The value used is called the *effective value* (Fig. 13-5). The effective value of an ac voltage is the value that produces the *heating* effect of a dc voltage of the same value. Whenever we speak of an ac voltage, it is understood we mean the effective value. The ac measuring instruments such as voltmeters and ammeters read in effective volts and amperes. The peak value of the ac sine wave is always higher than the effective value. In some applications, for example, control and protection circuits, the peak value of an ac voltage or current is important. To calculate peak values from the value of voltage or current measured, the following formula is used:

$$\text{Peak voltage} = \text{effective voltage} \times 1.414$$

In a 208-V ac system, the peak voltage reached on each half-cycle is

$$\text{Peak voltage} = 208 \times 1.414 = 294 \text{ V}$$

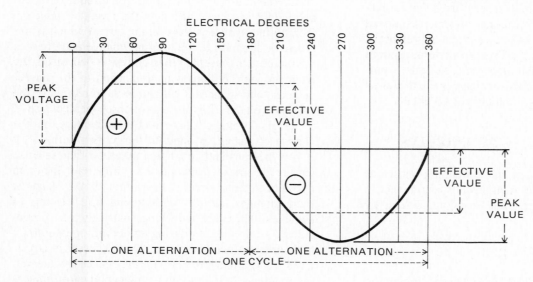

Fig. 13-5 Sine wave of an alternating voltage showing its peak (maximum) and effective values.

Some equipment, especially solid-state electronic controls, are affected by the peak-to-peak voltage. This is the sum of the positive peak and the negative peak. Since these two values are equal in the usual ac sine wave, the peak-to-peak value is *twice* the peak voltage found by the above formula. In the example given above, the peak-to-peak voltage in a 208-V ac system is

$$\text{Peak-to-peak voltage} = \text{Peak voltage} \times 2$$
$$= 294 \times 2$$
$$= 588 \text{ V}$$

that is, almost 600 V!

Another name for effective voltage is *root mean square*, or *rms*, *voltage*. The name comes from the mathematical method used to calculate effective values.

13-7 AC CURRENT

The discussion of effective voltage and peak voltage applies equally to ac current. To find peak current, use the formula

$$\text{Peak current} = \text{effective current} \times 1.414$$

Remember that most ac instruments read effective values. When an ammeter measures 10 A in a circuit, the current is actually reaching a peak of 10×1.414 or 14.14 A. The heating effect of this current, however, is the same as that produced by a dc current of 10 A.

13-8 AC TERMINOLOGY

Notice the apparent redundant term used for ac current. Since ac means alternating current, ac current becomes alternating current *current* and ac voltage is alternating current *voltage*. The proper terms should be *alternating current* and *alternating voltage*. Common usage in the field, however, have made the terms ac current and ac voltage completely acceptable.

13-9 RESISTANCE IN AC CIRCUITS

Direct current flows in a circuit containing resistance in accordance with the relationships expressed in Ohm's law. That is, current equals voltage divided by resistance.

$$I = \frac{V}{R}$$

Alternating current flows in a circuit containing only resistance in a similar way: The effective current is equal to the effective voltage divided by the resistance. The effectiveness of alternating current or direct current is therefore the same in resistance circuits, such as those containing incandescent lamps, coffee makers, toasters, ranges, soldering irons, room heaters, and other heat-producing equipment. All ac circuits also contain devices that have properties of inductance and capacitance. These devices do not behave the same in alternating as they do in direct current.

13-10 INDUCTANCE

Magnetism produced by a current changes with the current. If the current increases, magnetism increases. If the current decreases, magnetism decreases.

When current changes, the changing magnetic field cuts across adjacent conductors and generates a voltage in them. This property of generating a voltage by a change in current is known as *induction*.

The voltage produced by induction always opposes, or is opposite in polarity to, the voltage that produced the induction. Therefore, induction can be said to produce a *counterelectromotive force* (or *cemf*). When a conductor produces a cemf in itself, it is known as *mutual induction*. Conductors or devices that produce induction are said to have the property of *inductance*. The unit of inductance is the *henry*.

Inductance is present in dc and ac circuits. It is constantly in effect in all ac circuits, and it is more pronounced in coils of wire than in straight conductors. A device whose primary function is to add inductance to a circuit is called an *inductor*.

Inductance opposes any change of current in a circuit. When current is increased in a circuit, the magnetic field increases and cuts the circuit conductor, thereby inducing a voltage in the circuit which is opposite in direction to the original voltage and opposes the increase in current. When current is decreased, the magnetic field collapses in proportion to the current decrease and induces a voltage in the direction of the original voltage. This voltage opposes the decrease of the original voltage. This effect is momentary, but the change of current is opposed during this period.

It is the constantly changing magnetic field in an ac circuit that makes inductance an important factor to consider in these circuits. In dc circuits, the magnetic field changes during the opening and closing of switches. Therefore inductance must always be considered in dc circuits during switching operations.

Degrees of inductance are always present in any ac circuit. Even in a straight wire the constantly alternating magnetic field, cutting its own conductor, induces a voltage in the conductor which is opposite the original voltage. Inductance is practically negligible on short runs of straight wire in air, but it is a factor that must be considered in designing long transmission lines.

When wire is wound in a coil, inductance is increased. If an iron core is inserted through the center of the coil, the inductance is increased even further.

An examination of Fig. 13-6(a) will help explain the principle of inductance. A coil of wire is shown carrying an alternating current and producing an alternating magnetic field. All along the wire the magnetic field is expanding and collapsing, inducing a voltage in the wires in the field. The direction of the induced voltage is opposite to that of the original voltage.

If the turns of wire are widely spaced and wound around an air core, the effect is practically negligible, but where the sides of the wire are close to each other, the effect is more pronounced. The flux from each wire, in addition to cutting its own conductor, cuts the adjacent conductor, inducing a cemf in each. Thus the greater the number of turns in a coil, and the closer the spacing, the greater is the induced cemf.

In Fig. 13-6(b) an iron core is placed in the center of the coil in Fig. 13-6(a). Here the magnetism is concentrated because of the iron core, and the effective intensity of inductance has been greatly increased. Inductance in an ac coil is determined by the number of turns in the coil as well as other physical factors.

13-11   INDUCTIVE REACTANCE

The current in an ac circuit containing inductance is opposed by the cemf of self-induction. This opposition to current flow is called *inductive reactance*. Inductive reactance is measured in *ohms* since it has a limiting effect on the flow of current similar to resistance.

Since inductive reactance depends on the changing magnetic field, the rate at which the field changes affects its value. Inductive reactance can be found using the following formula:

$$X_L = 6.28 \times f \times L$$

where X_L = inductive reactance, ohms (Ω)
6.28 = constant
f = frequency, hertz (Hz)
L = inductance, henrys (H)

The inductance of a coil is constant for that coil and depends on the physical characteristics of the winding and core. Since most power and lighting circuits use 60 Hz, the above formula can be simplified to

$$X_L(60) = 377 \, L$$

where $X_L(60)$ equals inductive reactance at 60 Hz.

Problem 1

A coil has an inductance of 150 mH. The coil is connected to a 120-V ac, 60-Hz, line. What current will flow through the coil?

Solution

The circuit is shown in Fig. 13-7. Notice that the inductance value is 150 mH. The first step then is to convert this value in millihenrys to henrys. Since 1 mH = 0.001 H, 150 mH = 0.001 × 150 H = 0.15 H. In order to solve the circuit, we must find the inductive reactance.

$$X_L(60) = 377 \times L$$
$$= 377 \times 0.15$$
$$= 56.6 \, \Omega$$

In the problem given, $V = 2.12 \times 56.6 = 120$ V (which, of course, was the voltage applied across the coil originally).

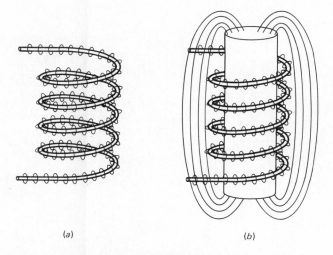

(a)   (b)

Fig. 13-6   Principle of self-inductance.

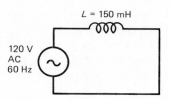

Fig. 13-7   A circuit with an inductance coil.

Ohm's law, with a slight change in symbols, can now be used to find the current in the circuit.

$$I = \frac{V}{X_L}$$
$$= \frac{120 \text{ V}}{56.6 \text{ }\Omega}$$
$$= 2.12 \text{ A}$$

The only difference in Ohm's law is the substitution of X_L for R. Thus the voltage across the coil can be found using the formula

$$V = I \times X_L$$

13-12 PHASE SHIFT IN INDUCTANCE

Inductive reactance produces a shift between voltage and current sine waves.

In a circuit containing only resistance, the voltage sine wave and the current sine wave coincide. That is, when the voltage reaches its peak value, the current also reaches its peak value. When the voltage is zero, the current is zero, and so on. The voltage and current are said to be *in phase* in a circuit containing only resistance.

In an inductive circuit, that is, a circuit containing inductance, the voltage and current do not peak at the same time nor do they pass through zero at the same time. Voltage and current are said to be *out of phase*. In relation to time, the current reaches its peak slightly later than the voltage does. In fact, each point on the current wave is reached slightly later than the relative point on the voltage wave. We therefore say that current *lags* the voltage in an inductive circuit.

At the instant the switch in Fig. 13-8(*a*) is closed, the inductance L acts as a short circuit. The inrush of current is limited only by the resistance R in the circuit. With $X_L = 0$, the voltage drop across L is therefore zero. This is shown by the 0° point in Fig. 13-8(*b*). The current through L begins developing a magnetic field that produces a cemf in L, and the voltage across L, that is, V_L, starts to increase. Its polarity, however, is opposite that of the current flowing through L, and the effect is to decrease I in the circuit. When the voltage has reached its peak value, the current will be at zero and beginning the positive half of its first cycle. Relatively speaking, we say that the current is lagging by 90°. As the cycles continue, the phase relationship between voltage and current will remain the same.

Current cannot flow without voltage. But two voltages are present in an inductive circuit. By way of explanation, it may be said that the original line volt-

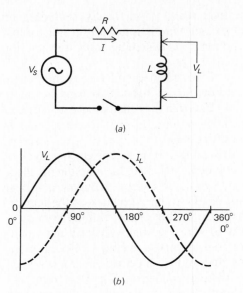

(a)

(b)

Fig. 13-8 Current lags voltage in an inductive circuit.

age uses part of its energy to build up a magnetic field. In collapsing, this magnetic field induces a second voltage in the circuit that causes current to flow after the original voltage has decreased. This can be more easily understood by studying the volt-ampere relationship in Fig. 13-8(*b*).

13-13 CAPACITANCE

Imagine two large sheet metal plates facing one another, close but not touching. Then connect each of the plates to a terminal of a battery such as in Fig. 13-9. The electrons from the negative terminal of

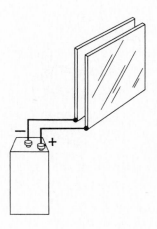

Fig. 13-9 A capacitor connected to a dc supply.

the battery will collect on one of the plates, while the electrons of the other plate will be drawn to the positive terminal of the battery. An electric field is now present between the plates, created by the negatively charged plate (where the electrons collected) and the positively charged plate (from which the electrons were drawn). The combination of the two conductors separated by a nonconductor is called a capacitor, and its effect on an electric circuit is called *capacitance*. In the above case the conductors were metal plates and the nonconductor (also called a *dielectric*) was air. Conductors can also be long strips of metal foil, chemicals, thin wires, and the like. Dielectrics can also be paper, plastic, oil, and the like.

The brief time during which the electrons are moving to one plate and leaving the other is called the *charging* period. Once the electrons have collected on one plate and left the other, we say the capacitor is *charged*. The size of the conducting surface and the type and size of the dielectric determine the amount of charge (or the capacity) of the capacitor.

If the battery were removed and the two plates short-circuited, the electrons would leave the plate that had an excess and move to the plate that had a deficiency of electrons. The electrons would continue to move until they were in balance at the two plates. At that point the capacitor would be *discharged*.

In effect a capacitor is a storage device. It will store an imbalance of electrons and thus a potential difference (voltage) for long periods of time. The voltage across the terminals of a charged capacitor is that of the source that charged the capacitor. For example, if a 120-V dc supply were used to charge a capacitor, the 120 V would hold across the capacitor terminals until it were discharged.

Capacitors are widely used in electronics and power systems. In power systems capacitors are commonly used to prevent sparking at contacts, increase the efficiency of circuits containing inductance, start motors, and smooth pulsating or irregular voltage and current waveforms.

The unit for capacitance is the farad, abbreviated F. In electronics very low value capacitors are used, in the order of several or hundreds of *microfarads* (μF). Large-power capacitors are generally specified in terms of the power they will safely handle. Capacitors must also be specified by the maximum voltage to which they will be connected.

13-14 CAPACITIVE REACTANCE

In a dc circuit a capacitor will allow current to flow only until the capacitor is charged. Once the voltage across the capacitor is equal to the applied voltage, current will cease to flow. Because of the periodic reversals of polarity in an ac circuit, a capacitor in an ac circuit will periodically charge, discharge, and recharge with opposite polarity. The result is that a capacitor allows current to flow continuously in an ac circuit. The constant charging and discharging, however, limits the amount of current that can flow. The effect is similar to that of inductive reactance. In the case of circuits containing capacitors, the impeding factor is known as *capacitive reactance*. Capacitive reactance is measured in ohms, and its behavior in an ac circuit obeys Ohm's law.

Capacitive reactance can be found using the following formula:

$$X_C = \frac{1}{6.28 \times f \times C}$$
$$= \frac{0.159}{f \times C}$$

where X_C = capacitive reactance, Ω
0.159 and 6.28 = constants
f = frequency, Hz
C = capacitance, F

In 60-Hz circuits this formula becomes

$$X_C(60) = \frac{0.159}{60 \times C}$$
$$= \frac{0.00265}{C}$$

where $X_C(60)$ is capacitive reactance at 60 Hz.

Problem 2

What is the capacitive reactance of a capacitor rated at 1000 μF and connected to a 60-Hz supply?

Solution

Since C is given in microfarads, it must be converted to farads.

$$1 \ \mu F = \frac{1}{1,000,000} \ F$$

$$1000 \ \mu F = \frac{1000}{1,000,000} = \frac{1}{1000} \ F \text{ or } 0.001 \ F$$

$$X_C = \frac{0.00265}{C} = \frac{0.00265}{0.001} = 2.65 \ \Omega$$

13-15 PHASE SHIFT IN CAPACITANCE

Like inductive reactance, capacitive reactance causes the voltage and current to be out of phase. However, in the case of capacitive reactance, the current *leads* the voltage. In fact, capacitors are often used in ac circuits because of their property of producing leading current.

In Fig. 13-10(a) a capacitor is connected in parallel across a load. The sine wave marked V in Fig. 13-10(c) represents the voltage across the load and the capacitor. If the load were pure resistance, the current through the load would be in phase with the voltage. At the same time, the current in the capacitor branch would be out of phase with the voltage and leading by 90°. The sine wave marked I in Fig. 13-10(c) is the current in the capacitor branch. The reason the current and voltage are out of phase is as follows. In Fig. 13-10(a) the top terminal of the line is negative and the capacitor is fully charged as shown. This represents the condition at 270° in Fig. 13-10(c). The current in the capacitor branch is zero because it is fully charged and its voltage is equal and opposed to the applied voltage. As the voltage in the top line becomes less negative, the capacitor voltage will be greater than the line voltage and the capacitor will begin discharging. That is, current will begin to flow in the capacitor branch. When the line voltage is zero, nothing will oppose the capacitor voltage and the current in the capacitor branch will reach its maximum value. This is represented by the 360° point in Fig. 13-10(c). Of course the 360° point represents both the

end of one cycle and the beginning (0°) of the next cycle. At 0° the voltage begins increasing in the same direction as the capacitor current until at 90° the line voltage has reached its peak in the positive direction. As the voltage increased, it charged the capacitor up to the applied voltage. This condition is shown in Fig. 13-10(b). Between 90° and 180° the capacitor is discharging. At the same time the line voltage is decreasing until it reaches zero at 180°. At the 180° point, the voltage reverses itself and has the same polarity as the discharging capacitor. This causes the capacitor to charge in the opposite direction until the voltage across the capacitor is equal, and opposite, to the line voltage. This brings it back to the condition at the 270° point from which the process repeats itself.

Since any two conductors separated by an insulator (including air) form a capacitor, most ac circuits contain capacitive reactance. In most cases its effect is negligible, but in long feeder runs and in transmission and distribution lines it could cause problems.

13-16 IMPEDANCE

The combined current-limiting effect of resistance and reactance in an ac circuit is known as *impedance*. Impedance is symbolized with the letter Z. It is measured in ohms. In an ac circuit current, voltage, and impedance obey Ohm's law in the following relationships:

$$Z = \frac{V}{I}$$

$$V = IZ$$

$$I = \frac{V}{Z}$$

The current/voltage phase relationships differ in resistances, capacitive reactances, and inductive reactances. For this reason we cannot simply add reactances to resistances to obtain the total impedance. However, capacitive and inductive reactances can be combined by merely finding the difference between their values. For example, the total reactance in a circuit that has 20-Ω inductive reactance in series with 15-Ω capacitive reactance is 20 − 15 = 5 Ω. Since the inductive reactance was greater than the capacitive reactance, the net effect is 5 Ω inductive reactance. If there was also series resistance in the circuit, it would have to be combined with the reactance to find the total impedance of the circuit. The formula for finding impedance in a series circuit is

$$Z = \sqrt{R^2 + X^2}$$

where Z = impedance
R = resistance
X = difference between X_L and X_C

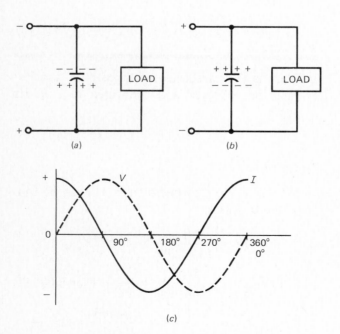

(a)   (b)

(c)

Fig. 13-10 Current leads voltage in a capacitive circuit.

Problem 3

Find the current in a series circuit containing a coil with 10-Ω inductive reactance, a capacitor with 7-Ω capacitive reactance, and a 4-Ω resistor. The applied voltage is 24 V ac.

Solution

The circuit is shown in Fig. 13-11. The net reactance in this circuit is

$$X = X_L - X_C$$
$$= 10 - 7$$
$$= 3\text{-}\Omega \text{ inductive reactance}$$
$$Z = \sqrt{R^2 + X^2}$$
$$= \sqrt{4^2 + 3^2} = \sqrt{16 + 9}$$
$$= \sqrt{25} = 5 \ \Omega$$

To find the current, use Ohm's law

$$I = \frac{V}{Z} = \frac{24}{5}$$
$$= 4.8 \text{ A}$$

Since the series circuit in this problem contained a net inductive reactance, the total current in the circuit will lag the voltage. However, the interesting thing about

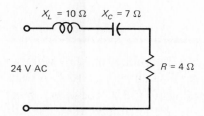

Fig. 13-11 A series circuit containing inductance, capacitance, and resistance.

phase relationships in ac circuits is that they can vary throughout the same circuit.

In the series circuit in the above problem the total current lagged the voltage. The current through the resistor, however, is in phase with the voltage across the resistor. The same *value* of current passing through the capacitor shifts and leads the voltage across the capacitor by 90°. Finally the same value of current will shift again as it passes through the coil. This time it will lag the voltage by 90° (Fig. 13-12).

13-17 POWER IN AC CIRCUITS

From the discussion of inductive and capacitive reactance, it can be seen that voltage and current are not always in phase. Since all ac circuits contain some reactance, the voltage and current are seldom exactly in phase. Under this condition the voltage and current will not reach their peak values at the same time. Thus we could not merely multiply the current by the voltage to obtain power in watts. The relationship $P = VI$ is true only when the voltage and current are in phase, such as across a resistor. If we placed a voltmeter and an ammeter in a circuit, they would read effective values but would not indicate phase relationships. Thus we cannot simply use voltmeter and ammeter readings to find ac power. But volts multiplied by amperes do provide some measure of behavior of an ac circuit so that the product is a useful figure. Volt-amperes, or more commonly VA, give some indication of the capacity of equipment to deliver current. For that reason it is often called apparent power, P_{app}. Some equipment with significant reactance can deliver more current than would be indicated by its power rating. Since current and voltage are always in phase across resistance, power in ac circuits can still be calculated using the formula

$$P = I^2R$$

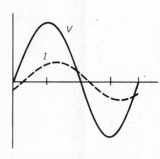

(a) LINE CURRENT LAGS LINE VOLTAGE

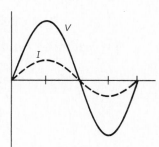

(b) CURRENT THROUGH RESISTOR IN PHASE WITH VOLTAGE ACROSS RESISTOR

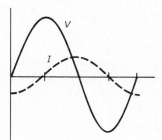

(c) CURRENT THROUGH INDUCTANCE LAGS VOLTAGE ACROSS INDUCTANCE BY 90°

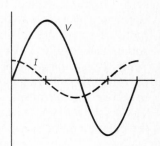

(d) CURRENT THROUGH CAPACITANCE LEADS VOLTAGE ACROSS CAPACITANCE BY 90°

Fig. 13-12 The current-voltage phase relationships for the circuit in Fig. 13-11.

Problem 4

Find the volt-amperes taken by the circuit in Fig. 13-11. What is the power consumed by X_L? What is the power consumed by X_C? What is the total power consumed by this circuit?

Solution

The volt-amperes taken by the circuit is $V \times I$. The voltage is 24 V. The current may be found as in the previous problem.

$$X = X_L - X_C = 10 - 7 = 3 \ \Omega$$
$$Z = \sqrt{R^2 + X^2}$$
$$= \sqrt{4^2 + 3^2} = \sqrt{25}$$
$$= 5 \ \Omega$$
$$I = \frac{V}{Z} = \frac{24}{5} = 4.8 \ A$$
$$P_{app} = 24 \times 4.8 = 115.2 \ VA$$

Since the voltage and current in X_L are 90° out of phase, no net power is consumed in X_L. Similarly, no power is consumed in X_C. The only power consumed in this circuit is the power consumed in the resistance.

$$P = I^2 R$$
$$= 4.8^2 \times 4$$
$$= 92.16 \ W$$

The important thing to remember about power in ac circuits is that only the resistive parts of the circuit consume power. Inductive reactance takes power from the source to build its magnetic field, but then when the magnetic field is collapsing, it returns the power to the source. The effect is that no net power from the source is consumed. Similarly capacitive reactance takes power from the source to charge its plates. During discharge the power is returned to the source. Again the effect is that no net power is consumed from the source. On the other hand resistance converts power to heat and the heat is lost. Thus the power used by the resistance is not returned to the source and there is a net consumption of power.

Many large-capacity ac circuits involve thousands of volts (kilovolts) and thousands of amperes. It is therefore common to speak of kilowatts (kW) and kilovolt-amperes (kVA) when specifying such circuits or the equipment that serves them.

13-18 POWER FACTOR

Energy consumption is related to power. Thus the electric utility bases its rates on watts or kilowatts.

But we saw from the preceding section that not all current is converted to power. We saw that kilowatts are not necessarily equal to kilovolt-amperes. These two units, however, are related by a ratio called *power factor* (*PF*). Power factor is an indication of the portion of volt-amperes that is actually power. To find power factor, the following formula can be used:

$$PF = \frac{\text{power}}{\text{volt-amperes}}$$
$$= \frac{P}{P_{app}}$$

The product of voltage and current is called *apparent power*. Wattage represents *true power*. The product of the voltage across a reactance and the current through the reactance is called *reactive power*.

Power factor is therefore true power divided by apparent power. It is commonly expressed as a percentage. When true power is equal to apparent power—such as in a circuit having only resistance or one in which the reactances exactly cancel one another—the power factor is equal to 1.00 or 100 percent. (To obtain percentage from the power factor ratio formula, multiply the answer by 100.)

Reactances consume no true power, and so the power factor of capacitance and inductance is 0 percent.

Problem 5

A voltmeter reads a line voltage of 200 V while an ammeter reads the line current of 10 A. A wattmeter gives a reading of 1800 W true power. What is the power factor of the circuit?

Solution

The apparent power is the product of the voltmeter reading and the ammeter reading.

$$P_{app} = 200 \times 10 = 2000 \ VA$$

The wattmeter reads 1800 W true power. Thus

$$PF = \frac{P}{P_{app}}$$
$$= \frac{1800}{2000}$$
$$= 0.9 \ \text{or} \ 90\%$$

A power factor less than 100 percent indicates that not all the current is doing useful work. Yet the conductors must be sized to carry the total current, and the capacity of the equipment in the system such as generators, transformers, switches, circuit breakers, etc., must be capable of handling this total current. Lower power factor thus puts an added strain on an electrical system while not contributing its share of work to the system.

Most electrical systems have a lagging power factor, meaning that the power factor is less than 100 percent owing to motors, coils, fluorescent lighting ballasts, etc. When the power factor of an entire system drops below 80 percent, measures must be taken to increase the power factor. Since low power factor makes increased demands on the power company without consuming power (on which the utility bases its bill), power companies often charge penalty fees for customers having excessively low power factors (below 70 percent lagging, for example).

At the engineering and design stages of an electrical installation, high power factor equipment may be specified wherever practical. Once a system has been designed, and in many cases installed, a low power factor condition may be unavoidable. In such cases, power factor correction equipment must be installed.

In most cases a low lagging power factor is corrected by adding capacitors across the main feeders or across the equipment feeders. Capacitors produce a leading power factor which partially neutralizes the lagging power factor. Capacitors used for power factor correction are usually large devices specified according to their operating voltage and their volt-ampere or kilovolt-ampere rating. Since capacitors act as storage devices, they present a serious shock hazard even when disconnected from a circuit. Before servicing or even handling a capacitor, make sure power to the capacitor has been disconnected. Then short the terminals of the capacitor with a heavy copper strap or bar attached to a wood or plastic pole. Never use a screwdriver or other tool. Wear protective eye covering. Heavy industrial capacitors used for power factor correction when shorted can produce very hot sparks capable of melting small wires and pitting tools.

Lagging power factor can also be corrected by using rotating equipment called *synchronous motors*. By varying the dc field of such motors, the power factor characteristics of the motor can be changed. If the field is adjusted to produce a leading power factor motor, the motor can be connected across the line to correct an existing lagging power factor condition.

These synchronous motors can be connected to loads so that they can be turning a load while also correcting power factor. Synchronous motors work best with constant load. They stall easily should the load suddenly increase. In addition, these motors have no starting torque and must be provided with special

windings or devices to make them self-starting. Like all rotating equipment they require periodic maintenance and lubrication. Synchronous motors are frequently used in pumping systems which are run continuously at a constant rate.

13-19 SINGLE-PHASE SYSTEMS

In dc systems, equipment is fed from two conductors, one positive (+) and the other negative (−). The negative path is sometimes ground (such as in the automobile electrical system), but there are always two paths going back to the source of power. In ac systems some equipment—but not all equipment—needs two paths also. Of course in ac systems one path will be positive and the other negative in one instant, and in the next instant their polarities will reverse. But there will be only two paths.

Such ac systems are called *single-phase* systems. Only one voltage wave and one current wave exist in single-phase systems (Fig. 13-13). Almost all residential wiring uses single-phase alternating current. The symbol for phase is ϕ (the Greek letter *phi*), and so on electrical drawings single phase is often abbreviated as 1ϕ.

Single phase is not usually generated directly. More often it is part of a system that generates more than one voltage wave. These are called *polyphase* systems. (*Poly* means "many.")

In the United States single-phase systems are usually rated at 110 to 125 V and 220 to 240 V. The most common voltages are 110, 115, 120, 220, 230, and 240 V at 60 Hz.

13-20 THREE-PHASE SYSTEMS

As noted above, a single-phase system is generally part of a polyphase system. In the United States two-

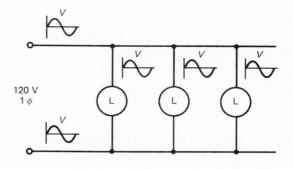

Fig. 13-13 A single-phase system serving a lighting load. Notice that the ac voltage wave is exactly the same throughout the system.

phase, three-phase, and six-phase systems are used. However, the most widely used polyphase system is three-phase. Almost all generation, transmission, and distribution systems used by utilities are three-phase.

Three-phase systems use three lines also called *phase legs* or *lines*. Unlike single-phase systems, the time relationship of the voltage wave in each line differs. Figure 13-14 shows a typical three-phase system serving a three-phase load. In this system a fourth wire, called the *neutral,* is used. The phase legs are denoted with the letters A, B, and C and the neutral with N. In a *balanced* system, the value of the voltages between A and B, B and C, and C and A are exactly equal. In Fig. 13-14 these voltages are 208 V.

$$V_{AB} = 208 \text{ V}$$
$$V_{BC} = 208 \text{ V}$$
$$V_{CA} = 208 \text{ V}$$

However, the waves in each leg are shifted so that the peak value is not reached at the same instant in each leg. In fact, in three-phase systems there is a shift of 120 electrical degrees between the waves in each leg. For example, in phase leg A the wave is shown starting at 0; in leg B the voltage at that instant is negative and increasing toward the negative peak value; in leg C at that instant the voltage is positive and decreasing toward 0. The three-phase loads are wired in such a way as to properly use these voltages.

The three-phase system shown in Fig. 13-14 is very common and is the standard system supplied by many electric utilities to commercial and industrial customers. It is very versatile since within the four-wire system shown it can deliver both single phase and three phase and both 208 V and 120 V without the use of a transformer.

Single phase may be obtained from this system by connecting a two-wire (single-phase) load across any two-phase legs. The voltage in that case will be 208 V, 1ϕ. Single phase may also be obtained by connecting a two-wire load between any phase leg and the neutral. In this case, the single-phase voltage will be 120 V, 1ϕ. The reason for this lies in the way the voltage wave shifts in each phase leg. This is discussed further in the chapter on transformers (Chap. 14).

The voltage between phase lines is usually referred to as the *line voltage,* while the voltage between any line and the neutral is referred to as the *phase voltage.* The system in Fig. 13-14 has a line voltage of 208 V and a phase voltage of 120 V.

The popularity of three-phase systems is due in part to its efficiency in the transmission and distribution of power. In terms of the conductors necessary to deliver a given amount of power, three-phase systems can deliver 1.73 more power for the same amount of wire than a single-phase system.

In addition to its advantage in delivering power, three-phase systems also permit the design of more efficient and dependable equipment. Three-phase motors, for example, have fewer starting problems and are more efficient than comparable single-phase motors. For that reason, larger-size ac motors are available only in three phase. Three-phase transformers can be made compact and versatile in that they can be connected to deliver both three-phase and single-phase power.

The three-phase, four-wire system shown in Fig. 13-14 is interesting because of the neutral conductor. Recall that 120 V single-phase loads may be served by connecting them between a phase leg and the neutral. The current path will then be the line and the neutral conductor. If the loads connected between each line and the neutral are equal, that is, if the system is *balanced,* no current will flow in the neutral conductor. The reason for this is the phase shift in voltage and current in each line. When these currents flow in the same line (the neutral), their sum *at every instant* is zero. An ammeter in each line would measure the effective value of current in each line (the ammeter cannot recognize any phase angle difference between lines). However, an ammeter placed in the neutral line will read zero since only the combined effect of the current will be registered on the meter.

Because of this property of three-phase four-wire systems, the National Electrical Code permits a reduction in the size of the neutral wire under certain conditions.

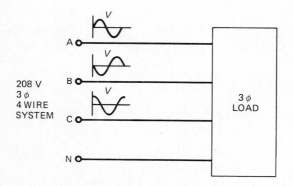

Fig. 13-14 **A three-phase system serving a three-phase load. Notice that the ac voltage wave is displaced in each phase leg.**

SUMMARY

1. Alternating current is the most widely used form of electricity for residential, commercial, and industrial electrical systems.

2. An ac voltage is produced by a conductor rotating in a magnetic field.

3. The right-hand generator rule is used to determine the polarity of an induced voltage and current. The thumb points in the direction of movement of the conductor, the index finger points in the direction of flux (north to south pole), and the middle finger points in the direction of current flow.

4. The periodic changes in value of an ac voltage and current can be represented by a sine wave.

5. The value of voltage induced in a conductor is related to the strength of the magnetic field and the speed with which the conductor cuts across the field.

6. One complete sine wave going through 360 electrical degrees is equal to one cycle.

7. The number of cycles occurring in 1 s is called frequency.

8. The unit of frequency is the hertz (Hz). The standard frequency used in the United States is 60 Hz. Other common frequencies are 50 Hz (used widely in Europe) and 25 Hz (used in some electrical railway systems and other industrial applications).

9. The highest voltage and current values of a sine wave are called maximum or peak values. The single value of voltage and current used to describe a varying ac wave is called the effective value.

10. Alternating-current systems are usually specified in terms of their effective values of voltage and current; the common ac ammeters and voltmeters measure effective values.

11. Although ac stands for alternating current, the terms ac voltage and ac current are acceptable and are in general use.

12. Resistance affects ac circuits in exactly the same way that it does dc circuits. Ohm's law applies to ac circuits containing resistance.

13. Conductors or devices that produce induction have inductance. The unit of inductance is the henry.

14. The opposition to ac current caused by inductance is called inductive reactance. The symbol for inductive reactance is X_L. The unit of inductive reactance is the ohm.

15. Inductive reactance can be found using the formula

$$X_L = 6.28\, fL$$

where X_L = inductive reactance, Ω
   6.28 = constant
   f = frequency, Hz
   L = inductance, H

16. If the frequency is 60 Hz, the formula for finding inductive reactance becomes

$$X_L\,(60) = 377\, L$$

17. A capacitor in an ac circuit produces capacitance. The unit of capacitance is the farad. The opposition to ac current caused by capacitance is called capacitive reactance. The symbol for capacitive reactance is X_C. The unit of capacitive reactance is the ohm.

18. Capacitive reactance can be found using the formula

$$X_C = \frac{0.159}{f\,C}$$

where X_C = capacitive reactance, Ω
   0.159 = constant
   f = frequency, Hz
   C = capacitance, F

19. If the frequency is 60 Hz, the formula for capacitive reactance becomes

$$X_C\,(60) = \frac{0.00265}{C}$$

20. Current through a resistance is in phase with the voltage across the resistance.

21. Current through an inductance lags the voltage across the inductance by 90°.

22. Current through a capacitance leads the voltage across the capacitance by 90°.

23. The combined opposition to ac current produced by resistance and reactance is called impedance. The unit of impedance is the ohm. Impedance is denoted by the letter Z.

24. Impedance, voltage, and current are related in an ac circuit by Ohm's law.

$$Z = \frac{V}{I} \qquad V = IZ \qquad I = \frac{V}{Z}$$

25. In an ac circuit, only resistance consumes power. This is called real or true power. Its unit is the watt.

26. The product of volts and amperes in an ac circuit is called apparent power and denoted by VA.

27. The ratio of real power to apparent power is called power factor.

$$PF = \frac{\text{power}}{\text{volt-amperes}} = \frac{P}{P_{app}}$$

28. Lagging power factor in a system can be improved by adding capacitance to the system.

29. An ac system that uses two wires and has a single sine wave is called single phase.

30. The most widely used polyphase system in the United States is the three-phase system.

31. In a 3-ϕ, 4-W system serving a balanced load, the neutral wire does not carry any current.

QUESTIONS

13-1 What advantages does alternating current have as compared with direct current?

13-2 How is the direction of current determined in a generator?

13-3 What is used in an ac generator in place of a commutator from which the voltage is transferred to brushes?

13-4 What type of voltage is always generated in a single coil rotating in a magnetic field, ac or dc?

13-5 What factors determine the value of voltage produced by a generator?

13-6 What is the voltage waveform produced by an ac generator called?

13-7 How many electrical degrees must a coil pass through to produce one cycle?

13-8 What term is used to specify the number of cycles produced in a second?

13-9 How many electrical degrees does a single coil pass through when making one complete revolution in a four-pole generator?

13-10 An eight-pole ac generator is required to produce 60 Hz. At what speed should it be rotated?

13-11 An ac voltmeter reads 110 V. What is the peak-to-peak voltage being measured?

13-12 What is another name for effective voltage?

13-13 The voltage across a 100-Ω resistor is 120 V ac, 1 ϕ, 60 Hz. What current is flowing through the resistor?

13-14 How many henrys in 150 mH?

13-15 An inductance of 60 mH is used in a 120-Hz circuit. Find the inductive reactance of the inductance.

13-16 If the inductance in question 13-15 were connected to a 120-V ac supply, find the current through the inductance.

13-17 Does current lead or lag voltage in an inductive circuit?

13-18 What are the basic components of a capacitor?

13-19 Does a constant current flow in a series circuit containing a resistor, an inductor, and a capacitor if the circuit is connected to 100 V dc?

13-20 What is the capacitive reactance of a 400-μF capacitor in a 60-Hz circuit?

13-21 A 10-Ω inductive reactance is in series with a 10-Ω capacitive reactance and the two are connected across a 120-V dc supply. What will happen?

13-22 The reactances in question 13-21 are connected in series across a 120-V, 60-Hz ac supply. What will happen?

13-23 A 10-Ω resistor is connected in series in the circuit of question 13-22. What current will flow in the circuit?

13-24 What is the relationship between current and voltage across a capacitance?

13-25 What is the total current-limiting effect in an ac circuit called?

13-26 What is power factor?

13-27 In general, is 1.00 PF desirable in an electrical system?

13-28 What is the PF of a resistance?

13-29 If the power factor is 100 percent, what must be true of the value of the reactances in the system?

13-30 Can the real power of a circuit ever exceed the apparent power of the circuit? Explain your answer.

13-31 A two-wire system with a single sine wave voltage is known as what type of system?

13-32 In a three-phase, four-wire system if the phase voltage is 120 V, what is the line voltage?

14
TRANSFORMERS

The transformer has led to alternating-current systems replacing direct-current systems for most residential, commercial, and industrial applications. Transformers can change the value of an ac voltage with very little loss of power. They are used throughout ac systems from the generating plant to the ultimate consumer of electricity. The basic transformer consists of two coils wound around an iron core. Alternating current supplied to one winding produces an alternating magnetic field which induces a voltage in the other winding. The value of the induced voltage is determined by the number of turns of wire in the two windings.

14-1 TRANSFORMER PRINCIPLES

The transformer winding that receives current from the line or other source is called the *primary winding*. The winding that delivers current to the load is called the *secondary winding*. In short, the primary winding is on the supply side of a transformer and the secondary winding is on the load side.

Two simple transformers are shown in Fig. 14-1. In some transformers, one coil is wound around the other instead of being separate as shown in the illustration. This illustration also shows the relationship between the voltage in the primary winding and the voltage in the secondary winding.

The relationship of the primary voltage to the secondary voltage is called the *voltage ratio*. If one winding has twice as many turns of wire as the other, it will have twice the voltage. When the ratio is given as $10:1$, it means that the high-voltage winding contains 10 times as many turns as the low-voltage winding. The higher value in the ratio pertains to the high-voltage winding, and the lower value (often 1) to the low-voltage winding. The ratio of the number of turns of wire in the primary to the number of turns of wire in the secondary is known as the *turns ratio*. It is therefore seen that the voltage ratio of a transformer is equal to its turns ratio. This can be written as a formula.

$$\text{Turns ratio} = \text{voltage ratio} = \frac{T_p}{T_s} = \frac{V_p}{V_s}$$

where T_p = number of turns in primary
T_s = number of turns in secondary
V_p = primary voltage
V_s = secondary voltage

Problem 1

A transformer has a voltage ratio of $5:1$. If there are 50 turns in the secondary, how many turns are in the primary?

Solution

$$\text{Voltage ratio} = \frac{5}{1} = \frac{T_p}{50}$$
$$T_p = 5 \times 50$$
$$= 250 \text{ turns}$$

In Fig. 14-1(*a*) the primary winding contains 50 turns of wire, and the secondary contains 50 turns of wire. Since both windings contain the same number of

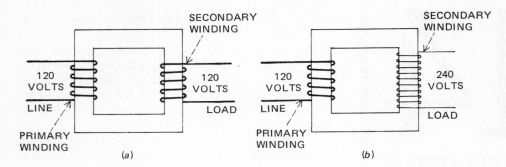

Fig. 14-1. Core and coil arrangements of simple transformers. (*a*) The primary winding and the secondary winding have the same number of turns. (*b*) The secondary winding has twice as many turns as the primary winding.

turns, the ratio is

$$\text{Turns ratio} = \frac{50}{50} = \frac{1}{1} \quad \text{or} \quad 1:1$$

In this transformer the load voltage will be the same value as the supply voltage. The primary is shown being supplied by 120 V. Therefore the secondary voltage can be found using the voltage ratio formula

$$\text{Voltage ratio} = \frac{V_p}{V_s}$$
$$\frac{1}{1} = \frac{120}{V_s}$$
$$V_s = 1 \times 120$$
$$= 120 \text{ V}$$

In Fig. 14-1(*b*) the turns ratio is 1:2. The load voltage is therefore twice the supply voltage. If the primary has 50 turns, the secondary will have 100 turns. With 120 V supplied to the primary, 240 V will be available on the secondary side.

$$\frac{50 \text{ turns}}{100 \text{ turns}} = \frac{120 \text{ V}}{240 \text{ V}}$$

The current in the two windings is in inverse proportion to the voltage ratio. That means the winding with the higher value in the voltage ratio will have the lower current. This can also be written as a formula.

$$\text{Voltage ratio} = \frac{V_p}{V_s} = \frac{I_s}{I_p}$$

where V_p = primary voltage
V_s = secondary voltage
I_s = secondary current
I_p = primary current

Problem 2

The load current of the transformer in Fig. 14-1(*b*) is 10 A. What is the primary current?

Solution

$$\text{Voltage ratio} = 1:2 = \frac{1}{2} = \frac{I_s}{I_p}$$
$$\frac{1}{2} = \frac{10}{I_p}$$
$$I_p = 2 \times 10$$
$$= 20 \text{ A}$$

The ratio does not affect the power or volt-amperes in the windings. Power is determined by the load. The power delivered to the load by the secondary winding is almost equal to the power delivered by the supply to the transformer primary. This is because power transformers are highly efficient devices—close to 100 percent efficient. The I^2R losses in the windings and certain magnetic losses account for most of the transformer losses.

For many calculations it is accurate enough to assume an ideal transformer, that is, one with 100 percent efficiency (no losses). For such a transformer

$$V_p \times I_p = V_s \times I_s$$

where V_p = primary voltage
I_p = primary current
V_s = secondary voltage
I_s = secondary current

Problem 3

The load on the secondary of a transformer draws 15 A at 120 V. If the transformer has a turns ratio of 4:1, what is the primary current? Assume an ideal transformer.

Solution

If the turns ratio is 4:1, the primary voltage V_p can be found with the formula

$$\text{Turns ratio} = \frac{4}{1} = \text{voltage ratio} = \frac{V_p}{120 \text{ V}}$$
$$V_p = 4 \times 120 \text{ V} = 480 \text{ V}$$

To find the current, use the formula for an ideal transformer.

$$V_p \times I_p = V_s \times I_s$$
$$480 \times I_p = 120 \times 15$$
$$I_p = \frac{120 \times 15}{480}$$
$$= \frac{1800}{480}$$
$$= 3.75 \text{ A}$$

Notice that this formula is merely another way of writing the current ratio formula.

$$\frac{V_p}{V_s} = \frac{I_s}{I_p}$$
$$\frac{480}{120} = \frac{15}{3.75}$$

To summarize the voltage, current, and power relationships in a transformer: voltage is in direct proportion to the windings or turns; current is in inverse proportion to the voltage (or turns); and power (or volt-amperes) is practically equal in each winding.

When the primary voltage of a transformer is higher than the secondary voltage, the transformer is called a *step-down* transformer. When the primary voltage is lower than the secondary voltage, the transformer is called a *step-up* transformer.

14-2 PRIMARY AND SECONDARY CURRENTS

A transformer automatically adjusts its input current to meet the requirements of its output or load current. Thus, when no current is being used from the secondary winding, no current flows in the primary except excitation current. Excitation current is the very small amount of current that is necessary to maintain the magnetic circuit.

If a transformer primary drew its full-load current continuously regardless of load, its operation would be uneconomical. But this is not the case since a trans-

former has an inherent and natural means of regulating its primary-current input in proportion to its load on the secondary.

For self-regulation, a transformer depends on counterelectromotive force generated in its primary winding by its own magnetism and an opposing magnetism produced by the current drawn by the load on the secondary winding. When the primary of a transformer is connected to its supply lines, current starts to flow, which produces a magnetic flux in the core. This magnetic flux, alternating with the current, induces a voltage in the winding in opposite direction to the original line voltage. This voltage is known as *counterelectromotive force* (abbreviated cemf) since it *counters*, or *opposes*, the *line voltage*.

The cemf in a well-designed transformer will nearly equal line voltage, and only a small excitation current will flow. (In the remainder of this discussion of self-regulation the excitation current will be neglected.)

Principles that afford self-regulation are illustrated with a loaded transformer in Fig. 14-2. For simplicity in discussion, the ratio will be considered 1:1. Thus the voltage and current in each winding are the same value.

When the primary is connected to the line with the load switches open, *no primary current flows because of high cemf in the primary winding*. When load switch 1 is closed, 5 A of load current will flow in the secondary winding. This 5-A load current will produce magnetism in the core of 5-A value in the *opposite direction* to the magnetism that is producing *cemf* in the primary winding which is limiting current flow in the primary. This counteracting effect is illustrated by arrows at the bottom of the core. Thus the load magnetism is opposing or countering the cemf of the primary, and *its opposition is in proportion to the load*. Accordingly, this 5-A load will produce magnetism *opposite to the cemf magnetism* and reduce it to the extent that 5 A will flow in the primary winding to furnish 5 A for the load.

If load switch 2 is closed, 10 A will flow in the secondary winding and will produce sufficient magnetism in the core to *counter enough cemf* to allow *10 A to flow in the primary* and supply its new load of 10 A.

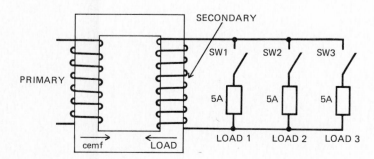

Fig. 14-2. Transformer self-regulating principle.

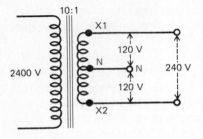

Fig. 14-3. A transformer providing three-wire single phase.

If load switch 3 is closed, 15 A will flow in the secondary and produce enough magnetism to counter the primary cemf and allow 15 A to flow in the primary. Thus the load current sufficiently regulates the primary current to supply the load, and the power input to the transformer practically equals the power output.

14-3 THREE-WIRE SINGLE-PHASE SYSTEMS

The transformers discussed to this point had single-winding primaries and single-winding secondaries. Each winding had two wires. These transformers operated from single-phase lines and delivered single phase to their loads.

A more common system is used for residential and small commercial electrical systems. The system was originally developed for direct current and is known as the *Edison three-wire* system. The system, however, is applicable to alternating current and in principle behaves the same way as in dc systems.

A typical three-wire system is shown in Fig. 14-3.

The primary and secondary windings could be only one part of a pole-mounted three-phase transformer. The transformer shown has a 10:1 voltage ratio overall, but the secondary contains a center tap so that half the secondary voltage is available between either of the outside lines and the center tap.

Assume a small commercial office building is supplied by this system and the loads are connected as shown in Fig. 14-4. Occupant X has 120-V lighting and receptacle loads and a single 240-V air-conditioning motor. Occupant Y has two small 120-V motors and 120-V lighting and receptacle loads. The various currents are shown in Fig. 14-4.

When all equipment is on, occupant X will draw 45 A and occupant Y will draw 35 A. The transformer secondary will need to deliver 45 A + 35 A = 80 A.

Let's see how this current is delivered to the respective loads. The air-conditioning motor (M3) is connected across lines A and B and draws 15 A at 240 V. Its 15 A flows in lines A and B. Lighting and receptacle load L1 is connected across line A and line N (the center tap). Notice that line N is connected to ground. Thus L1 draws its 30 A from lines A and N. Similarly L2 draws its current from lines B and N. At a given instant the polarities will be as shown in the figure. Line A is positive; line B is negative. Line N is shown with both a plus and a minus sign. Can it have two polarities at the same time? The answer is yes, if we are referring to two different reference points. Line N is less positive than line A; therefore it is negative with respect to line A. But at the same time it is less negative than line B so that line N is positive with respect to line B. Line N is called the *neutral* of the three-wire system.

This condition of polarity leads to an interesting situation if we trace current flow. Load L1 current starts from the transformer through line A through the load and back to the transformer by way of the neu-

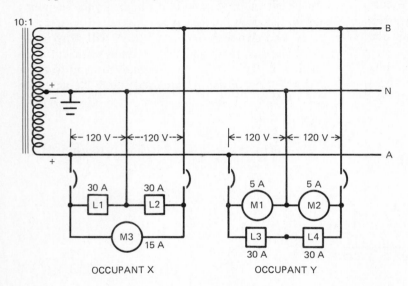

Fig. 14-4. A three-wire system serving two commercial loads.

tral. Now at that same moment load L2 is receiving its current from the neutral, through L2, and back to the transformer by way of line B. What we did was to trace 30 A going two ways through the same line, the neutral. Actually the two opposing currents cancel one another, and the current that serves L1 and L2 passes through lines A and B only. Note that this is true only when L1 = L2. For occupant B, M1 = M2 and L3 = L4. Thus no current will flow in the neutral serving occupant Y. This fact points up an important advantage of three-wire systems. In a balanced system—that is, one in which the load connected to line A is equal to the load connected to line B—no current flows in the neutral line. In an unbalanced system only the *difference* in the two loads will flow in the neutral.

With the neutral not carrying current, the 120-V loads are in effect connected in series across 240 V.

Problem 4

Find the current flowing in each line and the neutral of Fig. 14-4 under the following conditions:

(1) L1 = 10 Ω   L2 = 10 Ω
(2) L1 = 10 Ω   L2 = 20 Ω
(3) L1 = 10 Ω   L2 is open

Solution

(1) With the loads equal, no current will flow in the neutral.

$$I_N = 0$$

The two loads will in effect be in series and connected across 220 V. The current in lines A and B will be the same; by use of Ohm's law

$$I = \frac{V}{R} = \frac{220}{10 + 10} = \frac{220}{20}$$
$$= 11 \text{ A}$$

Therefore $I_A = I_B = 11$ A.

(2) In this case L1 and L2 are not equal and the difference in their currents will flow in the neutral. Loads L1 and L2 must be considered as being connected across 110 V. By use of Ohm's law

$$I_{L1} = \frac{V}{R_{L1}} = \frac{110}{10} = 11 \text{ A}$$

$$I_{L2} = \frac{V}{R_{L2}} = \frac{110}{20} = 5.5 \text{ A}$$

$$I_N = I_{L1} - I_{L2} = 11 - 5.5 = 5.5 \text{ A}$$

The current in line A serves L1, therefore $I_A =$ 11 A; line B carries the current for L2, therefore $I_B = 5.5$ A.

(3) With L2 open only the single load L1 will draw current, and line B will carry no current, or $I_B = 0$. The neutral will be the return path for I_{L1}. From the previous calculations $I_{L1} = 11$ A; therefore $I_N = 11$ A.

Power consumption is calculated as in previous single-phase systems.

Problem 5

Assume L1 and L2 are entirely resistive; find the total power consumed in each of the three parts of the previous problem.

Solution

(1) $P = V \times I$
        $= 220 \times 11$
        $= 2420 \text{ W}$

or $P = I^2R_{L1} + I^2R_{L2} = (11)^210 + (11)^210$
        $= 1210 + 1210$
        $= 2420 \text{ W}$

(2) $P_{L1} = V \times I$
        $= 110 \times 11 = 1210 \text{ W}$

$P_{L2} = V \times I$
        $= 110 \times 5.5 = 605 \text{ W}$

$P_T = 1210 + 605 = 1815 \text{ W}$

(3) $P_{L1} = 110 \times 11 = 1210 \text{ W}$

Notice that in part 1 the only current is 11 A since in effect this is a series current flowing through the load across the 220-V lines. In the other parts it was necessary to use the current drawn by the particular load using the 110 V across the individual loads.

14-4 NEUTRAL OF A THREE-WIRE SYSTEM

For satisfactory operation of a three-wire system, the neutral line must be continuous and never be allowed to open. That is why individual fuses, circuit breakers, or switches are never used in the neutral line. If the neutral line opens under certain conditions, electrical equipment can be seriously damaged.

Figure 14-5 shows the dangers of a broken neutral. Load L1 draws 10 A and Load L2 draws 5 A. With

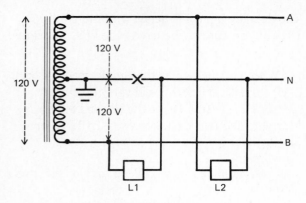

Fig. 14-5. Three-wire single-phase system with a broken neutral.

the neutral broken at point X, current for L1 cannot return to the transformer through the neutral. Instead it must go through the neutral to L2, through L2 and back to the transformer by way of line A. This could overload L2 if not protected. On the other hand, if L2 was switched off, L1 would be off also. If another appliance were to be switched on across line A and the neutral, it would be in series with L1 across the 240-V lines. Most likely the appliance would burn out.

It can be seen that the neutral line must be a continuous wire, in good condition, and properly connected. Many erratic, sometimes highly mystifying, problems, can be created by faulty, open, or poorly connected neutrals.

14-5 THREE-PHASE SYSTEMS

Three-phase systems can be obtained from three single-phase transformers connected in a number of different configurations that produce a variety of voltage and current combinations.

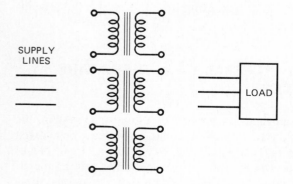

Fig. 14-6. Three single-phase transformers to be used in a three-phase system.

Three single-phase transformers are shown in Fig. 14-6. They are the same in all electrical characteristics. Their six primary terminals and their six secondary terminals must connect to the three (or four) supply lines and the three (or four) load lines.

The two most common connections for accomplishing this are the delta connection and the wye (or star) connection. The primaries need not be connected in the same configuration as the secondaries.

14-6 WYE (OR STAR) CONNECTION

If the same line from each of the single-phase transformers were connected together and the remaining three lines brought out to connect to the supply lines or load, we would have a *wye connection* as shown in Fig. 14-7. Each of the voltages of a three-phase system is out of phase with the other two voltages by 120°, and the voltage across each pair of lines is the same, in this case 208 V. The common connection (sometimes called the *star point*) is the neutral. The voltage measured across each line to neutral is called the *phase voltage*. The phase voltage therefore is actually the voltage across each winding. The 208 V measured across each pair of lines is known as the line-to-line voltage or simply the *line voltage*. The following relationships exist between line and phase voltages and current.

In a wye connection:

$$I_l = I_{ph}$$
$$V_l = 1.73 \times V_{ph}$$

where I_l = line current
I_{ph} = phase current
V_l = line voltage
V_{ph} = phase voltage

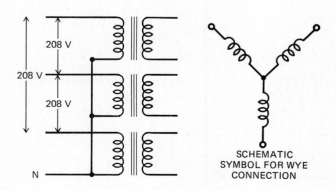

Fig. 14-7. The primaries of three single-phase transformers connected in a wye (or star).

Problem 6

A three-phase transformer is connected wye to a three-phase load. If the line voltage is 480 V, find the phase voltages.

Solution

In a wye connection

$$V_l = 1.73 \times V_{ph}$$
$$480 = 1.73 \times V_{ph}$$
$$V_{ph} = \frac{480}{1.73}$$
$$V_{ph} = 277 \text{ V}$$

This is one of the common three-phase voltages used in large commercial buildings. The 277 V is single phase and used for small motors and fluorescent lighting loads. The 480 V is used for larger three-phase motor loads.

14-7 DELTA CONNECTION

If the three single-phase windings (primary or secondary) in Fig. 14-6 were connected in series to form a loop, we would have a delta connection. This is shown in Fig. 14-8. The connection gets its name from the fact that the schematic symbol resembles the Greek letter delta (Δ).

The delta connection does not provide a single common point to serve as a neutral. The three lines are taken from the ends of the windings; therefore the line voltages are also the winding or phase voltages. The current in each line is drawn from two windings since two windings join to feed a single line. The voltage and current have the following relationships.

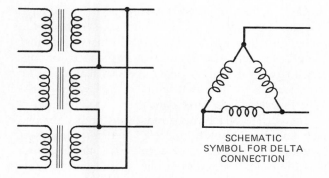

SCHEMATIC SYMBOL FOR DELTA CONNECTION

Fig. 14-8. The secondaries of three single-phase transformers connected in a delta.

In a delta connection:

$$I_l = 1.73 \times I_{ph}$$
$$V_l = V_{ph}$$

Problem 7

The primary of a three-phase transformer is connected as a delta. Find the line current and the line voltage if $I_{ph} = 17.3$ A and $V_{ph} = 480$ V.

Solution

In a delta connection

$$I_l = 1.73 \times I_{ph}$$
$$= 1.73 \times 17.3$$
$$= 30 \text{ A}$$
$$V_l = V_{ph}$$
$$= 480 \text{ V}$$

Although the three windings in a delta connection are connected in a closed loop, a short-circuit condition does not exist. That is because the voltages and currents in each winding or phase are displaced by 120° from one another. If the current in each winding was measured at any instant, the sum of all three would be zero.

14-8 TRANSFORMER IMPEDANCE

The primaries and secondaries of a three-phase transformer need not be connected in the same configuration. The possible combinations are wye-wye, wye-delta, delta-wye, delta-delta.

If three single-phase transformers are used to serve a three-phase system, the ratings of the individual transformers must be suitable for the three-phase connection.

Certain connections require that the voltage and impedance of the three transformers be the same to avoid circulating currents in the winding. Transformers connected delta-delta should have the same voltage and impedance, but the impedances can vary up to about 7 percent of the highest impedance. A transformer's impedance is always shown on the nameplate.

The impedance is determined by the total turns in the winding. This is not the same as the turns ratio. For example, a transformer with a 10:1 ratio might have 50 turns in one winding and 500 in the other, while another transformer with a 10:1 ratio might have 60 turns in one and 600 in the other. The transformer with the higher number of turns would probably have a higher impedance if the same size wire were used in both windings. Of course, the load on the primary is not likely to be the same as the secondary load, so that primary and secondary wire sizes are not likely to be the same.

Impedance is stated as a percentage. In transformers, the impedance is the percent of rated high voltage in the high-voltage winding that is necessary to produce full-load current in a short-circuited low-voltage winding. The technician must consider impedance when connecting transformers in combination or selecting control equipment for the system supplied by the transformers.

The amount of inrush current (with sufficient supply) that a three-phase transformer or bank can deliver to a system in which there is a short circuit may be calculated by dividing the impedance of the transformer into 100 and multiplying this by its rated current. For example, a transformer with 2 percent impedance can deliver 50 times its rated current under short-circuit conditions. A transformer with 4 percent impedance can deliver 25 times its rated current.

This can be expressed by the following formula.

$$I_i = \frac{100}{Z_t} \times I_R$$

where I_i = inrush current at short circuit
Z_t = impedance stated as a percent
I_R = rated load current

Problem 8

A three-phase transformer with 3 percent impedance has a rated load current of 120 A. If the secondary is short-circuited, what is the maximum current that will be drawn from the line?

Solution

Use the formula

$$I_i = \frac{100}{Z_t} \times I_R$$
$$= \frac{100}{3} \times 120$$
$$= 4000 \text{ A}$$

Remember these calculations find a fault current. Protective devices are expected to open the circuit to prevent damage to the transformer and the conductors. But protective devices are usually rated in terms of the maximum current they can handle in addition to the value at which they will open the circuit. If the maximum value is exceeded, the device may melt or be otherwise damaged and made inoperative. Thus the

short-circuit calculation just performed provides valuable information to the installer or maintenance electrician and technician.

14-9 TRANSFORMER CONNECTIONS

Three-phase transformers may have their windings connected as a wye or delta. Similarly, a bank of three single-phase transformers may be connected wye or delta.

When single-phase transformers are combined in a bank, the voltage ratios and impedances of the transformers should be the same, but these can vary to some degree in certain connections without disturbing the system. The transformers do not necessarily have to be of the same capacity. If there is a difference in capacity of the transformers, the total three-phase capacity of the bank will be limited to 3 times the capacity of the smallest transformer.

Problem 9

Three single-phase transformers with the following capacities: 25 kVA, 15 kVA, and 37.5 kVA are connected delta-delta to form a three-phase transformer. What is the rated capacity of the three-phase bank?

Solution

The smallest transformer is 15 kVA. Therefore the capacity of the bank is

$$3 \times 15 = 45 \text{ kVA}$$

When single- and three-phase loads are served from a three-phase delta secondary bank, the transformer serving the single-phase load is usually of greater capacity than the other transformers, but it should not be over twice as great.

Delta-delta–connected transformers should have the same impedance and voltage ratios, and in no case should the impedances of the transformers vary more than 7 percent from the highest impedance. In connecting delta-delta, it is recommended that a fuse be used across the last connection in closing the bank. The fuse should be large enough to carry the full exciting current of the transformers. If the fuse blows, the polarity of one or more of the transformers is wrong and will have to be corrected. If the fuse does not blow, it is safe to make the final connection permanent.

When either the primary or secondary side of a bank is connected delta, the voltage ratios of the transformers should be the same to prevent circulating current in the closed-circuit delta connection.

In wye connections, transformers should be matched in impedance and voltage ratio but not necessarily to prevent circulating currents, since a wye connection does not form a closed circuit in the windings.

For the sake of clarity, lightning (surge) arresters and overcurrent protection are not shown in the following transformer connection diagrams. But they are always connected between the high-voltage lines and the transformers in all ungrounded lines. Lightning arresters are always grounded, and transformer tanks should be grounded. Sometimes lightning arresters are also connected to the low-voltage lines.

Wye-Wye Figure 14-9 is a diagram of three single-phase transformers in a three-phase bank connected wye-wye. The H1 terminals of the transformers are connected to lines A, B, and C, and the H2 terminals are connected together on the primary side. If the primary is to be grounded, the grounding connection should be made at this point.

The secondary terminals X1 are connected to lines a, b, and c, and the X3 terminals are connected together. If the secondary is to be grounded, the grounding wire should be connected to this point.

The transformed line voltages will be determined solely by the transformer ratios, since a wye-wye connection does not change line voltage values. If the ratio is 10:1 and the primary line voltage is 2400 V, the secondary line voltage will be 240 V. Of course, if the primary line voltage is 2400 V, the phase voltage is 1387 V (2400/1.73). With a 10:1 voltage ratio, the secondary phase voltage will be 138.7 V. This results in a secondary line voltage of 240 V (138.7 × 1.73).

For a wye-wye connection the voltage ratios of these transformers should be the same, but the impedances do not necessarily have to be the same.

Wye-Delta Figure 14-10 is a diagram of three single-phase transformers in a three-phase wye-delta bank.

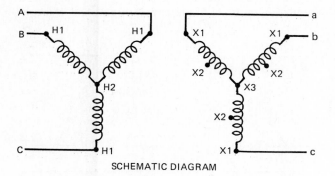

SCHEMATIC DIAGRAM

Fig. 14-9. Wye-wye connection.

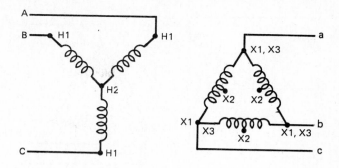

Fig. 14-10. Wye-delta connection.

The H1 terminals are connected to the high-voltage A, B, and C lines, and the H2 terminals are connected together. If the primary is to be grounded, the grounding wire should be connected to this point.

The secondary leads are connected delta to the secondary lines. Considering the X1 terminal of each transformer to be the start lead of the winding and the X3 terminals to be the finish leads, the connection is as follows. Terminal X1 of transformer 1 connects to terminal X3 of transformer 2. This connection point is then connected to line a. Terminal X1 of transformer 2 connects to terminal X3 of transformer 3. This connection point is then connected to line b. Finally, terminal X1 of transformer 3 is connected to terminal X3 of transformer 1, and this connection point is connected to line c. If the secondary is to be grounded, the grounding wire should connect to any one of the X2 terminals or phase lines.

The transformed line voltages are not determined only by the transformer ratios. If the ratio is 10:1 and the primary line voltage is 2080 V, the secondary line voltage will be 120 V. This is how the secondary line voltage was found. The primary line voltage of 2080 V resulted in a phase voltage of 1200 V (2080/1.73). With a 10:1 voltage ratio, the secondary phase voltage is 120 V (1200/10). Since in a delta winding the line voltage is equal to the phase voltage, the secondary voltage is 120 V.

Three-phase motors with only two protective units in the lines, operating on an ungrounded wye-delta or delta-wye system, might burn out if an open occurs in the primary supply.

If a primary circuit opens on these systems, a serious current imbalance occurs in the secondary lines. If the highest of these currents is in the unprotected motor line, a burnout, similar to single-phasing, is likely to occur in the motor. To protect against this condition, one fuse or circuit breaker in each of the three motor lines is necessary.

Wye-Delta Four-Wire System Figure 14-11 is a wye-delta connection that is commonly used for supplying

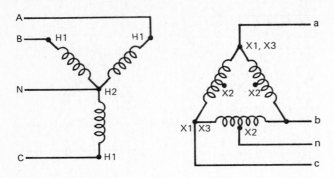

Fig. 14-11. A wye-delta four-wire system.

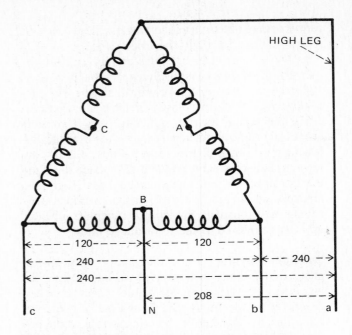

Fig. 14-12. A four-wire delta secondary showing the "high-leg" connection.

240-V three-phase and 120/240-V single-phase service. It is similar to the connection in Fig. 14-10, but a fourth line—a neutral—has been added to the primary and secondary sides. This neutral goes back to the transformers supplying the system, which is connected wye on the secondary side. It is grounded at both transformer banks and possibly at every third or fourth pole in overhead distribution systems. This system of grounding is controversial and not always accepted or approved. If the fourth wire is not grounded, it is known as the *common* wire.

The primary usually has 4160 V with standard 2400-V 10:1 ratio transformers. A connection between any one of the phase lines and the neutral line gives 2400 V for a single-phase transformer.

Single-phase transformers with 2400-V primaries, when connected wye in a three-phase bank, require 4160 line volts to give 2400 phase volts.

$$V_{ph} = \frac{V_l}{1.73}$$

$$= \frac{4160}{1.73} = 2400 \text{ V}$$

In the secondary, phase B transformer is on the bottom. This makes possible the neat connection shown in the illustration. The transformers are connected together, and then the connections are connected to the line in the manner shown. This is the way delta connections are commonly made.

The phase B transformer, which is to furnish 120-V single-phase loads, is usually of greater capacity than the other two transformers. Its secondary winding is tapped and connected to a fourth line which is also grounded and neutral.

This arrangement offers 240 V three-phase, 240 V single-phase, and the neutral. There are 208 V single-phase between line a and the neutral. This line is known variously as the "high leg," "wild leg," "red leg," or "stinger." So that a load will not be mistakenly connected across it and the neutral for 120-V cir-

cuits, this conductor must be identified by an orange outer finish or by clear tagging. It must also be identified at each point where a connection is made if the neutral is also present (NEC 250-5; 215-8). If a mistake is made, the circuit will be supplied with 208 V, which will burn out any standard 120-V equipment connected to it.

Figure 14-12 is a diagram of a four-wire delta secondary showing the line and phase voltages. Three-phase 240-V service can be obtained from lines a, b, and c. Single-phase 240-V service can be obtained from any two of the lines a, b, and c. Single-phase 120-V service can be obtained from either b to N or c to N. The high leg is a, with 208 V between it and the neutral line N.

Delta-Delta Three single-phase transformers are shown connected delta-delta in Fig. 14-13. To prevent circulating currents in the closed circuit that exists in

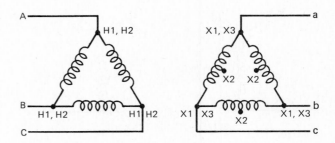

Fig. 14-13. Delta-delta connection.

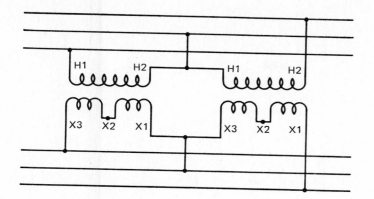

Fig. 14-14. Open-delta connection.

delta connections, it is necessary for all regulating taps, voltage ratios, and the impedances of the three transformers to be the same; however, the impedances can vary as much as 7 percent downward from the highest impedance before circulating currents are considered excessive.

If the primary has 2400 V and the transformer ratio is 10:1, the secondary voltage will be 240 V three-phase. The value of the secondary voltage in relation to primary voltage is determined solely by the ratio, since delta line and phase voltages are the same.

If 120-V single-phase service is desired, it can be obtained by connecting to one of the X2 taps and X1 or X3 of the same transformer. If three-wire 120/240-V single-phase service is desired, it can be obtained by connecting to the middle tap and to the two outside terminals of one transformer. For example, in the illustration, 120/240 V can be obtained by connecting to X2 of transformer B lines c and b. The connection at X2 should be grounded.

Open-Delta In case a transformer in a delta-delta bank becomes defective, it can be removed from the bank, and service can be normally continued without further adjustments if the load does not exceed 58 percent of the full-load rating of the bank. If one transformer is removed from a three-phase delta-delta–connected bank, the two remaining transformers will continue to operate in *open-delta*.

If the defective transformer in a bank is the one tapped at the X2 terminal for low-voltage single-phase service, the tap will have to be moved to one of the remaining transformers. If only the tap is changed, it will cause one of the other lines to become the high leg. The high voltage on it then would cause the destruction of all 120-V equipment connected to that circuit.

Since the original high-leg line is identified throughout the system where necessary, this line should remain the high-leg line, and a reconnection of lines at the transformer will have to be made. The

high-leg line is the line that does not connect directly to either side of the tapped transformer.

In a new installation for three-phase service where the load is expected to increase in the future, only two transformers need be initially installed, connected open-delta; a third transformer can be installed when future additional load requires it.

Two transformers are shown connected open-delta in Fig. 14-14. These two transformers can deliver 85 percent of the total full load of the two transformers. For example, if two 50-kVA transformers with a total capacity of 100 kVA are connected open-delta, they can deliver 85 kVA under full load.

Wye-Delta Open-Delta If a transformer in a wye-delta bank becomes defective, it can be removed and service can be continued at 58 percent of original full-load capacity, provided the primary system has a common wire that connects to the common point in the transformers at the other end of the primary system.

This connection is shown in Fig. 14-15. In effect, it is practically the same as two transformers operating open-delta. If the defective transformer is the one with the center X2 terminal tapped for single-phase service, it will be necessary to move the tap to one of the remaining transformers. This tap change will cause a different line to be the high leg. Since the original high-leg line is identified where required throughout the system, it will be necessary to change line connections to make the original high-leg line the same under the new conditions. If the system serves any three-phase motors, it may be necessary to interchange two lines other than the high leg to prevent reversing the motors.

Delta-Wye Three single-phase transformers connected delta-wye for three-phase are illustrated in Fig. 14-16. This connection is seldom used for service to a consumer but is often used as a substation transformer to go from a transmission system to a distribution system.

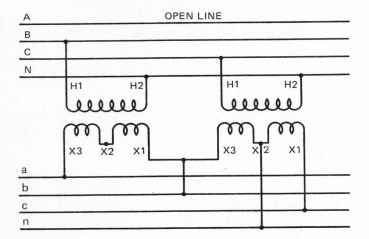

Fig. 14-15. Wye-open-delta connection.

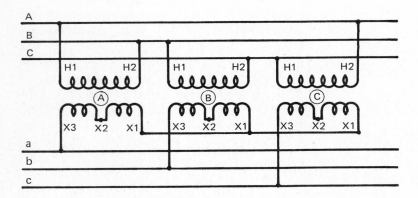

Fig. 14-16. Delta-wye connection.

Wye-Wye A commonly used connection for customer service requiring a heavy lighting or single-phase load as compared with a light three-phase load is shown in Fig. 14-17. This shows a four-wire primary and secondary wye-wye connection using 2400-V to 120/240-V transformers with the secondary windings connected parallel. The primary and secondary neutrals should be firmly tied together. On a 4160-V primary line the phase voltage is 2400 V, because of the wye connection.

The secondary windings, in parallel, deliver 120 V

between a line and the neutral. The wye connection delivers 208 V line to line. So 120-V single-phase circuits are connected from a line to neutral, 208-V single-phase circuits are connected to any two of the three lines, and three-phase 208-V circuits are connected to the three lines. The chief advantage of this connection is that the single-phase load can be well balanced among the three transformers. All three-phase and heavy single-phase equipment must be rated at 208 V. The more common 240-V motors or other equipment will not operate properly on 208 V.

Fig. 14-17. Four-wire wye-wye connection.

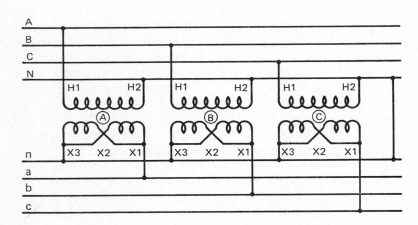

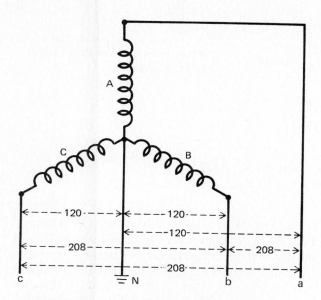

Fig. 14-18. Line and phase voltages in a four-wire wye connection.

This connection, with a different transformer ratio, is also commonly used for 277-V fluorescent lighting and 480-V power.

A schematic of the secondary of a four-wire wye connection showing voltages between the lines is given in Fig. 14-18. Voltage to ground in this system is 120 V, the same as from phase to neutral.

14-10 TRANSFORMER RATINGS

A transformer nameplate should give the name of the manufacturer, the rated kilovolt-amperes, frequency, phase, primary and secondary voltage, impedance, polarity, and type of coolant used.

A transformer, like an alternator, is rated in kilovolt-amperes (kVA), instead of kilowatts, which are used in dc ratings. Recall that a kilovolt-ampere is equal to a kilowatt only if the system is at 100 percent power factor. But 100 percent power factor is seldom obtained on ac systems; and in selecting power-producing equipment, the power factor of the system to be supplied must be considered in computing the kVA rating.

Usually, the electrical rating of a single-phase distribution transformer is given, as in the following example: rating—10 kVA, 2400/4160 Y to 120/240 V, single phase, 60 Hz. The first value in the voltage ratings is the primary-phase voltage. Therefore this transformer would require 2400 V single-phase per primary winding in a three-phase bank. The Y after 4160 indicates a wye (or star) three-phase connection.

A general-practice classification of transformers

is on the basis of *power* (or *transmission*) and *distribution*. A power transformer is one rated above 500 kVA, and a distribution transformer is rated 500 kVA and under.

14-11 TERMINAL IDENTIFICATION

Transformers, like dc generators, cells, or batteries, should, when connected in parallel or other combinations, have the polarities of the terminals identified. This makes it easy to connect transformers so that the voltages of the several units will be in the proper direction. In connecting a single-phase transformer alone, terminal identification has no significance, because connections can be made without regard to polarity.

Since the direction of alternating current is continually changing, polarity must be indicated for a given instant. To indicate polarity of a transformer, the terminals are identified by letters, while the direction of instantaneous voltage and current flow is indicated by numbers. Voltage direction and current flow are from low number to high. If a winding contains taps, the leads are numbered consecutively, in order of voltage values, beginning with 1. The high-voltage terminals are identified by the capital letter H, and the low-voltage terminals are identified by the letter X. This standard is used for the figures in this book.

In establishing the proper terminal markings on a single-phase transformer, the high-voltage terminal on the right, when facing the transformer on the side of the high-voltage terminals, is marked H1. The other high-voltage terminal is marked H2. If the instantaneous voltages in the two windings are in the directions shown in Fig. 14-19, the low-voltage termi-

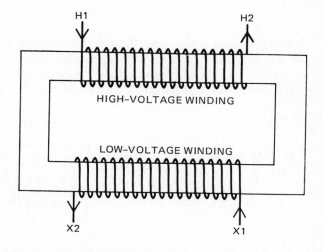

Fig. 14-19. Direction of instantaneous voltage in a transformer winding.

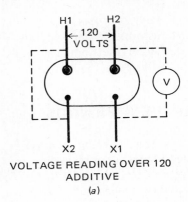

VOLTAGE READING OVER 120
ADDITIVE
(a)

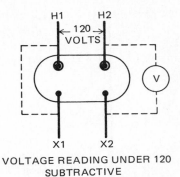

VOLTAGE READING UNDER 120
SUBTRACTIVE
(b)

Fig. 14-20. Testing transformer winding polarity with a voltmeter. (a) Additive. (b) Subtractive.

nal opposite H1 is marked X2, and the other terminal is marked X1. The voltage direction indicated here is in on X1 and out on X2.

Depending on their *polarity*, there are two classifications for transformers, *additive* and *subtractive*. If the instantaneous voltage in the two windings is in on H1 and out on X2, as shown in Fig. 14-19, the transformer is classed as additive. If the instantaneous voltage is in on H1 and out on the terminal now marked X1, the transformer is classed as subtractive, and, being opposite to that in Fig. 14-19, the markings would be interchanged with X2 on the right and X1 on the left to indicate this direction of voltage.

14-12 TESTING FOR VOLTAGE DIRECTION

A method of testing with a voltmeter for determining voltage direction in the windings is illustrated in Fig. 14-20. Any safe voltage is suitable for this test. This illustration shows 120 V being used; the voltmeter should have a range of about twice the testing voltage. The complete method of testing and establishing terminal markings is as follows.

When the transformer is faced on the high-voltage terminal side, the right-hand terminal is marked H1. The left-hand terminal is marked H2. A connection is made between H1 and the low-voltage terminal di-

rectly opposite H1, as shown by the dashed lines in Fig. 14-20. A voltmeter is connected between H2 and the low-voltage terminal directly opposite it, as shown by dashed lines in the illustration. The test voltage of 120 V is applied to the H1 and H2 leads. If the voltmeter reads *over 120 V,* the transformer is classed as *additive,* and the low-voltage terminal directly opposite the H1 terminal is marked X2. The other low-voltage terminal is marked X1, as shown in Fig. 14-20(a). This is standard marking for an additive transformer. It is termed an *additive transformer* because the voltages in the two windings add together and give a voltmeter reading above the test voltage.

If the voltmeter reads *less than the test voltage of 120 V,* the transformer is classed as a *subtractive* transformer, and the low-voltage terminal directly opposite H1 is marked X1, and the terminal directly opposite H2 is marked X2, as shown in Fig. 14-20(b).

Another method of testing transformer polarity is by checking an unknown transformer with one of known polarity. The connections for this test are shown in Fig. 14-21.

If the polarity of the unknown transformer is additive, no voltage will appear across the opening in the circuit of the unknown transformer.

If the polarity of the unknown transformer is subtractive, twice the normal voltage of the transformers will appear at the opening. The voltage condition at the opening can be tested with a voltmeter, a test light, or a small fuse. If no voltage is detected, the

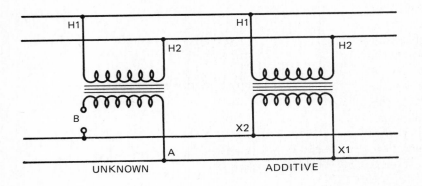

UNKNOWN   ADDITIVE

Fig. 14-21. Testing an unknown transformer for polarity using a transformer of known polarity.

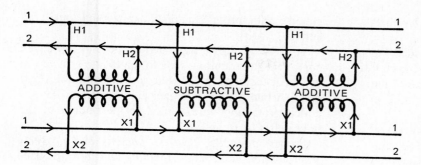

Fig. 14-22. Parallel connections of additive and subtractive transformers.

unknown transformer can be marked the same way as the known additive transformer.

The right-hand terminal on the high-voltage side should be marked H1, and the other terminal H2. The A terminal is X1 and the B terminal is X2. If a high voltage is present at the opening, the unknown transformer is subtractive, and terminal A should be marked X2, and terminal B should be marked X1.

If two unknown transformers are connected for operation together and an opening is made as shown in the illustration, and if no voltage is present at the opening, the transformers are properly connected and the opening can be permanently closed. If a voltage is present at the opening, it will be necessary to reverse the connections of one of the transformers' windings, on either the high- or low-voltage side.

The use of terminal markings is illustrated in Fig. 14-22. Two additive transformers are paralleled with one subtractive transformer. Correct polarity is attained when all the low-voltage terminals marked X1 are connected to line 1 and when all the terminals marked X2 are connected to line 2. The high-voltage terminals are connected H1 to line 1 and H2 to line 2. With these connections instantaneous voltages and current flow are in the proper direction in all transformers. This is indicated by the arrows showing voltage and current directions on the various circuits.

Single-phase transformers connected parallel must have the same voltage rating and impedance. The impedance of a transformer is indicated on the nameplate. If the voltages and impedances are not the same, a circulating current will exist between the windings at all times irrespective of a load on the transformers, and load current will not divide properly between the transformers.

Three-phase transformers contain three sets of windings mounted on a common core. The high-voltage terminals are marked H1, H2, and H3, and the low-voltage terminals are marked X1, X2, and X3. Proper polarity connections of the windings are made inside. It is also common practice to use three single-phase transformers connected in a bank for three-phase operation.

14-13 RATIO ADJUSTERS

The ratio of a transformer at times may not meet the exact requirements of certain loads. Occasionally, voltage drop in the feeder lines is excessive; when this occurs for a significant time period, some form of ratio adjustment in the transformer is required. In some transformers this adjustment is provided by taps, usually on the secondary winding. These taps are simply parts of the winding to which connections can be made.

Figure 14-23 shows a secondary winding with three terminal points for the load connection. In normal operation the switch connects terminal B to the load. With this connection assume the turns ratio is 1:10. If required, the switch can connect tap A to the load. This changes the turns ratio because fewer secondary turns are being induced by the same number of primary turns. The net effect will be for the secondary voltage to decrease. Similarly, if tap C were connected to the load, the voltage ratio would change. With more turns being induced by the primary, the secondary voltage would increase.

Some small transformers are equipped with a terminal block to which the ends of the secondary winding taps are connected. Changes in secondary voltage are

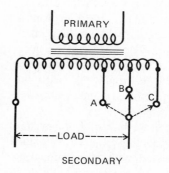

Fig. 14-23. Secondary taps provide a means for adjusting the turns ratio of a transformer.

made by changing tap connections at the terminal block.

Some transformers are equipped with no-load tap changers inside their housing, and others have a shaft brought through an airtight and watertight seal to the outside, where a hand wheel or lever can be used to change tap connections.

An external no-load tap changer contains an operating handle that can be locked in position. The secondary winding taps are connected to tin-plated copper studs, which are supported by molded insulators. A rotary contactor on the shaft under spring pressure makes contact with the studs for changing taps.

Large transformers are usually equipped with tap changers that can be changed under load. Some are manually operated, and others are automatic motor-driven changers. The actual switching of load currents is done under oil, but in a separate tank from the transformer tank to avoid deteriorating the coolant oil by the arc.

14-14 GROUNDING

In connecting a single-phase transformer alone, the polarity markings do not have any significance, since the transformer's high-voltage leads are simply connected to the high-voltage lines, and the low-voltage leads are connected in any convenient way to the low-voltage lines, with the center tap to the neutral line. But when one of the high-voltage lines is grounded and is therefore neutral, H2 is usually connected to it, and H1, with overcurrent protection and a surge (lightning) arrester, is connected to the hot line. The word *"neutral"* means *not charged electrically,* but in an electrical system the neutral line sometimes carries as much current as any of the other lines. The neutral line in a wiring system is the *grounded line* and is therefore *neutral to ground,* or earth. There is no voltage between the neutral line and the ground, since they are connected together and can be considered as one side of a circuit. The wire that connects the neutral line to ground is known as the *grounding wire.* This wire is also used to ground the cases and housing of the various equipment connected to the system.

There is a voltage between the hot lines of a system and the ground or any conducting objects in contact with the ground or connected with any conducting material that is in contact with the ground. Thus, if someone standing on the bare ground (or in contact with the grounded object) touches an ungrounded or hot line, the person will *make a circuit from the transformer through the hot line to the ground and through the neutral wire back to the transformer.* This completed circuit through the body can be (and usually is) lethal.

The neutral line is grounded at the transformer by connecting a grounding wire to it and to a ground rod driven at the base of the pole, a water pipe, or a butt plate. A butt plate is a copper plate (or about 30 ft of coiled grounding wire) nailed to the bottom of the pole before it is set.

Proper grounding *protects persons and property from possible hazards of lightning* and from the danger of the high-voltage lines accidentally coming in contact with the low-voltage lines and charging them with *high voltage.* A good system ground will divert lightning to the ground or cause a primary fuse or circuit breaker to open if the high- and low-voltage lines make contact. Grounding does not affect the operation of electrical equipment, which operates equally well whether or not the system is grounded.

The neutral line from the transformer to a customer's service entrance should be identified. This is usually done by stringing it on white insulators. At the service entrance, the NEC requires the neutral line to be grounded again, which can be done at one of three places. It can be grounded from the *service entrance cap,* the *meter socket,* or the *service cabinet.*

The grounding wire must be continuous and clamped with pressure connectors at each end. No soldered connections are allowed on the grounding wire. Such grounding on the premises of the customer, according to the Code, must always be to a *cold-water system if it is available on the premises.* If a cold-water system is not available, grounding can be done by connecting to a driven ground rod, buried metal plates, or some type of electrode. Grounding wires are identified with green insulation. They must not be used as circuit conductors.

The neutral line throughout the premises is identified by using wire with *white* or *gray* insulation. It should be continuous throughout the premises to the transformer, without any equipment that might open it, such as fuses, circuit breakers, or switches. A multipole switch can be used if it opens all hot lines and the neutral at the same time.

Grounding points of common transformer secondary service windings, and the location of overcurrent devices in lines in the service cabinet, as required by the Code, and voltages between secondary lines are illustrated in Fig. 14-24. It can be noticed that overcurrent devices are in all ungrounded lines and none are used in grounded or neutral lines.

14-15 PHASE SEQUENCE

In doing repair or replacement work on a three-phase system, it is possible inadvertently to reverse the

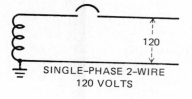

SINGLE-PHASE 2-WIRE
120 VOLTS

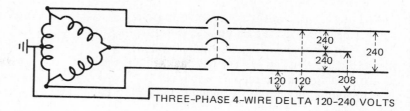

THREE-PHASE 4-WIRE DELTA 120-240 VOLTS

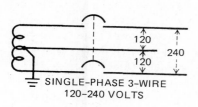

SINGLE-PHASE 3-WIRE
120-240 VOLTS

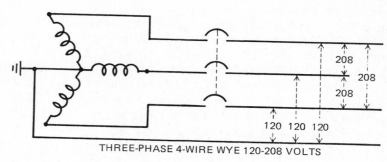

THREE-PHASE 4-WIRE WYE 120-208 VOLTS

Fig. 14-24. Grounding points for single- and three-phase transformer secondaries. Overcurrent protection can be provided by fuses or circuit breakers. Neutral lines are also grounded at the service entrance.

phase sequence of the system. Phase sequence is in the order in which each of the phase voltages reaches its peak and zero values. Remember that the phase voltages are 120° out of phase with one another. If two lines are transposed from their original positions, the phase sequence will be reversed. Since the direction of rotation of three-phase motors is determined by phase sequence, any change in the sequence will reverse the direction of rotation of the motors.

In many cases damage can result from the inadvertent reversal of three-phase motors, especially motors driving elevators, saws, planers, metalworking machines, etc. Therefore great care should be exercised in this respect when any changes are made on three-phase systems. Also, when paralleling three-phase transformers or three-phase banks, the phase sequence of each must be in the same order.

Several methods can be used to ensure proper phase sequence. The easiest way, usually, is to note and record the direction of rotation of a specific three-phase motor on the system before repairs or replacements on the system are started. After repairs are made, and before the entire system is connected, rotation of this specific motor should be carefully noted and compared with the record made before repairs. If both rotations are the same, phase sequence is the same for all other motors on the system. If the rotations before and after repairs are opposite, phase sequence is reversed, and it will be necessary to interchange any two of the three lines in the area where repairs were made.

Another method is to carefully mark and record connections and repair or replace only one part at a time, carefully checking the recorded connections during installation.

Still another method of maintaining proper phase sequence—and probably the most reliable—is to check and record the original phase sequence with a sequence indicator and check again after repair operations to determine if the sequence has been changed. Connections and use of a sequence indicator for secondary circuits of 480 V or less are illustrated in Fig. 14-25. This indicator consists of two neon lamps, A and B, connected as shown, with lamp A to lead 1 and lamp B to lead 3. The capacitor is connected to lead 2.

The leads of the indicator are marked 1, 2, and 3

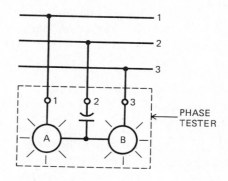

Fig. 14-25. Phase sequence tester consisting of two neon lamps and a capacitor. If lamp A is brighter, the phase sequence is 1-2-3. If B is brighter, the sequence is 3-2-1.

and can be connected in any order to the lines. The number of each lead should be placed on the line to which it is connected. This should be done on a portion of the line that can be disconnected from the load on the load side of the repair operation.

In operation, if lamp A is brighter than lamp B, the phase sequence is 1-2-3, or counterclockwise around the numbered terminals shown at the right end of the lines in the illustration. If lamp B is the brighter, the sequence is 3-2-1, or clockwise around the terminals. After repairs, a check should be made to see if conditions have been reversed. It should be made certain that the leads of the tester are connected to the right lines according to the lead numbers and the numbers previously placed on the lines. If the test shows that phase sequence has been reversed, interchanging any two of the three lines will correct the condition.

14-16   TRANSFORMER LOSSES

Power loss in a transformer is due chiefly to three factors—*eddy currents, hysteresis, and I^2R loss in the windings*. They all produce heat that must be dissipated.

Eddy currents are local short-circuit currents induced in the core by the alternating magnetic flux. In circulating in the core, they produce heat. They are minimized by laminating the core.

Hysteresis is the lagging of the magnetism in the core behind the alternating flux being produced. This lagging or out-of-phase condition is due to the fact that power is required to reverse magnetism; it does not reverse until the flux has attained sufficient force to reverse it. The reversing magnetism results in friction, and friction produces heat in the core which is a form of power loss. Hysteresis is minimized by the use of special steel alloys properly annealed.

The I^2R loss is sometimes referred to as *copper loss*. It is power lost when current flows in the windings. This represents the greatest loss in the operation of a transformer. The actual power lost can be determined in each winding by the formula

$$P = I^2R$$

where R = resistance of winding
I = current in winding
P = power loss in winding

Transformer efficiency varies with size and load. The larger size transformers are generally more efficient than the smaller sizes. A standard 2400-V 1 1/2-kVA transformer, operating at from one-fourth to full load, will vary in efficiency from 94 to 96 percent. A 2400-V 500-kVA transformer, operating under the same load conditions, will vary in efficiency from 98 to 99 percent.

14-17   CORES

Transformer cores are made of high-permeability steel alloys to conduct magnetism produced by the windings. High permeability ensures good magnetic linkage between the primary and secondary windings. Cores are made up of either coils of sheet steel or thin sheets of steel called *laminations*.

When an iron or steel core is placed in an alternating magnetic field, such as one produced by the primary winding of a transformer, the core will get hot. This heat represents a loss of power. The heat is produced by random currents, induced in the core by the primary current, and circulating freely throughout the core. These currents, shown in Fig. 14-26(a), are called *eddy currents*. By laminating the core, the paths of the eddy currents are reduced considerably as seen in Fig. 14-26(b), thereby reducing heat and power loss. The thinner the lamination, the greater will be the reduction in eddy currents. Eddy currents are prevented from flowing from lamination to lamination by a thin coating of insulating material on the flat surfaces of the lamination. This insulation results from the type of steel used (typically, silicon steel) or from a manufacturing process during the rolling of the sheet steel.

Laminations can be made from the special transformer sheet steel in a wide range of thicknesses. Typical laminations are 15 to 20 mils thick (0.015 to 0.020 in.). Cores are made by stacking the laminations to the desired thickness, then bolting or clamping the stack together tightly. Commercially stamped laminations are available in a range of shapes described as I's, E's, L's, F's, etc. Their shape is obvious from their designations. Figure 14-27 shows some

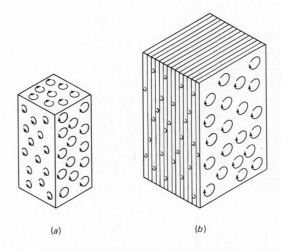

(a)   (b)

Fig. 14-26.   **Eddy currents in steel. (a) Circulating currents in a solid piece of metal. (b) Laminations reduce the path of the circulating currents.**

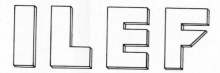

Fig. 14-27. **Typical lamination stampings.**

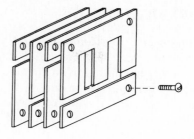

Fig. 14-29. **To reduce the air gap in the core, E's and I's are stacked alternately.**

of these shapes. In Fig. 14-28, the laminations are assembled to form typical cores. The placement of the windings is shown with broken lines.

A magnetic circuit must exist to permit the transformer to operate. The steel core provides the low-resistance (in magnetic circuits it is called *reluctance*) path similar to copper conductors in an electric circuit. However, wherever the edges of a lamination butt up against another lamination, a tiny space is formed. This space is called an *air gap*. Since air does not conduct the magnetic flux very efficiently, this gap presents a high reluctance path in the magnetic circuit which increases losses in the transformer. If the edges of the laminations could be superfinished, they would abut more solidly and the air gap would be greatly reduced if not completely eliminated. Since this is not possible from a practical standpoint, laminations must be stacked carefully and clamped tightly to keep the air gap to a minimum. In assembling laminations, alternate stacking is often used. Figure 14-29 shows how the E's and I's are alternated to reduce the air gap.

Cores of some medium-size transformers are made by winding a continuous strip of sheet steel on a rec-

tangular form with rounded corners. Such cores are known as *wound cores*. The spiral layers are bonded together, and after removal from the form, the core is sawed in half, making two U-shaped pieces. The sawed ends of each piece are finely machined to ensure a minimum air gap when the two pieces are assembled and clamped together in their original positions. The coil or coils are placed on the legs of the pieces when they are assembled. Wound-type cores are shown in Fig. 14-30.

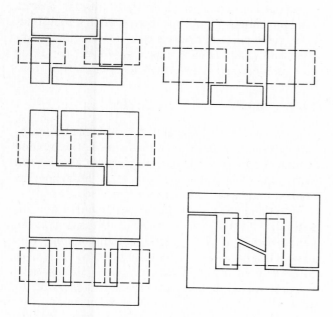

Fig. 14-28. **Laminations are arranged to form a continuous magnetic path.**

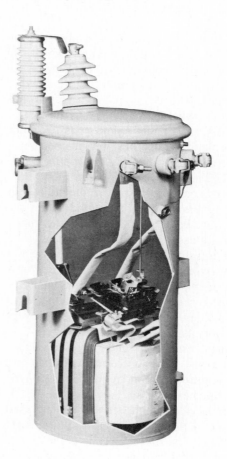

Fig. 14-30. **Wound-core transformer.**
(*McGraw Edison Company*)

14-18 TRANSFORMER COOLING

Like most electrical equipment, transformers are limited in capacity by their operating temperature. The life of class *A* insulation (organic materials) in an oil-immersed transformer is about halved for each 8° to 10°C rise in operating temperature. It is therefore essential that heat be carried away from the windings if normal life of insulation is expected.

Several methods are used to dissipate the heat produced in transformer cores and windings. Some transformers are designed to transfer the heat directly to the air surrounding the transformer. This type is known as *dry-type transformers*. They are designed with sufficient air spaces in and around the coils and core to allow sufficient air circulation for cooling. Some dry-type transformers depend on a blower for air circulation. The smaller sizes of dry-type transformers are not enclosed but are often coated with insulating material or covered with metal shells. Coils are sometimes encased in epoxy materials. Larger dry-type transformers are mounted in metal containers or tanks. Since the outside surface of the enclosure must dissipate the internal heat, the surface is often corrugated to increase surface area and provide greater cooling.

Many larger transformers, and those operating at very high voltages, use a liquid or vapor coolant for heat transfer. Oil is a commonly used coolant, but in applications where oil would present a fire hazard, other coolants must be used.

A popular nonflammable liquid used in the past is known generically as askarel. Askarel has excellent cooling and insulating properties, but it contains a chemical known as PCB (which stands for polychlorinated biphenyl). This chemical has been found hazardous to the environment, and therefore its use in electrical equipment is no longer permitted. Other liquids and vapors are being used as substitutes. Since the coolant in most cases is in direct contact with the windings of the transformer, it is absolutely essential that the coolant have excellent insulating properties. Silicone liquid (a synthetic polymer) meets many of the requirements for a good coolant. In addition to being nonflammable, it is odorless, clear, and very stable. Other coolants include mineral oil, high-fire-point petroleum distillates, and high-molecular-weight hydrocarbons. Vapor-cooled transformers use coolants such as nitrogen, nonflammable Freon-type liquids, and sulfur hexafluoride (SF_6).

A coolant conducts heat to the sides of the tank, and the tank conducts it to the outside air. To aid in the dissipation of heat, some tanks are equipped with fins to increase the radiating surfaces.

Large transformers may use vertical tubes around the tank as in Fig. 14-31. These tubes connect to the

Fig. 14-31. Large transformer with cooling tubes.
(*McGraw Edison Company*)

bottom of the tank and just below the coolant level at the top. The warm coolant in the tank enters the top of the tube and is cooled by the outside air. The cool coolant then circulates to the bottom of the tube and reenters the tank at a lower temperature than it entered at the top. This natural circulation is continued as long as the outside air is able to cool the coolant.

Water coils are sometimes immersed in the coolant and connected to a radiator on the outside with a pump for water circulation. In some cases fan-cooled radiators are used to cool the water.

Sufficient cooling enables a transformer to operate at a higher rating for the materials used in its construction, but a point is reached at which cooling equipment and its operational expenses will not offset the cost of additional transformer material.

14-19 CARE OF COOLANTS

Oil is a widely used coolant. It must be carefully selected, stored, installed, and maintained. The oil used in transformers is free of water, acids, alkalis, and sulfur. In order for it to circulate freely throughout the coolant system, it must have a low viscosity.

Water and oxygen are the greatest enemies of cooling oil. Oxygen will produce a chemical reaction called *oxidation*. The oxygen in the air combines with the coolant to produce a semisolid material that sinks to the bottom of the transformer tank. This material forms a sludge that reduces the effectiveness of the coolant and reduces its insulating properties.

The quality of transformer oil can be specified by a dozen or more factors. Generally the oil in industrial transformers is tested routinely for its dielectric strength and its acidity. Water in the oil usually causes a reduction in dielectric strength. The dielectric test uses a special cup containing two 1-in. electrodes spaced 0.1 in. apart. About a half pint of transformer oil is placed in the cup, and the voltage across the electrodes increased at the rate of 3000 V per second until the oil breaks down. At that point the voltage is read. The test is repeated a number of times to obtain an average reading. A breakdown voltage over 22 kV is considered acceptable, indicating that the oil is sufficiently dry.

The acidity test involves chemically neutralizing a sample of the oil to obtain a specific color. Although absolute standards of acidity do not exist, the recommendation of the manufacturer should be followed. Yearly acid tests are usually recommended, though sealed units could be tested less frequently.

If a customer is not equipped to test oil, this service is usually furnished by the supplier or manufacturer of the oil. Rigid rules regarding the taking of a sample of oil from a transformer should be observed. These rules should be obtained from the tester. It is essential that instruments used and the container for the sample oil be properly cleaned and kept clean. A minute amount of moisture introduced into the oil in taking a sample can cause the oil to fail the test.

Two of the most common ways for water to get into oil are through mishandling and by "breathing" of the transformer or storage barrels. Breathing is caused by pressure changes in a closed container which are caused by temperature changes. When the temperature of a container increases, it drives air out through even the smallest opening. When the container cools, the internal air pressure drops, and air from the outside is forced in. Moisture in the air forced into the container condenses to form water. Breathing can be prevented by keeping seals and plugs tight.

The same precautions that are taken with oil should be taken with other coolants in keeping them from water and foreign matter. Some of the coolants are harmful to the operator, especially the hands and some parts of the face. Special precautions recommended by the manufacturer should be observed in these cases.

14-20 SPECIALTY TRANSFORMERS

Transformers designed for specific applications other than serving as general sources of power are often called specialty transformers. Autotransformers, instrument transformers, signal and relay transformers, and constant-current transformers are specialty transformers of interest to industrial electricians and technicians.

14-21 AUTOTRANSFORMERS

An autotransformer *does not* refer to a transformer used in an automobile. It is a single-winding transformer that is mounted on an iron core. The primary and secondary currents flow in parts of the same winding. The ratio of an autotransformer is low, seldom exceeding 4:1. Two winding transformers are more economical for high ratios.

The main advantages of autotransformers are economy in construction and operating efficiency in low-ratio requirements. The main disadvantages are that the primary and secondary windings are connected together and their ratios are low. With the two windings connected together, the low-voltage winding is subjected to the high voltage in case of an accident. This and their low ratios make them unsuitable for distribution systems, but they are commonly used on transmission systems and single-phase electric railway systems in their higher ratings, and for reduced-voltage starting of ac motors in some of their lower ratings.

The operation of an autotransformer can be understood with an example. Figure 14-32 illustrates an

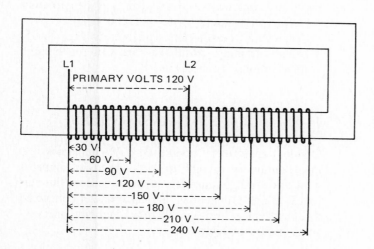

Fig. 14-32. An autotransformer with a 2:1 voltage ratio.

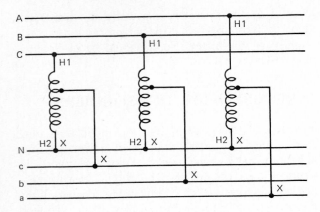

Fig. 14-33. An autotransformer used in the secondary of a four-wire system.

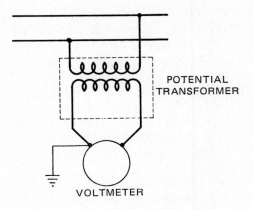

Fig. 14-34. Connecting a potential transformer to measure high-voltage lines.

autotransformer with a ratio of up to 2:1. The primary winding, between L1 and L2, is excited with 120 V. The leads are marked with voltages that are obtainable from them in both step-down and step-up connections. Voltages of 30 V, 60 V, 90 V, 120 V, 150 V, 180 V, 210 V, and 240 V are obtainable from this transformer.

If a load is placed across the 120-V lines, the primary current will flow directly from the primary lines into the secondary lines and to the load. Any current at less than 120 V will be primary current, and any current at more than 120 V will be primary current plus induced current from the winding outside the primary lines.

The voltage ratio is determined by the number of turns in the winding between the primary lines and by the number of turns in the winding between the secondary lines.

Connection of autotransformers for three-phase wye-wye three- and four-wire service is illustrated in Fig. 14-33.

14-22 POTENTIAL TRANSFORMERS

Potential transformers (often called *PTs*) are small two-winding transformers similar in construction to power or distribution transformers. They are used to transform high voltages to a low voltage for operation of voltmeters, wattmeters, watthour meters, and signaling and protective devices on high-voltage systems ranging from about 480 V and up. This facilitates wiring these devices and reduces the personal hazard of high voltages at panel and control boards.

Usually, the ratio is from the high voltage to 120 V on the secondary. If a 150-V meter is used, the meter reading is multiplied by a multiplier based on the ratio. Some meters are graduated on the scale to read the high voltage directly, although operating on the low voltage.

The connection of a potential transformer to a high-voltage line and a low-voltage voltmeter is shown in Fig. 14-34. A ground wire is shown connected to the secondary winding line at the meter. This is a very important ground connection that should be carefully made and maintained in good electrical condition. It protects the meter and operator against the high line voltage in case of an insulation failure in the transformer.

Most potential transformers operating on 13,800 V or less are dry type, while those for higher voltages are oil-insulated.

In connecting a potential transformer to a single device, polarity is of no consequence. However, polarity must be known when potential transformers are used in combination with certain types of equipment, such as wattmeters or watthour meters. They are marked to indicate polarity. Instantaneous voltages are assumed to be in on the marked terminal. The insulators for these terminals will be white, or a similar mark will appear at the base of each insulator, or H1 will appear at the high-voltage insulator base and X1 at the low-voltage insulator base.

The polarity markings are followed according to the installation instructions and drawings of the particular wattmeter or watthour meter being installed.

A potential transformer is pictured in Fig. 14-35. Two large bushings (insulators) are used for the high voltage primary connections.

14-23 CURRENT TRANSFORMERS

Current transformers (often called *CTs*) make possible measurement of current up to thousands of amperes with a low-range ammeter, or they measure the current in a high-voltage line with an ammeter connected to a low-voltage circuit, thereby removing the personal hazard of high voltage.

Fig. 14-35. Potential transformer for outdoor use rated at 34.5 kV. The two large bushings are used to connect the primary from line to line or from line to neutral. (*General Electric Co.*)

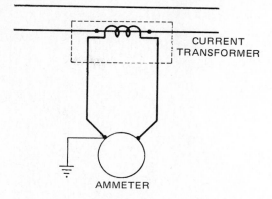

Fig. 14-36. Connecting a current transformer to measure high current values with a low-current ammeter.

Since it is not practicable to build self-contained ammeters in capacities above about 50 A, current transformers are used to step down high current values. The connection of a current transformer to a high-voltage line and ammeter is illustrated in Fig. 14-36. The ground will conduct the high voltage of the line to ground in event of insulation failure in the transformer. These transformers perform the same operational function as shunts do for dc ammeters. Current transformers are also used with wattmeters (Fig. 14-37) and watthour meters, and special types are used with signal and control devices.

Current transformers are constructed in several forms, but the primary is always in series with the power line—or the line itself is the primary in some cases. The primary is usually only one or two turns of a conductor which is connected in series with the power line; or it is a straight bar through the transformer core and is connected in series with the power line; or the power line may be run through the secondary of the transformer.

The secondary winding contains the proper number

of turns to provide the required current ratio for the meter. In some types this requires a large number of turns for the secondary winding. A large number of turns in the secondary of a current transformer creates an extremely dangerous condition if the secondary leads are open. A very high voltage, capable of producing a fatal shock, builds up in the secondary winding when it is open. For this reason the secondary leads should always be connected to an ammeter or kept short-circuited if the meter is removed. Many current transformers are equipped with a short-circuiting device. Since the secondary is wound with many turns of small wire, the impedance is high and very little current will flow even under short-circuit conditions. Great care must be exercised in removing the meter from the circuit. The secondary leads should be short-circuited before removal of the meter is attempted. The secondary winding should also be kept well grounded, especially when the transformer is operating on high voltages.

Current transformers can be connected to a single instrument without regard to polarity, but when connected in combination with other devices, such as wattmeters and watthour meters, it is necessary to know the polarity. Polarity is commonly indicated by

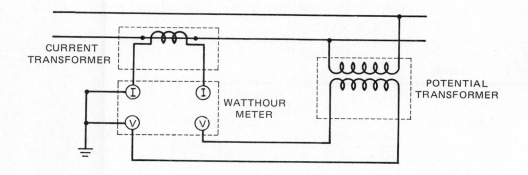

Fig. 14-37. Connecting a potential transformer and a current transformer to a wattmeter. The connection shown is for a single-phase system.

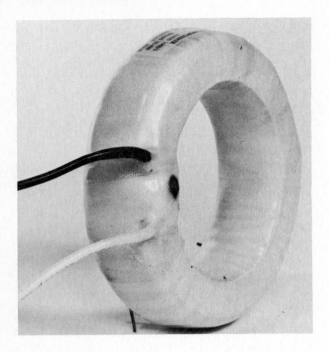

Fig. 14-38. A current transformer. (*General Electric Co.*)

marking one primary and one secondary lead. Instantaneous current flow is assumed to be into the marked primary lead and out of the marked secondary lead.

A current transformer that is used with the power line running straight through it is pictured in Fig. 14-38. The polarity is indicated by the black and white leads. This transformer is sometimes called a *doughnut-type transformer.*

A 5-A ammeter is generally used with a current transformer, and the transformer ratio is usually stated as 100:5, 250:5, or 500:5, or any value to 5 as the case may be. Some transformers and meters are designed to operate together, and the meter is calibrated to read current directly.

14-24 REMOTE-CONTROL AND SIGNAL TRANSFORMERS

Small two-winding transformers, of about 100-VA capacity, are used for operating doorbells, buzzers, chimes, remote-control house wiring, temperature-control units, etc. These transformers contain sufficient impedance to limit the secondary current to the extent that a short circuit of the secondary will not burn them out or create a fire hazard.

Transformers, to meet Class 2 NEC specifications, have their secondary current limited according to Table 14-1. Transformers meeting these specifications can be used in Class 2 wiring (NEC 725-31).

There are few limitations on Class 2 wiring methods on the secondary side of a transformer in the usual residential installation. Such wiring must be done with insulated wire; it must be kept 2 in. or more from light or power wiring, and a barrier must separate it from power wires if it enters a box containing power wires.

Doorbells usually require from 6 to 8 V, chimes from 8 to 20 V, temperature-control equipment about 24 V, and solid-state electronic equipment 5 to 15 V.

Control and signal transformers are sometimes mounted on an outlet box cover with the primary leads brought through the cover opposite the side containing the transformer. Connection of the primary leads to the power lines can thus be made in a box, as required by the Code. The secondary leads are usually connected to screw terminals on the side of the transformer. These terminals are usually marked to show the secondary voltage.

14-25 CONSTANT-CURRENT TRANSFORMERS

A constant-current transformer furnishes an unvarying current to a load although the load requirements may vary. A typical example of the use of these transformers is series street-lighting systems.

A constant-current transformer consists of two coils, a primary and secondary, mounted one above the other on an iron core. The bottom coil is stationary, and the top coil is movable on the core. The top coil is also counterbalanced so that it will remain stationary at any point on the core.

An increase in current flow in the secondary due to any cause increases the strength of the magnetic poles and pushes the coils further apart. The increased air space between the coils increases magnetic leakage

TABLE 14-1
Power Limitation for Class 2 AC Circuits

Maximum circuit voltage, V	0–20	21–30	31–150
Maximum circuit current, A	8.0	8.0	0.005
Maximum nameplate rating, VA	5 × max. voltage	100	0.005 × max. voltage
Maximum nameplate current, A	5.0	100 × max. voltage	0.005

between the coils and reduces current flow in the secondary coil. Thus a practically constant flow of current is maintained in the secondary coil and its circuit. In a well-designed transformer, current will remain within about 1 percent of the set value over the entire load range of the transformer.

While series street lighting is no longer common, a typical system had about 73 lamps rated at 32.8 V, drawing 6.6 A on a 2400-V line. Each lamp socket was equipped with a film-disk insulator across its terminals. If the lamp burned out and opened the series circuit, the high voltage that resulted across the socket terminals was enough to break down the film disk and short-circuit the terminals. The remaining lights continued to burn but drew increased current owing to decreased resistance in the circuit. The increased current flow caused the secondary coil of the constant-current transformer to move away from the primary and reduce the current flow in the secondary circuit. By this automatic adjustment to varying load conditions, a constant-current transformer was able to maintain the 6.6-A flow on the line required by the lamps.

14-26 PROTECTIVE DEVICES

Transformers are protected against overloads and short circuits by fuses or circuit breakers connected on the primary side. Fuses usually are of the indicating type and will blow on an overload. When the fuse blows it swings down into an open position. This type of cutout is shown in Fig. 14-39. The fuse contains a renewable element.

The fuse is provided with a ring for manually disconnecting the circuit with a hot-line stick. Another type of cutout is enclosed in a housing with a door that opens when the fuse blows.

Fig. 14-39. Indoor open disconnect-type high voltage fuse cutout. (*Westinghouse Electric Co.*)

Distribution transformer primary fuses are usually rated at 200 to 300 percent of full-load primary current. Transformers serving heavy-starting motor loads might require higher ratings or time-delay fuses. Primary fuse ratings for transformers rated 600 V and over within the scope of the Code and having only primary protection are limited to a maximum of 250 percent of the primary full-load current (NEC 450-3).

Circuit breakers, usually submerged in oil, are of the manual or automatic reclosure types. The manual breaker opens and remains open until it is reclosed by an operator. The automatic or repeater breaker opens momentarily and recloses a predetermined number of times before it remains open if the fault on the line is not cleared. Frequently a short circuit will blow itself clear, relieve the line of the fault, and permit continued operation without further attention when reclosure-type breakers are used.

14-27 SURGE ARRESTERS

The function of surge arresters is to pass abnormal voltage to ground and offer continuous insulation for normal voltage to ground. For maximum efficiency, arresters should be connected as close as possible to equipment they are to protect and should be well grounded.

Nearly all arresters are connected to the line at the top of the arrester; the grounding wire is connected to the bottom of the arrester. An arrester should be installed on all incoming, ungrounded, and high-voltage lines to a transformer. The transformer tank should be grounded to the arrester grounding wire.

Basically, all arresters used for lightning protection have about the same principle of operation and construction. The circuit through them is composed of the connecting line at the top, an air gap, a semiconducting material known as the valve element sealed in a porcelain or glass body, and the grounding connection at the bottom.

The air gap serves as the insulating medium between the line side and the grounded side of the arrester. It effectively insulates normal line voltages but allows lightning to pass. The semiconducting material in the body of the arrester provides a comparatively low resistance circuit for lightning and other abnormally high voltages; it also aids in quenching the arc at the air gap when the high voltages are dissipated, thereby stopping the flow of line current to ground.

Lightning arresters are of two types—expulsion type and valve type. In the valve type the element conducts high voltages but offers high resistance to line voltages, thereby aiding the air gap in quenching

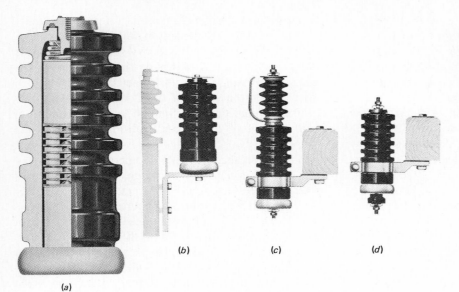

(a)

(b)   (c)   (d)

Fig. 14-40. Examples of surge arresters. (a) Cross-section of valve-type arrester. (b) Transformer mounted arrester with spark gap. (c) Conventional air gap arrester. (d) Arrester used to isolate ground lead. *(Ohio Brass Co.)*

the arc. The expulsion arrester is designed so that heat generated by the current through the arrester creates gases in the chambers in the body that expel conducting gases out through the bottom of the arrester, thereby aiding the air gap in stopping line current flow to ground.

By far, the most commonly used arresters are the valve type. Figure 14-40 shows examples of surge arresters in common use for distribution transformer protection.

Some single-phase distribution transformers are equipped with lightning arresters mounted on and grounded to the tank and with circuit breakers in the secondary winding to open on excessive overloads as in Fig. 14-40(b).

Lightning rods are often used in conjunction with surge protection systems. Solid-state equipment sensitive to surges and transients are often protected by surge arresters similar to lightning arresters. Such devices react much faster than lightning arresters, though they operate on much lower voltage levels.

SUMMARY

1. A transformer is an electrical device that transforms an ac voltage from one value to another circuit.

2. The basic transformer consists of two coils wound on an iron core. The iron core provides a low-reluctance path for the magnetic circuit developed by the current in the windings.

3. The winding of a transformer that receives current from the supply is the primary winding. The winding connected to the load is the secondary winding.

4. The degree of transformation of ac voltage in a transformer is determined by the ratio of turns of wire in the two windings. The voltages of the two windings are in direct proportion to the ratio, while the currents are in inverse proportion to the ratio.

5. Transformer cores are laminated to minimize eddy currents. Eddy currents are local short-circuit currents induced in the core and are a form of power loss.

6. A transformer automatically adjusts its energy intake in proportion to its energy output. This is accomplished as the magnetism from the secondary, or load, winding counters the effects of cemf produced by magnetism in the primary, or supply, winding.

7. Transformer terminals are marked for polarity to facilitate connecting them in combinations with other transformers.

8. Power loss in a transformer is due chiefly to the I^2R loss in the windings, eddy currents in the core, and hysteresis. Hysteresis is power required to reverse the magnetism in the core.

9. Power losses occur when energy is converted to heat. Excessive heat is injurious to the insulation and coolant in a transformer and must be dissipated.

10. Various means are used to dissipate heat, including fins on the tank, a coolant such as oil, cooling tubes, water coils in the coolant, and radiators or fans.

11. Water in coolant oil reduces its dielectric strength. Transformer oil with sufficient dielectric strength should withstand 22,000 V or more between 1-in. disks spaced 0.10 in. apart without breaking down.

12. The grounded line from a transformer is known as the neutral line, and the ungrounded lines are known as hot lines. The wire that connects the grounded line to ground is known as the grounding wire.

13. Grounding is done solely to protect persons and property against lightning and excessive line voltages. Electrical equipment operates equally well with or without grounding of the system.

14. Customer service systems, operating at less than 150 V to ground, are required by the Code to be grounded at the transformer and at the service entrance. One element of a complete grounding system is connection to an underground water system if it is available on the premises.

15. The neutral line on a three-wire single-phase system from the service head cap to the equipment it serves should be continuous, i.e., without switches, fuses, etc., and must be identified with gray or white insulation.

16. In connecting single-phase transformers in three-phase banks, it is necessary in some cases that the impedance and voltage ratios of the transformers be matched.

17. If the phase sequence on a three-phase system is reversed, the direction of rotation of all three-phase motors on the system will be reversed. This, under certain conditions, can be highly dangerous to persons and property.

18. In a wye connection, $V_l = 1.73 V_{ph}$. The phase and line currents are the same.

19. In a delta connection, $I_l = 1.73 I_{ph}$. The phase and line voltages are the same.

20. An autotransformer is a one-winding transformer, usually with a low ratio, seldom over 4:1. In large sizes, they are used chiefly on transmission and railway systems.

21. Potential transformers (PTs) are used to transform high voltages to comparatively harmless low voltages for metering and control purposes.

22. Current transformers (CTs) are used to step down heavy currents for metering purposes.

23. Fuses or circuit breakers are connected in the primary lines at transformers to serve as disconnects and protect transformers against excessive overloads and short circuits.

24. Surge arresters are used to conduct lightning from line to ground. They should be connected to all ungrounded outdoor lines but are usually not used in secondary service lines. Arresters should be well grounded.

QUESTIONS

14-1 What winding of a transformer is connected to the line or power source?

14-2 What winding of a transformer is connected to the load?

14-3 There are 65 turns of wire in the primary of a transformer and 13 turns in the secondary. What is the turns ratio (in its simplest terms)?

14-4 What is the voltage ratio of the transformer in question 3?

14-5 The load current for the transformer in question 3 is 10 A. What is the primary current?

14-6 A step-up transformer has a voltage ratio of 1:10. The load is rated at 138 kVA. If the primary voltage is 1380 V, what is the primary current?

14-7 In an ideal transformer if no current is taken from the secondary, no current will flow in the primary. However, in an actual transformer current does flow in the primary even when no current is drawn from the secondary. What is this primary current called? What is its purpose?

14-8 Explain some of the advantages of a three-wire single-phase system.

14-9 A single-family house is served with a 120/240-V three-wire single-phase system. Each wire has an ampacity of 100 A. Show how a total load of 200 A can be connected to this system.

14-10 A three-wire system has two hot legs, A and B, and a neutral conductor, N. (**a**) If A is carrying 10 A and B is carrying 15 A, what is the neutral carrying? (**b**) If A is carrying 20 A

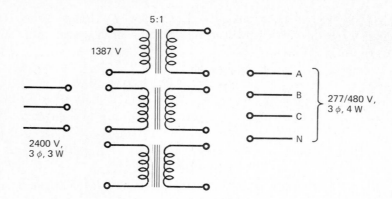

5:1

1387 V

2400 V,
3 φ, 3 W

A
B
C
N

277/480 V,
3 φ, 4 W

Fig. 14-41. Three single-phase transformers for question 14-22.

and B is carrying 20 A, what is the neutral carrying?

14-11 Explain the difference between a three-wire single-phase system and a three-wire three-phase system.

14-12 What are the two most common three-phase connections?

14-13 Which one of the connections in question 12 can have a true neutral?

14-14 The phase voltage of a certain three-phase system is 120 V and the line voltage is 208 V. How are the phases connected?

14-15 If the phase current in the system in question 14 is 17.3 A, what is the line current?

14-16 The phase voltage of a certain three-phase system is exactly the same as the voltage measured across any two lines. How are the phases connected?

14-17 The phase current in the system in question 16 is 17.3 A. What is the line current?

14-18 What does a transformer's turns ratio indicate about the impedance of the windings?

14-19 How is percent impedance of a transformer found?

14-20 A three-phase wye-wye–connected transformer has 5 percent impedance. The low-voltage windings are short-circuited. The high-voltage winding is connected to a varying supply voltage. Starting from 0 V, the voltage is slowly increased on the high-voltage winding until the full-load current is flowing in the low-voltage winding. At that point the voltage measured across the high-voltage winding is 24 V. What is the rated voltage of the high-voltage windings?

14-21 Three single-phase transformers are to be connected delta-delta. Should they all have the same voltage ratio? Should they all have the same impedance? Explain your answer.

14-22 Figure 14-41 shows three single-phase transformers. Each primary winding is rated at 1387 V. The turns ratio is 5 : 1 and the transformers are step-down types. The supply lines are 2400 V, 3 φ, 3 W. The load requires 277/480 V, 3 φ, 4 W. Redraw the transformers showing how the primaries and secondaries are to be connected to provide the necessary service.

14-23 Explain the high-leg or wild-leg connection. What are the advantages and disadvantages of this type of connection?

14-24 What is an open delta? Why would it be used?

14-25 How are transformers rated?

14-26 Under what conditions is it important to establish the relative polarity of transformer windings?

14-27 Explain the use of ratio adjusters.

14-28 What are two major purposes of grounding systems?

14-29 Why is it important to maintain proper phase sequence when making repairs in three-phase systems?

14-30 What are the major power losses in transformers? Indicate whether the losses are in the core or the windings.

14-31 How are transformers cooled? Describe three methods.

14-32 Does an autotransformer have separate primary and secondary windings? Explain your answer.

14-33 What are potential transformers and why are they necessary?

14-34 What are current transformers and why are they necessary?

14-35 Explain the principles of operation of a constant-current transformer.

14-36 What is a surge arrester and why is it important?

15
SINGLE-PHASE MOTORS

Single-phase motors operate on a single waveform from a two-wire alternating-current system. Practically all single-phase current in commercial use is obtained from two lines of a three-phase supply system.

Alternating single-phase current produces a varying magnetic field. This varying field, when applied directly to the main field windings of a typical single-phase motor, cannot produce torque for starting. Series motors contain an armature winding in series with the main winding and therefore do develop sufficient starting torque. All other single-phase motors require some provisions for producing a starting torque.

15-1 TYPES OF SINGLE-PHASE MOTORS

General-purpose single-phase motors are classified in types by the methods used for starting them. They are selected for service principally by the amount of starting torque they can develop. There are five general types of single-phase motors. In order of starting torque, beginning with the least, the five are *shaded pole, split phase, capacitor, repulsion,* and *series.* Because the shaded-pole motor has a very low comparative starting torque, it is used only in small sizes and in applications such as timers, small fans, light-duty kitchen appliances, etc.

Average starting torques of single-phase motors, expressed as a percentage of the running torque of each motor, are as follows:

Shaded pole	100%
Split phase	200%
Capacitor	300%
Repulsion	400%
Series	500%

(All figures are approximate.)

Shaded-pole, split-phase, and capacitor motors are classed as *induction* motors, since rotor current is induced in the rotor from the stator winding. They use a *squirrel-cage* rotor, which contains no windings.

Repulsion and series motors use wound armatures similar to dc armatures. Repulsion motor armatures receive current by *induction* from the stator winding. Series motor armatures, in series with the fields, receive current *directly from the line* through the fields.

15-2 SQUIRREL-CAGE ROTORS

Split-phase, capacitor, and shaded-pole induction motors use squirrel-cage rotors. The name is derived from the appearance of the conductors in these rotors.

The conductors of a squirrel-cage rotor consist of large copper bars inserted in slots in the core, and a copper ring at each end to which the bars are welded. Some motors are aluminum bars and rings to reduce the weight of the rotor. The assembly of bars and rings has the general appearance of a squirrel cage. The conductors are uninsulated. In ac motors, current is induced in the squirrel-cage conductors by alternating magnetism from the stator fields.

A typical squirrel-cage rotor is shown in Fig. 15-1. This is a die-cast rotor. During manufacture, the rotor core is placed in a die, and molten aluminum alloy is

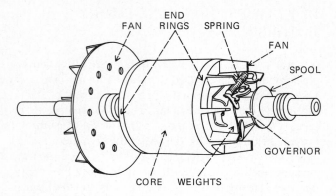

Fig. 15-1. A simplified squirrel-cage rotor. For clarity, the core laminations are not shown.

cast in the die and holes through the rotor core to form the bars and end rings. The fins on the end rings help radiate heat in the rotor and also serve as fan blades to cool the motor.

Some single-phase rotors contain copper bars soldered together or to an end ring. If they get too hot, the solder melts and is thrown from the rotor by centrifugal force when running. Symptoms of lost solder are excessive heating (the rotor will get hotter than the remainder of the motor), low starting torque, and noisy operation. Rotors can be tested for lost solder and broken die-cast bars by placing them in an armature growler and sprinkling iron filings on the core with the growler excited. Filings will cling to the core only over bars that are intact and properly connected to the end rings.

15-3 CENTRIFUGAL SWITCHES

Most split-phase and capacitor motors use a centrifugal switch as part of their starting and running systems. One part of the switch is mounted on the rotor as shown in Fig. 15-1.

The governor weights shown in the figure are held inward toward the shaft by springs. However, the weights are pivoted so that they can move away from the center of the shaft. When the rotor turns, centrifugal force tends to push the weights outward. At about 75 percent of full-load speed the weights are pushed outward. When the rotor is at rest, the spool presses against the contacts of a stationary switch mounted on the frame of the motor. The pressure of the spool on the switch contacts keeps the switch closed. When the rotating weights are pushed outward, the spool moves close to the rotor core and relieves the pressure on the stationary switch contacts. The contacts then open.

As the rotor slows, centrifugal force diminishes and the spring pulls the weights close to the shaft again. This pushes the spool away from the core and against the stationary switch contacts, closing the contacts. When the rotor finally stops, the weights and spool are in position to help start the motor again.

To adjust a centrifugal switch, the rotor core is centered in the stator core and end play is taken out with thrust washers on the shaft. The switch is then adjusted for slight overplay on closing.

There are scores of designs for switches that operate by centrifugal force. Replacement springs for

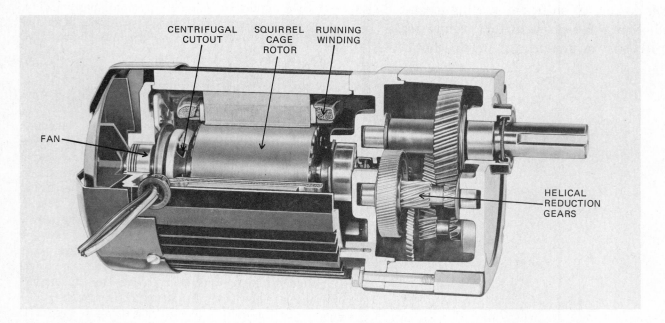

Fig. 15-2. A typical small single-phase squirrel-cage induction motor. (*Bodine Electric Company.*)

these switches must be identical in tension with the original to operate at the proper speed.

15-4 SPLIT-PHASE MOTORS

A single-phase split-phase motor (Fig. 15-2) operates with single-phase current, but it divides, or *splits,* the single-phase current into what appears to be *two-phase* current. The two phases produce a *rotating magnetic field* in the stator for starting.

Single-phase current is split by two windings, the *main running* winding and an *auxiliary* winding, the *starting* winding, which is displaced in the stator 90 electrical degrees from the running winding. The starting winding is connected in series with a switch, centrifugally or electrically operated, to disconnect it when the starting speed reaches about 75 *percent of full-load speed.*

Phase displacement is accomplished by the difference in inductive reactance of the two windings and the physical displacement of the windings. The starting winding contains *comparatively few turns of small wire.* Few turns mean reduced inductive reactance, and the small wire means increased resistance to limit current flow. Thus limited current flow and few turns reduce *inductive reactance* in the starting winding to a *comparatively low value.*

The *running winding* contains a *comparatively large number of turns of large-size wire,* and *inductive reactance is comparatively high.*

The way in which the two windings in a split-phase motor produce a rotating magnetic field in the stator is shown in Fig. 15-3. A two-pole stator, with starting and running windings, is shown. The sine waves indi-

cate the time displacement of current in the starting and running windings.

In Fig. 15-3(*a*) and point *a* on the waveform, the current in the starting winding has already begun to increase. It leads the current in the running winding because its inductive reactance is much lower. Since the magnetism produced by these currents follows the same wave pattern, the two sine waves can be thought of as the waveforms of the electromagnetism produced by the two windings.

In Fig. 15-3(*b*) and point *b* on the waveforms, the magnetic field of the starting winding is beginning to weaken while the magnetic field of the running winding is increasing toward its maximum value. This appears to shift the position of the poles developed in the stator, as shown in Fig. 15-3(*b*).

In Fig. 15-3(*c*) and point *c* on the waveforms, the magnetic field of the starting winding has dropped to zero while the magnetic field of the running winding is near its maximum value. The positions of the poles are now as shown in Fig. 15-3(*c*).

As the alternations in current (and magnetism) continue, the position of the poles changes in what appears to be a clockwise rotation. Naturally the poles do not physically move. It is the periodic buildup and weakening of the poles that give the appearance of rotation.

The important fact is that the stationary conductors in the squirrel-cage rotor are influenced by the constantly changing magnetic field. The *rotating magnetic field cuts the squirrel-cage conductors* of the rotor and *induces current* in them. This current creates *magnetic poles* in the rotor which *interact* with the *poles* of the *stator rotating magnetic field* to produce *torque* by the rotor.

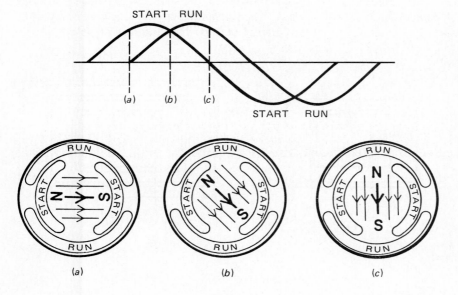

Fig. 15-3. The difference in inductive reactance between the running and starting windings produces a rotating magnetic field.

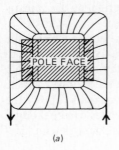

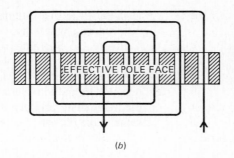

(a)   (b)

Fig. 15-4. Comparison of dc and ac motor windings. (*a*) Concentrated pole windings used in dc motors. (*b*) Concentric windings used in ac stator windings.

The direction of rotation of a split-phase motor is *from a starting pole of one polarity toward the adjacent running pole of the same polarity*.

Starting current of the average split-phase motor is about 5 to 7 times full-load current. If the motor does not have built-in protection, it should be protected by a time-delay fuse or circuit breaker with an interrupting setting of about 125 percent of motor full-load current. Full-load current is given on the motor nameplate. Split-phase motors are available in size up to about 1/2 hp.

15-5   SPLIT-PHASE WINDINGS

Split-phase motor poles are formed usually by *concentric* coils. They differ from dc motors, which use *concentrated* windings placed on *solid pole pieces*. Concentric and concentrated windings are shown in Fig. 15-4. The coil in Fig. 15-4(*a*) is a concentrated winding, that is, one coil. In Fig. 15-4(*b*) the winding consists of coils wound in slots of a stator to form a pole. The winding is distributed concentrically among adjacent slots as shown. This is also known as a *distributed winding*. While most ac motors use distributed windings in their stator, both types of windings are used in the different types of single-phase motors.

15-6   WINDING A CONCENTRIC POLE

The development of a concentric-coil–distributed-winding pole is illustrated in Fig. 15-5. For simplicity the coil shown consists of only two turns of wire. The beginning wire is called the *start* lead. The wire is not cut between coils.

In Fig. 15-5(*a*) the two turns fill two slots separated by two empty slots. The next slots to the left and right, respectively, are filled with two turns as shown in Fig. 15-5(*b*). Finally the next two slots to the left and right are filled with the two turns [Fig. 15-5(*c*)]. The end of the winding is known as the *finish* lead. The three concentric windings of two turns each make up a single pole in the pattern shown.

15-7   CONNECTING CONCENTRIC POLES

The start and finish leads of a completed winding are connected to produce proper magnetic polarity in each pole, and for connection to the line leads. The polarity is determined by the right-hand coil rule. A single-circuit connection is illustrated in spiral form in Fig. 15-6.

In the illustration, the start lead of pole A forms line lead L1. The finish lead of pole A is connected to the finish lead of pole B. The start lead of pole B is con-

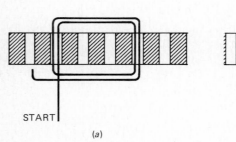

START

(a)

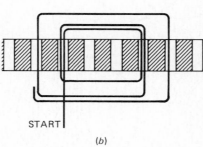

START

(b)

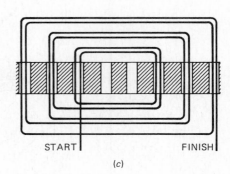

START   FINISH

(c)

Fig. 15-5.   Winding concentric coils in a stator.

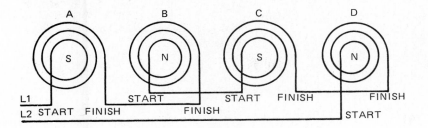

Fig. 15-6. **Connecting four concentric stator windings to produce alternate north-south poles.**

nected to the start lead of pole C. The finish lead of pole C is connected to the finish lead of pole D. Finally the start lead of pole D forms line lead L2.

Note that finish leads of adjacent poles are connected together and start leads are connected together to form alternate magnetic poles.

Four poles formed by concentric coils in stator slots in a flat form are shown connected for alternate magnetic polarity in Fig. 15-7. The start and finish leads are designated, respectively, by the letters S and F. This is the way an actual winding in a stator would be connected.

15-8 REWINDING SPLIT-PHASE MOTORS

A motor winding should be studied thoroughly before rewinding is attempted. Data on all parts of the winding should be carefully recorded on a data form so that there will be no guessing about data after the winding is stripped from the motor.

A typical data form is shown in Fig. 15-8. This form is filled in with data taken from the stator in Fig. 15-9. This winding will be stripped and rewound, and the step-by-step process of rewinding will be discussed and shown in the accompanying figures.

Most of the electrical and temperature data are taken from the nameplate of the motor. The boxes in the horizontal table represent slots in a stator. In this method of recording data, the fill of any one slot, with either or both running and starting winding turns, can be determined at a glance.

The squares shown under this table are used to indicate the connection of the running windings. The circles to the right of these squares are used to record lead positions. The *outer circle* represents the *running winding*, the *next circle* represents the *starting winding*. When the position of the leads is known and recorded, the winding, after being completed, can be laced so that the leads will emerge from the winding in the proper places. Otherwise it will be necessary to partially assemble the motor to determine lead positions.

Running winding data come next. The *pole span* is the number of stator slots occupied and spanned by the *complete pole*. The space *Do Poles Lap?* should be answered "yes" if the outer coil of a pole laps in the same slot with the outer coil of another pole.

In the space *Coils Fit* indicate whether the winding is "tight" or "loose." Occasionally it is necessary to omit a few turns from a coil that is too tight and include them in another coil. If such a change is necessary, it should be made in all poles and recorded in the *Number of Turns per Slot* table in the form.

In the starting winding section, the data in the space marked *Total Turns per Pole* are used for changing from *concentric* to *skein* coils. For example, if a pole of three coils has a total of 54 turns, it can be skein-wound with 18 turns in a skein. The skein is produced by wrapping wire around two pegs to obtain a single coil. The skein is made long enough to be twisted into smaller coils.

15-9 PREPARATION FOR REWINDING

A four-pole stator to be rewound is shown in Fig. 15-9(*b*). Data for this motor are recorded in the data sheet shown in Fig. 15-8. The coils are cut to facilitate

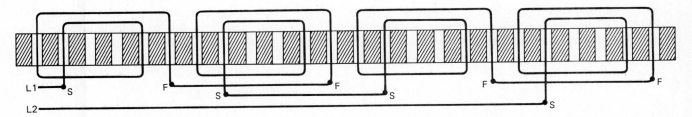

Fig. 15-7. **Concentric stator windings shown in flat layout.**

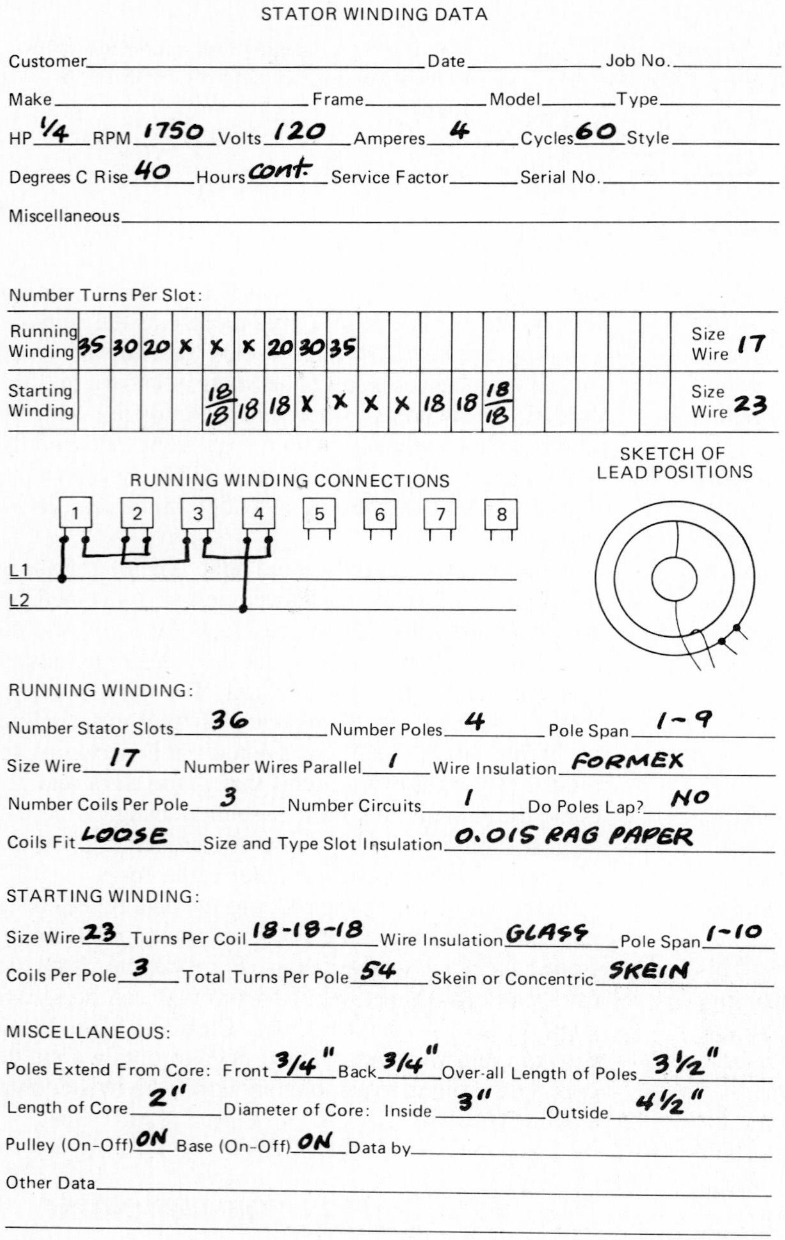

STATER WINDING DATA — STATOR WINDING DATA

Customer_____ Date_____ Job No._____

Make_____ Frame_____ Model_____ Type_____

HP **¼** RPM **1750** Volts **120** Amperes **4** Cycles **60** Style_____

Degrees C Rise **40** Hours **cont.** Service Factor_____ Serial No._____

Miscellaneous_____

Number Turns Per Slot:

Running Winding	35	30	20	X	X	X	20	30	35									Size Wire **17**
Starting Winding				$\frac{18}{18}$	18	18	X	X	X	X	18	18	$\frac{18}{18}$					Size Wire **23**

RUNNING WINDING CONNECTIONS

SKETCH OF LEAD POSITIONS

1 2 3 4 5 6 7 8

L1
L2

RUNNING WINDING:

Number Stator Slots **36** Number Poles **4** Pole Span **1~9**

Size Wire **17** Number Wires Parallel **1** Wire Insulation **FORMEX**

Number Coils Per Pole **3** Number Circuits **1** Do Poles Lap? **NO**

Coils Fit **LOOSE** Size and Type Slot Insulation **0.015 RAG PAPER**

STARTING WINDING:

Size Wire **23** Turns Per Coil **18-18-18** Wire Insulation **GLASS** Pole Span **1-10**

Coils Per Pole **3** Total Turns Per Pole **54** Skein or Concentric **SKEIN**

MISCELLANEOUS:

Poles Extend From Core: Front **¾"** Back **¾"** Over-all Length of Poles **3½"**

Length of Core **2"** Diameter of Core: Inside **3"** Outside **4½"**

Pulley (On-Off) **ON** Base (On-Off) **ON** Data by_____

Other Data_____

Fig. 15-8. A typical data form for a single-phase stator.

counting the number of turns and stripping from the stator.

An examination of the winding disclosed it to be connected as shown by the circular diagram in Fig. 15-9(a). The running winding is four-pole, one-circuit, and the starting winding is four-pole, one-circuit. These pole connections and the lead positions are recorded in the data sheet.

In Fig. 15-9(b) each pole of the winding is shown cut preparatory to completing the data. At this stage the turns can be easily counted and the remainder of the data taken. No part of the winding should be destroyed until all data have been taken.

After complete data have been taken, the stator is heated until all varnish is removed or softened and the winding can be easily stripped from it. Heating can be done with a torch or in an oven. The stator can be held in a vise, and the wires easily pulled from it with pliers. Varnish is also softened chemically. Extreme care must be taken in handling the stator, and all chemicals must be removed before rewinding. Care must also be taken when removing coils and cleaning the stator to prevent separating laminations. Avoid scratching or otherwise damaging the inside surface of the stator since this could affect the air gap between stator and rotor. Extreme localized heat on the stator

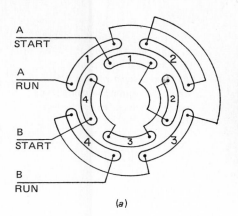

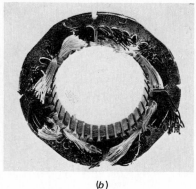

Fig. 15-9. A single-phase stator. (*a*) Circular winding diagram. (*b*) Actual stator with running and starting windings in place.

can permanently distort the metal laminations. Although rewinding a stator is practical, restacking a laminated stator that has been damaged is not. The cleaned stator is shown in Fig. 15-10. The stator teeth have been straightened and aligned, and burrs have been filed from it preparatory to insulating the slots. A convenient shop-built holder for rewinding is shown in Fig. 15-11. The base is 10 × 12 in., and the upright piece is 4 × 4 in., 6 in. high, with a V slot in the top as shown. Two-inch leather strapping will wrap easily around the stator without slipping. A door spring will provide sufficient tension to hold the stator down. A commercial stator holder is shown in Fig. 15-12.

In Fig. 15-13(*a*) the stator is shown insulated with slot insulation. This insulation must extend completely to the underside of the slot teeth and at least 1/8 in. out from each end. In Fig. 15-13(*b*) the first coil of a running pole has been wound. This beginning lead, known as the starting lead, is *sleeved,* that is, covered with plastic tubing that serves as insulation. The sleeving extends about 1 in. into the slot. Dowels are placed in the three center open slots to keep the

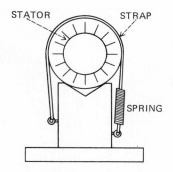

Fig. 15-11. A shop-built stator holder for rewinding.

wires down at the ends. The dowels are left in the slots until the running winding is completed.

Electrically, it makes no difference where the winding is started in an induction motor. There may, however, be a mechanical reason for winding poles in the same position in the stator as the original winding. In brush-type motors, poles must be in the original positions.

Fig. 15-10. Stator after removal of windings and cleaning.

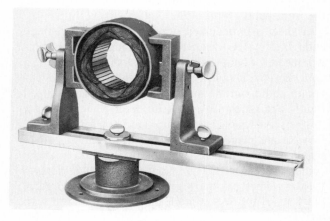

Fig. 15-12. A commercial stator holder. (*Crown Industrial Products Co.*)

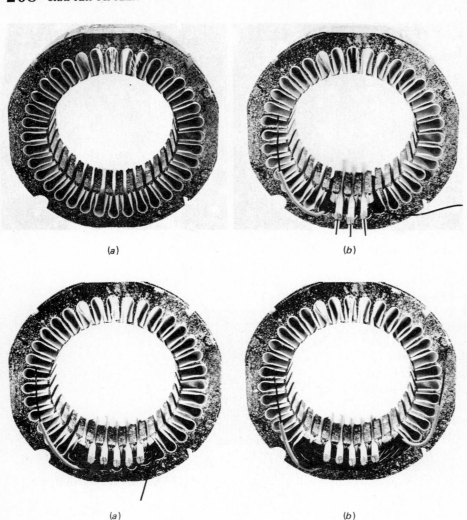

(a)

(b)

Fig. 15-13. The first running winding. (a) All slots are insulated and ready for winding. (b) Dowels keep coils in place. When the winding is completed, wedges are inserted to close off the slot.

(a)

(b)

Fig. 15-14. Completing the first pole. (a) The second coil of the pole is completed with dowels and wedges in place. (b) The final coil is completed and the finish lead is insulated with a plastic sleeve.

In winding, the wire is drawn from its spool with the right hand and looped. The left hand reaches through the stator bore from the back to grasp the loop and pull it through the stator and into the left slot. The wire is then laid in the right slot and drawn tight with the right hand.

A sheet of vulcanized fibre about 4 × 4 in., thick enough to closely clear the slot opening, is a convenient means of tamping wire in the slots. In beginning a coil, the wire should be wound clockwise and in layers up and down the inside of the teeth, then tamped tightly against the inside of the teeth.

The second coil has been wound in Fig. 15-14(a). Dowels are placed in the slots with the first coil to maintain room for the starting winding and hold the coil ends down as they are wound. Since, according to the data sheet, no starting winding coils are to go in the slots of the second coil, the slots are closed with wedges.

In Fig. 15-14(b) the third and final coil has been wound, and the slots are closed with wedges. The

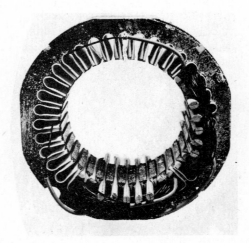

Fig. 15-15. The second pole complete. Dowels and wedges are in place. Note that the stator has been rotated in its holder so that winding is done with the slots at the bottom.

Fig. 15-16. Stator with the four complete running windings.

Fig. 15-17. With dowels removed, a length of cord is used to determine the size of a skein coil.

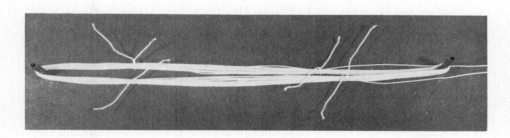

Fig. 15-18. An improvised skein form made of two nails. The bundle of wires is tied with string in a few places preparatory to being installed in the stator slots.

end, or finish, lead is sleeved. This completes one running winding pole.

In Fig. 15-15 the second pole is wound. Dowels are in place to hold the winding down. The second pole is shown completed in the manner of the first pole. The stator is turned in the holder as winding progresses, with the pole being wound at the bottom of the stator.

All the running winding poles have been wound in the same manner as the first pole, and leads are bent back out of the way in Fig. 15-16.

15-10 SKEIN WINDING

In Fig. 15-17 dowels have been removed from two adjacent poles. A length of cord is placed in the slots that will contain the starting winding. This cord is also used for measuring the length of a skein coil for the starting winding. The cord is adjusted for proper fit at the ends and is tied, as shown. Skein winding is practical for No. 22 wire and smaller. The cord is threaded in the slots in the manner of installing the skein coil, as shown in the illustrations.

Skein coils can be wound on adjustable forms or on an improvised jig as shown in Fig. 15-18. This consists of spools on nails spaced for the length of the coil. When the skein is wound on a form, it is tied as

shown at four to six places to hold it when it is removed from the form and placed in the slots. The number of turns of wire in a skein coil is the total number of turns in a pole divided by the number of times the skein is to be threaded through the slots.

A skein starting-winding pole is shown started in Fig. 15-19. The dowels have been removed from the slots to be occupied by the winding. The leads are

Fig. 15-19. The first coil of the starting winding is placed in the slots. The rest of the skein can be seen hanging down from the back of the stator.

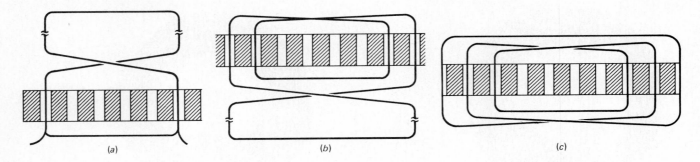

Fig. 15-20. Continuing the starting winding. (*a*) Flat representation of the stator with the starting winding inserted in the slots. The coil is twisted in the back and (*b*) brought forward, inserted in the slots, and the balance of the coil twisted in front and (*c*) inserted in the slots.

sleeved, and the end of the skein containing the leads is laid in the inner slots of the pole. Wedges are inserted to close the slot. The remainder of the skein is hanging from the back of the stator.

Figure 15-20(*a*) shows a slot representation of the stator with the starting winding in place. The skein is half-twisted at the back and brought forward and placed in adjacent slots and wedged. The remainder of the skein is half-twisted at the front [Fig. 15-20(*b*)] and laid back over the adjacent slots it is to occupy to form a three-coil pole [Fig. 15-20(*c*)]. In Fig. 15-21 two coils of the skein have already been inserted in the slots. The third coil will complete one of the starting winding poles.

The starting poles are shown completed in Fig. 15-22. This completes the entire winding, and the poles are ready to be connected. In connecting the poles, the location of the line leads is determined by reference to the "sketch of lead positions" on the data sheet. Pole leads are selected for line leads to provide this position, and the remaining pole leads are cross-connected to give alternate polarity of the poles.

The connected starting winding is shown in Fig. 15-23(*a*). The running winding is also connected, and the line leads are connected to both windings as shown in Fig. 15-23(*b*). Line leads should be of a size sufficient for mechanical strength requirements and at least one size larger than the total cross-sectional area of the winding leads to which they are connected. Figure 15-23(*b*) also shows the winding spirally laced and completed. It is now ready for varnishing and baking.

Half-hitching is sometimes used in lacing, but circular lacing as shown provides a lighter and more solid lacing job. A winding should be laced tightly to allow insulating varnish to effectively seal and hold the winding solidly to avoid vibration of individual wires.

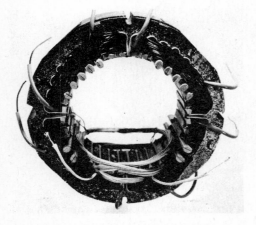

Fig. 15-21. The stator with the skein winding in place.

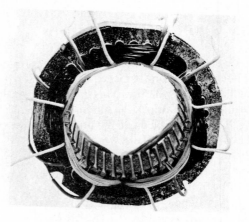

Fig. 15-22. All poles completed and ready for connection.

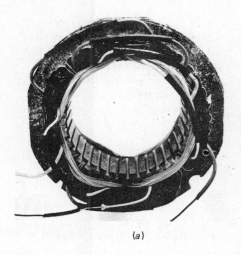 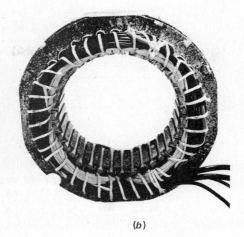

(a) (b)

Fig. 15-23. Connecting the completed windings. (*a*) Starting winding poles connected. (*b*) All poles connected, line leads connected, and entire winding spiral laced and ready for varnish and baking.

15-11 SPEED OF AN AC INDUCTION MOTOR

The speed of an ac induction motor—such as split-phase, capacitor, shaded-pole, and most three-phase motors—is determined by the line frequency and the number of poles in the motor.

To determine the speed in revolutions per minute, the line frequency in hertz is multiplied by 60 and the product divided by the number of pairs of poles in the motor. The result will be *synchronous speed,* which is the speed that is in step with the frequency or the changing magnetic field. But induction-motor rotors are unable to keep up with the rapidly changing field. There must be some slip between the rotor speed and the magnetic field in order for the field to cut rotor conductors and induce a current in them to support rotation. The difference between synchronous speed and actual motor speed is known as *slip.* Induction motor slip is usually about 3 or 4 percent of synchronous speed. Thus the approximate speed of the average four-pole induction motor connected to a 60-Hz system is about 1750 rpm. This figure can be calculated using the following formulas.

$$S_s = \frac{f \times 60}{pp}$$

where S_s = synchronous speed, rpm
f = line frequency, Hz
pp = no. of pairs of poles

and $S_a = S_s \left(1 - \frac{S_{sl}}{100}\right)$

where S_a = actual speed at no load, rpm
S_s = synchronous speed (from previous formula), rpm
S_{sl} = slip, %

Problem 1

Calculate the actual speed of a single-phase induction motor operating at 208 V, 60 Hz. The motor has four poles. Use an average 3 percent slip.

Solution

Use the formula

$$S_s = \frac{f \times 60}{pp}$$

$$= \frac{60 \times 60}{2}$$

$$= 1800 \text{ rpm}$$

Notice that the speed is independent of the applied voltage. The frequency is 60 Hz, and four poles are converted to two pairs of poles. The 1800 rpm is the synchronous speed, but the rotor has 3 percent slip. Therefore

$$S_a = 1800 \left(1 - \frac{3}{100}\right)$$

$$= 1800 \ (0.97)$$

$$= 1746 \text{ or nominally 1750 rpm}$$

The rated speed of an induction motor is its full-load speed. Slip decreases with load. The difference between no load and full load is so small that induction motors are sometimes considered as constant-speed motors.

15-12 MULTISPEED SINGLE-PHASE MOTORS

A multispeed motor is a motor capable of running at *two or more* constant speeds. Since the number of

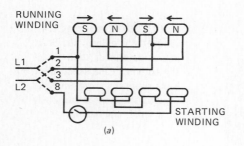 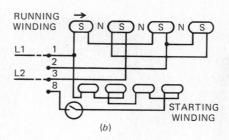

Fig. 15-24. Connections of a multispeed split-phase motor. (*a*) Four salient poles for high speed. (*b*) Four salient poles and four consequent poles for low speed.

poles and line frequency determine the speed of an induction motor, a change in the number of poles will change the speed.

There are two commonly used methods of changing the number of poles in a motor—the *multiwinding* and the *consequent-pole* methods. In the multiwinding method, a motor is equipped with two or more *separate windings* with the desired numbers of poles in each. A selector switch is used to connect the winding that will produce the desired speed.

In the consequent-pole method, one winding is connected in such a way that one connection produces *salient poles of alternate polarity,* while another connection produces *salient and consequent poles,* thus doubling the number of poles and halving the speed. A salient pole is one produced by a current-carrying coil would around an iron core, such as in the stator of an ac motor or the field of a dc motor. A consequent pole is a pole that is automatically created or formed as a consequence of the *existence of a salient pole.*

The principles of a consequent-pole multispeed winding in a single-phase motor are illustrated in Fig. 15-24. In Fig. 15-24(*a*), with leads 1 and 3 of the running winding connected to L1, and leads 2 and 8 connected to L2, four alternate salient poles are formed for high speed. This will be seen by tracing the two circuits through the poles from L1 to L2.

In Fig. 15-24(*b*), lead 1 is connected to L1 and lead 3 to L2 for a series circuit which will produce four south poles. Four north consequent poles will develop between the south poles to produce a total of eight

poles, which will result in one-half the speed of the four-pole winding for slow speed.

A speed-selector switch makes the proper connections for the desired speed, which, under the consequent-pole method, must be either one-half or twice that of the other. A *variable-speed* motor is any motor that can operate satisfactorily over a comparatively *wide speed range* with suitable controls. Series, shaded-pole, and permanent-split capacitor motors are generally classed as *variable-speed motors.* Speed control of these motors is accomplished by any of several means of reducing line voltage. Single-phase motors are also controlled by electronic speed controllers. Such devices are described in Chap. 24.

15-13 REVERSING SPLIT-PHASE MOTORS

Split-phase motors are reversed by reversing polarity of either the starting or running winding. The standard practice in reversing is to *reverse the starting winding.* Figure 15-25 is a diagram with standard lead identification for a single-voltage reversible split-phase motor. Terminal board connections are for ccw and cw rotation.

In all single-phase motors, standard rotation (counterclockwise as viewed from opposite the drive end) is obtained when the running winding lead T4 and starting winding lead T5 are connected together on the same line.

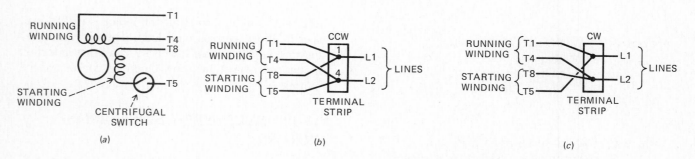

Fig. 15-25. Split-phase reversible motor connections. (*a*) Lead identification. (*b*) Terminal board connections for counterclockwise rotation. (*c*) Terminal board connections for clockwise rotation.

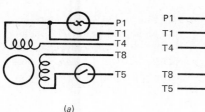 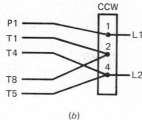 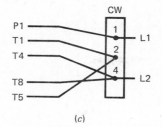

(a)   (b)   (c)

Fig. 15-26. Split-phase reversible motor with built-in (internal) thermal protection. (*a*) Lead identification. (*b*) Terminal board connections for counterclockwise rotation. (*c*) Terminal board connections for clockwise rotation.

When colors are used instead of numbers, they are T1, blue; T4, yellow; T5, black; T8, red. For motors with built-in thermal protection, standard lead identification and terminal connections for ccw and cw rotation are shown in Fig. 15-26.

In motors with built-in internal thermal protection, running winding lead T1 is connected to lead T8 and is *in series* with T8. Both are connected *in series* with the built-in thermal-protection device. If the thermal protection opens (as it would during an overload condition), both running and starting winding circuits are broken.

In motors without internal thermal protection, leads T1 and T8 are both connected to the line. Thus they are *in parallel* with one another. A break in either circuit would not necessarily break the other circuit.

15-14 CAPACITOR-START MOTORS

A capacitor-start motor is a type of split-phase motor with a *capacitor in series with the starting winding*. Alternating current can flow into and out of a capacitor but not through it. A capacitor in series with a starting winding limits the flow of current in the starting winding, which permits a larger-size wire to be used than can be used in a split-phase starting winding. A capacitor also introduces capacitive reactance to neutralize inductive reactance, which permits more turns in the starting winding than can be used in a split-phase starting winding.

With larger wire and more turns, a capacitor starting winding delivers more starting torque than a split-phase winding. The starting torque of the average capacitor-start motor is about 300 to 350 percent of full-load torque.

Capacitor-start motors are available in sizes up to about 2 hp. They are extensively used for applications requiring more starting torque than that provided by split-phase motors. Typical applications include compressors, pumps, and air conditioners. The larger sizes are extensively used at voltages ranging from about 200 through 250 V in areas where three-phase current is not available.

15-15 TYPES OF CAPACITORS

Electrolytic capacitors are commonly used in capacitor-start motors. Electrolytic motor-starting capacitors are built in sizes up to 800 μF. Because of rapid movement of the electrons in the capacitor's dielectric, heat is rapidly generated; and so the time of operation must be limited. They can be used for starting only and are usually limited to 20 starts per hour, with an average of 3-s duration per start. A relief plug is usually built into a capacitor to relieve internal pressure from heat and prevent an explosion in case of overheating. Typical electrolytic motor-starting capacitors are shown in Fig. 15-27.

Capacitors are rated in microfarads. Some capacitors are rated with a double number as a minimum and maximum rating, while others contain only a single number for an average rating.

In replacing capacitors, the exact required size should be used because a capacitor too large or too small will result in the same adverse effect of reducing starting torque.

If a capacitor is good, it will hold a strong charge and spark proportionately when the terminals are shorted for discharge. A severe electric shock can be delivered by a charge capacitor, and *care should be used in working with capacitors*. If it does not hold a charge, it is defective. If a test light is connected in series with the capacitor, failure to light indicates an opening in the capacitor (not a common defect). A short in a capacitor is a more common defect.

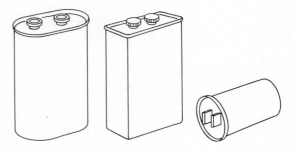

Fig. 15-27. Typical electrolytic motor-starting capacitors.

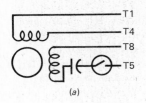

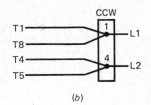

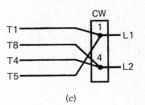

(a)   (b)   (c)

Fig. 15-28. Reversible capacitor-start motor. (a) Lead identification. (b) Terminal board connections for counterclockwise rotation. (c) Connections for clockwise rotation.

If a capacitor is defective, it should be replaced with another capacitor of equal rating. Capacitors are rated according to voltage, capacitance, power factor, duty cycle, and temperature. The specified voltage is the maximum working voltage and is usually higher than the line voltage or motor operating voltage. The specified capacitance is the minimum value; commercial capacitors may have as much as 20 percent more capacitance available than the value stamped on the case. Capacitors have extremely low leading power factors ranging from 5 to 10 percent, depending upon size. Power factor increases sharply as temperatures decrease. As noted above, motor-starting capacitors are rated for intermittent duty. Capacitors are normally rated at 25°C (77°F), but they can operate up to 65°C (150°F).

If the rating of a defective capacitor is unknown, the safest practice is to obtain the correct voltage and capacitance ratings from the manufacturer of the motor.

Capacitors deteriorate and lose capacity with shelf life, or age, and misuse. A formula for testing capacitors for 60-Hz operation is as follows. For 60-Hz applications:

$$C = \frac{2650 \times I}{V}$$

where C = capacitance, μF
      I = current through capacitor circuit, A
      V = voltage across capacitor, V

The constant 2650 is derived from the formula for capacitive reactance (see Chap. 4).

$$\text{Constant} = \frac{10^6}{2 \times 3.14 \times 60} = \text{approx. } 2650$$

To make the test, first check the capacitor for short circuit by connecting it across the line in series with a small fuse of sufficient capacity to carry normal current. A shorted capacitor will blow the fuse. If the capacitor is free of short circuits, an ammeter is connected in series with it to determine the current. The value thus obtained and the test voltage are applied in the formula; the result is compared with the capacitor rating to determine the condition of the capacitor.

15-16 SINGLE-VOLTAGE REVERSIBLE CAPACITOR-START MOTORS

Diagrams for standard connections for single-voltage reversible capacitor-start motors, *without built-in thermal protectors,* are given in Fig. 15-28. At Fig. 15-28(a) is shown standard terminal identification, at Fig. 15-28(b) terminal board connections for ccw rotation, and at Fig. 15-28(c) connections for cw rotation.

Rotation of capacitor motors, like rotation of split-phase motors, is *reversed by reversing polarity* of either the *starting winding* or the *running winding.* It is standard practice to reverse the polarity of the *starting winding* for reversing rotation.

Diagrams for standard connections for single-voltage reversible capacitor-start motors, *with built-in thermal protectors,* are given in Fig. 15-29. It will be noticed that these motors contain five leads. At Fig. 15-29(a) is shown standard lead identification, at Fig. 15-29(b) terminal board connection for ccw rotation, and at Fig. 15-29(c) connections for cw rotation.

15-17 DUAL-VOLTAGE CAPACITOR-START MOTORS

Capacitor motors are usually wound for dual voltage operation. This is provided for basically by arranging

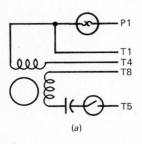

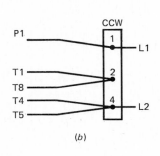

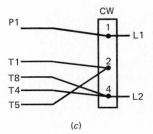

(a)   (b)   (c)

Fig. 15-29. Reversible capacitor-start motor with thermal protector. (a) Lead identification. (b) Terminal-board connections for counterclockwise rotation. (c) Connections for clockwise rotation.

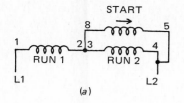

(a)   (b)

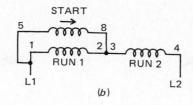

Fig. 15-30. High-voltage connection of a dual-voltage reversible capacitor-start motor. (*a*) The running windings are in series while the starting winding is in parallel with one of the running windings. (*b*) The direction of rotation is reversed by connecting the starting winding across the other running winding.

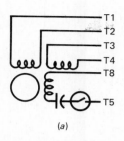

(a)

ROTATION	L1	L2	JOIN
HIGH VOLTAGE			
CCW	T1	T4-T5	T2-T3-T8
CW	T1	T8-T4	T2-T3-T5
LOW VOLTAGE			
CCW	T1-T3-T8	T2-T4-T5	
CW	T1-T3-T5	T2-T4-T8	

(b)

Fig. 15-31. Connections for a dual-voltage reversible capacitor-start motor. (*a*) Lead identification. (*b*) Rotation and dual-voltage connection chart.

the running winding in two circuits that can be connected in series for high-voltage operation, or parallel for low-voltage operation. The starting-winding circuit, including the capacitor, is one circuit for low voltage only. It is therefore connected in parallel with one of the two running sections for high voltage and is thus in series with one of the running sections for high voltage.

This high-voltage connection is illustrated in Fig. 15-30. To reverse, one starting lead is interchanged with the line. In Fig. 15-30(*a*) 5 connects to L2 for one direction of rotation. Current is from 8 to 5, as shown by the arrow. In Fig. 15-30(*b*) 5 is connected to L1 for rotation in the opposite direction. Current is from 5 to 8, which is a direction opposite to that in Fig. 15-30(*a*). Some dual-voltage motor terminal boards are equipped with links for changing connections.

Diagrams for standard lead identification and terminal board connections for ccw rotation, and a rotation chart for dual-voltage reversible capacitor-start motors, are shown in Fig. 15-31.

15-18 CAPACITOR MOTORS WITH THERMAL PROTECTION

Most capacitor motors are equipped with a built-in thermal protector. Line current flows as shown in Fig. 15-32(*a*), through P1 and terminal 1 in the protector, through thermally operated contacts and terminal 2. Here it divides, according to the protective system which is identified in three groups, and part of the current goes through the heater element between terminals 2 and 3, and to the motor. If excessive current flows in this circuit, because of overload or any other cause, the heater causes the thermally operated contacts to open and break the line circuit.

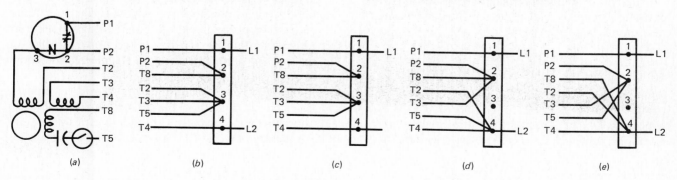

Fig. 15-32. Group I dual-voltage reversible capacitor-start motor with thermal protector. (*a*) Lead identification. (*b*) Terminal board connections for high-voltage ccw rotation. (*c*) Connections for high-voltage cw rotation. (*d*) Connections for low-voltage ccw rotation. (*e*) Connections for low-voltage cw rotation.

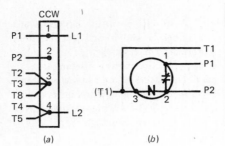

ROTATION	L1	L2	JOIN	JOIN	INSULATE
HIGH VOLTAGE					
CCW	P1	T4-T5	T2-T3-T8		P2-T1
CW	P1	T4-T8	T2-T3-T5		P2-T1
LOW VOLTAGE					
CCW	P1	T2-T4-T5	P2-T3	T1-T8	
CW	P1	T2-T4-T8	P2-T3	T1-T5	

(a)　　　(b)　　　(c)

Fig. 15-33. Group II dual-voltage reversible capacitor-start motor with thermal protector. (*a*) Terminal board connections for high-voltage ccw rotation. (*b*) Thermal protector connections. (*c*) Rotation and dual-voltage connection chart.

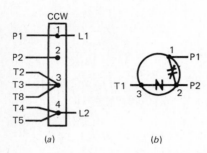

ROTATION	L1	L2	JOIN	INSULATE
HIGH VOLTAGE				
CCW	P1	T4-T5	T2-T3-T8	P2
CW	P1	T4-T8	T2-T3-T5	P2
LOW VOLTAGE				
CCW	P1	T2-T4-T5	P2-T3-T8	
CW	P1	T2-T4-T8	P2-T3-T5	

(a)　　　(b)　　　(c)

Fig. 15-34. Group III dual-voltage reversible capacitor-start motor with thermal protector. (*a*) Terminal board high-voltage connections for ccw rotation. (*b*) Thermal protector connections. (*c*) Rotation and dual-voltage connection chart.

For dual-voltage motors, three methods are commonly used in connecting thermal protectors. These three methods are designated by Groups I, II, and III.

Diagrams of standard lead identification and terminal board connections for Group I dual-voltage reversible capacitor-start motors are given in Fig. 15-32.

Diagrams of standard connections for Group II dual-voltage reversible capacitor-start motors and a connection chart are given in Fig. 15-33.

Diagrams of standard connections and a connection chart for Group III dual-voltage reversible capacitor-start motors are given in Fig. 15-34.

15-19 PERMANENT-SPLIT CAPACITOR MOTORS

A permanent-split capacitor motor has a small oil capacitor permanently connected in series with the starting winding and does not contain a centrifugal switch. It starts and runs with the capacitor in the circuit. The capacitor is therefore rated for continuous duty. This motor is suitable only for low-starting-torque requirements.

Figure 15-35 shows standard terminal identification and terminal board connections for rotation for permanent-split single-voltage reversible capacitor motors.

An arrangement for switch reversing of a permanent-split capacitor motor is shown in Fig. 15-36. The motor contains two identical windings displaced from each other in the stator. A forward-reverse selector switch transfers the capacitor to the proper winding to make it the starting winding for the desired direction of rotation.

In Fig. 15-36(*a*), with the switch in forward position, the top winding is the running winding and the bottom winding, with the capacitor in series with it, is the starting winding for forward. In Fig. 15-36(*b*) the

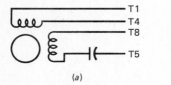

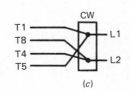

Fig. 15-35. Permanent-split single-voltage reversible capacitor motor. (*a*) Lead identification. (*b*) Terminal board connections for ccw rotation. (*c*) Connections for cw rotation.

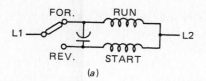

Fig. 15-36. Method of reversing a permanent-split capacitor motor with two displaced identical windings. (a) With the switch in the forward position, the top winding serves as run winding. (b) With the switch in the reverse position, the top winding serves as the start winding.

switch is in reverse position. The top winding with the capacitor in series with it is the starting winding for reverse. Typical applications of this system are operation of valves, signal devices, shaft-mounted fans, television antenna rotators, etc. In the antenna rotator application, the capacitor is located in the control box. Permanent-split capacitor motors range up to 3/4 hp for fans and blowers, but generally are less than 1/3 hp.

15-20   TWO-VALUE CAPACITOR MOTORS

A two-value capacitor motor starts on one capacitor connected in the starting winding circuit and runs with a lower-value capacitor in the same circuit. These motors usually have an oil capacitor permanently connected in series with the starting winding. An electrolytic capacitor with a starting switch in series with it is connected in parallel with the oil capacitor.

The motor starts with both capacitors in parallel; as the motor comes up to speed, the switch opens the electrolytic capacitor circuit, leading the oil capacitor in series with the starting winding for running. This motor is used for high starting-torque applications.

A terminal identification diagram showing the connection of the two capacitors and the starting switch is given in Fig. 15-37(a). The diagrams in Fig. 15-37(a) through (c) are for two-value single-voltage reversible capacitor motors. The terminal board connection at Fig. 15-37(b) is connected for ccw rotation, while the connections in Fig. 15-37(c) are for cw rotation.

15-21   REPULSION MOTORS

Repulsion-type motors have a higher starting torque and draw proportionately less starting current than any of the other types of constant-speed single-phase motors. Repulsion motors develop torque by the *repulsion of like magnetic poles.* An *induced current is opposite in direction to the inducing current,* and resultant magnetic poles of the two currents are of *like polarity* and *face each other.* Repulsion motors have a field winding similar to the running winding of a split-phase motor and a wound armature similar to a dc armature. The brushes on the commutator are short-circuited together by shunts and a common uninsulated brushholder. The armature is not connected to the line. It receives its current by induction from the field windings.

The brushes allow induced current to flow in the armature winding and produce armature magnetic poles that are repelled by the poles of the field winding to produce torque. These principles are illustrated in Fig. 15-38.

The shaded arcs in Fig. 15-38(a) represent the stator poles of a two-pole motor. The circle represents the circuit formed by the armature winding. If the poles are excited, voltage will be induced in the armature winding, or circle, in the directions of the arrows during one half-cycle.

The voltage in the two sides is from the dot at the bottom of the circle toward the dot at the top of the circle. Thus both voltages "buck" at the top and no current flows. In Fig. 15-38(b) a circuit is shown by a

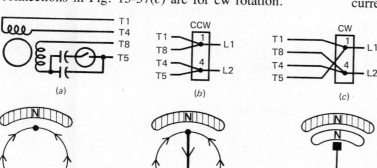

Fig. 15-37. Two-value single-voltage reversible capacitor motor. (a) Lead identification. (b) Terminal board connections for ccw rotation. (c) Connections for cw rotation.

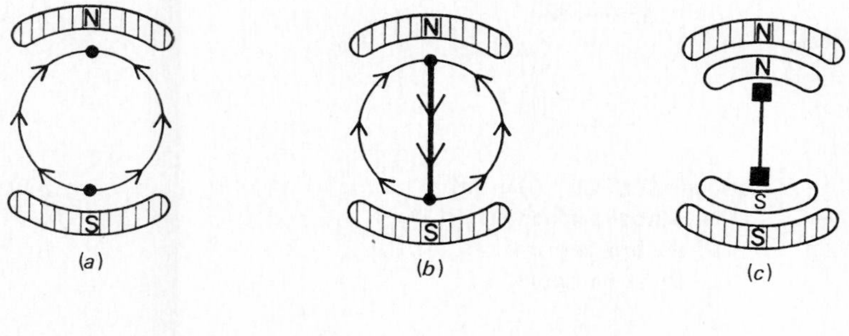

Fig. 15-38. Principles of a repulsion motor. (a) Direction of induced voltage indicated by the arrows. (b) A jumper provides a current path. (c) Brushes in hard neutral; no torque is developed.

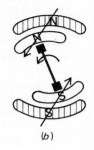

(a)   (b)

Fig. 15-39. Working neutral for (a) clockwise rotation and (b) counterclockwise rotation.

jumper from the top dot on the circle to the bottom dot. Current can now flow in the direction of the arrows.

In Fig. 15-38(c) short-circuited brushes are shown to represent the jumper, and the white arcs represent magnetic poles formed on the armature by the induced current flowing in the armature winding, brushes, and jumper.

No torque is developed in Fig. 15-38(c) because the brushes cause the pattern of current flow to form like poles directly under the main poles. Repulsion now is toward the center, or axis, of the armature. Hence *rotational torque is not possible*. In the present position the brushes are in *hard neutral* position. To develop torque, the brushes must be shifted to *working neutral* to form armature poles slightly to one or the other side of the main poles.

The illustrations in Fig. 15-39 show the effect of shifting the brushes. In Fig. 15-39(a) the brushes have been shifted *clockwise*. The armature poles are now forming in the direction of the brush shift, slightly clockwise from the main poles. Repulsion between the north pole of the main field at the top and the north pole at the top of the armature is now in the direction of the arrow shown across the top poles, and torque is *developed clockwise*, as shown by the inside arrow. Likewise, torque is developed *clockwise* by the south main and armature poles at the bottom of the figure. The brushes are now in *working neutral* position.

In Fig. 15-39(b) the brushes are shown shifted slightly *counterclockwise* from hard neutral, and torque is being developed *counterclockwise*.

15-22 REVERSING REPULSION MOTORS

Torque and direction of rotation of a repulsion motor are determined by the relative positions of the brushes and the main poles. To reverse the direction of rotation of a repulsion motor, the brushes are shifted, usually from 10 to 15 electrical degrees from hard neutral, in the direction of the desired rotation.

In rewinding repulsion motor fields, the poles should occupy the same slots as those occupied by the original winding poles.

In Fig. 15-40, the brushes are shown shifted *90 degrees from hard neutral*, and the armature poles are forming midway between the main poles where *no torque* can be produced. This is the *soft neutral* position for the brushes. No current flows in the armature with the brushes in soft neutral position, and the fields draw only a small excitation current that produces only a soft hum in the fields. If the brushes are shifted from soft neutral, rotation will be opposite the direction of the shift.

To summarize the conditions of brush positions, the *position of the brushes determines where the armature poles form*. At *hard neutral* the armature poles are *directly under the main poles and no torque is produced*. The motor will issue a hard, harsh noise with no rotation.

Working neutral is a brush position from 10 to 15 electrical degrees either side of hard neutral. *Working neutral produces maximum torque*. It can be located accurately by measuring starting torque with a prony brake. A convenient method of locating working neutral is to find the brush position that provides fastest acceleration from start to full speed under load. The soft neutral position of the brushes produces no torque, and the motor issues only a soft hum.

15-23 TYPES OF REPULSION MOTORS

There are three types of repulsion motors—*repulsion*, *repulsion-induction*, and *repulsion-start induction-run*. The repulsion-start induction-run, better known as *repulsion-start induction*, is the most commonly used of the three types.

The field winding, similar to the running winding of a split-phase motor, is the same in all three types of repulsion motors. The armature winding, similar to dc

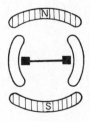

Fig. 15-40. The soft neutral position of the brushes produces no torque.

armature windings, is the same in all three types of repulsion motors.

In starting torque, the three types are the same. The straight repulsion motor has *no-load runaway* characteristics similar to those of a series motor. It runs continuously by its brushes. If the brushes are raised while running, this motor will stop. It uses an axial, or horizontal, commutator.

The repulsion-induction motor is a repulsion motor with a *squirrel-cage winding* on the armature *under the regular winding*. This is an induction winding. It serves to *stabilize* the no-load speed by cutting magnetic lines of force to afford the effect of *regenerative braking* at speeds above synchronous speed. Variation of speed from no load to full load is about 15 to 25 percent. Heavy currents flow in the induction winding at no-load speeds, and this motor usually heats more at no-load than at full-load. If the brushes are raised while running, the motor will continue to run on the induction winding as an induction motor. It uses an axial (horizontal) commutator.

15-24 REPULSION-START INDUCTION MOTORS

A repulsion-start induction motor starts as a repulsion motor, and at about 75 percent of full speed *automatically converts to an induction motor*. This conversion is accomplished by a mechanism operated by a cen-trifugal switch which short-circuits the commutator bars and converts the *wound* armature in effect into a *squirrel-cage induction armature*. The motor then operates as an *induction motor*. Thus, the high starting torque of a repulsion motor and the stable running torque of an induction motor are combined in one motor.

A cutaway view of a repulsion-start induction motor with parts labeled is shown in Fig. 15-41.

In operation, a repulsion-start induction motor starts with the brushes on the commutator as a repulsion motor. At about 75 percent of full speed the governor weights open and force back the push rods, which move the collar or barrel containing the short-circuiting "necklace" into the commutator.

The necklace is composed of copper segments strung at the inner end on a copper retaining wire and free at the outer end. Because of centrifugal force, the necklace segments firmly contact a brass or metal ring inside the commutator and the commutator bars, which short-circuits all the bars. In effect, the winding is converted to a squirrel-cage winding, and the motor accelerates from this point to full speed and runs as an induction motor.

The commutator is of the radial type, and the brush-holder is pushed from the commutator by the necklace barrel in its movement. To reduce wear and noise, the brushes are removed from the commutator during running periods, since the commutator is short-circuited and the brushes are ineffective and not needed. All motors with radial commutators lift the brushes after

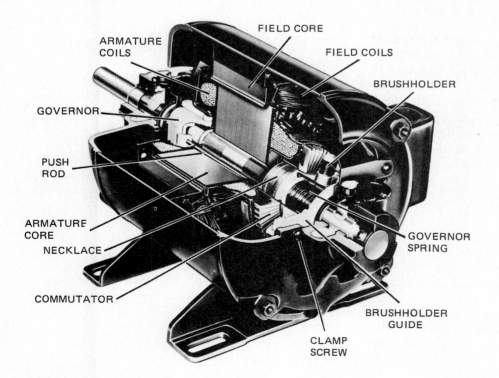

ARMATURE COILS
FIELD CORE
FIELD COILS
BRUSHHOLDER
GOVERNOR
PUSH ROD
ARMATURE CORE
NECKLACE
COMMUTATOR
CLAMP SCREW
BRUSHHOLDER GUIDE
GOVERNOR SPRING

Fig. 15-41. Repulsion-start induction motor.

Fig. 15-42. Radial commutator and short-circuited brushes used in a repulsion-start motor.

starting. Only a few repulsion-start induction motors with axial commutators lift the brushes after starting.

For reversing rotation, the brushholder is retained by the brushholder guide, which is rotated and clamped to hold the brushes in proper position for the desired rotation.

The same care is required by the commutator as by a dc commutator. The surface contacted by the necklace must be kept clean and free from burned or pitted areas. Poor contact between these points will cause armature current to circulate from the brushholder to the motor frame, bearings, and shaft and back to the armature. This current flow pits the bearings and shaft and results in rapid bearing wear.

A repulsion-start induction armature with a radial commutator and short-circuited brushes is shown in Fig. 15-42. The slot in the side of the brushholder allows the latter to move axially in alignment with the guide. When the brushes are badly worn, the motor in starting will seesaw or alternate from a condition of repulsion to induction, since the brushes are raised from the commutator before the necklace can complete conversion to induction conditions. Short brushes also create a smutty surface on the commutator. If a commutator in this condition is rubbed with the fingers, a smutty appearance of soot will result, indicating short or stuck brushes.

15-25 DUAL-VOLTAGE REPULSION MOTORS

Most repulsion motors are arranged for dual-voltage operation. The fields are connected in two groups, and the two leads of each group are identified and brought out of the motor for external connections. The fields are connected in series for high voltage, or in parallel for low voltage. A motor and controller voltage change may require a change of thermal overload element.

A lead identification diagram and a chart showing high- and low-voltage connections are given in Fig. 15-43(a) and (b). All three types of repulsion motors use the same method of connections for dual voltage.

15-26 ELECTRICALLY REVERSIBLE REPULSION MOTORS

Repulsion motors are commonly reversed by shifting the brushes. This changes the relationship of current flow and poles of the armature and the poles of the main fields.

For applications where the mechanical shifting of the brushes is not practical, two main field windings, displaced from each other, are used in the stator. One

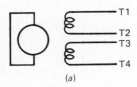

(a)

VOLTAGE	L1	L2	JOIN
HIGH	T1	T4	T2-T3
LOW	T1-T3	T2-T4	- - -

(b)

Fig. 15-43. Dual-voltage repulsion motor. (a) Lead identification. (b) Dual-voltage connection chart.

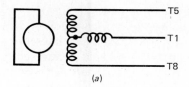

ROTATION	L1	L2	OPEN
CCW	T1	T5	T8
CW	T1	T8	T5

(a)   (b)

Fig. 15-44. Electrically reversible repulsion motor. (a) Lead identification. (b) Rotation connection chart.

winding forms main poles to one side of the armature poles for one direction of rotation, while the other winding forms main poles on the other side of the armature poles for opposite rotation. Thus the main field poles can be electrically shifted for reversing instead of the brushes. A selector switch is used to excite the winding for desired rotation. The lead identification and connection for reversing of this type of electrically reversible motor are given in Fig. 15-44.

15-27 TESTING REPULSION ARMATURES

Some repulsion motor armatures are equipped with equalizers behind the commutator. To test either the armature or the equalizers without removing the armature from the motor, it is only necessary to raise or remove the brushes from the commutator and excite the fields. If the armature can be freely turned, the armature is free of short circuits. If the armature locks or tends to lock in certain places as it is turned, it is shorted.

15-28 SHADED-POLE MOTORS

A shaded-pole motor is a type of split-phase induction motor. It employs a short-circuited band, or winding, on one side of each pole to produce a shifting or rotating magnetic field to start rotation (Fig. 15-45). Small shaded-pole motors generally have two poles with a single field coil mounted between them. The field winding is of the concentrated coil type. The salient poles are split or slotted about one-third of the way across the pole face as shown in Fig. 15-45. A single band of copper is mounted around the smaller pole

created by the slot. In effect the copper band is a shorted field coil.

In operation, magnetism will build up in the left or unshaded portion of the core as illustrated in Fig. 15-45(a). When it starts in the shaded or banded area, it induces a current in the band. This current produces a magnetic field which opposes the original magnetic field and delays the original in building up in the shaded area of the core.

In time, the opposition of the banded area will be overcome, and the complete pole will be magnetized as in Fig. 15-45(b). At the end of a half-cycle the magnetism in the unshaded area of the core will disappear, but when magnetism in the shaded area begins to collapse, it induces a current in the band. This current produces a magnetic field which is of the same polarity as the original. Thus magnetism with the original polarity is maintained for a period of time. This remaining magnetism is shown in Fig. 15-45(c).

An examination of all three illustrations shows that during one half-cycle magnetism in Fig. 15-45(a) is in the left area of the pole at the beginning. Magnetism has shifted to all areas later as shown in Fig. 15-45(b); and it has shifted to the right banded area at the completion of the half-cycle.

In the stator, the *shifting* of magnetism in all the poles produces a *rotating magnetic effect* similar to that of a split-phase motor. A rotating magnetic field induces a current in the rotor which produces magnetism in the rotor. The rotor magnetism and the rotating-field magnetism interact to produce rotation.

Rotation of a shaded-pole motor is *in the direction of the rotating magnetic field;* thus rotation is from the unshaded area toward the shaded area, or toward the banded side of a pole.

Rotation of a shaded-pole motor can be reversed by removing the stator from the motor and turning it

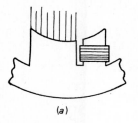
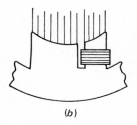
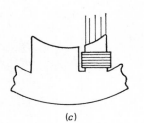

(a)   (b)   (c)

Fig. 15-45. Principle of the shaded-pole motor.

Fig. 15-46. Series motor. *(Baldor Electric Co.)*

around, end for end, during reassembly. This places the banded areas opposite the original positions and reverses rotation.

Speed control of shaded-pole motors is accomplished by reducing the line voltage to the motor. This is done with a variable resistor or impedance coil in series with the motor, or by increasing inductive reactance in the motor by adding turns to the main winding in the form of an auxiliary winding between taps.

Starting torque of shaded-pole motors varies from 75 to 100 percent of full-load torque. Motor sizes range up to $\frac{1}{4}$ hp. They are widely used on applications requiring only low starting torque, such as fans and blowers.

15-29 SERIES MOTORS

Alternating-current series motors are similar to dc series motors in construction and operating characteristics, except that the core of ac motors must be laminated.

A series motor that can operate on either alternating or direct current at the same voltage is known as a *universal series motor*. Otherwise, series motors are identified for either ac or dc operation.

A typical small series motor is shown in Fig. 15-46. Series motors can deliver higher horsepower for their weight than any other type of motor. Small series motors are used on portable hand-held commercial equipment such as saws, drills, polishers, grinders, and sanders, as well as on household equipment such as food mixers, vacuum cleaners, and sewing machines.

Variable speed control is accomplished with variable-resistance, impedance coil, tapped-winding controls, or solid-state electronic controls.

Excessive no-load speed of a series motor can be controlled with an automatic centrifugal speed governor. This system is illustrated in Fig. 15-47. A resistor and capacitor are shown paralleled with a centrifugal switch in the line. The centrifugal switch is adjusted for the desired speed range. When the motor exceeds the maximum of the range, the centrifugal switch opens, placing the motor in series with the resistance to lower the speed. The capacitor serves to reduce sparking and damage to the governor contact points. This type of speed control operates best in ranges between 2000 and 6000 rpm. No-load to full-load speed can be controlled within 5 percent.

A *compensated series motor* contains a main winding and a compensating winding which is displaced 90° from the main winding. Both windings, containing the same-size wire, and the armature are connected in series. Schematic and circular diagrams of a compensated series motor are given in Fig. 15-48.

A compensating winding performs the same function as interpoles in dc motors in reducing sparking at

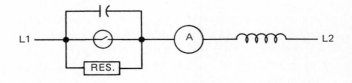

Fig. 15-47. Series motor with built-in centrifugal governor for speed control.

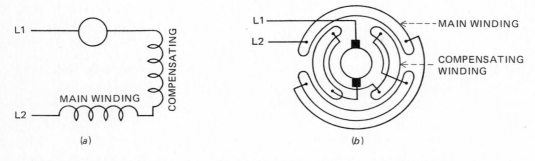

Fig. 15-48. Compensated series motor. *(a)* **Schematic diagram.** *(b)* **Winding connections.**

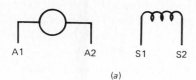

ROTATION	L1	L2	JOIN
CCW	A1	S2	S1–A2
CW	A1	S1	A2–S2

(a)   *(b)*

Fig. 15-49. Reversible universal (series) motor. (a) Lead identification. (b) Rotation connection chart.

the brushes and permitting reversing without the necessity of shifting the brushes to compensate for field distortion and reduce sparking. Polarity of a compensating winding can be determined by the M-G rule explained in Sec. 8-11. The compensating winding, like interpoles, must be reversed when rotation is reversed.

The most common source of trouble in hand-held portable series motors is overloading, which causes excessive heating of the armature. This heat is sometimes sufficient to melt solder in the commutator. Centrifugal force throws molten solder from the commutator, creating an open circuit. Current crosses the mica between the bars connected to the open coil. This chars the mica and pits the bars.

15-30 REVERSING SERIES MOTORS

Not all series motors can be reversed in practice. Motors with stationary brushholders, designed for equipment requiring only one direction of rotation, such as saws, drills, sanders, etc., have the armature leads connected to the commutator one or more bars from neutral to compensate for field distortion and reactance. If these motors are reversed, severe sparking will occur at the brushes, since lead swing is for one direction only.

Standard terminal markings for reversible universal series motors and a chart giving reversing connections are shown in Fig. 15-49. Reversing either the field or armature will reverse a series motor.

SUMMARY

1. Single-phase motors operate on two-wire single-phase alternating current.

2. The five principal types of single-phase motors are shaded pole, split phase, capacitor, repulsion, and series.

3. Squirrel-cage rotors, used in shaded-pole, split-phase, capacitor, and three-phase motors, receive current by induction from the main fields. These motors are classified as induction motors. The rotor winding consists of uninsulated bars embedded in the rotor core and connected to a ring at each end of the core.

4. Split-phase, capacitor, and repulsion motor fields are usually wound with concentric coils in distributed slots.

5. Single phase produces an oscillating magnetic field which alone cannot produce starting torque, except in the series motor.

6. Single-phase motors, except the series motor, must be provided a means of starting. They are typed according to the means of starting and are selected for service according to starting torque.

7. Starting torque of single-phase motors, in round numbers, in percent of full-load torque, is as follows: shaded pole, 100; split phase, 200; capacitor, 300; repulsion, 400; series, 500.

8. Split-phase motors, with two windings having different inductive reactances, split the single-phase current into two-phase current which, flowing in the two windings physically displaced from each other, produces a rotating magnetic field for starting. Rotation is from a starting pole of one polarity toward the adjacent running pole of like polarity. Reverse rotation is obtained by reversing the polarity of one of the windings, usually the starting winding.

9. A split-phase starting winding is disconnected at about 75 percent of full speed by a centrifugal or electrically operated switch.

10. A capacitor motor is a type of split-phase motor with a capacitor in series with the starting winding to increase starting torque.

11. There are two types of motor-starting capacitors: (a) electrolytic, for high starting torque, and (b) oil, for low starting torque and continuous operation.

12. The size of the capacitor needed for a motor is determined by the design of the motor. A capacitor of the proper size should never be replaced with one of different size as a substitute.

13. A two-value capacitor motor starts on one value of capacitance and runs with a lower value in the starting winding.

14. Repulsion motors operate on the principle of repulsion between like poles. The position of the brushes determines the physical relationship between the field and armature poles and the direction of rotation.

15. There are three types of repulsion motors—repulsion, repulsion-induction, and repulsion-start induction. The starting torques of the three are practically the same. The running characteristics vary widely.

16. A repulsion motor has no-load runaway characteristics. A repulsion-induction motor varies in speed about 15 to 25 percent between no load and full load. A repulsion-start induction motor has a practically constant speed.

17. Repulsion motor stator poles induce like poles in the armature. Brushes displace the armature pole slightly from a centerline between the stator pole and the armature axis. Rotation results from repulsion of like poles and is in the direction of displacement. Rotation is reversed by shifting the brushes in the direction of desired rotation.

18. For starting, a shaded-pole motor employs a shading band mounted to enclose about one-third the area of one side of a pole. This produces a shifting or rotating effect of the field magnetism that cuts rotor bars to induce current in them for starting. Rotation is toward the banded side of the poles.

19. Series ac motors develop torque by the same principle employed in dc series motors. The direction of rotation can be reversed by reversing the armature or field current. A universal series motor operates equally well on either ac or dc voltage of the same value.

QUESTIONS

15-1 Arrange the following single-phase motors in the order of decreasing torque, with the highest torque first: split phase, repulsion, shaded pole, series, capacitor.

15-2 A small electric kitchen timer probably uses what type of single-phase motor?

15-3 Which of the following motors uses a wound rotor: series, repulsion, split phase?

15-4 What instrument is used to test a squirrel-cage rotor for a broken rotor bar?

15-5 Explain how a centrifugal switch opens and closes the starting winding circuit.

15-6 Explain how the starting winding makes a single-phase induction motor self-starting.

15-7 If the full-load current of a typical split-phase motor is 4 A, what would the starting current likely be?

15-8 Explain the difference between concentric and concentrated windings.

15-9 What is the source of most electrical and temperature data about a motor?

15-10 Define *pole span*.

15-11 Calculate the synchronous speed of a single-phase induction motor with eight poles operating at 120 V, 60 Hz.

15-12 The motor in question 11 is rewound so that it will run twice as fast. Should it have more or fewer poles?

15-13 How does slip affect a motor's speed? What is a typical value of a slip for a single-phase induction motor?

15-14 How can the direction of rotation of a split-phase motor be reversed?

15-15 How does a standard split-phase motor differ from a capacitor-start motor?

15-16 What are the two types of capacitors used for motor starting?

15-17 How many starts per hour are motor capacitors limited to?

15-18 What are two types of capacitor defects? Which of the two is more common?

15-19 Do capacitors have leading or lagging power factors?

15-20 A capacitor is rated at 100 μF. A test of the capacitor shows 4.5 A in the capacitor circuit with 120 V across the capacitor. Is the capacitor shorted? Is it open? Do the test results prove the capacitor is defective? Explain your answer.

15-21 What winding of a capacitor is generally reversed to reverse the direction of rotation of the motor?

15-22 What type of capacitor is used for continuous duty in a capacitor motor?

15-23 What are the functions of the capacitors in a two-value capacitor motor?

15-24 Which of the following motors generally draws the least starting current: split phase or repulsion?

15-25 Explain how a repulsion motor develops its starting torque.

15-26 Does hard neutral or working neutral produce starting torque in a repulsion motor?

15-27 How is the direction of rotation of a repulsion motor reversed?

15-28 Name the three types of repulsion motors. Which of the three is the most common?

15-29 Explain the operation of a repulsion-start induction motor.

15-30 Where is a short-circuiting *necklace* used?

15-31 What is meant by a *shaded pole*?

15-32 Can a shaded-pole motor have its direction of rotation reversed? If not, why not? If so, how can it be reversed?

15-33 How does the core of a dc series motor differ from that of an ac series motor?

15-34 What is the name of the series motor that is used both on alternating as well as direct current?

15-35 How is a series motor protected against runaway speeds?

15-36 What is the function of compensating windings in an ac series motor?

15-37 How are series motors reversed?

16
THREE-PHASE MOTORS

Three-phase motors are the workhorses of industry. They have the ruggedness and reliability that make them the most dependable of all types of motors.

Three types of ac motors are the *squirrel-cage induction motor,* the *wound-rotor induction motor,* and the *synchronous motor.* The stationary-field windings (called *stators*) of each are basically identical.

In ac motors, three-phase current in the stator produces a *rotating magnetic field* although the stator coils themselves remain stationary. This rotating magnetic field makes three-phase motors *self-starting.* No auxiliary starting windings or devices are needed to start them. Moreover, three-phase motors, because of the self-induced counterelectromotive force (cemf) in their windings, do not necessarily require protection against excessive starting currents. However, starting currents sometimes can be excessive and require current-limiting devices to protect the line. The self-starting and self-protection characteristics of three-phase motors contribute to their ruggedness, reliability, and economy of operation.

16-1 THREE-PHASE ROTATING MAGNETIC FIELDS

Three-phase current in a three-phase winding produces a rotating magnetic field in the stator of a motor. This rotating magnetic field, cutting conductors in the rotor, induces a current in the rotor conductors which produces a magnetic field in the rotor. The north and south poles of the stator's rotating field interact with the magnetic field of the north and south poles of the rotor and attract and repel the rotor poles in the direction of field rotation, producing torque in the rotor.

Although we speak of the field as rotating, there is no actual movement of the field. The periodic changing of field strength and polarity gives the *appearance* of movement.

A step-by-step illustration of the process of rotation of the magnetic field of a two-pole three-phase motor is given in Figs. 16-1 and 16-2. A series of three sine waves produced by three-phase current and depicting the relative strength and polarity of magnetism in a motor at intervals of 30 electrical degrees is shown in Fig. 16-1. The horizontal line represents zero magnetism. The portion of the sine wave above this line indicates north polarity, while the wave below this line indicates the polarity is south. The magnetic effect produced at each point in a two-pole stator is shown in Fig. 16-2. Each instant of the sine wave is numbered to correspond with the numbers under each stator. (The lettering on the magnetic sine wave and stators is in the order in which phase poles actually appear in a motor, 60 electrical degrees apart. On three-phase voltage or current sine-wave curves, phases usually are lettered in the order in which they reach positive maximum, which is 120° apart.)

At point 1 on the sine wave, following the dashed line downward, it is seen that phase B is at maximum strength north, phase A is north but decreasing in strength, and phase C is north but increasing. In Fig. 16-2, stator 1, this condition is illustrated by a large N for phase B and a small N for phases A and C.

At point 2 on the sine wave, A is zero, B is north and decreasing in strength, and C is north but increasing. In Fig. 16-2, stator 2, phase A is zero, and B and C, of equal strength, produce a north pole, the strongest part of which is between the two phases. Thus, from point 1 to point 2, the north pole has moved 30 electrical degrees clockwise.

At point 3 on the sine wave, C is maximum strength north, B is north and decreasing, while A is increasing, but A has changed polarity to south, indicated by A being below the horizontal line. Stator 3 shows this magnetic condition. Phase pole C has a large N for

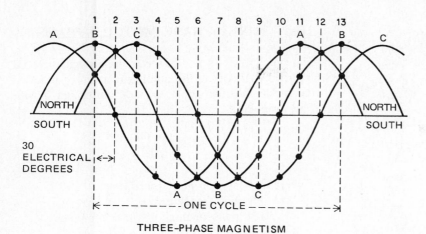

Fig. 16-1 The magnetic field strengths developed by three-phase currents.

maximum strength north; B has a small N, and A is shown with a small S. Thus the main north pole has shifted 30° more clockwise from point 2 to point 3.

An examination of the sine wave at any point in the cycle will show the location of the poles in the stator of the corresponding point. For example, point 6 on the sine wave shows A and B south and C at zero. In stator 6, the north pole has shifted 150° clockwise from its position in the stator in point 1. At the end of the cycle the north pole will have rotated 360°, or one revolution.

A three-phase motor will run but will not start on single-phase current. If a fuse or circuit breaker opens while the motor is running, the motor usually will continue to run if loaded below 60 percent of full load without any apparent ill effects, but it will not start under these conditions.

16-2 DIRECTION OF ROTATION

Figure 16-3 shows a three-phase system with each phase using two wires, connecting the power source phases with the motor phase poles. By reversing phase B of the source and changing other connections, three-phase current can be transmitted over three wires instead of six wires as shown. In actual practice, three-phase is transmitted and distributed over three wires, although frequently a common neutral or fourth wire is used.

Since the center or phase B pole of an alternator is reversed, the sine waves illustrating the values and polarities of a generated three-phase voltage would appear as in Fig. 16-4. It shows that phases A and C in the first alternation are positive, and phase B is negative. This condition will not produce a rotating field in

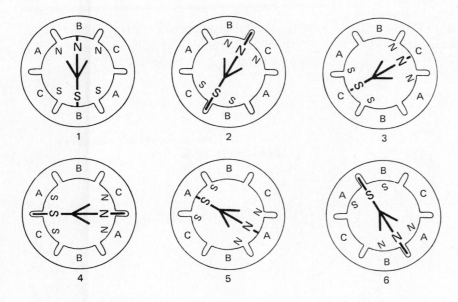

Fig. 16-2 The rotating magnetic field produced by the three-phase current in Fig. 16-1.

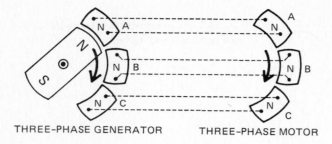

THREE-PHASE GENERATOR   THREE-PHASE MOTOR

Fig. 16-3 Generating three-phase power for a three-phase motor.

a three-phase motor if all the phase poles in the motor are connected for the same polarity. For phase B magnetism to be of the same polarity as that of phases A and C, the phase B pole in a motor is connected opposite in direction to A and C, since phase B current is opposite to A and C. This produces field magnetism of the same polarity in all three phases. This is illustrated in Fig. 16-1, which shows the effect of reversing phase B to make the magnetism of all three phases the same polarity. Phase B magnetism is shown as being north and on the same side of the time line as A and C.

In all three-phase motors operating on a three-wire three-phase system with the phase poles lettered A, B, and C, the center, or B phase, is always connected in *opposite polarity in relation to the A and C phases.* This is true of all motor and alternator windings and should be remembered when connecting these windings. The only apparent exception to this requirement is discussed in Sec. 16-16 on consequent-pole motors.

A motor will run in the direction of rotation of its magnetic field. A three-phase motor is reversed by *interchanging any two of the three supply lines* anywhere between the motor and the source of supply. This reverses direction of rotation of the magnetic field.

16-3 THREE-PHASE MOTOR SPEEDS

The synchronous speed of the rotating magnetic field is found by dividing the number of pairs of poles of the motor into the line frequency multiplied by 60. The formula is

$$S_s = \frac{f \times 60}{PP}$$

where S_s = synchronous speed, rpm
f = line frequency, Hz
PP = pairs of three-phase stator poles

In a synchronous motor, the rotor will lock onto the speed of the rotating magnetic field. Therefore this formula can be used to calculate the speed of a synchronous speed motor. Asynchronous motors (that is, motors that rotate at less than synchronous speed) try to reach synchronous speed but cannot. The difference between synchronous speed and the actual speed is called *slip*. Slip is usually stated as a percentage. A typical squirrel-cage motor might have a slip of 3 percent. The actual speed of a motor can be calculated using the following formula.

$$S_a = S_s\left(1 - \frac{S_{sl}}{100}\right)$$

where S_a = actual speed, rpm
S_s = synchronous speed (as found using previous formula), rpm
S_{sl} = slip, %

Problem

Find the full-load speed of a three-phase motor with 24 actual poles and a slip of 4 percent. The motor is rated at 25 hp, 480 V, 60 Hz.

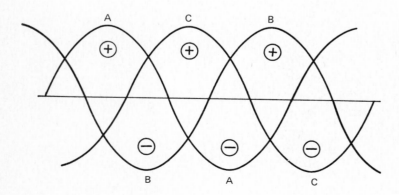

Fig. 16-4 Three-phase voltages in the system shown in Fig. 16-3.

Solution

A 24-pole three-phase motor has

$$\frac{24}{3} = 8 \text{ poles per phase}$$

The motor is therefore considered an *eight-pole,* three-phase motor. Therefore the motor has 4 pairs of poles.

The synchronous speed can be found using the formula

$$S_s = \frac{60 \times 60}{4} = \frac{3600}{4}$$
$$= 900 \text{ rpm}$$

$$S_a = 900\left(1 - \frac{4}{100}\right) = 900\left(\frac{96}{100}\right)$$
$$= 864 \text{ rpm}$$

Notice that the horsepower rating and voltage do not enter into the speed calculations, but the frequency does.

16-4 SQUIRREL-CAGE INDUCTION MOTORS

Three-phase squirrel-cage motors (Fig. 16-5) are the most commonly used motors for industrial applications. This is especially true for motors rated 5 hp and higher and voltages over 200 V. The construction of the rotor is essentially that of the single-phase squirrel-cage motor discussed in Chap. 15. Three-phase motors, however, are self-starting and therefore require no starting windings, capacitors, or other separate starting methods. This simplicity in design, the absence of brushes, slip rings, or commutators, and consequently its low maintenance requirements are some of the reasons for this motor's popularity. One of the drawbacks of this type of motor in the past was the difficulty in controlling its speed. Modern solid-state devices, however, are able to provide effective accurate speed control by varying the voltage and frequency applied to the motors.

Squirrel-cage motors have good speed regulation and therefore are considered constant-speed motors. However, speed does vary with load, and if 0 percent speed regulation is required, a synchronous motor is usually chosen.

The stator winding of squirrel-cage motor is its only winding. It serves as the primary of a transformer by inducing a voltage and current in the copper or alumi-

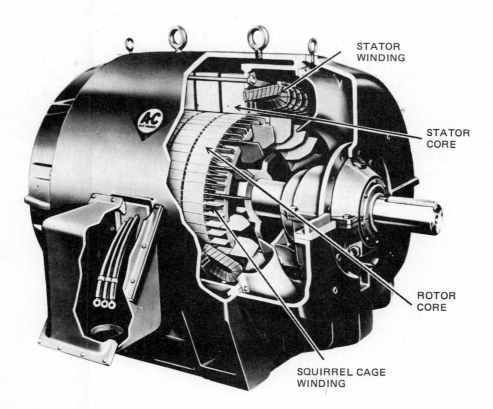

Fig. 16-5 Three-phase squirrel-cage induction motor. (*Siemens-Allis, Inc.*)

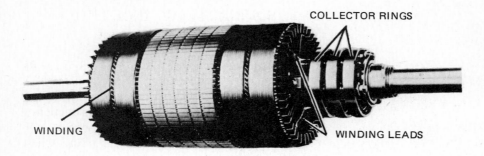

COLLECTOR RINGS

WINDING

WINDING LEADS

Fig. 16-6 Rotor for a wound-rotor induction motor. (*Siemens-Allis, Inc.*)

num bars of the rotor. The magnetic poles produced by the rotor current and the magnetic poles of the stator interact to produce the motor's torque. To maintain its torque characteristics and efficiency, three-phase motors must be supplied from an essentially balanced three-phase system. Even a small unbalance (3 to 4 percent) can reduce efficiency to as much as 80 percent of the motor's rated value.

16-5 WOUND-ROTOR MOTORS

A *wound-rotor motor,* commonly called *slip-ring motor,* is an induction-type three-phase motor with an insulated winding on the rotor. This winding, known as the *secondary winding,* with the same number of poles as the stator, is wound lap or wave and connected as either a wye or a delta three-phase winding. The three starting leads of the rotor winding are connected to three collector rings. A rotor for a wound-rotor motor is shown in Fig. 16-6.

The *stator* winding, which is the *primary* winding, is a three-phase winding, connected either wye or delta, with the same number of poles as the rotor, and usually arranged for dual-voltage connections.

Brushes on the rotor collector rings conduct current, induced in the rotor winding by the stator, to a control system of variable resistance. This arrangement makes it possible to vary the resistance and current in the rotor circuits. As in the case of a trans-

former, changes in the secondary current produce proportional changes in the primary current. This makes possible the control of inrush stator current during starting and provides for varying the starting torque and running speed.

Speed control is obtained by increasing the resistances in the rotor circuits which decreases rotor current and increases slip at the slower speeds. At no load, a wound-rotor motor runs near synchronous speed. Although resistances in the rotor circuits do not necessarily have to be equal at all times, the usual speed control consists of a variable three-phase rheostat that varies the resistance equally in each rotor circuit.

The power circuits of a manual wound-rotor motor control with the motor are shown in Fig. 16-7.

16-6 SYNCHRONOUS MOTORS

A synchronous motor is an ac motor that operates at synchronous speed within its load range regardless of the variation of its load. It is more efficient than any other electric drive. It is especially suited for heavy low-speed loads served by direct drives.

A synchronous motor is equipped with a regular *three-phase winding* in its stator. The rotor is equipped with two collector (or slip) rings and *dc poles,* which are sometimes called *rotating fields.* Direct current is supplied through the collector rings

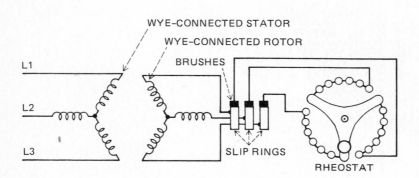

WYE-CONNECTED STATOR

WYE-CONNECTED ROTOR

BRUSHES

L1

L2

L3

SLIP RINGS

RHEOSTAT

Fig. 16-7 Simplified diagram of a manual speed control for a wound-rotor motor.

COLLECTOR RINGS

D-C LEADS

D-C POLES

INDUCTION STARTING WINDING

WELDED BARS AND END RINGS

Fig. 16-8 Synchronous motor rotor. *(Siemens-Allis, Inc.)*

to the dc poles, which maintain a *fixed polarity*. The rotor is also equipped with a *squirrel-cage winding* which functions as a *starting winding*. Except for the induction winding in some cases, a synchronous motor and an ac generator or alternator are of the same construction.

A synchronous motor rotor is shown, with principal parts named, in Fig. 16-8. A small exciter dc generator, sometimes mounted on the shaft of the synchronous motor, is used to supply direct current during running periods.

In operation, three-phase current is supplied the stator winding, and the motor starts as a squirrel-cage induction motor on the squirrel-cage winding. At about 95 to 98 percent of synchronous speed, direct current is supplied to the dc poles. The poles lock the rotor in step with the three-phase rotating magnetic field which "rotates" at synchronous speed. Attraction of *unlike three-phase rotating magnetic poles* and *dc rotor poles* supplies running torque. Stator starting equipment for synchronous motors, where it is desired to limit starting inrush current, is usually one of the types used to start regular squirrel-cage induction motors.

Rotor starting equipment can consist only of a type of double-throw switch to supply direct current to the rotor for running and short-circuit the rotor dc windings during starting or stopping. The dc winding is shorted during starting or stopping to prevent buildup of destructively high voltages by induction from the ac winding.

In most installations the rotor control system consists of several pieces of protective and regulating equipment. A synchronizing relay is used to excite the dc rotating fields at the proper instant for maximum pull-in torque. A thermally operated relay is used to stop the motor if it is caused to drop below synchro-

nous speed and run on the starting winding. The relay is actuated by a thermal element placed in the starting winding and connected through collector rings to the relay.

Synchronous motors, with or without load, are frequently used for power-factor correction. Overexciting the dc fields can create a leading power factor which is used to neutralize the lagging power factor produced by inductive loads. When a synchronous motor is used only for this purpose, it is known as a *synchronous condenser*. For power-factor correction it is usually equipped with an ac ammeter in its primary circuit, a dc ammeter in the rotor circuit, a power-factor meter on the ac system, and a field rheostat in the exciter controls.

16-7 MOTOR SPECIFICATIONS

The National Electrical Manufacturers Association (known as NEMA) publishes standards of motor design and performance. Standard designs are designated by the letters A through D. Other NEMA designations define motor enclosures as open, totally enclosed, or sealed. Motor specifications also include insulation types which are directly related to the temperatures at which the motors can safely and continuously be operated.

NEMA Design Types A *design A motor* has high starting torque for starting under heavy loads. It can run at almost twice its full load before reaching its breakdown point. The breakdown point is the point at which additional load produces an abrupt drop in speed. Ultimately the motor will stall and the windings burn from the excessive current. A *design B motor* is similar in some respects to the design A motor, but its overload characteristics are not as good; its breakdown point is lower than design A. It has good starting torque and average starting current. It has good efficiency especially at the higher horsepower ratings. Typical applications include fans, blowers, and compressors. Because it represents what might be called an optimum design, the design B motor is perhaps the most popular motor for general-purpose applications using across-the-line starting. A *design C motor* has high starting torque but lower than average starting current. It has high efficiency and is especially suitable for starting and accelerating rapidly under very heavy loads. Typical applications include conveyor belts, compressors, crushers, and pumps. A *design D motor* is suitable for very high starting loads such as elevators, cranes, hoists, and presses. This type of motor has high slip and relatively low efficiency, especially at the higher slip percentages.

Motor Enclosures   The NEMA standards designate dozens of enclosure types, of which a few will be described here. In the *open* motor category are the following:

1. General purpose. These have ventilating openings that permit external cooling air to pass over and around the windings of the motor.
2. Drip-proof. These motors have ventilating openings that allow the motors to continue operating despite the fact that liquids or particles strike the enclosure at an angle as much as 15° from the vertical.
3. Weather protected. A type I weather-protected motor has ventilating openings that keep the entrance of rain, snow, and other airborne particles to a minimum. The type II motor is so constructed that any rain, snow, or particles that enter the motor because of winds will be blown out again before they can enter the internal ventilating openings that lead to the windings or terminals.

The *sealed windings* motor is a squirrel-cage motor with specially wound coils and an insulation system that seals off the windings and electrical connections from any external contaminants.

The important motors in the *totally enclosed* category include the following:

1. Nonventilated. Though not completely airtight, this type of motor does not allow external air to enter the enclosure. No external cooling method is used.

2. Waterproof. This motor is designed to withstand a steady stream of water from a hose and not allow any to enter except leakage around the shaft. However, water must not enter the oil reservoir. These motors must also have a means of draining any water that does enter the enclosure.

3. Explosion-proof. Motors with this designation are built to contain an explosion of gases that may occur within the enclosure. In addition they will not ignite any gases that may be present externally.

Insulation Types   The ability to run continuously at full load and occasionally overload is directly related to the motor's ability to handle the increased temperatures produced by the motor. Thus the means of dissipating heat and the ability of the insulations used in the motor to withstand high temperatures determine, in part, the horsepower rating of a motor. Motor insulation is classified by letter according to the temperatures they are capable of withstanding without a serious deterioration of their insulating properties. The temperature ratings are limits based on the sum of the ambient temperature (the temperature of the air in the surroundings) plus a temperature rise resulting from operation of the motor. For example, the four insulation classes are

Class A 105°C (221°F)
Class B 130°C (266°F)
Class F 155°C (311°F)
Class H 180°C (356°F)

A motor with Class A insulation located in a room having an ambient temperature of 85°F (29.4°C) can tolerate a rise in temperature of almost 76°C (137°F). Consider that water boils at 100°C (212°F), and the temperatures under which motors operate can be appreciated. While most insulation will not burn or melt if the limit is exceeded, the useful life of the insulation is seriously reduced. For every 10°C rise above the limit, the life of the insulation is reduced by half. If a motor is ordered without stating an insulation class, more likely than not a Class B insulated motor will be assumed.

16-8   THREE-PHASE STATOR WINDINGS

An understanding of three-phase windings enables an electrician to rewind motors or determine where and how many shorted, grounded, or damaged coils can be cut out of a winding to continue operation in an emergency. In cutting out a defective coil, cut the

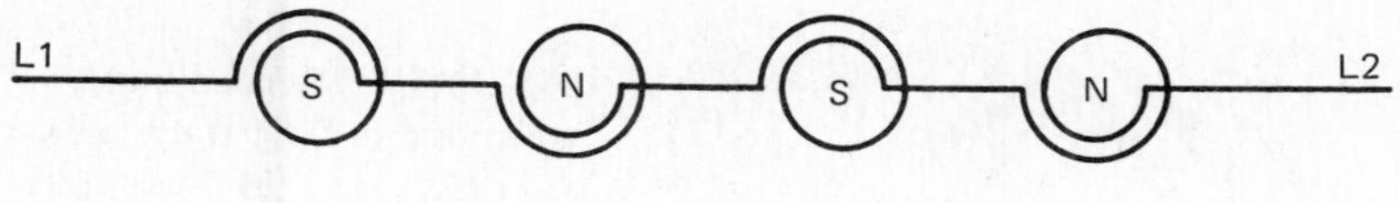

L1 —(S)—(N)—(S)—(N)— L2

Fig. 16-9 Single-phase concentric winding for a four-pole stator.

leads and install a jumper to complete the circuit thus opened. Cut the defective coil in two at the back to avoid induced current in it from the remainder of the winding.

The stator windings of all three-phase rotating equipment are either wye- or delta-connected, with modifications in some cases. Actually, a three-phase winding is three separate single-phase windings in one stator. The way the finish leads of the three single-phase windings are connected determines whether the finished winding is wye or delta. This is illustrated in Figs. 16-9 and 16-10.

A simple single-phase concentric-coil four-pole connection with poles of alternate polarity is shown in Fig. 16-9. Three of these windings in a stator, displaced 60 electrical degrees from each other, and with polarity of the center winding reversed, will make a three-phase winding.

16-9 WYE CONNECTIONS

If the finish leads of the three windings are connected *together,* it would be a *wye* winding. If the finish leads are *connected to certain start leads,* the winding would be a *delta* winding, as shown in Fig. 16-10.

Beginning at the top of the diagram, SA, the start of A phase, is T1, or terminal 1, of the winding; FA, the finish of phase A, is connected to FB and FC to form a star point for this four-pole wye winding; SB is T2; and SC is T3. This winding is known as a *four-pole single-circuit wye winding.* (Wye windings and con-

nections are also known as *star windings* and *connections.*)

The number of poles in a motor is the number of poles per phase. It requires three phase poles to make one motor pole. In Fig. 16-10(*a*) there are 12 *phase poles* but only 4 *motor poles.* The number of circuits in a winding is the number of circuits in *one phase.* In Fig. 16-10(*a*) there is only one circuit from the start of a phase to the wye or star point.

16-10 DELTA CONNECTIONS

A delta-connected three-phase winding is shown in Fig. 16-10(*b*). The *finish leads* of phases are connected to the *start leads of phases in alphabetical order.* The finish of A is connected to the start of B. The finish of B is connected to the start of C. The finish of C is connected to the start of A. These connections form the winding terminal leads T1, T2, and T3 as shown in the figure.

Briefly, the connections are as follows: *finish of A to the start of B, finish of B to the start of C,* and *finish of C to the start of A.*

A wye-connected winding requires 1.73 of the voltage of a delta-connected winding. A delta-connected winding requires 58 percent of the voltage of a wye-connected winding. It can be readily seen that voltage requirements are greater between T1 and T2 of the wye winding in Fig. 16-10(*a*) than the requirements between T1 and T2 of the delta winding in Fig. 16-10(*b*).

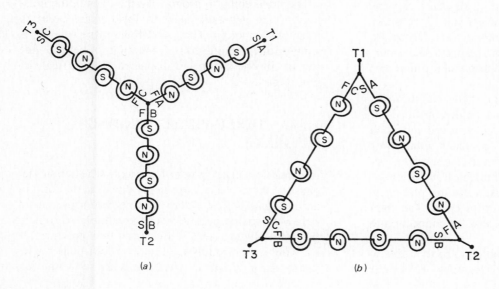

Fig. 16-10 Three-phase connections of single-phase windings. (*a*) Four-pole wye. (*b*) Four-pole delta.

(a)

(b)

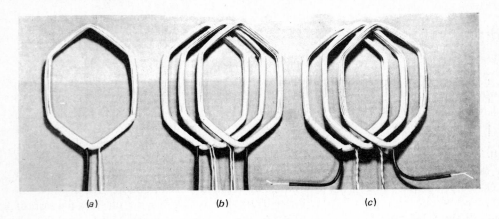

(a)   (b)   (c)

There is no difference between the operational characteristics of the two connections, but they allow greater latitude in designing windings. Voltage and current requirements determine the number of circuits and turns and the size of wire in coils for a winding. Wye and delta connections require different voltages and currents. A designer selects the connection and winding that afford the most favorable mechanical and electrical arrangement of a winding.

A comparison of data for wye- and delta-connected windings for a given motor for the same voltage, horsepower, and speed follows. For a delta winding, as compared with wye, the coil turns are 1.73 of a wye and the wire size is 58 percent of a wye. For a wye winding, as compared with a delta, the coil turns are 58 percent of a delta, and the wire size is 1.73 of a delta. The current, voltage, and power required for each winding are the same.

16-11 THREE-PHASE COILS

Occasionally, when end space is limited, a three-phase stator is wound with three single-phase concentric-coil windings. But practically all three-phase windings are composed of diamond coils. These are coils wound on a form and later installed in the stator. Coils are also made on forms that make round end coils.

Diamond coils, named for their shape, are shown in Fig. 16-11. At Fig. 16-11(*a*) is a single coil, made of magnet wire wound on a form and removed and taped at the lead end. The left lead is insulated with tubing to form a pole lead, also known as a *phase coil* because it contains a specially insulated lead for cross-connection to other poles. Phase coils are placed on each side of a pole phase group of coils. The right lead, usually sleeved, is a *stub* lead for connection to other coils in the pole.

Three coils are shown assembled to form a pole in Fig. 16-11(*b*). At Fig. 16-11(*c*) the stubs of the three

coils are connected to complete a continuous circuit through them. Making these stub connections is known as *stubbing* the winding. The group of three coils stubbed to form a pole is known as a *pole phase group*. Small to medium-size coils are usually wound on a multicoil form, and the wire is not cut between coils for a pole phase group. This is known as *gang winding*. Stubbing is eliminated by gang winding.

Some single-phase motor windings are composed of diamond coils. A single-phase winding of diamond coils is shown in Fig. 16-12. In Fig. 16-12(*a*), the coils are shown placed in the slots. These are semiopen slots, and coils must be capable of being "mushed" or "fanned out" to place them in the slots. This type of coil is known as a *diamond mush* coil.

In Fig. 16-12(*b*), the three coils per pole are shown stubbed to form a complete pole. Cross-connections of the pole leads are shown made to form alternate polarity, as pictured in Fig. 16-9, to complete the single-phase winding. These cross-connections are sometimes called *series connections* or *jumpers*. Line leads are connected to the two pole leads shown extending into the bore of the stator.

Diamond coils, in groups like the ones in the illustration, are generally used to form phase poles in three-phase motors. The usual three-phase winding is simply three single-phase windings, like the one shown, in one stator and displaced 60 electrical degrees apart.

16-12 THREE-PHASE WINDING DIAGRAMS

In general practice, poles of a motor, whether single-phase or three-phase, are usually drawn in the form of an oval or diamond figure. Each figure represents a pole but with no indication of the number of coils in the pole. Figure 16-13 illustrates two practices in showing a connection diagram of a four-pole single-phase motor circuit. Diamond coils are shown in Fig.

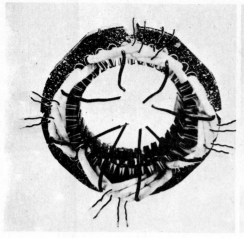

(a)   (b)

16-13(a) and ovals in Fig. 16-13(b). The arrows indicate the direction of current flow through the poles. The right-hand coil rule can be used to determine the resultant polarity.

In ac equipment the polarity is changing with each alternation; accordingly polarity is taken to mean the relative polarity of one pole in relation to others, or the polarity of all poles at a given instant.

In Fig. 16-13, current entering on L1 of either diagram will produce a south pole in the pole at the bottom of the figure and alternate poles as it flows through the remainder of the circuit to L2. In using the oval method to illustrate a three-phase winding, it is only necessary to show the phase poles of each of the three phases, and connections. If it is desired, the number of coils in a pole phase group can be written in each oval.

The development of a drawing for a three-phase four-pole single-circuit winding is shown in Fig. 16-14(a). Twelve phase poles, lettered according to phase, are shown with the A phase poles connected in series, or single circuit. The start of phase A is marked SA, and the finish of phase A is marked FA.

The arrows show the required direction of instantaneous flow of current for proper polarity.

A complete illustration of a three-phase four-pole single-circuit winding, which can be either wye or delta, depending on how the finish leads are connected, is shown in Fig. 16-14(b). The starting leads of each phase are marked S and the finishing leads of each phase are marked F. It will be noted that the polarity of the center or B phase is opposite that of the A and C phases. The B phase *starting lead* is the *right lead,* while the A and C phase *starting leads* are the *left leads.* If FA, FB, and FC are connected together, the winding will be wye-connected. For a delta connection, FA is connected to SB, FB is connected to SC, and FC is connected to SA; SA will be T1, SB will be T2, and SC will be T3; T is for terminal.

16-13   THREE-PHASE STATOR DATA

All necessary data should be taken from a winding before it is stripped from the stator. The most important data are the number of poles, the coil span, the

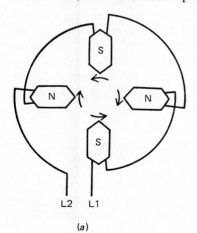
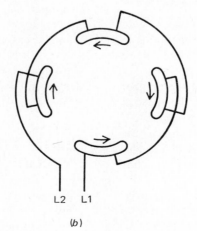

(a)   (b)

Fig. 16-13 Wiring diagram of a single-phase winding. (a) Diamond coils. (b) Oval coils.

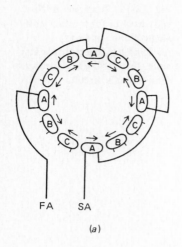

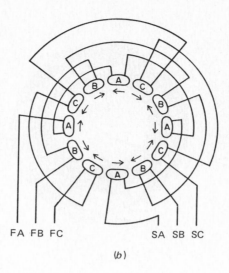

FA   SA

(a)

FA FB FC   SA SB SC

(b)

Fig. 16-14 Development of a four-pole single-circuit three-phase winding. *(a)* The phase A winding. *(b)* All coils for each of the phases are connected. To obtain a wye connection, FA, FB, and FC are connected together. To obtain a delta connection, FA is connected to SB, FB to SC, and FC to SA.

number of turns and wire size in the coils, and the connection. Other important information is listed in a three-phase stator data form shown in Fig. 16-15.

Most of the data for the first five lines of the form are taken from the nameplate. In the spaces for "coils per pole phase group," the number of coils per pole phase group is recorded in the proper square in each column. Some windings, known as *odd-group* windings, do not have the same number of coils per pole phase group. The number of coils per pole phase group is determined by dividing the total number of coils by the number of phase poles.

Odd-Group Windings. A four-pole motor with 12 phase poles and a total of 36 coils will have three coils per pole phase group, which is a whole number. But a six-pole motor with 18 phase poles and a total of 48 coils will have an average of 2 2/3 coils per pole phase group. An uneven grouping would have to be made in this case. The coils per pole phase group would be:

2-3-2-3-2-3-3-3-3-2-3-2-3-2-3-3-3-3

for a total of 18 pole phase groups and 48 coils. These numbers would be placed in the squares beginning at the left in the phase A column with No. 1 pole phase group.

In a well-balanced odd-group winding the number of coils in a pole phase group should be the same as the group in the same phase diametrically opposite it in the stator. In the grouping previously shown, the group of two coils of pole phase 1 in phase A is balanced by a two-coil group in pole phase A10, which is opposite it in the stator.

There are several methods of organizing a balanced odd-group winding, but none of them is general enough to serve in all cases. So the arrangement of

coils of an odd-group winding should be recorded before the winding is stripped from the stator.

In the pole connection section of the data form, the connection of one phase is recorded. Occasionally, for mechanical or other obvious reasons, a winding is connected in a certain way that should be duplicated in the new winding, and this connection should be recorded as a guide. Several modifications of connections that produce the same electrical and magnetic effects will be discussed later.

Skeleton wye and delta figures are shown in the center of the data form. These figures are to be used to show the circuit arrangement of the phase poles of a winding. For example, a single-circuit four-pole winding would be shown by drawing one line with four small circles on it in the blanks containing the phase identification letter.

Most of the remainder of the data form is self-explanatory. The coil span is the number of slots occupied and spanned by a coil. The number of poles can be calculated by dividing the coil span into the total number of slots and taking the nearest whole even number. The number of circuits is the number of circuits in one phase. In a dual-voltage winding, the number of circuits is determined by the number of circuits when the winding is connected for the highest voltage. However, sometimes the number of circuits is given for both the high and low voltages, such as one- and two-circuit.

16-14 THREE-PHASE REWINDING PROCEDURES

The step-by-step procedures for rewinding a three-phase motor are discussed and illustrated in Figs. 16-16 through 16-24.

THREE-PHASE STATOR DATA

Customer_____ Date _____ Job No. _____

Manufacturer_____Frame_____Model_____Type_____

HP_____rpm_____ Rated Voltage_____ Rated Current_____ Phase_____ Frequency_____

Degrees C Rise_____ Hours_____ Service Factor_____ Serial No. _____

Miscellaneous _____

COILS PER POLE PHASE GROUP

A	B	C	A	B	C	A	B	C	A	B	C	A	B	C	A	B	C	A	B	C	A	B	C	A	B	C	A	B	C
1	2	3	4	5	6	7	8	9	10	11	12	13	14	15	16	17	18	19	20	21	22	23	24	25	26	27	28	29	30

Pole Connections
for Phase A: (1) (2) (3) (4) (5) (6) (7) (8) (9) (10) (11) (12)

Line 1_____

Line 2_____

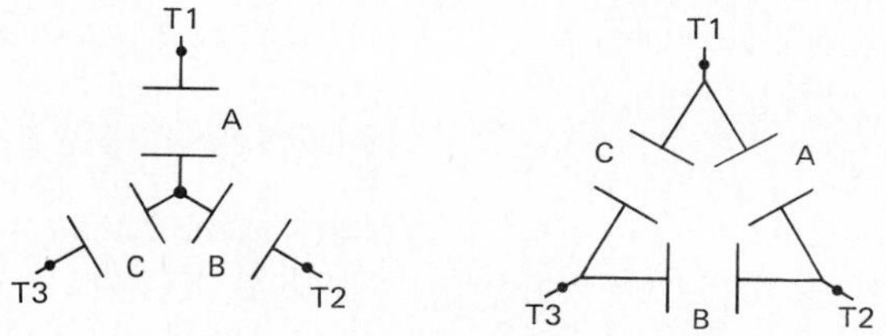

Number of Poles_____ Number of Coils_____ Number of Slots_____

Coil Span_____ Type Connection_____

Number Pole Phase Groups_____ Number Circuits (high voltage)_____

Turns Per Coil_____ Wire Size_____ Wires Parallel_____

Wire Insulation_____ Slot Insulation_____

Poles Extend from Core: Front_____ Back_____ Overall Length of Poles_____

Length of Core_____ Diameter of Core: Inside_____ Outside_____

Pulley-On-Off_____ Base-On-Off_____ Inspection Plate-On-Off_____

Coils Fit_____ Size of Lead Wire_____ Pounds of Scrap Wire_____

Miscellaneous_____ Data by_____

Fig. 16-15 Three-phase stator data form.

Fig. 16-16 A stator cleaned, slot insulation in place, and ready to be rewound.

In beginning the rewinding, data were recorded according to the preceding explanations of a data form. The stator to be rewound, in Fig. 16-16, was found to have a four-pole single-circuit wye winding with a coil span of 1–9. It was stripped, cleaned, and insulated according to instructions for preparing single-phase stators.

A whole sample coil should always be carefully taken from the old winding for measurements for new coils. Before insulating a stator, the slots should be carefully inspected and freed of all dents, burrs, misaligned teeth, and copper deposits due to grounds in previous windings.

Diamond coils are inserted in the stator as illustrated in Fig. 16-17(a). The coil is "fanned out" or "mushed" to allow it to slide into the slot. Very thin guide paper is usually placed in the slot to guide the coil through the opening in the teeth and prevent damage to coil insulation. Any convenient place around the stator bore can be selected to begin winding.

Only one side of each coil for the number of the span is inserted in the beginning. A piece of insulation is laid over the teeth to protect the coil sides left out of slots. The winding in the illustration will progress clockwise around the stator bore; toward the end, coils will be inserted under the raised coil sides of the span until all coils are in the stator. The other coil sides of the span are then inserted in the slots. This type of winding is known as a *span-up* or *throw-up* winding and is most frequently used. If both sides of beginning coils are inserted as winding progresses, the winding is known as a *span-down* or *throw-down* winding.

In Fig. 16-17(b) a pole phase group of three coils has been inserted. The beginning coil is a right-hand phase coil, the center coil is a stub coil, and the last coil is a left-hand phase coil. When the stubs are connected, these three coils will form a *pole phase group*.

A second pole phase group of three coils has been inserted in Fig. 16-18. In Fig. 16-19 the third pole phase group of three coils has been inserted for a total of nine coils, which is equal to the pole span of 1–9. Both sides of the ninth coil are placed in slots, and the sides of the remaining eight coils are left out until the winding progresses all the way around the stator and all coils are in the stator.

The top side of the ninth coil is inserted in the slot containing the bottom side of the first coil, and the slot is closed with a wedge. Before the wedge is inserted, a piece of insulation, known as *phase insulation,* is placed between the second and the third pole phase group to extend from the stator teeth to beyond

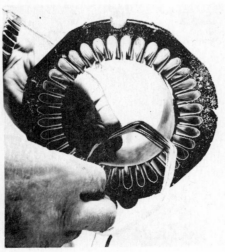

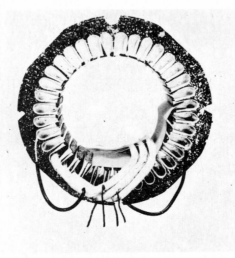

Fig. 16-17 The beginning of a three-phase winding. (a) Installing a diamond mush coil in a slot. (b) First pole phase group, with span up, in place.

(a)　　　(b)

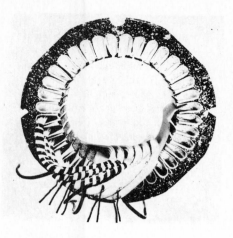

Fig. 16-18 Coils of two pole phase groups inserted in the stator slots.

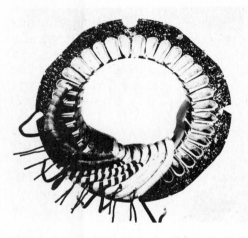

Fig. 16-19 Coils of three pole phase groups inserted in slots forming a single motor pole. The pole span is 1–9, and both sides of the ninth coil are in the slots.

the ends of the coils and thus insulate the phases. Phase insulation is placed between all pole phase groups as winding progresses. Three pole phase groups, composing one motor pole, have now been inserted in the stator.

In Fig. 16-20(a), winding has progressed until all the coils are in the stator. The span was raised and tied out of the way to wind under the span. Phase insulation between all pole phase groups is shown, except the insulation for the span. The next step is to release the span and insert the top sides of these coils and the wedges. This is shown completed in Fig. 16-20(b). The winding is completed in the slots and wedged.

The next step is to partially shape the winding and arrange the stubs and pole leads in an orderly manner, as shown, preparatory to making stub and cross-connections between pole phase groups.

In Fig. 16-21(a) the pole phase group with white ends has been selected as phase A, and the phase leads of this phase have been bent into the stator bore. The next step is to connect, solder, and insulate the stubs of the entire winding. Stub solder joints can be insulated by taping with electrical tape and slipping a piece of tubular insulation (called *spaghetti*) over them.

In Fig. 16-21(b) operations on all stubs are completed, and the cross-connections between pole phase groups in phase A have been made, soldered, and insulated. To insulate cross-connections, a large piece of spaghetti can be slipped well back on one lead and moved over the connection after it is made and soldered, as shown in the illustration.

The cross-connections of phase A are made, and the start and finish leads are lettered according to the schematic diagram in Fig. 16-14(a). The start lead of phase A is lettered SA, and the finish lead of phase A is lettered FA.

(a)

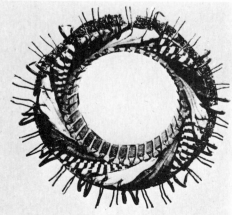

(b)

Fig. 16-20 All coils are in the stator. (a) The span tied out of the way. (b) The top sides of the coils are inserted and the slots wedged.

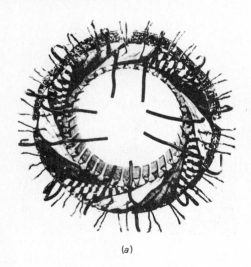

(a)

(b)

Fig. 16-21 Arranging and connecting stubs and pole leads.
(a) Start leads and finish leads of phase A bent into the stator bore before connecting phase A.
(b) Stubs and phase A leads connected. The start of phase A is labeled SA and the finish is marked FA.

A pole phase group on either side of the phase A pole can be selected as phase B. The phase lead of this pole that will give opposite polarity of the adjacent phase A pole is selected and marked SB for the start lead of phase B. Phase B is opposite in polarity to phases A and C. Phases can be selected and connections made either clockwise or counterclockwise around the stator. The start and finish leads of phase B are marked, and phase B cross-connections have been made in Fig. 16-22.

Phase C, with start and finish leads marked, has been cross-connected in Fig. 16-23. The starts of all phases are marked, and the finishes of all phases are marked. All the cross-connections for this four-pole single-circuit wye winding are shown in the schematic diagram in Fig. 16-14(b). The winding is now ready for a wye or delta connection. At this stage it can be connected either way.

Wye (or Star) Connections The data on this winding are single-circuit wye. To make a wye connection,

finish leads FA, FB, and FC are connected together, and the winding is completely connected except for the terminal leads. Terminal lead wire should be of a size sufficient for required mechanical strength and at least one size larger than the total cross-sectional area of the winding leads it is connected to. Flexible lead wire will be connected to the start lead of each phase for the terminals; SA will be T1; SB will be T2; and SC will be T3.

The completed winding is shown in Fig. 16-24(a). The cross-connections and terminal leads have been firmly laced, the coils have been shaped, and the completed job is ready for drying, varnishing, and baking. The back end of the winding, with the coils shaped and phase insulation trimmed, is shown in Fig. 16-24(b).

A large stator is shown being rewound in Fig. 16-

Fig. 16-22 Phases A and B completed.

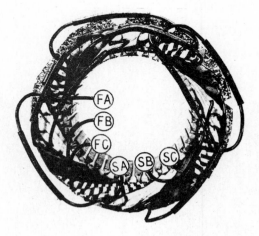

Fig. 16-23 All pole phase groups connected. Leads SA, SB, and SC will be connected to terminal leads. Leads FA, FB, and FC will be connected together to form a wye-connected stator winding.

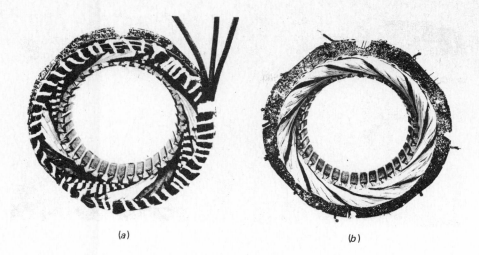

Fig. 16-24 Completed stator winding. (*a*) Terminal leads connected and cross-connections and terminal leads securely laced. (*b*) The back of the stator after shaping and trimming phase insulation.

(a) (b)

25. This is a span-up winding with diamond-formed coils. The stubs and cross-connections are plainly visible. There are 54 pole phase groups for this 18-pole winding. Operation on 60 Hz for this 18-pole motor will produce a synchronous speed of 400 rpm. Owing to slip, the actual speed will be 3 to 5 percent less.

Multicircuit Wye Windings The number of circuits in a winding is determined by the number of circuits through one phase. The number of circuits in all phases of a winding is the same.

A schematic diagram of a four-pole two-circuit three-phase winding is shown in Fig. 16-26(*a*). There are two circuits in each phase from the terminal lead to the wye point. Two separate wye points, with three phases in each wye point, could be used instead of the common wye point shown in the diagram. If two are used, they should be connected with a jumper to minimize the flow of unequal currents in the two groups of the winding.

A diagram for a four-pole four-circuit wye is shown in Fig. 16-26(*b*). Each pole in this winding forms a circuit.

So far only a four-pole wye winding has been discussed. Circuits of windings with any number of poles are always in accord with the same principles; that is, each circuit always contains the same number of poles.

The important things in making connections in all three-phase windings are as follows: Keep all phase

Fig. 16-25 Winding a large 18-pole three-phase stator. (*National Electric Coil Div., McGraw-Edison Service.*)

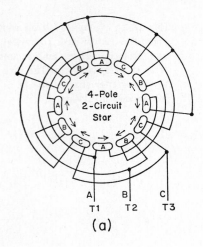 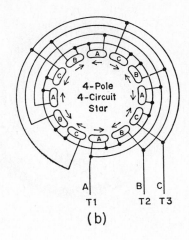

A   B   C
T1   T2   T3

(a)

A   B   C
T1   T2 T3

(b)

Fig. 16-26 Multicircuit windings. (*a*) Four-pole two-circuit wye. (*b*) Four-pole four-circuit wye.

poles in their proper phase circuits; divide phase poles evenly in phase circuits; and give all poles the proper magnetic polarity.

Skip-a-Group Connections Two additional methods of beginning connections or making cross-connections of three-phase windings are shown in Fig. 16-27. The starting leads of each phase are selected by the skip-a-group method. There is a pole phase group between the beginning pole phase groups.

The 1-and-4 and 1-and-7 Connections The cross-connections in Fig. 16-27 are made between adjacent pole phase groups of the phase. This connection is known as a *1-and-4*, or *short-throw*, connection; that is, a pole phase group is cross-connected to the *fourth* pole phase group from it. A *1-and-7*, or *long-throw*, connection is shown in Fig. 16-28. Beginning at SA and counting *pole phase group A*, it is connected to the

seventh pole phase group from it. There is no operational difference in a 1-and-4 and a 1-and-7 connection.

Dual-Voltage Wye Windings Most three-phase windings are arranged to permit operation on one of two voltages. This is accomplished by equally dividing the circuits of each phase into two sections and arranging them to be externally connected, series for the high voltage or parallel for the low voltage. Usual voltages are 240 and 480 V. When voltage changes are made on a motor and its controls, it is necessary to change the overload relays.

The division of phase circuits and the ANSI standard terminal numbering system for a wye winding are shown in Fig. 16-29. The numbering is in consecutive order clockwise around the phases, beginning with 1 for the start of phase A.

A chart for external connections of the numbered

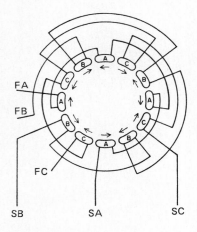

FA

FB

FC

SB   SA   SC

Fig. 16-27 Skip-a-group, or short-throw, connection.

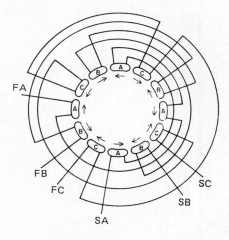

FA

FB

FC

SA

SC

SB

Fig. 16-28 A 1-and-7, or long-throw, connection.

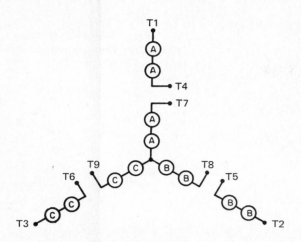

Fig. 16-29 Four-pole three-phase dual-voltage wye-connected winding with ANSI standard terminal numbering system.

16-31(*a*) would be *one- and two-circuit* and Fig. 16-31(*b*) would be *two- and four-circuit*. The procedure for dual-voltage arrangements for four poles is the same for a motor of any number of poles.

Delta Connections A delta connection is made simply by connecting the finish leads of each phase to the proper starting lead of each phase. A four-pole one-circuit delta connection is shown in Fig. 16-32(*a*). The connection is FA to SB, FB to SC, and FC to SA. A four-pole two-circuit delta connection is shown in Fig. 16-32(*b*). In this winding there are two circuits through each phase, with the end of each circuit connecting to the start of the next phase.

In all delta-connected windings, regardless of the number of poles, the same order of phase circuits and final phase connections is followed.

Dual-Voltage Delta Windings. In dual-voltage delta connections, as in dual-voltage wye connections, the phase circuits are divided into two equal sections, and leads from each section are brought externally to the motor and numbered for identification. The leads are connected in proper combinations to series or parallel the two sections for high or low voltage.

A schematic diagram of the arrangement of a four-pole one-circuit winding for dual voltage is shown in Fig. 16-33. This winding is one-circuit when connected for high voltage and two-circuit when connected for low voltage.

Phase A is divided in the center, and the ends of the

terminals for high or low voltage is given in Fig. 16-30. The heading "Tie together" in the chart means that the terminals are to be connected and insulated.

In expressing the number of circuits in a dual-voltage winding, the number of circuits in one phase when connected for *high voltage* is given. A diagram of a single-circuit dual-voltage four-pole wye winding, with terminals numbered, is shown in Fig. 16-31(*a*). A two-circuit winding is shown in Fig. 16-31(*b*). Sometimes the number of circuits is given for both *high and low* voltage. In this case, Fig.

DUAL VOLTAGE WYE CONNECTIONS

VOLTAGE	L1	L2	L3	TIE TOGETHER
LOW	1-7	2-8	3-9	4-5-6
HIGH	1	2	3	4-7, 5-8, 6-9

Fig. 16-30 Line connections to a wye-connected stator.

Fig. 16-31 Four-pole dual-voltage wye-connected windings. (*a*) Single circuit, high voltage. (*b*) Two-circuit winding.

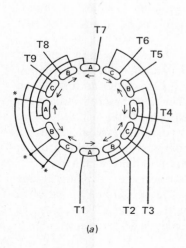

(*a*)

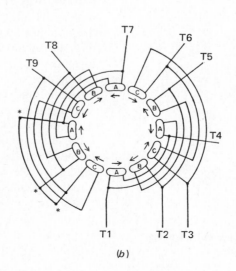

(*b*)

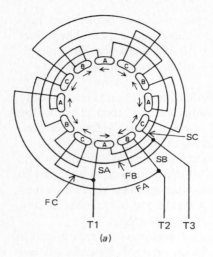

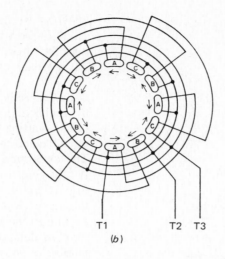

Fig. 16-32 Four-pole delta-connected stator winding. (a) Four-pole one-circuit. (b) Four-pole two-circuit.

sections are numbered; B and C phases are likewise divided and numbered. It will be noticed that, beginning with 1 at the start of phase A, the order of the numbers is numerical clockwise on the lead nearest the starting lead in each phase. Or, considered in a clockwise spiral form, 1, 2, and 3 would be clockwise in the first loop of the spiral; 4, 5, and 6 would be on the next loop; and 7, 8, and 9 on the next loop.

A schematic diagram of a four-pole one-circuit dual-voltage delta winding is shown in Fig. 16-34(a). When 4 and 7 are connected for high voltage, there is one circuit through the phase which makes it a single-circuit winding. A four-pole two-circuit dual-voltage delta winding is shown in Fig. 16-34(b).

A chart for connections of dual-voltage delta terminals for high and low voltage is given in Fig. 16-35.

The high-voltage connections for wye and delta windings are the same. Low-voltage connections for these two windings are not the same. A comparison of the wye and delta connection charts in Fig. 16-30 and Fig. 16-35 will show the difference.

In case of doubt as to whether a dual-voltage winding is wye or delta, a test light or ohmmeter can be used for a continuity test. If a circuit is indicated for three groups of three leads each, the winding is connected delta. If a circuit is indicated for four groups, three groups with two leads and one group with three leads, the winding is connected wye.

Occasionally, a motor is arranged for dual-voltage use by bringing the six start and finish phase leads out to be connected wye for high voltage or delta for low voltage.

16-15 MULTISPEED THREE-PHASE MOTORS

If more than one speed is required in an induction motor, the other speeds can be obtained by changing the number of poles in the motor. For multispeed motors, the number of poles can be changed by two methods.

One method is to wind the motor with two or more complete and separate windings, with each winding containing the proper numbers of poles to furnish one of the desired speeds. When two or more windings are used, they are connected wye. If the windings were connected delta, the active winding would induce a voltage in the idle windings which would result in circulating currents in the idle delta windings.

An arrangement of connections of poles in one phase for *alternate* or *consequent-pole* polarity is shown in Fig. 16-36. With T1 and T2 connected to one line and current entering on T4, *alternate polarity*

Fig. 16-33 Four-pole three-phase dual-voltage delta winding with ANSI standard terminal numbering system.

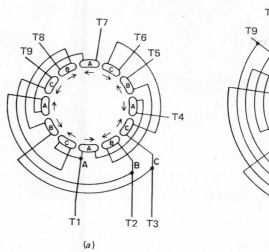

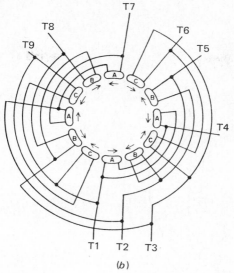

Fig. 16-34 **Four-pole dual-voltage delta-connected windings.**
(*a*) Single circuit. (*b*) Two circuit.

(*a*)   (*b*)

DUAL VOLTAGE DELTA CONNECTIONS

VOLTAGE	L1	L2	L3	TIE TOGETHER
LOW	1-6-7	2-4-8	3-5-9	– – – – –
HIGH	1	2	3	4-7, 5-8, 6-9

Fig. 16-35 **Line connections to a delta-connected stator.**

is produced by the *salient poles* as indicated by the top arrows. With T4 open, and current entering on T1 to T2, *north poles are produced by all the salient poles, and consequent poles develop between them, which doubles* the total number of poles.

A schematic diagram of a winding for a four-pole three-phase motor, with pole phase connections of 1-and-7, is shown in Fig. 16-37. This winding, with T1, T2, *and* T3 *connected together*, and T4, T5, *and* T6 *connected to the lines*, forms a *two-circuit wye winding* to produce *four salient poles of alternate polarity* for the *high speed*. The inside arrows show alternate polarity of the salient poles when the winding is traced as a two-circuit wye winding from T4, T5, and T6.

With T1, T2, *and* T3 *connected to the lines* and T4, T5, *and* T6 *left open*, the winding forms a *single-circuit delta* to produce the *same polarity* in all the *salient poles*. Consequent poles develop between the salient poles, *doubling the total number of poles* in the motor for the *slow speed*.

A controller can be used to make the necessary connections for the desired speed. Each obtainable speed is one-half or double the other speed.

In a consequent-pole multispeed motor containing more than one delta-connected winding, provisions must be made to open the *closed delta circuit of the idle winding to avoid induced circulating currents*. For this purpose, the delta connection is usually opened at T1, and a terminal established and numbered T7. In case of four-speed consequent-pole motors, one winding is numbered T1 to T7, and the corresponding terminals of the other winding are numbered T11 to T17.

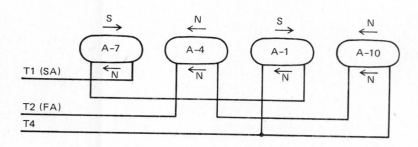

Fig. 16-36 **Arrangement of windings of one phase for alternate or consequent-pole polarity.**

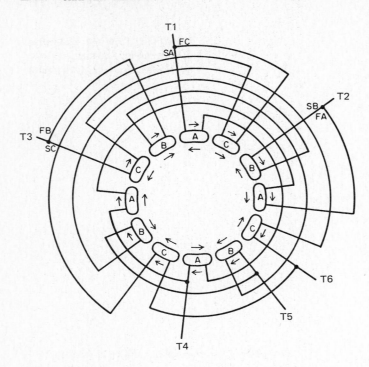

Fig. 16-37 Multispeed four-pole three-phase winding.

16-16 THREE-PHASE CONSEQUENT-POLE MOTORS

Some three-phase single-winding motors are originally and permanently connected for consequent-pole operation. In this case, all salient poles of all phases are connected for the same polarity. The formation of consequent poles reverses the B phase and provides alternate polarity for all poles.

16-17 WINDINGS WITH SEVERAL POLES

In this chapter illustrations of three-phase stator windings have been confined chiefly to four-pole windings. Windings with other numbers of poles are on the same principles as the four-pole windings. A six-pole winding, connected for dual-voltage wye operation, is shown in Fig. 16-38. This winding contains one more

Fig. 16-38 Six-pole three-phase stator connected for a dual-voltage wye.

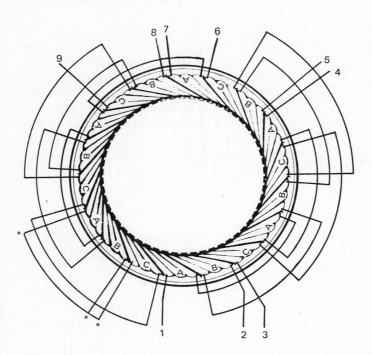

phase pole per circuit than the four-pole dual-voltage wye shown in Fig. 16-31(a). A winding with more poles would simply contain more poles per circuit. A two-pole dual-voltage winding would contain only one pole per circuit.

16-18 IDENTIFICATION OF THREE-PHASE MOTORS BY LEADS

Types of motors and windings can often be identified by the number of leads and the numbering system on the leads. Proper identification is necessary before a motor can be selected and installed. Three leads indicate a single-voltage, one-speed motor (Fig. 16-26).

Six leads, numbered 1 through 6, provided a test light or ohmmeter test indicates continuity between all leads, indicate two-speed consequent-pole single winding (Fig. 16-27); if there is continuity between pairs of leads only, the six leads indicate wye-delta start-run arrangement. If a six-lead motor is marked dual-voltage, it is arranged to connect the winding wye for high voltage, or delta for low voltage. Circuits between 1, 2, and 3 and between 11, 12, and 13 indicate a multispeed two-winding motor (two wye windings).

Nine leads, numbered 1 through 9, indicate dual-voltage arrangement for operation on one of two voltages (Fig. 16-31) or an arrangement for part-winding starting.

Ten leads, numbered 1 through 7 and 11 through 13, indicate a three-speed two-winding motor. The winding numbered 1 through 7 is arranged for two speeds by the consequent-pole method shown in Fig. 16-37, with the seventh lead for opening the closed delta circuit when this winding is idle.

Fourteen leads, numbered 1 through 7 and 11 through 17, indicate a four-speed two-winding consequent-pole motor.

In addition to the above, an extra lead can be tapped to a winding for obtaining low voltage between the tap lead and one line lead for operation of a lamp, signal device, or control system, or two leads can be connected to a thermostat embedded in the winding or located elsewhere in the motor. In these cases the extra leads are usually smaller wire than the regular motor leads. An understanding of windings, with the aid of an ohmmeter or test light, will help to determine these conditions.

SUMMARY

1. The three major types of three-phase motors are the squirrel-cage induction motor, the wound-rotor induction motor, and the synchronous motor.

2. Three-phase windings produce a rotating magnetic field that permits three-phase motors to be self-starting.

3. In order to create the rotating magnetic field, the B phase winding must always be reversed.

4. The apparent speed of the rotating magnetic field can be found with the formula

$$S_s = \frac{f \times 60}{PP}$$

where S_s = synchronous speed or speed of the rotating magnetic field, rpm

f = frequency, Hz

PP = no. of pairs of three-phase stator poles

This formula also gives the synchronous speed of a motor.

5. A nonsynchronous motor will rotate at less than synchronous speed due to slip. Slip is usually specified as a percentage.

6. Squirrel-cage motors are the most common motors for industrial applications.

7. Three-phase squirrel-cage motors are popular because they are self-starting, operate without brushes, slip rings, or commutators, and do not have a winding in the rotor.

8. Squirrel-cage motors have good speed regulation and for many applications may be considered constant-speed motors.

9. Wound-rotor motors contain a three-phase winding on their rotor. The winding is known as a secondary winding. It contains the same number of poles as the stator.

10. Wound-rotor motors require three slip rings and brushes in order to connect speed control devices to the rotor circuits.

11. An external rheostat controls the induced current in the rotor which, in turn, controls the speed of the motor.

12. Synchronous motors operate at only the single synchronous speed. If the load varies, the speed will remain constant. If the load becomes excessive, the motor will stall.

13. Synchronous motors contain a regular three-phase stator winding. The rotor is wound to develop fixed polarity. The rotor is fed from a dc supply through two collector rings and brushes.

14. Synchronous motors are not self-starting unless provided with additional starting devices or windings. The rotor of a synchronous motor often is equipped with a squirrel-cage winding which serves as a starting winding.

15. A small dc generator mounted on the synchronous motor shaft supplies the rotor field current.

16. By variation in the strength of the dc field, the synchronous motor can have either leading or lagging power factor. Synchronous motors are therefore often used for power factor correction.

17. The National Electrical Manufacturers Association (NEMA) publishes standards of motor design and performance.

18. NEMA design types are designated by the letters A, B, C, and D.

19. NEMA standard enclosure types include open, sealed, and totally enclosed.

20. Insulation classes are A, B, F, and H. These specify the temperature limits to which the insulation can be subjected. The maximum temperature consists of the ambient temperature plus a temperature rise due to the operation of the motor.

21. Stator windings are either wye- or delta-connected.

22. Most stators are wound with diamond coils.

23. Each pole of the motor consists of three-phase windings. A four-pole motor would have twelve windings.

24. Before rewinding a stator, all data should be recorded. A sample coil from the old winding should also be saved.

25. The nameplate of a motor contains valuable information on horsepower, rpm, voltage, current, phase, frequency, temperature rise, service factor, and manufacturer's specific information on model number, style, serial number, type of enclosure, frame size, and the like.

26. Many types of motors and windings can be identified by the number of leads, the lead markings, and certain continuity tests.

QUESTIONS

16-1 Describe the physical and electrical differences between the three major three-phase motor types.

16-2 How does the rotor of a wound-rotor induction motor obtain its current?

16-3 What makes a three-phase induction motor self-starting?

16-4 How is the direction of rotation of a three-phase motor reversed?

16-5 Will a three-phase motor start if connected to a single-phase system? Explain your answer.

16-6 While running on three-phase, one line opens so that an unloaded three-phase motor is supplied with only single-phase. Will the motor continue to run? Why?

16-7 Compare the voltage requirements of a wye-connected winding and a delta-connected winding.

16-8 What is a pole phase group?

16-9 What important data are found on a motor nameplate?

16-10 What is the speed of an eight-pole, 480-V, 60-Hz, 3-Φ synchronous motor?

16-11 What would happen to the speed of the motor in question 16-10 if it were overloaded?

16-12 A four-pole squirrel-cage induction motor rated at 208 V, 60 Hz, 3Φ has a 3 percent slip. What is its speed?

16-13 How is the power factor of a synchronous motor varied?

16-14 How are synchronous motors made self-starting?

16-15 Which of the three major motors listed in question 16-1 requires brushes?

16-16 How are the windings of a three-phase induction motor connected for dual-voltage operation?

16-17 How are the windings of a three-phase induction motor connected for multispeed operation?

16-18 How is the speed of a wound-rotor motor controlled?

16-19 What types of three-phase motors are indicated by nine leads numbered 1 through 9?

16-20 Why must the B phase winding be reversed in a three-phase motor?

17

ALTERNATING-CURRENT MOTOR CONTROLS

Alternating-current motor controls are of the manual or automatic (magnetic or solid-state) types and are classified by the method of starting they afford. There are two general methods of starting an ac induction motor—*across-the-line* and *reduced voltage*. In across-the-line starting the full line voltage is applied to the motor upon starting. Alternating-current motors can be started on full voltage without damage to the motor.

Although the motor itself could withstand full-voltage starting, the electrical system might not. In these cases, reduced starting voltage is necessary. In reduced-voltage starting a lower voltage is applied to the motor upon starting. As the motor accelerates, the voltage is increased in one or more steps until the motor is up to full speed at which point full line voltage is applied for running the motor. This chapter deals with manual and magnetic starters and controls.

17-1 SINGLE-PHASE MOTOR MANUAL CONTROLS

Controls for small single-phase motors usually consist of a simple on-off toggle or knife switch, with or without built-in overload protection. Some single-phase motors have built-in thermal overload protection. A simple control diagram for a single-phase motor is illustrated in Fig. 17-1(a). Overload protection is required either in the line or in the motor. In Fig. 17-1(b), a thermal overload element is shown in the switch. In operation, overload current produces excess heat which actuates a device that unlatches the contacts and opens both lines to the motor.

In a grounded single-phase system, the thermal element should be in L1, the ungrounded conductor. Generally, a switch must break all ungrounded conductors in any system.

A toggle switch of the type generally used for manual control of single-phase motors is shown in Fig. 17-2.

17-2 SPLIT-PHASE MOTOR REVERSING CONTROLS

Split-phase and capacitor motors are reversed by reversing either the starting winding or the running winding. Drum switches are commonly used to start and reverse split-phase motors.

A typical wiring connection of a split-phase motor connected to a reversing drum control is shown in Fig. 17-3. The heavy lines indicate the circuits through the drum control for forward and reverse. This control also has an off position.

In forward position [Fig. 17-3(a)], L1 is connected so that current flow in the start and run winding to L2 is in the direction shown by the arrows. In the reverse position [Fig. 17-3(b)], current flow in the start winding is reversed. Since ac reverses direction in every half cycle, the arrows indicate *relative* direction at a particular instant. Figure 17-4 shows a typical drum switch with its cover removed.

17-3 PRINCIPLES OF MAGNETIC CONTROLS

In across-the-line magnetic controls for motors, the main line closing coil is the heart of the system. All functions of starting, stopping, and protection by the auxiliary controls are accomplished by acting on the closing coil. An illustration of this is given in Fig. 17-5. A three-phase boiler-stoker motor for a coal-

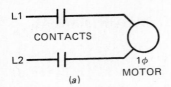

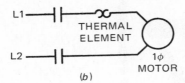

Fig. 17-1 Single-phase motor starters. (*a*) **Double-pole on-off switch.** (*b*) **On-off switch with thermal overload element.**

fired heating system with a two-wire control is shown connected with its controls.

The controls of this system consist of (1) a thermostat for controlling temperature of the heated area, (2) a kindling control to operate the stoker occasionally to maintain the fire, (3) a high-pressure switch for the boiler which opens to stop the stoker if boiler pressure is too high, (4) a low-water switch to stop the stoker if water in the boiler is too low, (5) a shear-pin switch to stop the motor if a jam occurs in the feed screw or stoker mechanism and shears a pin for protection, and (6) a set of overload-relay contacts.

All the controls are connected in series with the closing coil. The pilot controls are the thermostat and kindling controls. If the thermostat does not run the stoker often enough to maintain the fire in the boiler, the fire will go out. The kindling control is a clock or other time-operated control, connected in parallel with the thermostat. In order to keep the fire kindled, the kindling control periodically operates the stoker, usually for a short interval each hour. If a fault in the heating system occurs, any one of the remaining controls, in series with the closing coil, will open and deenergize the closing coil to stop the motor.

The motor in Fig. 17-5 is a three-phase motor. In a three-phase system, as shown in the illustration, the control circuit begins on L1 and ends on L2, accord-

ing to standard practice. The closing coil should always be connected to L2 with the controls on the L1 side of the circuit. If the system is grounded, L2 should be the grounded conductor. Under this arrangement, an accidental ground in the control system circuit will not start the motor. However, if a ground should occur between the coil and L2, the closing coil would be energized and it would start the motor, but this part of the circuit, located in the control cabinet, is not likely to become grounded.

17-4 SINGLE-PHASE MAGNETIC CONTROL

Magnetic control of single-phase motors is accomplished with a magnetically operated switch consisting of a set of main contactors in the line operated by an electromagnet, a thermal overload relay if the motor is not otherwise protected, and two-wire or three-wire control. Two-wire control is generally used for automatic operation.

Automatic operation of a motor is usually accomplished by using pressure switches, float switches, thermostats, limit switches, thermocouples, photoelectric cells, time switches, and various types of electronic controls.

Automatic control requires two wires from the motor switch to the pilot switch, permitting control from remote locations. Unless interlocks or other auxiliary devices are used, two-wire control will restart a motor after restoration of voltage following a power outage.

17-5 ALTERNATING-CURRENT CONTACTORS AND RELAYS

Alternating-current power contactors and control relays differ in some respects from dc contactors and relays in construction and operation.

Alternating-current contactors are noisier than dc contactors. To reduce noise and produce smoother operation, shading bands are used on ac contactor cores.

The coil of an ac contactor, containing fewer turns than a dc contactor for the same voltage, depends on inductive reactance in addition to resistance to limit

Fig. 17-2 Toggle switch starter for small single-phase motors. (*Westinghouse Electric Co.*)

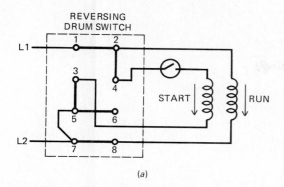

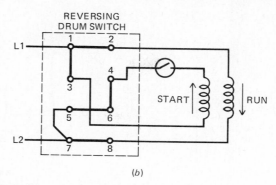

Fig. 17-3 Reversing drum switch connected to a split-phase motor. (a) Forward position. (b) Reverse position.

current flow in the coil. If an ac contactor fails to close completely, an air gap will exist in the magnetic circuit. This air gap will reduce the ability of the coil to generate sufficient cemf to protect itself. Overheating or complete burnout of the coil will result.

The operating parts of an ac contactor must be kept *clean* and *free* to operate properly to avoid burnouts. The faces of the magnetic core at the openings should be kept *slightly* moistened with a rust preventive to inhibit the accumulation of rust and dirt which can cause air gaps. A high-grade *very light machine oil* and dry graphite are the *only* lubricants permissible for lubrication of controls to *avoid gumming* and *sticking* of contactors.

When an ac control circuit is run a long distance, for example, several hundred feet or more, the two conductors add a significant amount of capacitance to the circuit. This may make it difficult to stop a motor quickly since the capacitance will permit current to flow for a period of time. This problem is usually solved by connecting a high resistance across the main coil. The effect of this resistor is to dissipate the en-

ergy in the conductor capacitance, thus stopping the flow of current.

The proper value of resistance is usually found through trial and error. A 10,000-Ω resistor may be tried first to see the effect when the stop button is pressed. Additional resistance can be added in steps of 1000 Ω until pressing the stop button brings the motor to a proper stop.

Thermal relays serve the same purpose in ac control circuits as they do in dc control circuits. The same two types of operating devices, bimetal and molten alloy, are used in both ac and dc thermal relays.

Three-Phase Overload Elements Many three-phase motor controllers are equipped with only two thermal overload elements. Occasionally, on this type of control, a three-phase motor will burn out with apparent evidence of single-phasing, that is, operating with one feeder line open.

Under certain conditions of a system fed by a wye-delta or delta-wye connected transformer, a severe overload can be created on one secondary line if a

Fig. 17-4 Reversing drum control with cover removed showing the arrangement of cams and contactors. (*Eaton Corp.*)

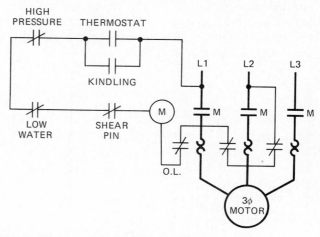

Fig. 17-5 An example of a magnetic starter with auxiliary contacts.

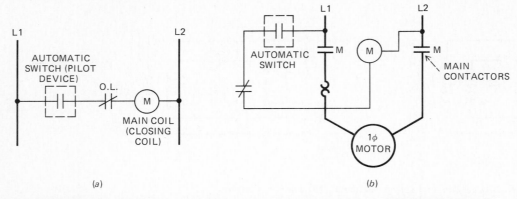

Fig. 17-6 Two-wire control for a single-phase motor. (*a*) Control diagram. (*b*) Wiring diagram for control and power.

primary line is opened. If this secondary line is the one not equipped with an overload unit, the motor can be damaged without opening the overload relays in the other two lines.

In the case of a small three-phase 1-hp motor operating on a line with a 10-hp motor, an open in the supply secondary line can cause a current that is as much as 145 percent of normal in one line of the 1-hp motor. If this line does not contain an overload unit, the motor can be destroyed. Three overload elements, or other approved means of protection, are required for three-phase motors by NEC Table 430-37.

17-6 TWO-WIRE CONTROL

Automatic operation of a single-phase motor with two-wire control is illustrated in Fig. 17-6. Figure 17-6(*a*) is a two-wire control diagram. The control circuit starts at L1 and includes an automatic switch (known as a pilot device) such as a float switch, pressure switch, thermostat, etc., and overload contacts and the main closing coil. The closing coil is also known as a holding coil and is usually identified by the letter M. The contacts operated by this coil also are designated by the letter M.

The power circuit as shown in Fig. 17-6(*b*) contains the main line contactors and an overload relay element. The control works as follows. When the pilot contacts are closed (for example, by a rise in the water level or an increase or decrease in temperature), the closing coil M is energized and the main line contactors are closed; the motor starts. When the pilot contact opens, the coil is deenergized and the main line contactors open, thus stopping the motor.

If an overload occurs, the thermal contacts open, deenergizing the coil which opens the line contactors, stopping the motor.

17-7 THREE-WIRE CONTROL

Three-wire control, as the name implies, requires three wires from the motor controller to the push-button station. A typical three-wire control diagram is shown in Fig. 17-7(*a*). The power and control circuit wiring diagram for a single-phase motor are shown in Fig. 17-7(*b*).

The control circuit contains a start-stop push-button station and a set of auxiliary contacts which are housed with the main line contactors. This set of contacts is known as *sealing-circuit contacts, maintain-*

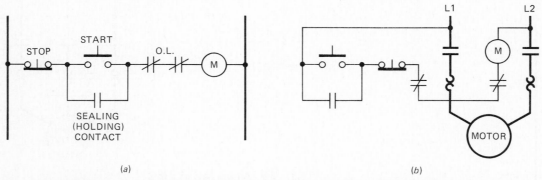

Fig. 17-7 Three-wire control for a single-phase motor. (*a*) Control diagram. (*b*) Wiring diagram for control and power.

ing contacts, auxiliary contacts, or *holding contacts.* They will be referred to here as *sealing* or *maintaining contacts.*

Sealing contacts are always connected across the start button. When closed, they act as a bypass around the start button to maintain a circuit when the start button is released. The start button is returned to an open position by a spring. Similarly the stop button is kept in the closed position by a spring and is *opened* only by pressing the button.

The operation of a three-wire control system as shown in Fig. 17-7(*b*) is as follows. The start button is pressed and held for a few seconds. This completes the circuit from L1 through the start button, the stop button, the overload (O.L.) contacts, the main coil, and back to L2. The main coil is thus energized, closing the main line contactors and the sealing contacts. Closing the sealing contacts shorts out the start button so that the main coil remains energized even after the start button is released.

The motor is now running. To stop the motor, the stop button is pressed, breaking the main coil circuit and deenergizing the coil. This opens the main line contactors, stopping the motor. It also opens the sealing contact so that even when the stop button is released, the circuit to the main coil remains open. The start button must be pressed again to restart the motor.

In case of overload while the motor is running, the O.L. contacts which are NC will open, deenergizing the main coil and stopping the motor. To restart the motor after the overload has been cleared requires pressing the start button again. The overload contacts will close automatically or require pressing a reset button before pressing the start button.

As additional protection, the main coil will deenergize at no voltage and low voltage (70 to 80 percent of full voltage). Upon restoration of full voltage, the start button will have to be pressed to start the motor.

17-8   PUSHBUTTON STATIONS

Four commonly used switches in a pushbutton station are illustrated by their symbols in Fig. 17-8. A normally open (NO) pushbutton used as a start button is shown in Fig. 17-8(*a*). This switch makes a circuit when it is pressed and returns to its open position when the button is released. The switch in Fig. 17-8(*b*) is a normally closed (NC) pushbutton switch which opens when it is pressed. It is used as a stop switch. A break-make switch is shown in Fig. 17-8(*c*). It is used as either a start switch or a stop switch, or for interlocking controls. In this switch the top section is NC while the bottom section is NO. When the button is pressed, the bottom contacts are closed as

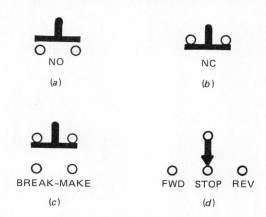

Fig. 17-8 Symbols of common pushbutton switches. (*a*) Momentary contact start. (*b*) Momentary contact stop. (*c*) Break-make. (*d*) Selector switch.

the top contacts open. There is no off position on this switch. When the button is released, the switch returns to its original position. The switch in Fig. 17-8(*d*) is a selector switch for switching from one circuit to another, as from forward to reverse. Selector switches usually have an off position. Other applications include "auto-manual-off" operations. Selector switches remain in their set positions until changed manually.

Occasionally, it is necessary to design and assemble a control system on the job. Various kinds of push button and selector switches, with nameplates and enclosures, are available for this purpose. Several types of control stations assembled from available standard units are shown in Fig. 17-9. Pilot lights are used in the control station to indicate conditions of various circuits. Pilot lights with line voltage or low-voltage filaments or neon lights with transformers or resistors built in are available.

17-9   MULTIPLE-CONTROL STATIONS

On many occasions more than one point of control is required for a job. In these cases two or more pushbutton stations are used. Two-point control is shown in Fig. 17-10. Only three wires are used between the motor switch and the first control station, and three wires are used between the first and second control station. Regardless of the number of control stations, only three wires are used in wiring between them.

All the start switches are connected in parallel, and the sealing-circuit contact is connected in parallel with the combination. Start switches are always connected in parallel with the sealing-circuit contacts. The stop switches are all connected in series.

Fig. 17-9 Some common pushbutton types.

This diagram shows that closing any of the start switches will make a circuit from L1 to L2. The sealing-circuit contact will bypass all start buttons. Any stop button will break the circuit to the closing coil.

17-10 CONTROL ACCESSORIES

Many types of switches, sensors, transducers, and similar devices are used for various automatic control functions. Special control features are often needed on a job. A knowledge of what is available and how it is used will aid the industrial electrician in these problems. A wealth of control information can be obtained from catalogs and bulletins from the various manufacturers of control equipment.

Some commonly used control items are pictured in Fig. 17-11. Figure 17-11(a) is an extremely sensitive overcurrent relay. This relay is sometimes used for quick, automatic stopping of a machine in case of overload to prevent breakage or damage. One motor power line is connected through the relay winding, and excessive current through the relay causes it to open its contacts, which are in series with the motor closing coil. This stops the motor. The contacts must be bypassed to start the motor.

An example of the use of an overcurrent relay is shown in Fig. 17-12, which is a diagram of a control system for a grinding mill driven by a 25-hp motor. A feed conveyor driven by a 1-hp motor supplies raw material to the mill on a continuous basis. When the large motor is overloaded, the excessive current through the overcurrent relay causes it to open its contacts, which are in series with the feed conveyor motor closing coil. This stops the feed motor. When the load on the large motor is reduced, the relay closes its contacts and restarts the conveyor motor. Thus the supply of material to the mill is regulated by the load on the mill motor.

A dashpot time-delay switch is shown in Fig. 17-11(b). This is used for time-delay operations. It consists of a cylinder containing a special fluid and a piston connected to an armature through the coil. When energized, the coil draws the armature upward, and the piston rises. The rise is slow because the piston must pump fluid through a small orifice from the top to the bottom of the cylinder. When the armature reaches the top, it operates the control points. Pneumatic time-delay switches use air instead of liquid to produce the time-delay effect.

Figure 17-11(c) is a zero-speed switch. A rotating element in the switch is operated by connection to a rotating shaft such as the motor armature shaft. The rotating element is magnetically coupled to an assembly that operates the contacts. The contacts can be arranged to open or close when the motor starts or stops.

Drum controllers are used as master switches to operate magnetic controls and for direct control of small motors. A drum control circuit for reversing a single-phase motor is shown in Fig. 17-3. The control itself is shown in Fig. 17-4.

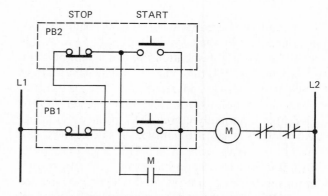

Fig. 17-10 Multiple-point control with a three-wire pushbutton.

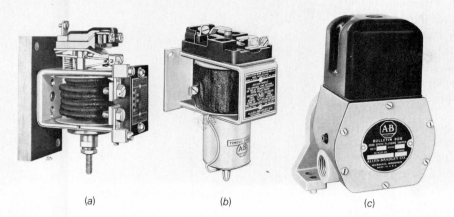

(a) (b) (c)

Fig. 17-11 Automatic control switches and sensors. (*Allen-Bradley.*)

A control relay is shown in Fig. 17-13(*a*). A relay of this type is shown in use as a jogging relay JR in Fig. 17-13(*b*); these relays are also used for magnetic starters for small motors. A limit switch is shown in Fig. 17-14. This switch is operated by movement of a piece of equipment, such as a machine or elevator, to limit travel. The switch is activated by the movement of a lever, roller, push button, or other device. The switch itself is usually sensitive to small movements and light pressure. Most switches are of the momentary contact type. It is extensively used in automatic machines, often as a sensor for safety purposes.

17-11 MULTISPEED CONTROL

A multispeed motor is a motor that can operate at two or more constant speeds. To change the speed of a squirrel-cage induction motor, it is necessary to change either the frequency of the power supply, the number of poles in the motor, or the amount of slip. The poles of an induction motor can be changed by equipping the motor with two or more separate windings or by arranging one or more windings for salient and consequent-pole operation.

Two-Winding Multispeed Control In a two-winding two-speed motor each of the windings is wound to form the number of poles necessary to produce the desired speed. Any two induction motor speeds can be produced by the two-winding method. Only the winding producing the desired number of poles is energized. A two-speed multispeed magnetic switch is shown in Fig. 17-15. This switch consists of two sets of main line contactors and two sets of overload relays.

A diagram of a typical two-winding multispeed system with controls, main line contactors, and motor is shown in Fig. 17-16. The control system is composed of a stop switch and electrically interlocked high- and low-speed start switches as shown in Fig. 17-17. The main contactors consist of two sets of contacts, each with an auxiliary sealing-circuit contact. The motor contains two separate windings with the desired number of poles in each winding.

For high-speed operation, the high button is pressed, which forms a circuit from L1 through the stop switch, contacts 5 and 6 of the low switch, contacts 4 and 3 of the high switch, coil H and overload contacts to L2. Coil H closes its contactors to energize

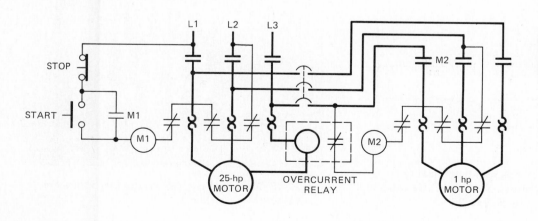

Fig. 17-12 Overcurrent relay used to interlock two motor systems.

the winding in the motor with the fewer number of poles for high-speed operation. After the sealing-circuit contact is closed, the high control circuit is through low contacts 5 and 6, the sealing contact, coil H, and the overload contacts to L2. Pressing the low button will break this circuit between 5 and 6.

For low-speed operation, pressing the low button makes a circuit from L1 through the stop switch, contacts 1 and 2 of the high switch, 8 and 7 of the low switch, coil L and the overload contacts to L2. Coil L closes its line contactors to the winding in the motor with the greater number of poles for slow-speed operation.

One-Winding Multispeed Control A one-winding multispeed motor contains a winding suitable for two ways of connection to the line. One of the connections forms salient poles for high speed, and the other connection forms salient and consequent poles for low speed. This method of obtaining multispeed operation is known as the *consequent-pole method*. In this method the low speed is one-half of the high speed. A starter for a single-winding multispeed motor is similar to the multispeed starter shown in Fig. 17-15, except that there are five main line contactors in one of the sets instead of three.

A typical one-winding multispeed starter wiring diagram of the starter, controls, and motor is shown in Fig. 17-18. A control diagram is shown in Fig. 17-19.

Fig. 17-14   Limit switch.   (*Eaton Corp.*)

Operation of the multispeed system illustrated in Figs. 17-18 and 17-19 is as follows. Pressing the high-speed button makes a circuit from L1 through the stop switch, contacts 5 and 6 of the low switch, contacts 4 and 3 of the high switch, coil H and two sets of overload contacts to L2.

Coil H closes its five high-speed contactors, which connects the motor terminals T4, T5, and T6 to the line, and connects motor terminals T1, T2, and T3 together to form a two-circuit wye connection of the motor winding. This connection produces four salient poles in each phase of the motor to make it a four-pole motor for high speed.

(a)

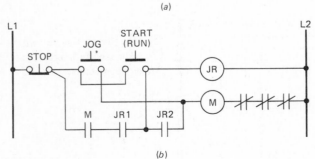

(b)

Fig. 17-13   Control relay. (a) This relay is intended for mounting in a protective enclosure. (*Eaton Corp.*) (b) Jog circuit using a control (jog) relay marked JR.

Fig. 17-15   Multispeed magnetic control used with a two-speed motor.   (*Eaton Corp.*)

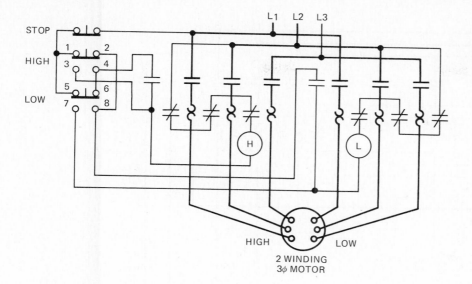

17-16 Typical wiring diagram of a multispeed magnetic control used with a two-speed three-phase induction motor.

For low-speed operation, pressing the low button makes a circuit from L1 through the stop switch, contacts 1 and 2 of the high switch, contacts 8 and 7 of the low switch, coil L and the two sets of overload contacts to L2. Coil L closes its main line contactors, which connects motor terminals T1, T2, and T3 to the line. This connection forms a single-circuit delta winding in the motor with all poles of each phase of the same polarity. Consequent poles develop between the salient poles to double the number of motor poles for slow-speed operation of the motor.

The controls are electrically interlocked since pressing the high or low buttons will break one set of contacts while making another set. The high and low contactors are mechanically interlocked to prevent both contactors from closing at the same time and forming a short circuit across the lines.

Usually, the two-circuit wye high-speed and one-circuit delta low-speed winding is used for constant-torque applications, and the one-circuit delta high-speed and two-circuit wye low-speed winding is used for constant-horsepower applications. For these rea-

sons the two-circuit wye connection is not always high-speed.

17-12 REVERSING CONTROLS

A three-phase motor is reversed by interchanging any two of the three lines. Standard practice is to interchange L1 and L3. Figure 17-20 shows the lines and motor terminals connected to a three-pole double-throw switch. The blades are shown with heavy lines. Jumpers are installed between terminals of the switch as shown in the drawings.

When the switch blades are closed to the right for forward motor rotation, as indicated in Fig. 17-20(a), L2 is connected to T2, L1 is connected to T1, and L3 is connected to T3.

When the switch blades are closed to the left for reverse rotation, as indicated in Fig. 17-20(b), L2 is connected to T2 as in forward position, but L1 and L3 are interchanged. Line L1 is connected to T3, and L3 is connected to T1 for reverse rotation of the motor.

To reverse a three-phase motor with a magnetic starter, two sets of main line contactors are used with suitable controls.

A typical circuit diagram of a magnetic reversing starter, with controls, connected to a three-phase motor is shown in Fig. 17-21. When the forward push button is pressed, the main line contactors connect L1 to T1, L2 to T2, and L3 to T3. When the reverse push button is pressed, the main line contactors connect L1 to T3, L2 to T2, and L3 to T1.

The dashed line between the forward and reverse contactors means they are mechanically interlocked. The mechanical interlock usually consists of a bar mounted between the contactors in such a manner that

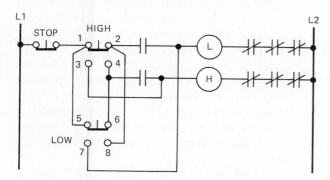

Fig. 17-17 Control diagram for the magnetic control in Fig. 17-16.

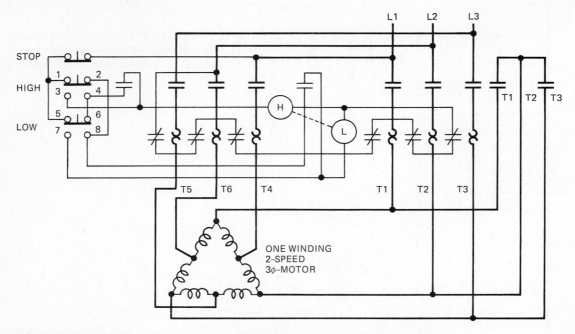

Fig. 17-18 Typical wiring diagram for a two-speed single-winding three-phase motor.

when one set of contactors is "in," the other set is blocked "out." This interlock eliminates the possibility of both sets of contactors closing at the same time and creating a short circuit between L1 and L3.

The push buttons are electrically interlocked to eliminate the possibility of energizing both the forward and reverse coils at the same time. In the electrical interlocking system shown in Fig. 17-21, the circuit to the forward push button contacts 3 and 4 and forward coil goes through the top NC contacts 5 and 6 of the reverse switch. Pushing the reverse button will break the circuit to the forward coil before it makes a circuit to the reverse coil. Likewise, pushing the forward button will break the circuit at its contacts 1 and 2 to the reverse switch and coil before it makes a circuit to the forward coil.

The control diagram for Fig. 17-21 is given in Fig. 17-22. For forward operation, the control circuit is from L1 through the stop switch, reverse contacts 5 and 6, forward contacts 4 and 3 (when the forward button is pressed), coil F and overload contacts to L2. Coil F closes the forward main line contactors and the forward sealing contact. When the forward button is released, the control circuit is from the stop switch, through contacts 5 and 6 of the reverse switch, the sealing contact, coil F and overload contacts to L2.

Pushing the reverse button makes a circuit from L1 through the stop switch, forward contacts 1 and 2, reverse contacts 8 and 7, coil R and overload contacts to L2. Coil R closes the main line reverse contactors, to start the motor in reverse, and its reverse sealing contact. The control circuit now is through the stop switch, forward contacts 1 and 2, through the reverse sealing contact to coil R and the overload contacts to L2.

17-13 START-JOG-STOP CONTROL

Some driven machines require a control system that will allow them to be moved or driven slightly forward or reverse for repair or adjustment. This kind of control is known as *jogging*. For accuracy in movement, the sealing circuit must be disconnected to prevent the motor continuing to run after the point of desired travel is reached.

A control arrangement for jog operations is illustrated in Fig. 17-23. This figure shows a control station equipped with start and stop buttons and a selector switch. When the selector switch is set on "run," the sealing circuit is connected in the circuit around the start switch for continuous running, and the stop switch is used to stop the motor.

When the selector switch is set on "jog," the sealing circuit is opened and the motor is started and stopped by pressing and releasing the start switch only. This is an inexpensive jog system, but it is not completely safe because of the human element involved in determining the position of the selector switch, especially when two or more persons are involved on a job.

The jog system illustrated in Fig. 17-13(*b*) overcomes some of these problems. This system employs stop, jog, and start switches and a jog relay. The jog

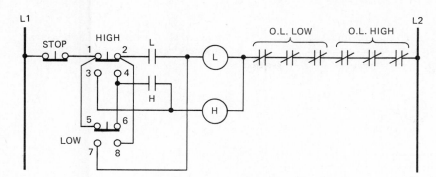

Fig. 17-19 Control diagram for the magnetic controller in Fig. 17-18.

relay, marked JR in the diagram, is a two-pole control relay. The two pole contacts are marked JR1 and JR2 in the diagram. Although called a jog relay coil, it is not used for jogging. It is energized only during the running operation.

Pressing the jog button closes the jog switch to form a circuit from L1 through the jog switch, coil M and overload contacts to L2. The closing coil M closes the main line contactors to start the motor. The sealing-circuit contact is closed by the closing coil, but the contacts of the jog relay are open, which keeps the sealing circuit open. The motor will stop when the jog button is released.

Pressing the start switch completes a circuit from L1 through the stop switch, start switch, and jog relay coil to L2. This energizes the jog relay coil. The jog relay closes its contacts JR1 and JR2. Closing of JR2 completes a circuit from L1 through the stop switch, start switch, JR2, and the closing coil M. The closing coil M closes the main line contactors, to start the motor, and closes the sealing-circuit contact. A circuit to coil M is now established from L1 through the stop button, sealing-circuit contacts, jog relay contacts JR1 and JR2, and coil M to L2. This permits continuous operation of the motor even with the start or jog buttons open. Pressing the stop button will deenergize coil M to stop the motor.

17-14 REDUCED-VOLTAGE STARTER

Alternating-current motors, unlike dc motors, do not necessarily need protection against heavy inrush starting current. Inductive reactance limits the flow of current in ac motor windings. However, conditions other than motor protection sometimes require that the inrush starting current be limited. Conditions may require a smooth, shockless start for driven machines, minimum line-voltage disturbance, and protection of fuses, circuit breakers, and other equipment.

The most commonly used methods of reduced-voltage starting are primary resistance, reactor, part-winding, wye-delta, and autotransformer starting.

Wound-rotor motors with starting controls are also used where starting currents must be limited.

Open and Closed Transition Starters Some reduced-voltage starters disconnect the motor from the line during the transition period when changing from reduced voltage to full voltage for running. This type of starter is known as an *open transition starter*.

Because open transition starting causes a severe line-voltage disturbance at the transition period, it cannot be used in some cases. Most open transition starters can be equipped to minimize disturbance satisfactorily during the transition period.

Closed transition starters *do not open the power circuit* at any time after the motor starts.

Primary Resistance Starters A primary resistance starter starts a motor with resistance in series with the motor windings during the starting period. A timing device then closes a set of contactors that short out the resistance, thus applying full voltage to the motor.

A primary resistance starter is shown in Fig. 17-24(a). This starter consists of two sets of contactors, a timer, and thermal overload relays. Resistors are mounted behind the panel.

A wiring diagram of a typical resistance starting system connected to a three-phase motor is shown in Fig. 17-24(b). A control diagram is shown in Fig. 17-24(c). When the start button is closed, current flows from L1 through the closing coils to L2. Coil S

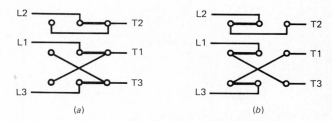

Fig. 17-20 Using a three-pole double-throw switch to reverse a three-phase motor. The heavy lines indicate the switch blades. (a) Forward. (b) Reverse.

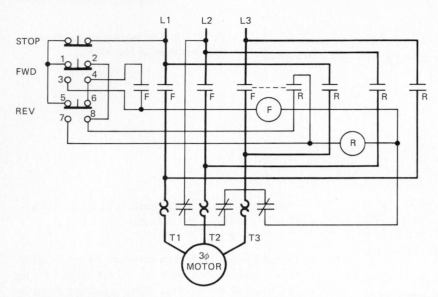

Fig. 17-21 An example of a three-phase magnetic reversing circuit.

closes the starting contactors, which starts the motor on reduced voltage with line current through the resistance.

Timer T is connected across the line when the starting contactors are closed. At the preset time contact T is closed, energizing coil R which closes the run contactors. The run contactors bypass the resistors in the power circuits to the motor. This places the motor across the line on full voltage for running.

Primary resistance starters are closed transition starters and provide comparatively smooth acceleration. The resistors cause a large voltage drop at starting. Since current decreases as motor speed increases, the voltage drop decreases, which increases the voltage to the motor for increased torque during acceleration.

Several steps of starting can be provided by the addition of timer contacts and resistors in the power circuit.

Reactor Reduced-Voltage Starters A reactor starter is similar to a primary resistance starter but with reactor coils replacing the resistors.

Reactor starters are more expensive and must be selected very carefully for a given job. They are employed chiefly in cases where the use of resistance would be difficult because of high voltage or current.

Autotransformer Starters An autotransformer starter uses autotransformers to reduce the voltage for starting. Once the motor comes up to speed, the autotransformers are dropped out of the circuit.

A manually operated autotransformer starter is shown in Fig. 17-25. This starter contains two autotransformers, a thermal overload relay with a reset and stop button, and stationary and movable contacts. The movable contacts are operated by a handle. The movable contacts, in starting position, connect the autotransformer to the line and the motor to the autotransformer for starting on reduced voltage. As the motor comes up to speed, the handle is pulled to the running position, and the movable contacts disconnect the autotransformer and connect the motor across the line for running.

A three-winding autotransformer starter connected to a three-phase motor for starting and running is

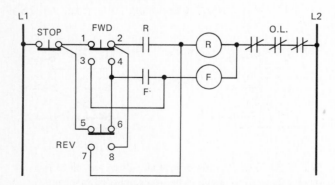

Fig. 17-22 Control diagram for magnetic reversing of a three-phase motor.

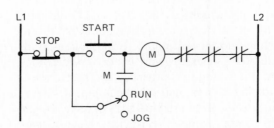

Fig. 17-23 A simple jog control using a selector switch.

shown in Fig. 17-25(b) and (c). In Fig. 17-25(b), the heavy lines show the switch position for starting through the autotransformers. In Fig. 17-25(c), the switch is in the run position. The autotransformer is out of the circuit, and full voltage is applied to the motor directly from the line.

When the handle is moved to the running position, the holding coil H is connected across L1 and L2 and energized to hold the movable contacts in its running position against a spring which is exerting force to open the contacts. The holding-coil circuit contains overload contacts and a stop button. In case of overload, excessive current in the thermal O.L. elements will open the overload contacts, deenergizing the holding coil; the spring then opens the main line contacts to stop the motor. The stop button stops the motor by breaking the holding circuit and deenergizing the holding coil.

The transformers in the diagrams are shown with taps on the windings. Three taps, affording voltages to the motor of 50, 65, or 85 percent of line voltage, are frequently provided. The tap allowing maximum torque at minimum current is selected for the job.

Autotransformer starters, which are open transition starters, provide more starting torque per ampere than any other reduced-voltage starter. Motor current, at reduced voltage, often exceeds line current. A momentary excessive line current occurs during transition from start to run. Some autotransformer starters are equipped with a signaling device to aid the operator in determining transition time.

Wye-Delta Starters In wye-delta starting, a three-phase motor winding is connected as a wye for starting and reconnected as a delta for running. In this method of starting the voltage across phases is reduced to 58 percent during the starting period. A motor arranged for this type of starting has the starting and finishing leads of each phase brought outside the motor and numbered. The start lead of phase A is T1 and the finish lead is T5. The start lead of phase C is T3 and the finish lead is T6.

The principles of making and switching from a wye to a delta connection, using a three-pole double-throw knife switch, are illustrated in Fig. 17-26. With the switch blades thrown to the right as shown in Fig. 17-26(a), T4, T5, and T6 are connected together to form a wye for starting. Each phase receives 58 percent of line voltage for starting.

With the switch thrown to the left, FA is connected to SB, FB to SC, and FC to SA to form a delta connection. Each phase now receives full line voltage for running.

A wye-delta magnetic starter is shown in Fig. 17-27. This starter contains three sets of main line contactors, a timing device, and an overload relay.

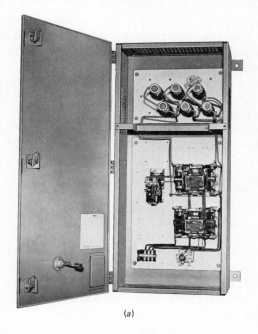

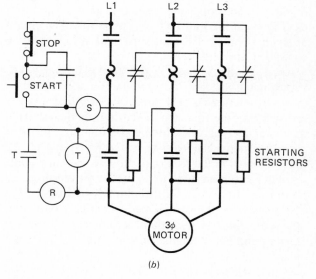

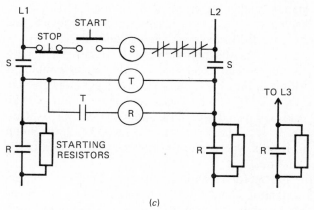

Fig. 17-24 A primary resistance starter. (*a*) **Starter mounted in a cabinet.** (*Eaton Corp.*) (*b*) **Wiring diagram.** (*c*) **Control diagram.**

A typical wiring diagram of a wye-delta starter connected to a three-phase motor with three-wire control is shown in Fig. 17-28. The control diagram for this circuit is shown in Fig. 17-29.

Pressing the start button forms a circuit from L1 through the stop and start switches and the time opening (TO) contacts to coil S, and through the overload contacts to L2. Coil S closes its contactors, which connects 4, 5, and 6 of the motor winding together to form a wye connection. Coil S also opens NC contact S2 to coil 2M and closes contacts S1, which energizes coil 1M.

Coil 1M closes its main line contactors to start the wye-connected motor. Coil 1M also closes sealing-circuit contacts 1M and its auxiliary contacts 1M, which starts the timer T. The timer in time opens the TO contact to coil S. Coil S, deenergized, opens its contactors to open the motor wye connection and also allows NC contacts S2 to close and energize coil 2M, which closes its contactors. With 1M closed, this forms a delta connection of the motor winding for running. When coil S was deenergized, it opened contacts S1 (near the start button), but coil 1M had closed its sealing contact 1M across the start switch to allow continued operations.

Part-Winding Starters A part-winding starter starts a motor by energizing only part of the winding (usually one-half) for the first step in starting, then energizing the remainder of the winding for running. In some cases the motor cannot start on the first step, but the time involved in the first step allows a system voltage regulator sufficient time to adjust for the starting current on the first step. This minimizes starting-current disturbances on the second step when the motor actually starts. Such a procedure is known as *increment starting* and is acceptable to some utilities in certain cases.

A part-winding starter is shown in Fig. 17-30. This starter contains two sets of main contactors, one for each part of the motor winding, a timing device, and two overload relays with reset buttons.

A wiring diagram of typical circuits for a part-winding starter connected to a three-phase motor is shown in Fig. 17-31(a). A control diagram is given in Fig. 17-31(b). The motor contains windings with leads numbered according to the dual-voltage system. Leads 4, 5, and 6 are connected together at the motor to form a wye connection for the first section of the winding.

The motor is started by pressing the start button, forming a circuit from L1 through closing coil S and the overload contacts to L2. Coil S closes its main contactors, energizing the first section of the motor winding and starting the motor. Coil S also closes two

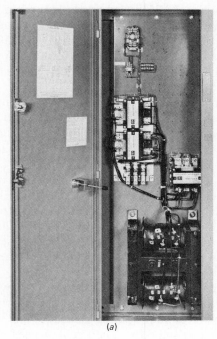

(a)

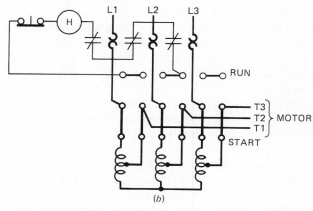

(b)

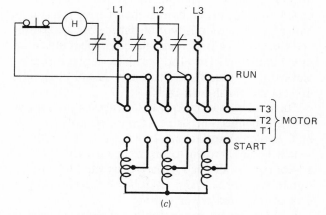

(c)

Fig. 17-25 Reduced-voltage motor starting with an autotransformer. (*a*) Autotransformer mounted in a cabinet. (*Eaton Corp.*) (*b*) Starting the motor with the autotransformer in the circuit. (*c*) When the motor comes up to speed, the switch is thrown and the autotransformer is disconnected from the motor circuit.

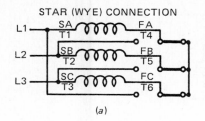

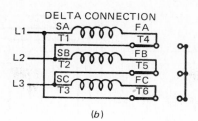

(a)   (b)

auxiliary contacts—the sealing contact S and timer contact S. The timer eventually closes contact T to coil R. Coil R closes its main contactors to energize the second half of the motor winding for running. Some (but not all) standard dual-voltage motors can be used for part-winding starting since their windings are divided into two sections.

For smoother starting, some part-winding starters start a motor in three steps with resistors in series with the first part of the winding for the first step, bypass the resistors for the second step, and energize the second part of the winding for the third step for running.

Part-winding starting affords about 45 percent of full-load torque for starting torque at about 60 to 70 percent of locked-rotor current.

17-15 WOUND-ROTOR MOTOR CONTROL

A wound-rotor motor is a type of three-phase induction motor with an insulated three-phase rotor winding, connected either wye or delta. The three winding leads are brought out and connected to three collector rings. The rotor winding is known as the *secondary winding*. The stator winding, known as the *primary winding*, is a regular three-phase winding.

Resistance in the rotor or secondary circuit can be varied by a control system connected to the collector rings of the rotor through brushes on the rings. This arrangement makes possible a wide variation in starting and running torques which is not obtainable from a squirrel-cage motor.

A wound-rotor motor produces more starting torque per ampere than any other ac motor. In some wound-rotor motors, 150 percent of full-load inrush current will produce 150 percent of full-load torque in starting, as compared with about 500 percent inrush current for 150 percent starting torque for an average squirrel-cage motor.

Figure 17-32 shows the torque curve of a squirrel-cage motor started across the line, together with five different torque curves of a wound-rotor motor with five different values of resistance in the rotor circuit. With no resistance in the wound-rotor circuit, the motor's performance would be similar to that of the

squirrel-cage motor. With resistance $R1$ in the rotor circuit, starting torque is increased but running torque is decreased. Resistance $R2$ increases starting torque but decreases running torque. Resistances $R3$ and $R4$ also increase starting torques and decrease running torques. Finally, resistance $R5$ decreases both starting and running torques. Its value is too high and therefore would not be used.

If the value of resistance $R4$ is used to start on the first step, $R3$ on the second step, $R2$ on the third step, $R1$ on the fourth step, and if the rotor circuit is shorted on the fifth or final step, starting and running torques will be about 250 percent up to about 85 percent of full speed.

A manual wound-rotor motor secondary circuit controller is shown in Fig. 17-33(a). The bank of three sets of resistors is connected to the contact segments on the face plate.

Fig. 17-27 Wye-delta starter. (*Eaton Corp.*)

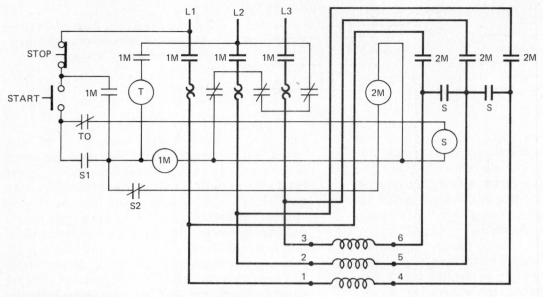

Fig. 17-28 Wiring diagram for an automatic wye-delta starter.

A typical circuit for a wound-rotor motor with a manual controller and magnetic across-the-line starter is shown in Fig. 17-33(*b*). Pressing the start button makes a circuit from L1 through the closing coil M and the overload contacts to L2. Coil M closes its main line contactors to energize the stator winding, and the motor starts with all resistance in the rotor circuit. Resistance is gradually cut out as the motor accelerates to full speed, when all the resistance is cut out of the circuit and the three lines are shorted.

When the resistance controller is in its full resistance position, it mechanically closes a switch in the start switch circuit. With the controller in any other position, the start circuit is opened, which avoids starting with insufficient rotor resistance.

A circuit diagram of an automatic wound-rotor motor starter connected to a motor is shown in Fig. 17-34(*a*) and (*b*). Pressing the start button makes a circuit from L1 through closing coil M and overload contacts to L2. Coil M closes its auxiliary sealing-

circuit contacts and the main line contactors, thus starting the motor with all resistance in the rotor circuit.

Closing the start switch also energized the timer T, which in time closes the contact T1 to coil R1. This coil closes the R1 contactors in the resistor circuit which bypasses resistors *R*1 and leaves the motor running on resistors *R*2. Eventually, the timer closes its contact T2 to coil R2, which closes its contactors R2 in the resistor circuit. This bypasses all resistance in the rotor circuit and connects the lines together. Under these conditions the motor runs as a squirrel-cage motor.

Any desired number of steps in starting can be provided for by the addition of resistors and contactors in the power circuit and contacts on the timer.

Speed control of wound-rotor motors is usually accomplished by the use of manually operated drum controllers. As resistance is added in the secondary circuits of a loaded motor, slip increases and speed decreases.

Wound-rotor motors are usually started across the line with full line voltage, but they can be used with reduced-voltage starting equipment when necessary to minimize line disturbance. Reversing is accomplished only by interchanging any two of the line supply leads. This reverses the direction of rotation of the rotating magnetic field and therefore reverses rotation of the rotor.

The circuit conductors from the collector rings to the controller can be connected in any order, since the rotor circuits do not affect direction of rotation. According to the National Electrical Code (Art. 430-23), these conductors for continuous duty shall have a cur-

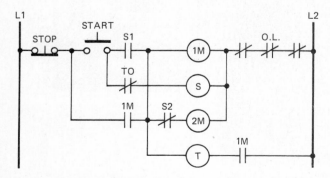

Fig. 17-29 Control diagram for the starter in Fig. 17-28.

Fig. 17-30 Part-winding starter. (*Eaton Corp.*)

rent-carrying capacity of 125 percent or more of the full-load secondary (or rotor) current shown on the nameplate of the motor. Conductor sizes for other than continuous duty are given in NEC Table 430-22(*a*). Conductor sizes for the resistor circuits, if apart from the controller, are given in NEC Table 430-23(*c*).

The frequency of the induced alternating current in the rotor circuit decreases as speed increases. At full speed the frequency is less than 1 Hz. An electrically operated clamp-on ammeter cannot accurately measure ac current in wound-rotor circuits because of the low frequency. Moving-vane meters are used for this purpose.

17-16 INDUCTION MOTOR BRAKING

There are four general methods commonly used to brake the speed of an induction motor and its load. The four methods are *dynamic braking*, *regenerative braking*, *plugging*, and *mechanical braking*. Dynamic braking of induction motors consists in applying direct current to two of the supply lines of the motor after the main line contactors are opened. The direct current in the motor field winding produces magnetic poles in the stator. The magnetic field induces a voltage in the rotating rotor bars. Since the bars are short-circuited, a heavy current flows in them. This current produces a magnetic field that opposes movement through the original field, and this action produces a braking effect on the rotor. The braking principles involved are similar to the braking effect of a load on a generator.

Fig. 17-31 A typical part-winding circuit. (*a*) Wiring diagram. (*b*) Control diagram.

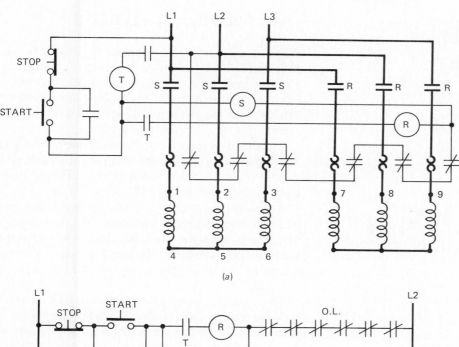

(*a*)

(*b*)

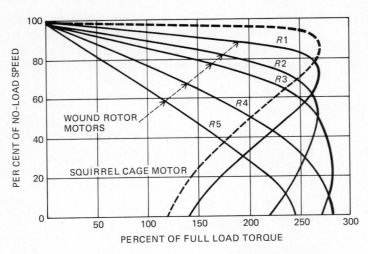

Fig. 17-32 Torque curves of a squirrel-cage motor and a wound-rotor motor with various resistances in the rotor circuit.

A simple control diagram of a variable dynamic-braking system is shown in Fig. 17-35. This system consists of the start-stop pushbutton, coil DB and contactor, the motor circuit, a transformer, a full-wave rectifier to produce direct current, and a rheostat.

The main coil is equipped with two NC contactors connected in the dc circuit. The motor can be stopped by pressing the stop button part way. This opens the main contactors to stop the motor and closes the NC contactors. Pressing the stop button all the way energizes the DB coil, which closes the DB contactor in

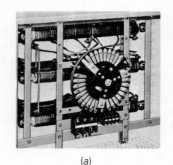

(a)

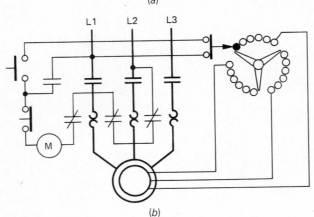

(b)

Fig. 17-33 Manual wound-rotor motor starter and control. (a) Control with cover and handle removed. (Allen-Bradley.) (b) Wiring diagram.

the dc circuit. This puts direct current across lines L2 and L3 and produces dynamic braking. The intensity of braking can be regulated by the rheostat.

Regenerative braking is the braking naturally afforded by a motor when the load drives an induction motor above its synchronous speed. The rotor bars are forced to cut field magnetism, thereby inducing current in the short-circuited bars. The rotor is thus loaded and produces a braking effect similar to that of dynamic braking.

17-17 BRAKING BY PLUGGING

Plugging a motor means reversing it while it is running in order to bring it to a quick stop. For plugging, a starter must have contactors of sufficient capacity for this service, and the driven equipment must be able to bear the shock resulting from plugging the motor.

Automatic plugging can be accomplished by the installation of a zero-speed plugging governor and control contacts on a reversing starter. This arrangement is illustrated in Fig. 17-36.

A zero-speed governor is a rotating device that uses magnetic coupling to automatically open or close a set of contacts when the governor either starts or stops rotating. For automatic plugging service it is connected to the motor, preferably mounted directly on the shaft.

The normally closed governor contacts are operated with the main line contactors. When the stop switch is pressed, the main line contactors open and the governor contacts close to form a circuit through the governor and coil P. Energizing coil P closes the reverse contactors to reverse the motor and stop it. When the motor brakes to a stop, the governor opens the circuit to coil P, which opens the reverse power contactors before the motor can start rotating in reverse.

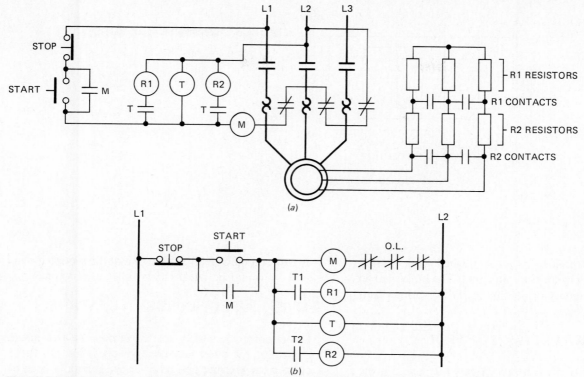

Fig. 17-34 Automatic wound-rotor motor three-step starter. (*a*) Wiring diagram. (*b*) Control diagram.

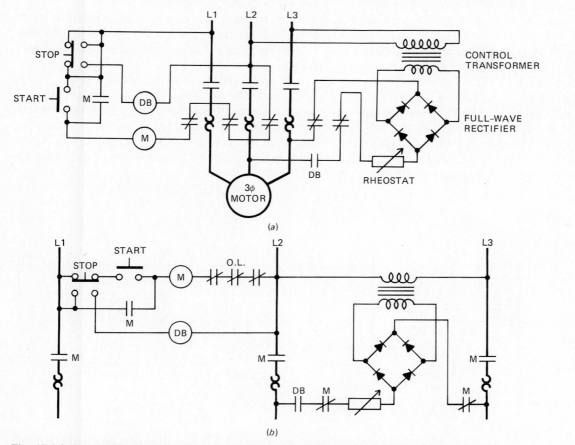

Fig. 17-35 Dynamic braking of a three-phase motor. (*a*) Wiring diagram. (*b*) Control diagram.

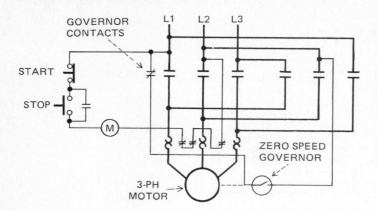

Fig. 17-36 Automatic plugging starter.

Mechanical braking is accomplished by a friction brake operated by an electromagnet. The brake can be controlled manually or automatically by various control means.

17-18 LOW-VOLTAGE CONTROL

Some motor voltages are too high to be safe for the operator at the control station. In these cases a control transformer is used to reduce the control voltage to a safe value. The reduced voltage is usually 115 or 120 V.

A control transformer is shown connected in a control circuit in Fig. 17-37. The primary of the transformer is connected to L1 and L2. The secondary is connected through a fuse to the control circuit. The sealing-circuit contact on the main line coil is used as usual. In addition to the safety afforded by the low voltage, this system also provides a control that is not likely to start a motor accidentally because of a ground in the control system. Two grounds in the con-

trol circuit between the fused end of the secondary and coil M would be necessary to start the motor.

17-19 DIRECT-CURRENT CONTROL

Direct-current controls are sometimes used in an ac controller for quiet and positive operation. In these cases a transformer and rectifier are generally used to supply low-voltage direct current. A dc control system is illustrated in Fig. 17-38.

When the start button is pressed, a circuit is made from the fused end of the transformer through the stop switch, start switch, relay coil CR, and overload contacts to the transformer.

The relay CR closes three sets of contacts, CR1, CR2, and CR3. Contact CR2 is the sealing-circuit contact. Contact CR1 connects the transformer to the rectifier. Contact CR3 connects the main line closing coil M to the rectifier. This energizes closing coil M with direct current to close the main line contactors and start the motor.

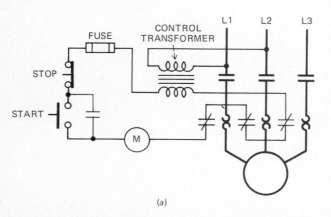

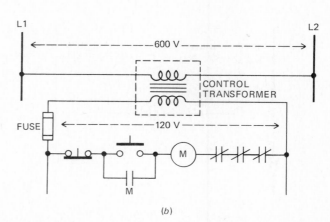

Fig. 17-37 Use of a control transformer. (a) Wiring diagram. (b) Control diagram.

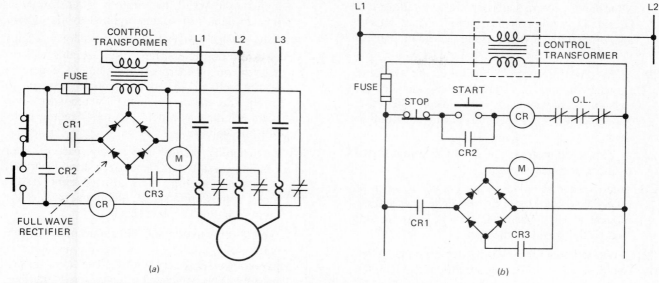

Fig. 17-38 Direct-current control. (*a*) **Wiring diagram.** (*b*) **Control diagram.**

SUMMARY

1. The two general methods of starting ac induction motors are across-the-line and reduced voltage.

2. Alternating-current motors can be started across-the-line at full voltage, without damage to the motor, because of the high reactance of the windings.

3. Reduced-voltage starting is used to protect the circuits and equipment connected to the motor.

4. Small single-phase motors are usually started across-the-line using simple on-off toggle switches rated for motor loads.

5. Small motors often have thermal overload protection built into the motor or as part of the motor starting switch.

6. Single-phase motors are reversed by reversing the leads of either the starting winding or the running winding.

7. A magnetic controller consists of a main coil that operates the main line contactors and various auxiliary relays and sensors.

8. The main coil of a magnetic control is normally connected across L1 and L2 of a three-phase system.

9. Alternating-current contactors and relays differ from those used in dc systems, in that the ac coils rely on reactance to limit current and must operate with a varying magnetic field.

10. To reduce the dangers of single-phasing three-phase motors, overload relays should be provided in all three lines feeding the motor.

11. Automatic operation of motors can be provided by a two-wire control in series with the main coil of a magnetic starter.

12. Automatic operation of motors using manual start-stop pushbuttons requires three-wire control.

13. The typical start-stop pushbuttons are momentary contact switches; the start button is normally open while the stop button is normally closed.

14. To maintain a closed circuit after the start button is released, a sealing contact is wired across the start button in three-wire control.

15. Four common control switches used in motor control circuits are the momentary contact, normally open (NO) switch, used as a start button; the momentary contact, normally closed (NC) switch, used as a stop button; the make-break switch which is a combination NO-NC switch; and the selector switch, used as an auto-manual-off switch.

16. Switches can be provided with pilot lights to indicate the operating condition of a circuit.

17. Multiple-control stations can be used to operate equipment from more than one location.

18. Multiple three-wire start-stop control stations can be added by wiring all start buttons in parallel and all stop buttons in series. A single sealing contact is wired across the parallelled start buttons.

19. Many types of control accessories are available for use with magnetic starters. These include various types of switches; mechanical, chemical, and electrical sensors; transducers and similar devices.

20. A multispeed motor is one that can operate at two or more constant speeds.

21. Rated motor speed depends on the number of poles, the frequency of the power source, and the amount of slip. Changing any one of these will change the speed of the motor.

22. A two-winding motor has its stator wound so that two different sets of poles are produced. Thus the motor will run at two different speeds.

23. A one-winding motor can be made to run at two different speeds by changing the connections of the individual windings.

24. A three-phase motor is reversed by interchanging any two lines feeding the motor. Standard practice is to interchange L1 and L3.

25. Starting and stopping a motor quickly so that it turns only slightly forward or reverse is called jogging. Jogging controls use a start-stop button but not a sealing contact when jogging is required.

26. Although full voltage usually will not damage an ac motor, reduced-voltage starting is used to protect other equipment or the external electrical system.

27. The most common reduced-voltage starting methods are primary resistance, reactor, part winding, wye-delta, and autotransformer starting.

28. If the motor circuit is broken during the transition from reduced voltage to full voltage, the starter is known as an open transition starter.

29. If the motor circuit remains closed during the transition from reduced voltage to full voltage, the starter is known as a closed transition starter.

30. Primary resistance starters start a motor at reduced voltage by adding series resistance in the lines feeding the motor. When the motor comes up to speed, the resistance is bypassed and full voltage is applied to the motor.

31. Reactor starters substitute coils for resistors to produce the reduced voltage at starting.

32. Autotransformer starting produces a reduced voltage from a tapped winding. When the motor comes up to speed, the entire autotransformer is bypassed and full voltage applied to the motor.

33. A wye-delta starter connects the stator windings as a wye during starting. In effect, this will reduce the voltage across the winding to 58 percent of full voltage. When the motor comes up to speed, the starter automatically reconnects the stator winding as a delta, thereby feeding it the full line voltage.

34. Part-winding starters have only part of their winding (usually half) energized during starting. Series resistors limit the starting current. As the motor comes up to speed, additional parts of the motor winding are energized.

35. Wound-rotor motors can be started under a wide variety of conditions by varying the resistance in its rotor (or secondary) circuit. The stator (or primary) can be connected directly across the line, although it is sometimes connected for reduced-voltage starting.

36. Alternating-current induction motors can be brought to a quick stop using any one of four common methods: dynamic braking, regenerative braking, plugging, and mechanical braking.

37. Dynamic braking is produced by applying direct current to two of the motor lines after the main power has been disconnected from all lines. The motor begins acting as a generator under load, and the magnetic fields bring the motor to a halt.

38. Regenerative braking is produced naturally when a motor is driven above its synchronous speed by its load. The process of braking is similar to that of dynamic braking.

39. Plugging is the process of reversing a motor while it is running. This produces a strain on the motor and the driven equipment. Plugging can be used only when the shock of reversing the motor will not damage the equipment or the system.

40. Mechanical braking is accomplished by friction brakes that can be applied to a drum or disk and operated manually or automatically.

41. When the motor operating voltage is too high to be used safely in control circuits, a step-down transformer is used. Typical control circuit voltages are 115 and 120 V.

42. Because dc controllers are quieter and more positive acting than ac controllers, control circuits are sometimes fed from low voltage direct current. The direct current used in these circuits is usually obtained by connecting a step-down control transformer across two lines and then wiring a full-wave bridge rectifier across the transformer secondary.

QUESTIONS

17-1 What are the two general methods for starting ac induction motors?

17-2 What type of starting device is used with small single-phase motors?

17-3 How are split-phase motors reversed? Draw a diagram showing how this can be done using a manual switch.

17-4 What component of a magnetic controller operates the contactors that connect the line to the motor?

17-5 What part of a three-phase magnetic control circuit should always be connected to L2?

17-6 Name six types of devices that are used with magnetic controllers to provide automatic starting and stopping of motors.

17-7 Why can ac closing coils have fewer turns than dc closing coils with the same rated voltage?

17-8 What electrical characteristic other than voltage drop and power loss is of concern when motor control wires are run very long distances?

17-9 Why should regular lubricating oil be avoided in magnetic controllers?

17-10 Explain the operation of the two main methods used in thermal overload devices.

17-11 Why is it good practice to wire overload devices in each leg of a three-phase motor circuit?

17-12 Draw a control circuit showing a start-stop pushbutton, a main closing coil, and three overload contacts. Will this control circuit permit continuous running of the motor?

17-13 If your answer to question 17-12 was yes, indicate the part of the control circuit that ensures continuous running. If your answer to question 17-12 was no, show how the circuit could be made suitable for continuous running.

17-14 What are other names for the sealing contact?

17-15 List four types of switches used for three-phase motor control.

17-16 How are start switches wired in multiple-control station systems using three-wire control?

17-17 How are stop switches wired in multiple-control station systems using three-wire control?

17-18 Explain the operation of a dashpot time delay switch.

17-19 What is a multispeed motor?

17-20 What two factors affect the speed of an ac induction motor?

17-21 Which of the two factors named in question 17-20 is usually changed to vary the constant speed of a motor?

17-22 Describe the electrical differences between a one-winding and a two-winding multispeed motor.

17-23 How is the direction of rotation of a three-phase motor reversed?

17-24 What is meant by jogging a motor?

17-25 Why is it important to prevent continuous running of a motor that is being jogged?

17-26 Under normal conditions, does full-voltage starting damage an ac motor?

17-27 What are the four main methods used for reduced-voltage starting?

17-28 Explain the effect on the electrical system of closed transition starters and open transition starters.

17-29 Which arrangement of windings is usually used for starting in a wye-delta starter?

17-30 What part of a wound-rotor motor is called the primary winding and what part is called the secondary winding?

17-31 What winding of a wound-rotor motor, primary or secondary, contains the speed regulator?

17-32 How is a zero-speed switch used in a motor control circuit?

17-33 What is meant by plugging a motor?

17-34 List four common methods of braking an ac motor.

17-35 Why is direct current sometimes used in ac motor control circuits?

17-36 What is the purpose of a control transformer?

18

ALTERNATING-CURRENT POWER SYSTEMS

The major source of electric energy in the United States is the alternating-current generator. These generators produce electricity by making a magnetic field cut across a series of conductors. (The theory of generation is covered in Chap. 6.) The bulk of electric energy required by industrial, commercial, and residential users is supplied by utility companies from off-site generating plants. Many modern electrical installations also require on-site generating facilities for emergency supplies as required by Code; examples are hospitals, supplemental sources for economic peak loads, and uninterruptible supplies for critical commercial uses such as computers and data processing equipment.

18-1 MECHANICAL AND ELECTRICAL CHARACTERISTICS

Alternating-current generators, also called *alternators* and *synchronous* generators, produce electricity by rotating a magnetic field across stationary coils of wire or by rotating coils of wire in a stationary magnetic field (Fig. 18-1). The rotating part is called a *rotor* while the stationary part is called the *stator*. The magnetic field is supplied by direct current, usually from a small generator. Large alternators normally use the rotating magnetic field arrangement. This is because it is easier to insulate the coils for the very high voltages they will handle if the coils are stationary. Similarly the slip rings and brushes supplying the magnetic field coils need carry only the relatively low dc voltages used by the field. In addition, as the rotor turns, extremely high mechanical forces (called centrifugal forces) act on the conductors in the rotor to pull them outward. To overcome these forces, it is necessary to use restraints to hold the numerous coil sides in place. This adds weight to the rotor. Since mechanical energy must be employed to turn the rotor, additional weight would require the input of additional mechanical energy.

To reduce windage losses in large alternators, especially those running at thousands of revolutions per minute, the rotors containing the magnetic field winding are constructed very long relative to their diameter. The smaller diameter also helps reduce the centrifugal forces pulling at the field windings.

18-2 TYPES OF GENERATORS

The mechanical energy required to turn an alternator is supplied by machines or devices called *prime movers*. Alternator systems are generally named for their prime movers. The large power company generators are turned by *steam turbines* such as shown in Fig. 18-2. In areas where rivers and lakes can be used, water-driven turbines and waterwheels turn the generators. Steam, however, is the primary source of energy for turbines. The steam is produced in large boilers by burning fuel oil, coal, or gas. Nuclear power plants employ radioactive materials to heat the water in the boilers. Except for the fact that nuclear power plants are carefully monitored and tightly controlled closed systems, the basic electric generating components are similar to those using coal, oil, or other fuels.

Smaller alternators, often for on-site emergency or standby service, use internal-combustion engines (Fig. 18-3). Diesel oil, gasoline, and gas engines are

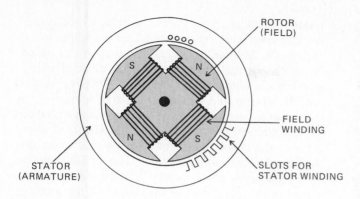

Fig. 18-1 A simplified alternator.

the most common types of prime movers for on-site generating plants. Some utilities employ jet engines as prime movers during peak load periods. Though no longer in common use, reciprocating steam engines are also used to turn alternators.

Wind power is being harnessed to turn medium-size alternators, though such installations are generally experimental and serve as supplements to standard generating plants.

18-3 VOLTAGE AND FREQUENCY

Large alternators used by power companies generate voltages as high as 23 kV. Smaller alternators, especially on-site generating equipment, are able to use lower voltages because they are closer to the place where the energy is required. Power companies must transmit their service over great distances. By em-

ploying very high voltages, they are able to reduce the line current, power loss in the lines, and the size of the conductors themselves. As a trade-off, the very high voltage systems require special expensive equipment, highly trained personnel, tall transmitting towers, and large transformers, switches, circuit breakers, and the like.

On-site alternators generally produce the system voltage used by the plant or building in which they are located. Small local transformers, usually of the dry type, are used to obtain special higher or lower voltages.

In the United States the most common generated frequency is 60 Hz. Some specialized applications (such as transportation systems) use 25 Hz. While 50-Hz systems are not common in the United States, they are standard in Europe and elsewhere around the world. In certain areas of the United States, for example, parts of the west and southwestern states, 50-Hz systems still exist. It is not unusual to specify equipment in these areas for 50/60-Hz operations. In the long run, these systems will probably convert to 60 Hz.

Since the frequency produced by an alternator is dependent on the speed of the rotating field, constant speed of the rotor is critical to maintaining a constant frequency. Recall that the formula for frequency is

$$f = \frac{PP \times N}{60}$$

where f = frequency, Hz
PP = pairs of poles in field
N = speed of rotor, rpm

Fig. 18-2 A modern steam turbine of the type employed by power companies. (*General Electric Company.*)

Fig. 18-3 An emergency diesel-driven alternator.
(*Caterpillar Tractor Co.*)

For example, a four-pole rotor turning at 1800 rpm will produce a frequency of

$$f = \frac{PP \times N}{60}$$
$$= \frac{4/2 \times 1800}{60} = \frac{3600}{60}$$
$$= 60 \text{ Hz}$$

Notice that the four poles of the rotor appear in the formula as two pairs of poles (4/2).

18-4 FIELD EXCITATION

Alternators require a dc source to feed the field coils. The field coils in turn provide the magnetic flux to generate the ac voltage.

Several methods are used to obtain direct current. One of the oldest methods is to mount a small dc generator on the alternator's shaft. As the alternator is rotated by the prime mover, the dc generator is also generating a voltage. The dc voltage is produced in the armature and delivered to the commutator. The dc generator brushes are connected to the alternator's field coils (rotor) by slip rings and brushes as shown in Fig. 18-4. This system, while relatively simple, does require maintenance and control. Producing and delivering the dc power involves sliding contact and friction between brushes and commutator and slip rings; it involves brush wear and replacement. The dc generator introduces an additional set of bearings that must be maintained. Since the dc generator is

mounted directly on the alternator shaft, its speed cannot be varied independently to adjust the output to the field coils. Similarly changes in the alternator speed could change the dc output and thus affect the field strength and output of the alternator. Therefore a system of controls and regulators must be used.

In another method, a small alternator is mounted on the main alternator shaft. The field of the small alternator is stationary and mounted on an insulated shaft within a permanent magnet rotor. The field is supplied by a separate dc power supply. The ac voltage is produced in the stator by the interaction of the rotating permanent magnet field and the stationary dc field. The ac output is fed to rectifiers mounted on the alternator itself. This arrangement is similar to that used in an automobile's electrical system. The small alternator does not use slip rings or brushes. The output of the stator terminates in insulated sections on the alternator frame. Contacts, directly connected to the rectifiers, tap the voltage from the insulated sections. The output of the small alternator and therefore the field strength of the main alternator can be controlled very easily by varying the dc supply of the small alternator. Since the generated voltage of the main alternator is dependent upon field strength, the rheostat used to vary the current to the field of the small alternator in effect can be used to vary the output voltage of the main alternator.

While this method of supplying the field does eliminate a set of brushes and permits control of the field strength, it still uses a rotating machine and requires careful maintenance.

High-power solid-state (electronic) rectifiers offer a means of providing direct current under very controlled conditions without separate rotating machinery. Two systems are in common use. In one system, part of the normal ac output of the alternator is fed to

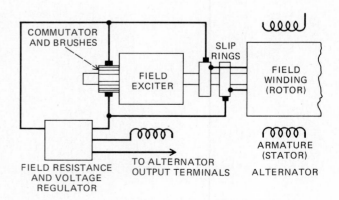

Fig. 18-4 A typical brush-type field exciter system.

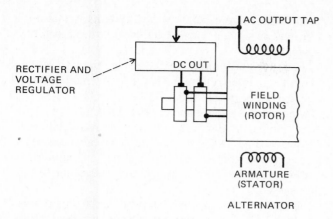

Fig. 18-5 The alternator output voltage used as a source of direct current for field excitation.

As a result a magnetic field will be created around the stator windings. The magnetic field thus created will oppose the rotation of the rotor and increase the load on the prime mover. Either manually or, as in most cases, automatically the prime mover will be throttled up to produce more torque and restore the speed of the alternator to its original set value.

Although the generated voltage and frequency of the alternator will remain constant at constant speed, the terminal voltage will decrease. This is because of the internal impedance of the alternator itself. The stator windings have resistance and inductive reactance.

There is a voltage drop and power loss in the stator due to IR and I^2R. Because of the inductive reactance in the stator windings, the terminal voltage is decreased even further owing to the IZ drop.

As in the case of motors, the magnetic field created by current in the stator windings tends to shift the normal magnetic field produced by the rotor. This effect, called *armature reaction,* further reduces the terminal voltage.

The difference between the terminal voltage and the generated voltage divided by the generated voltage is known as *voltage regulation.* Written as a formula, it is

$$VR\ (\%) = \frac{V_G - V_t}{V_G} \times 100$$

where $VR\ (\%)$ = voltage regulation as a percent
V_G = generated voltage, V
V_t = terminal voltage, V

a rectifier bank which in effect then serves as a dc power supply. The voltage output of this supply is precisely regulated and controlled by the power supply circuitry (Fig. 18-5).

In the other system, separate windings are embedded in the stator slots as shown in Fig. 18-6. The alternating current generated in these windings is fed directly to the rectifiers and treated as in the first method. Both of these methods have generally replaced the separate generator methods, discussed above, in large modern turbine-driven alternators.

18-5 VOLTAGE REGULATION

The output voltage of an alternator is dependent upon the speed with which the field is rotating, the strength of the field, and the number of turns in the stator. Generally it is not feasible to change the number of turns in the stator.

When a load is connected to the output terminals of the alternator, current will flow in the stator windings.

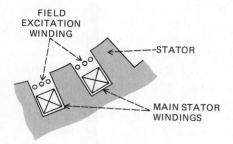

Fig. 18-6 Field excitation winding embedded in the main stator slots.

Most industrial and commercial loads are inductive. As the lagging power factor is lowered, the effect on the terminal voltage gets worse and the voltage regulation decreases.

Circuits with resistance and both inductive and capacitive reactance can have voltage drops that exceed the applied voltage. Similarly, an alternator with inductive internal impedance that is feeding a capacitive load can have a terminal voltage greater than the generated voltage. The increase is due to the effects of leading current on the internal resistance and reactance of the alternator.

Armature reactance also varies with load and specifically the power factor of the load. Inductive load currents tend to produce magnetism in the stator windings that opposes the dc field of the rotor. Capacitive load currents produce stator magnetism that strengthens the rotor field, thus producing a higher generated voltage. As the load becomes more capacitive, the field will increase in strength and the generated voltage will also increase.

Problem 1

The generated voltage of an alternator is 21,000 V. At full load the voltage at the alternator terminals is 19,300 V. What is the nameplate voltage regulation of this alternator?

Solution

The voltage regulation indicated on an alternator's nameplate is based upon full-load characteristics. In this case $V_t = 19,300$ V and $V_G = 21,000$ V. Using the formula

$$VR\ (\%) = \frac{V_G - V_t}{V_G} \times 100$$

$$= \frac{21,000 - 19,300}{21,000} \times 100$$

$$VR\ (\%) = 8.1\ \%$$

18-6 ON-SITE GENERATING SYSTEMS

On-site alternators serve a number of purposes, and the selection, installation, and maintenance of these alternators is largely determined by their purpose.

Emergency Systems Alternators installed for the purpose of providing legally required emergency service are covered in NEC Art. 700. Emergency service is generally defined and required by local, state, or federal law. The main reason for emergency service is to safeguard human life. Continuous lighting and power are essential in places of public assembly, theaters, hotels, sports arenas, health care facilities, and the like. The NEC requires that the emergency system be applied within no more than ten seconds after the interruption of normal service. The only NEC require-

Fig. 18-7 A small portable alternator used on a construction site. (*Onan Corp.*)

ment concerning capacity of the systems is that it be adequate to serve the connected load of the emergency system.

Engine-driven alternators are commonly used to provide emergency power. The engines may use diesel, gasoline, or gas fuel.

Generator systems are rated from a few hundred watts for small portable systems to several hundred kilowatts (Fig. 18-7). In some large installations multiple units are employed with parallel operation providing the increased energy requirements of the building served.

In selecting the proper systems to serve a particular application, consideration must be given to the initial cost as well as the cost of installation (especially when extensive structural work is involved), cost of fuel and fuel storage, maintenance and reliability, and starting characteristics.

While the NE Code requires merely "adequate" capacity to serve the emergency connected load, a good rule of thumb calls for 25 percent more capacity than the maximum connected emergency load. This extra capacity not only serves as a safety measure for unexpected excess demand, it also allows for heavy demand during motor starting. It may also provide some of the extra capacity that would be needed for future expansion.

The conditions under which the alternator must operate must be given careful consideration (Fig. 18-8). Outdoor installations, in particular, must be sheltered properly. In cold climates starting aids may be necessary. At high altitudes or in hot climates the capacity may need to be derated. Where dampness can cause trouble, it may be necessary to install dehumidifiers or heaters. Noise and vibration are important factors that are especially critical for on-site alternators.

Standby Systems The National Electrical Code covers two types of standby systems—those that are required by local, state, or federal laws and those that are optional and not legally required. Article 701 of the NEC, Legally Required Standby Systems, covers those services not already covered by the provisions of Art. 700 on emergency systems. Whereas Art. 700 addresses itself directly to safeguarding human life, Art. 701 is concerned indirectly with hazards to human life and other related matters. For example, standby power may be legally required for equipment which aids in rescue or fire-fighting operations. In this category might be ventilating and smoke-removal systems. While not directly life-threatening, certain industrial processes and sewage disposal and treatment operations could create hazards to health if electric services to them were interrupted.

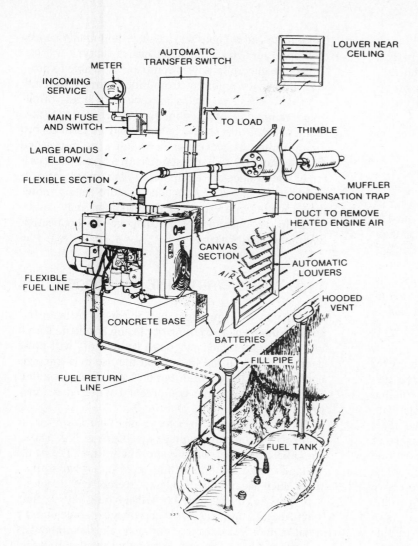

METER
INCOMING SERVICE
MAIN FUSE AND SWITCH
LARGE RADIUS ELBOW
FLEXIBLE SECTION
FLEXIBLE FUEL LINE
CONCRETE BASE
FUEL RETURN LINE
AUTOMATIC TRANSFER SWITCH
TO LOAD
LOUVER NEAR CEILING
THIMBLE
MUFFLER
CONDENSATION TRAP
DUCT TO REMOVE HEATED ENGINE AIR
CANVAS SECTION
AIR
AUTOMATIC LOUVERS
HOODED VENT
BATTERIES
FILL PIPE
FUEL TANK

Fig. 18-8 In locating standby alternators, fuel supplies and cooling must be carefully considered. (*Onan Corp.*)

Optional standby systems, NEC Art. 702, are intended to provide uninterrupted service for the protection and continuity of business operations and safeguarding property. Such systems may be found in industrial plants, office buildings and stores, farms, and homes. Heating and refrigeration equipment is often connected to standby service especially where spoilage of food is a problem or maintenance of temperature is critical as, for example, in ovens, melting furnaces, and heat treatment operations. A class of service called *uninterruptible power supplies* (UPS) is especially important in data processing and communications facilities, computer installations, and information networks.

18-7 ON-SITE PRIME MOVERS

Prime movers for on-site alternators may be gasoline, diesel, or gaseous fuel engines. Turbines powered by diesel fuel or gas are also used but not as extensively as engine-driven systems.

Diesel Engines Diesel-driven alternators are quite common not only for on-site emergency and standby power but as a supplemental source of energy during periods of high demand. They are often used by smaller utility companies for just such a purpose.

The large diesel engines used for this purpose are rugged and reliable. Although the initial cost of diesel engines is higher than that of gasoline or gaseous fuel engines, diesel maintenance costs are lower and their useful life is longer. Because diesel engines are extremely heavy, installation of large units requires special foundations, as well as sound- and vibration-reduction systems. Diesel fuel does not present an explosive hazard, and so fuel storage is relatively straightforward. If the fuel is stored in a cold area, it must be warmed to prevent sluggish engine starting. Like gasoline and gaseous fuel engines, an exhaust stack must be provided to expel the fumes of combustion. Most engines (except the very small gasoline engines) also require a circulating water system for cooling.

Gasoline Engines Gasoline engines are typically used for alternators under 100 kW. Although less rugged than diesel engines, gasoline engines have a relatively low initial cost and reliable fast-starting capability. Storage of large quantities of gasoline over long periods, however, does present a safety hazard. The highly volatile fuel is explosive and should be kept in tanks buried in earth. Off-site burial is safest.

Gaseous-Fuel Engines Engines are available that can run on a wide variety of gaseous fuels. Typical gasses used are natural gas, LPG, and propane. Propane in liquid or gaseous form can be stored in tanks outside the building and above ground. Natural gas as supplied by pipes by a gas utility is convenient but means a dependency on the gas supplier for a constant, reliable, and immediate source of fuel. Fuel stored in tanks under pressure has a lower heat (btu or calorie) content and therefore does not produce the rated horsepower of the engine. This in turn means a lower kilowatt output from the alternator.

Turbine Prime Movers The large alternators used by power companies are turned by steam-powered turbines. Smaller turbines using gaseous fuel are suitable for loads in excess of 500 kW. Turbines are relatively small and lightweight when compared with the alternator load they are capable of delivering. Because of this, and because the high-speed turbine is almost free of vibration, it can be mounted on-site on the roof or upper floors of a building without requiring excessive damping. The principle of operation is similar to the operation of a jet engine inasmuch as the products of combustion create the power to turn the turbine blades. These same combustion products must, however, be exhausted to the atmosphere, thus requiring large stacks or chimneys and perhaps gas treatment equipment. Just as jet noise is a problem, so too is the high sound level produced by the turbine. Soundproofing is essential if the turbines are to be used in any populated areas. Turbines, by nature, are slow-starting machines. This starting characteristic is unacceptable for emergency systems. Therefore, turbines that must come up to capacity within the NEC-required 10 s are provided with auxiliary independent starting means. A common system uses an air supply under pressure connected to a pneumatic motor. The compressed-air motor is directly connected to the turbine shaft and is capable of bringing the alternator up to speed in 10 s while the turbine is being fired.

18-8 POWER LINE PROBLEMS

Most power used by industrial, commercial, and residential buildings is supplied by local power companies. These companies are required by law to provide continuous and reliable service including maintenance of proper voltage, frequency, and capacity.

Though generally power companies are able to provide the service required, from time to time problems do occur. These problems may be very brief, often going unnoticed by customers. At other times the problems may last a few minutes and be noticed by the customer. However in these cases the trouble is cleared quickly with no serious effects on the customer or connected equipment. The most serious problems are those of a sustained nature lasting more than 15 min. In these cases the problems may cause hardship to customers and damage to connected equipment and systems.

18-9 MOMENTARY LINE PROBLEMS

A momentary line problem may be caused by an operation as innocent as opening or closing a switch. Turning electric tools and machines on and off will produce problems in the line. Even the use of a kitchen appliance can have a brief but noticeable effect on the line and any sensitive equipment connected on the line.

These momentary problems take the form of a "spike" of very high voltage as shown in Fig. 18-9. In some cases the voltages are many times the peak voltage of the system. While most industrial equipment such as motors, heaters, overcurrent devices, and the like are capable of withstanding these transient, that is, momentary, high voltages, some equipment cannot. Electronic equipment and instruments, as well as computers, are extremely sensitive to transients, and operation of these devices can be seriously disrupted even by a spike lasting less than a complete cycle of the ac sine wave. Consumer electronic equipment such as television receivers, high-fidelity stereo amplifiers, video cassette recorders, and home computers sometimes contain built-in surge protector devices.

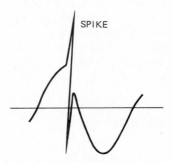

SPIKE

Fig. 18-9 A high-voltage "spike" can damage sensitive electronic equipment.

High-voltage spikes caused by lightning can be extremely damaging and disruptive. Surge suppressors (see Chap. 19) are essential for all incoming service to a building. However the best that can be expected from a surge suppressor is a lowering of the level of the spike. In many cases this still not afford full protection to all electronic equipment. It takes a little less than 17 ms (0.017 s) for a 60-Hz voltage to go through one cycle. A single spike caused by lightning may reach as high as 50,000 V but last only half that time. It is time enough to burn out critical equipment on the line.

Transients affect data-processing and computer equipment in another way. Even spikes of a few thousand, and in some cases a few hundred, volts can introduce so-called glitches or errors in data stored in computers. In addition, voltage spikes can disrupt computer programs by changing or eliminating instructions within the program itself. This will cause the program to run incorrectly or not at all.

18-10 SHORT-TERM PROBLEMS

Power line problems lasting only a few minutes can be caused by on-site load changes or power company switching. They may also result from natural or accidental damage to parts of the electrical system.

When an electrical device is connected across the line, there is a sudden surge of current through the device and in the line. This is particularly noticeable with motor loads but occurs with most loads, including lamps. When heavy loads are switched on across the line, the very high current in the line produces a momentary, but high, voltage drop in the line. This is usually noticed as a brief dimming of lights or momentary slowing down of motors. The initial surge soon passes and the rated current is drawn by the device. The effect on the electrical system of starting heavy loads is a short-term reduction in line voltage.

When heavy loads are disconnected from an electrical system, a short-term disruption in voltage also takes place. Most loads, particularly motors, fluorescent lighting, transformers, and the like, are inductive. Inductive loads have the property of wanting to maintain the condition of their circuit. If there is no current in an inductance and the inductance is connected across the line, current will not flow immediately. Similarly, if current is flowing through an inductance and the switch is opened, the inductance develops a high voltage and attempts to have the current "jump" the gap across the open switch. Evidence of this is the spark that appears across switch, circuit breaker, and relay contacts. Thus unlike the closing of an inductive circuit which can cause an undervoltage disruption in the system, opening an inductive circuit can cause an overvoltage disruption.

Overvoltage disruptions in the form of high-voltage spikes are safety hazards as well as equipment hazards. Sensitive electronic devices and instruments are often damaged because they are connected to electrical systems to which high inductive loads (such as motors) are also connected.

Short-term and momentary line problems are similar in many ways. Local switching or power company load operations may cause both. The difference in the two problems, however, is the duration of the disturbance. In most cases the momentary problem does not have far-reaching consequences on the other loads connected to the system, and in fact often goes unnoticed. The short-term disruption, on the other hand, does last long enough to cause inconvenience and sometimes damage.

18-11 EXTENDED POWER DISRUPTIONS

When service in an electrical system is disrupted for an extended period, say, in excess of 15 min, widespread inconvenience, damage to equipment, and life-threatening conditions may occur.

The most dangerous extended service disruption is the system failure, usually referred to as a *blackout*. Blackouts can occur in small blocks of a system or can be more widespread, even to a power company's entire system. Loss of power in small sections of a system usually occurs suddenly and totally. Fuses in a distribution power line may open because of a short circuit, a small distribution transformer may burn out, or a pole carrying the distribution lines may topple and break the conductors. The result will usually cut off power immediately, and the outage will last an extended period.

Disruptions of this type are common. Some electric utilities, systems, and geographical locations are more susceptible than others. Overhead systems are, of course, more prone to damage to their lines than underground systems. Poles, switches, fuses, and transformers are exposed to adverse weather, accidents, and vandalism.

When blackouts are caused by, or originate at, central distribution or transmission points, the effect is usually not an immediate stoppage of service. In some cases rerouting the distribution and readjusting loads will permit the power company to restore service throughout the affected area. In other cases the outage can be localized so that the smallest number of consumers are inconvenienced.

Major outages are characterized by an initial black-out, a brief restoration of service as the power company attempts to adjust loads and isolate the problem, and then, when that fails, a complete and extended blackout.

Total blackouts lasting more than 15 min present the greatest danger both to life and equipment. The lack of light, heat, and signal systems obviously creates conditions that affect safety. While separate emergency and standby systems are mandated by law in special occupancies such as hospitals and places of public assembly, most residences and commercial and industrial buildings depend on the power company for regular service. Very often the emergency service in these buildings is supplied by a set of alternate feeders connected to the same transformer as the regular feeders. This is intended as an emergency system in the event that the regular feeders or on-site switchgear providing regular service is damaged. It cannot serve as an emergency system if the power company itself loses service.

18-12   BROWNOUTS

A significant reduction in line voltage for extended periods is commonly referred to as a *brownout*. Brownouts can be intentional or accidental.

Intentional brownouts are produced when the power company reduces the voltage delivered to the system or some given part of the system. It is a method used by the company to force a reduction in power consumption. For example, during a heat wave in a large city, the demand for power to operate air-conditioning equipment may tax the capacity of the electrical system to deliver additional load. Since there is no easy way to prevent additional load from coming on-line (short of disconnecting entire sections from the system), the power company may choose to lower the voltage to all customers, or at least to significant areas of heavy usage.

The effect of lowered voltage is an immediate reduction in power demand. This can be seen from the power formula

$$P = \frac{V^2}{R}$$

where P = power, W
V = voltage, V
R = resistance, Ω

Since the resistance of a particular electrical device or appliance does not change in most cases, a reduction in voltage will lead to a reduction in power.

Problem 2

An electric heater is rated at 1500 W when operating at its rated voltage of 120 V. To conserve power, the power company reduces the system voltage by 10 percent. What will be the power reduction? Give the answer in watts and percent.

Solution

Assume that the resistance of the heater remains constant. The heater resistance can be calculated from its rated voltage and power.

$$R = \frac{V^2}{P}$$
$$= \frac{(120)^2}{1500} = \frac{14,400}{1500}$$
$$= 9.6 \ \Omega$$

A reduction in voltage of 10 percent means that the new voltage will be

$$V_r = 120 - 0.1 \times 120$$
$$= 120 - 12$$
$$= 108 \ V$$

The reduced voltage V_r is therefore 108 V. When the heater is connected across 108 V, the power consumed will be

$$P = \frac{V^2}{R}$$
$$= \frac{(108)^2}{9.6} = \frac{11,664}{9.6}$$
$$= 1215 \ W$$

This will be the new power consumed by the heater when operating under the reduced voltage. To find the percent reduction in power, divide the reduction in power by the rated power and multiply by 100.

$$\text{Difference} = 1500 - 1215 = 285 \ W$$
$$\% \text{ power reduction} = \frac{285}{1500} \times 100 = 19\%$$

Thus with a 10 percent reduction in voltage, the power company was able to lower the power demand by 19 percent.

Brownouts may often go unnoticed, especially in residential areas where the only indication is a slight dimming of incandescent lighting. In homes with air-conditioning units there may be a noticeable decrease in the effectiveness of the unit to cool. Toasters may take slightly longer to toast; electric ovens and coffee makers may take a little longer to bake or broil or make coffee, but in most cases these appliances will continue to function. In rare cases, the voltage may be lowered to a point where the compressor motors in air conditioning units and refrigerators and freezers will be unable to start.

Brownouts do have a significant impact on commercial and industrial customers. Electronic equipment used in data processing, computers, and industrial controls is extremely sensitive not only to voltage changes but also to the actual voltage level. In large installations where continuity and reliability are critical, electrical and electronic devices called *voltage regulators* are connected in the line to maintain a constant rated voltage. These regulators, however, operate over a relatively narrow range of under- and overvoltage conditions. Thus brownouts can seriously affect such installations.

Industrial loads consist mostly of induction motors. These motors run slower and hotter when operated under reduced voltage. Such operation over long periods of time can significantly lower the life of the motor by speeding the deterioration of the motor's various insulations. The hotter operation can also break down the lubricating oils and greases used in the motor and therefore damage bearings and gears. The net result is earlier and more frequent motor failures.

To prevent the damage that might result from low-voltage operation, many motor controls contain low-voltage coils which automatically open the motor contacts when the voltage falls below a certain value. Should the brownout voltage fall below this value, the motor circuit will be opened and the motor will stop. When the full voltage is restored or at least raised to a higher value, the motor will remain stopped. It will be necessary to restart the motor manually. In cases of two-wire motor control, under low-voltage conditions startup may be automatic when the normal voltage is restored. Therefore special precautions must be followed when investigating outages of equipment operated by two-wire controls.

Intentional brownouts are not uncommon and in many cases may involve a voltage decrease of only a small percentage. Accidental brownouts, on the other hand, may result in a greater drop in voltage, but such drops are usually restored much quicker. Accidental brownouts are generally caused by a failure of the power company's equipment or transmission and distribution lines.

18-13 LINE DROP PROBLEMS

Transmission and distribution power lines contain inductance and capacitance in addition to resistance. Alternating current also tends to concentrate near the surface of conductors as shown in Fig. 18-10. This characteristic is known as *skin effect*. While direct current tends to distribute itself uniformly throughout the cross-sectional area of the conductor, the central portion of an ac cable carries little or no current. The net result is that ac lines present considerable impedance to the flow of current which leads to a high line voltage drop.

To compensate for this drop and to ensure that adequate voltage is available throughout the system, the power company specifies a base voltage as well as the highest overvoltage and the lowest undervoltage it will allow during regular peak load operations. For example, 208 V may be set as the base voltage with 200 V the minimum voltage allowed and 216 V the maximum voltage.

Since the load on the power company's line is ordinarily not distributed evenly, a theoretical load center can be calculated. The load center is considered that point at which a concentrated load representing the effects of the other line loads is located.

The procedure for finding the load center is as follows. The distance from each load to the distribution transformer is multiplied by the load at that point. All the products are added, and the result is divided by the total load. The load center need not occur at the site of an actual load, but it does represent the effects of the entire load on the system.

DIRECT CURRENT DISTRIBUTED UNIFORMLY

(a)

ALTERNATING CURRENT CONCENTRATED NEAR THE SURFACE OF THE CONDUCTOR

(b)

Fig. 18-10 Skin effect. (*a*) Current distribution in the cross section of a conductor carrying direct current. (*b*) The same conductor carrying alternating current.

Problem 3

Find the load center of a distribution system if the distances and loads are as follows:

Load 1 = 7.5 kVA at 900 ft from the transformer

Load 2 = 33.0 kVA at 1250 ft from the transformer

Load 3 = 15.0 kVA at 2500 ft from the transformer

A one-line diagram of the system is shown in Fig. 18-11.

Solution

Each load is multiplied by its distance from the transformer and the products are added. The total load is also found.

$$\text{Load 1} \times \text{distance 1} = 7.5 \text{ kVA} \times 900 \text{ ft}$$
$$= 6,750 \text{ kVA-ft}$$
$$\text{Load 2} \times \text{distance 2} = 33.0 \text{ kVA} \times 1250 \text{ ft}$$
$$= 41,250 \text{ kVA-ft}$$
$$\text{Load 3} \times \text{distance 3} = 15.0 \text{ kVA} \times 2500 \text{ ft}$$
$$= 37,500 \text{ kVA-ft}$$
$$\text{Total load} = 55.5 \text{ kVA}$$
$$\text{Total kVA-ft} = 85,500 \text{ kVA-ft}$$

Divide the total kVA-ft by the total load to find the load center.

$$\text{Load center (LC)} = \frac{\text{kVA-ft}}{\text{kVA}} = \frac{85,500 \text{ kVA-ft}}{55.5 \text{ kVA}}$$
$$\text{LC} = 1540 \text{ ft}$$

The load center as found in this problem represents the point at which, theoretically at least, the entire load would be acting. To make sure the voltages in this system keep within the minimum and maximum limits, the base voltage must be established at the load center. Obviously any load located a greater distance from the load center will have a higher voltage drop and therefore a lower voltage at its incoming service panel. The loads closer to the transformer will have a voltage higher than the base voltage.

Ordinarily in urban areas, especially very large cities, the loads served by a transformer are concentrated in a relatively small area, and distances between transformer and even its farthest load are short. In rural areas, however, the distances are greater, and the loads spread out to a greater extent.

The problem above might represent a rural system being served by a single distribution transformer.

Problem 4

The previous problem represents the demand loads of three farms. The base voltage is 240/120 V, three-wire, single-phase. A 75-kVA distribution transformer serves this system. The three-wire distribution line has a resistance of 0.04 Ω per 1000 ft. (Neglect any other line impedances.) Figure 18-12 is a schematic diagram of this system. Find the voltages at the secondary terminals of the transformer and the actual voltages at each of the three farms when they are drawing their demand loads.

Solution

The base voltage of 240 V can be considered at the load center which is 1540 ft from the transformer. We will assume that the load is balanced and is at 100 percent power factor. In a balanced system only the two hot legs carry current. The distance from the load center to the transformer is 1540 ft. Figure 18-13 shows the distance to the load center and loads 1, 2, and 3. The current in each line is shown as I_1, I_2, I_3, and I_{LC}.

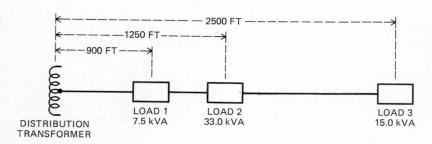

Fig. 18-11 One-line diagram of a distribution system with three loads.

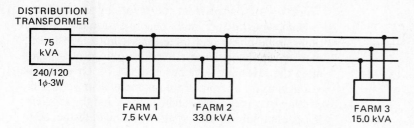

DISTRIBUTION TRANSFORMER

75 kVA

240/120 1ϕ-3W

FARM 1 7.5 kVA

FARM 2 33.0 kVA

FARM 3 15.0 kVA

Fig. 18-12 A rural distribution system.

$$I_1 = \frac{55.5 \text{ kVA}}{240 \text{ V}} \times 1000$$

$$= 231 \text{ A}$$

$$I_2 = \frac{(55.5 - 7.5) \text{ kVA}}{240 \text{ V}} \times 1000$$

$$= 200 \text{ A}$$

$$I_{LC} = \frac{15.0 \text{ kVA}}{240 \text{ V}} \times 1000$$

$$= 62.5 \text{ A}$$

$$I_3 = I_{LC} = 62.5 \text{ A}$$

Next, the resistance in each length of line can be determined.

$$R_{I_1} = \frac{0.04 \text{ }\Omega}{1000 \text{ ft}} \times 900 \text{ ft} = 0.036 \text{ }\Omega$$

$$R_{I_2} = \frac{0.04 \text{ }\Omega}{1000 \text{ ft}} \times 350 \text{ ft} = 0.014 \text{ }\Omega$$

$$R_{I_{LC}} = \frac{0.04 \text{ }\Omega}{1000 \text{ ft}} \times 290 \text{ ft} = 0.012 \text{ }\Omega$$

$$R_{I_3} = \frac{0.04 \text{ }\Omega}{1000 \text{ ft}} \times 960 \text{ ft} = 0.038 \text{ }\Omega$$

Naturally since there are two conductors in each section, the resistance of that section of line is twice the above value, as shown in Fig. 18-14.

The voltage drops can now be calculated for each run of the line.

$$V = I \times R$$
$$V_{R_1} = 231 \times 0.072 = 16.6 \text{ V}$$
$$V_{R_2} = 200 \times 0.028 = 5.6 \text{ V}$$
$$V_{R_{LC}} = 62.5 \times 0.024 = 1.5 \text{ V}$$
$$V_{R_3} = 62.5 \times 0.076 = 4.75 \text{ V}$$

The voltage at the load center is 240 V; therefore, at Load 3 the voltage is 4.75 V less, or

$$V_3 = 240 - 4.75 = 235.25 \text{ V}$$

At Load 2 the voltage is higher than at the load center by 1.5 V, or

$$V_2 = 240 + 1.5 = 241.5 \text{ V}$$

At Load 1 the voltage is higher than at Load 2 by 5.6 V.

$$V_1 = 241.5 + 5.6 = 247.1 \text{ V}$$

And, finally, at the transformer the voltage is 16.6 V higher.

$$V = 247.1 + 16.6 = 263.7 \text{ V}$$

This problem shows the effect of line drop on the actual voltage received by the customer of the power company. The customer closest to the transformer continually operates equipment and appliances at a 3 percent overvoltage, while the customer farthest from the transformer continually works with a 3 percent undervoltage. Since it would be uneconomical for the power company to provide equipment to bring the voltage up to the base value at each customer's location, the power company uses voltage regulators to maintain the base voltage at its proper level.

18-14 VOLTAGE REGULATORS

Voltage regulators are used in electrical systems to maintain a relatively set voltage. For example, the base voltage discussed in the previous section can be kept fairly constant if a voltage regulator is connected across the line. The location of the voltage regulator is determined by the locations of the loads and the critical voltages in the system.

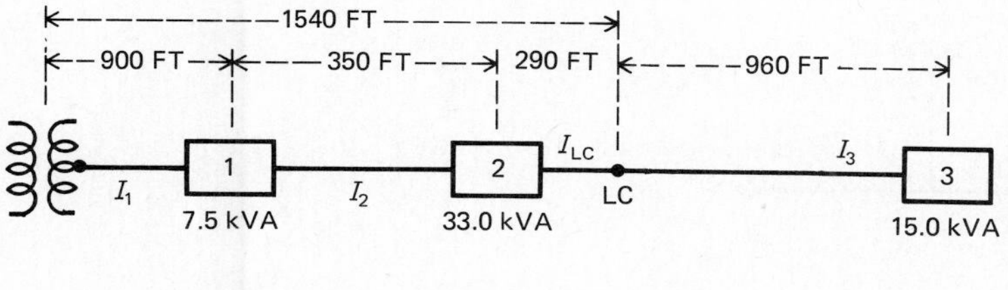

Fig. 18-13 The system of Fig. 18-12 with load currents and distances to the transformer.

1540 FT

900 FT — 350 FT — 290 FT — 960 FT

1   I_1 7.5 kVA I_2 2 I_{LC} 33.0 kVA   LC I_3 3   15.0 kVA

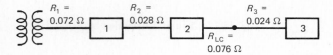

Fig. 18-14 Line resistances for the system of Fig. 18-12.

A voltage regulator is an autotransformer with two windings as shown in Fig. 18-15. This is actually a simplification of a commercial regulator. In practice, regulators contain numerous control devices, automatic switches, and meters.

The principle of operation of a regulator is based on the boost and buck effect of the small winding which is in series with the line. Assume the voltage ratio between the primary and secondary windings is 10:1 and the base voltage of the distribution transformer is 240 V. The secondary winding of the autotransformer will then have 24 V which could be added to the 240 V or subtracted from 240 V, depending upon the position of the reversing switch. The lines to the load would therefore have a voltage ranging from 240 + 24 = 264 V to 240 − 24 = 216 V.

If the distribution transformer voltage should drop to 216 V because of loading, the reversing switch could be set to the boost position and 21.6 or 22 V would be added to the distribution transformer voltage: 216 + 22 = 238 V. If the distribution transformer voltage rose to 270 V because of a sudden removal of load, the switch could be set to the buck position and 27 V would buck the 270 V, and the line would receive 270 V − 27 V = 243 V. So while the distribution transformer might swing between 216 and 270 V, the line voltage swings would be only between 238 and 243 V.

The voltage regulator described above had a single nonadjustable voltage ratio. Some commercial regulators are available with taps in the secondary winding. These taps permit changing the voltage ratio between the primary winding which is connected across the line and the secondary winding in series with the line serving the load. As with the single ratio regulator, the voltage produced in the secondary winding may be added or subtracted from the incoming line voltage.

Both types of regulators described are available in both single-phase and three-phase units. Most commercial regulators have automatic equipment for operating the reversing switches and taps for changing the voltage ratio. Automatic equipment also senses the cycling of voltage rises and drops so that the regulator is not continuously adjusting the voltage. Rather, only sustained drops or rises are compensated. Solid-state electronic devices monitor the amount and duration of voltage changes and react according to programmed settings of the control equipment.

Voltage regulators are rated according to the percentage change they can effect on the circuit. For example, if the voltage can be raised or lowered by 10 percent (as in the case given above), the rating of the voltage regulator would be 10 percent of the system load rating. If the system were rated at 240 V × 200 A = 48 kVA, then the voltage regulator would be rated at 48 kVA × 0.1 = 4.8 kVA. In general the rating of the regulator would be the full-load current multiplied by the voltage that can be added or subtracted by the secondary winding when the primary winding has its rated voltage.

18-15 PARALLELING ALTERNATORS

Electric generating stations rarely contain a single alternator. The load demands usually exceed the capacity of the alternators that the utility can economically provide. In addition it is unwise to have all the necessary capacity requirements tied up in one machine. Thus more often than not more than one alternator must serve a system, and this calls for parallel operations of alternators.

To operate two alternators in parallel, the following conditions must be present:

1. The terminal voltages must be equal
2. The frequencies must be equal
3. The voltages must be in phase and in the same phase sequence

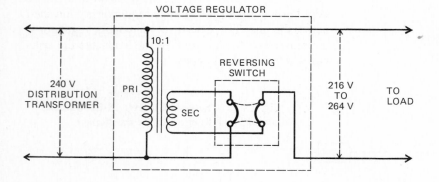

VOLTAGE REGULATOR

240 V DISTRIBUTION TRANSFORMER

10:1

PRI

SEC

REVERSING SWITCH

216 V TO 264 V

TO LOAD

Fig. 18-15 A simplified voltage regulator used to stabilize line voltage.

When these three conditions have been met, the alternators are said to be *synchronized*.

The first item is obvious and easily tested. The two alternators must be producing the same voltage at their output terminals, and a voltmeter can make that determination.

Similarly, the frequency can be measured with a frequency meter or frequency counter.

Although a simple phase sequence indicator can be constructed with lamps as described in Chap. 14, Sec. 14-15, a more accurate measurement is usually made by use of a phase sequence meter. A *synchroscope* is also used to compare the frequencies and phase relationships of the two alternators that are to be connected in parallel.

Basically the synchroscope is a small induction motor driven by the voltages of the two alternators that are being paralleled. The shaft of the motor has a pointer which passes over the face of a dial as in Fig. 18-16. The difference in frequencies of the two alternators causes the shaft of the synchroscope to rotate rapidly in either the clockwise or counterclockwise direction. If the incoming alternator, that is, the alternator being added to the system, is too slow, the pointer will rotate in the counterclockwise direction. As the difference in speed of the two alternators is reduced and the frequencies and phase relationships get closer to synchronization, the pointer will rotate more slowly in the direction indicating whether the incoming alternator is rotating too slowly or too rapidly. Adjustments are carefully made to the speed of the incoming alternator until the pointer remains steady at the black (index) mark. At this point the two alternators are synchronized in phase and frequency.

Parallel operation also requires a division of the load so that the two alternators are serving the system as efficiently as possible. Alternating current loads consist of a real portion (the kilowatt power) and a reactive portion (the kilovar power). Both portions of the load should be divided proportionally between the two alternators.

The real power delivered to the system depends on the turbines or engines driving the alternators. Adjustments in these prime movers are made by speed governors which automatically regulate the amount of fuel or steam going to the prime mover.

Fig. 18-16 A synchroscope.

Reactive power and, coincidentally, the power factor at which the two alternators operate are controlled automatically by changes in the field excitation of the alternators. As discussed with synchronous motors, field excitation variations are capable of changing the value of the power factor. They can also change the power factor from lagging to leading and vice versa.

18-16 ALTERNATOR CONTROL SYSTEMS

While technical details of most alternator controls are beyond the scope of this text, three control systems are fundamental to alternators and other electric machines and therefore will be discussed here. These controls are voltage regulators, field excitation controls, and speed governors. A one-line diagram showing these elements is in Fig. 18-17.

Voltage Regulators Like the voltage regulators used to maintain a relatively even line voltage, alternator voltage regulators are used to keep the output voltage of the alternator within a given range. They do this automatically when load changes produce a sustained change in the output voltage. The regulator acts quickly when it senses a sudden drop in output voltage by increasing the alternator's field current. But this current increase is beyond what would normally be required to bring the voltage back up to its rated value. Its purpose is to bring the voltage up to normal very quickly. In fact, if left alone, the new field current would increase the output voltage *above* the rated value. The alternator output voltage rises rapidly, and as it nears the rated value the regulator again senses

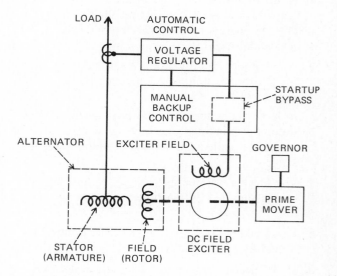

Fig. 18-17 One-line diagram of an alternator and its basic control systems.

that the voltage will overshoot this value and automatically decreases the excitation current. Once the output voltage is within range, automatic fine adjustments are made and the output voltage and field current settle into their new values to serve the increased load.

During normal start-up of an alternator, the voltage regulator is bypassed and the voltage buildup results from the residual magnetism in the alternator field. Once the output voltage builds up to the level where the regulator would normally take over, further voltage leveling is effected by the regulator.

Field Excitation Controls The alternator requires a source of direct current to produce a constant magnetic field. The field coils are normally in the rotor, and a number of methods are used to deliver direct current to these coils. Basically the source of direct current is a generator. The two types in use are designated as either *brushless exciters* or *brush-type exciters*. Both types are usually connected to the alternator field through a voltage regulator which contains a sensing circuit connected to the alternator's output. This circuit can detect changes in the output characteristics which would then be translated into signals to the voltage regulator. The regulator in turn will change the field excitation to produce the effect needed to adjust the alternator output.

Governors These are devices that sense and control the speed of the prime movers. As noted earlier, an increase in load on the alternator will cause the alternator to decrease in speed, thus affecting the output voltage and frequency. Real electric power output in kilowatts would also be reduced. Since the loaded alternator presents an increased mechanical load to the prime mover, a method must be used to increase the prime mover's power and restore the alternator's speed. The governor does this automatically by sensing a decrease in speed and reacting by adjusting the fuel or steam to the prime mover. Similarly a sudden decrease in the alternator's load would reduce its mechanical load on the prime mover and lead to an increase in speed. The governor would then act to reduce fuel and steam in order to decrease the prime mover's speed.

Governors are either mechanical or electrical. The mechanical governor operates by means of a rotating flywheel. Changes in centrifugal force are produced by changes in speed. Hydraulic actuators are linked to the fuel controls and either increase or decrease fuel as required. Electrical governors are either small generators or frequency sensors connected to the alternator's output. The signal from the governor activates a solenoid which is again linked to the fuel controls.

While these governors sense speed, other governors are available that react to load variations. Since load changes take place before the alternator and prime mover can react by changing speed, load governors are able to set speed-changing devices sooner than speed governors.

SUMMARY

1. Alternating-current generators, commonly known as alternators, are the major source of electric energy for commercial, industrial, and residential uses.

2. Alternators generally produce electricity by rotating a magnetic field past a stationary set of windings.

3. In large alternators the field is in the rotor. The stationary windings in which the voltage is induced are called the armature which is in the stator.

4. The mechanical energy required to turn alternators is supplied by machines called prime movers.

5. Large alternators used by electric utilities are turned by steam turbines.

6. Hydroelectric generating plants use water power to turn the turbines.

7. Steam is produced in generating plants by burning coal, gas, or oil. Nuclear fuel is used to produce steam in closed and shielded systems.

8. Smaller alternators are driven by internal combustion engines which use diesel fuel, gasoline, or gas.

9. Alternate sources of energy for turning alternators include wind and tides.

10. Large alternators generate voltages as high as 23 kV. Transformers step up the generated voltages to hundreds of thousands of volts for transmission.

11. High voltages are used for transmission in order to reduce line loss and permit using smaller-size conductors.

12. The standard frequency used in the United States is 60 Hz.

13. Alternators require a source of direct current to provide the rotor's field excitation.

14. Field excitation is usually provided by a small dc

generator mounted on the alternator's shaft. The output is fed directly to the alternator's field winding through collector rings or brushes.

15. The difference between the alternator's generated voltage and its terminal voltage is an indication of its voltage regulation *VR*.

$$VR\ (\%) = \frac{V_G - V_t}{V_G} \times 100$$

16. On-site generating systems are used to provide emergency or standby power as required by the NEC.

17. On-site alternators generally use gasoline, diesel, or gaseous fuel engines as prime movers.

18. Small on-site alternators sometimes use gaseous fuel turbines as prime movers.

19. Power line problems are usually characterized by their duration, such as momentary, short-term, and extended power disruptions.

20. Momentary disruptions last for only a few seconds or perhaps only part of a cycle. Their presence usually goes unnoticed except that certain sensitive electronic equipment may be affected or damaged unless properly protected.

21. Short-term problems are those which can be corrected in less than 15 min. They can cause disruptions and inconvenience but because of their short duration rarely produce major problems for users.

22. Extended power disruptions are a threat to life and property. Such outages are usually referred to as blackouts.

23. A brownout is a reduction in system voltage, often intentional but sometimes accidental. The purpose of an intentional brownout, or voltage reduction, is to effect an immediate reduction in power demand.

24. Line drop problems are common and most noticeable in systems in which customers are spread out over relatively long distances.

25. Line drop problems are characterized by voltages higher than a base voltage for customers close to the distribution transformer and voltages under a base voltage for customers farthest from the transformer. Only customers located near the system's load center receive voltages close to the base voltage.

26. Voltage regulators are autotransformers used to automatically maintain line voltages close to the established base voltage.

27. Power companies generally use more than one alternator to serve their system. Two or more alternators may be connected in parallel to serve the same system.

28. In order to parallel alternators, it is necessary that their (a) terminal voltages be equal, (b) frequencies be equal, and (c) voltages be in phase and in the same phase sequence.

29. A synchroscope is used to compare the frequencies and phase relationships of two alternators being connected in parallel.

30. Alternator control systems include voltage regulators, field excitation controls, and speed governors.

QUESTIONS

18-1 What are two other names for an alternating-current generator?

18-2 Why do large alternators have their armature winding in the stator?

18-3 What is a prime mover?

18-4 List three types of fuels used to operate internal combustion prime movers.

18-5 What is the prime mover in a nuclear power plant?

18-6 In general, is the output of a power company's large alternator fed to a step-up or step-down transformer before transmission? Why?

18-7 What is the standard frequency generated in the United States? List two other common frequencies in power systems and state where they are used.

18-8 At what speed must a six-pole single-phase alternator be turned to produce a frequency of 60 Hz?

18-9 Describe three methods used to supply direct current to the field of an alternator.

18-10 How does a brushless generator operate?

18-11 What is voltage regulation as applied to an alternator?

18-12 The generated voltage of an alternator is 11 kV. When the alternator is delivering its full-load current of 1000 A, the terminal voltage is found to be 10.7 kV. Find the voltage regulation of this alternator.

18-13 What is the main reason for mandated emergency systems?

18-14 What type of prime mover is generally used with emergency systems?

18-15 What is the basic difference between emergency systems and standby systems?

18-16 What type of alternate electrical system would be suitable for a large data processing installation, emergency, standby, or optional standby? Why?

18-17 List two advantages and two disadvantages of an on-site diesel engine–driven alternator.

18-18 List two advantages and two disadvantages of an on-site gasoline engine–driven alternator.

18-19 Describe a method used to bring a small gas turbine–driven alternator up to speed quickly.

18-20 List three general classifications of power line problems based on duration.

18-21 What is a voltage spike and how is it caused?

18-22 What causes the large arc when opening a switch in a motor circuit?

18-23 Why are electrical systems with overhead distribution lines more prone to extended outages than underground distribution systems?

18-24 Why do power companies sometimes reduce the system's voltage intentionally?

18-25 The power company has reduced its voltage from its normal 120 V to 110 V. What will be the percentage power reduction of a heater rated 1000 W at 120 V? State the assumption you made in solving this problem.

18-26 Why can brownouts increase maintenance bills in an industrial plant?

18-27 What is the name of the characteristic property of alternating current traveling along the surface of a conductor rather than uniformly through the entire cross-section of the conductor?

18-28 Calculate the load center for the following distribution system:

No.	Load, A	Distance from Transformer, ft
1	200	150
2	100	300
3	100	500
4	150	1200

18-29 Calculate the voltage drops in each of the feeders shown in Fig. 18-18 if the loads are as given in question 18-28 and the conductor resistance is 0.02 Ω per 1000 ft.

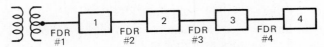

EACH FEEDER CONSISTS OF TWO CONDUCTORS

Fig. 18-18 Circuit for question 18-29.

18-30 What is the rating of a voltage regulator that can raise or lower the line voltage 5 percent if the full-load current is 200 A and the rated voltage is 480 V?

18-31 A 7 percent voltage regulator is operating in a system with a base voltage of 208 V. When the voltage in the system drops to 200 V, to what value can the load voltage be raised?

18-32 List three conditions that must be present before two alternators can be synchronized.

18-33 Why must a synchroscope be used with another type of instrument before synchronized conditions can be assured?

18-34 What part of an alternator's system must be regulated to aid in distributing the real kilowatt load between two machines?

18-35 What are three major control systems used to adjust voltage and frequency in an alternator system?

19
RESIDENTIAL WIRING

Residential wiring is a subject of interest to everyone. Most of us live in houses or apartments and most of us work in buildings of some kind. We use and depend on the electrical system constantly to a degree seldom realized and appreciated until "the power goes off."

A well-planned and wired electrical system for a house will serve efficiently and safely, and it will allow for reasonable additional future loads. A system that is poorly planned and wired does not allow for future loads; it is inefficient and expensive to use, and it can result in destruction of property or injury and death.

Fortunately the National Electrical Code provides standards for planning and installation of reasonably safe electrical systems. Every electrician should have a Code book and be able to apply its provisions in planning and installing a job. A copy of the Code should be used in connection with this chapter and the next on industrial wiring.

19-1 LOCAL CODES AND UTILITY REGULATIONS

Before any electrical work is started, electricians should acquaint themselves with the requirements of any local codes that apply in the area in which the work is to be done. They should also be fully acquainted with regulations of the utility company that will supply electric energy to the system.

Most areas covered by a local code require a permit and inspection fee for a job, and in some cases an examination and license for the electrician are required. However, in nearly all instances an owner is allowed to do work on his or her own property with only a permit and inspection fee, and possibly a written examination.

In this chapter we will use the word "shall" the same way it is used in the Code. It will mean that the condition referred to is required by the National Electrical Code. In most cases when a Code requirement or regulation is discussed in this chapter, the number of the section, paragraph, or table in the Code referring to the subject is given *in parentheses following the discussion*.

19-2 WIRING TOOLS

The tools used by an electrician in residential wiring are few and simple. Figure 19-1 illustrates a practical set of tools usually meeting all needs for a job. The name of each and its principal use are as follows: (1) claw hammer, for nailing straps and boxes; (2) brace, to hold extension bit, for boring holes for cables; (3) hacksaw, for cutting armored cable; (4) folding rule, 6-ft; (5) wood chisels, for notching framing; (6) locking pliers, for holding split-bolt connectors while tightening, also used as a substitute for a pipe wrench; (7) lineman's pliers, for general wire work; (8) wire cutter and stripper; (9) electrician's knife; (10) screwdriver; (11) conduit bender, for bending conduit and electrical metallic tubing. Other tools used by the electrician include a keyhole saw, for sawing small openings for boxes; an electric drill and bits; fish tape, for pulling wire and cable through conduit and other raceways; and a crimping tool, for attaching lugs to wire.

19-3 DEFINITIONS OF TRADE TERMS

The following terms are commonly used in the Code and trade. A complete list of definitions can be found in NEC Article 100.

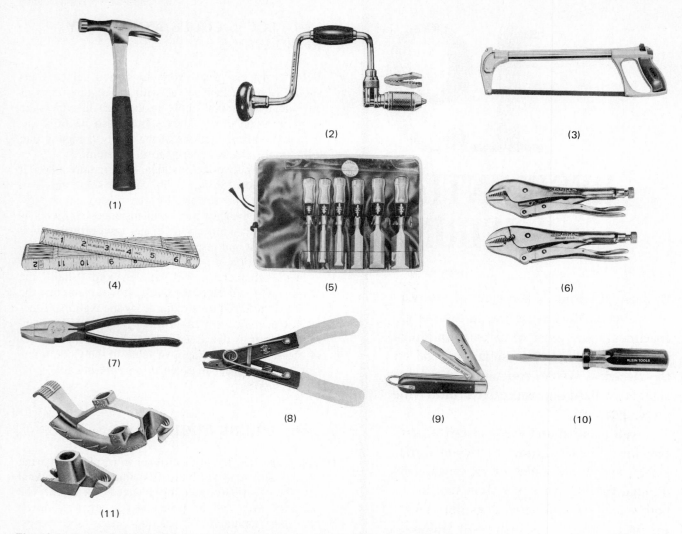

Fig. 19-1 Common tools used in house wiring. (*Klein Tools, Inc.; Ideal Industries, Inc.*)

Accessible (as applied to wiring methods): Capable of being removed or exposed without damaging the building structure or finish, or not permanently closed in by the structure or finish of the building (*See* Concealed.)

Accessible (as applied to equipment): Admitting close approach because not guarded by locked doors, elevation, or other effective means. (*See* Readily Accessible.)

Ampacity: Current-carrying capacity expressed in amperes.

Appliance: Utilization equipment, generally other than industrial, normally built in standardized sizes or types, which is installed or connected as a unit to perform one or more functions such as clothes washing, air conditioning, food mixing, deep frying, etc.

Approved: Acceptable to the authority enforcing this Code.

Branch circuit: That portion of the wiring system between the final overcurrent device protecting the circuit and the outlet(s).

A device not approved for branch-circuit protection such as a thermal cutout or motor overload protective device is not considered as the overcurrent device protecting the circuit.

Branch circuit, appliance: A circuit supplying energy to one or more outlets to which appliances are to be connected, such circuits to have no permanently connected lighting fixtures not a part of an appliance.

Branch circuit, general purpose: A branch circuit that supplies a number of outlets for lighting and appliances.

Branch circuit, individual: A branch circuit that supplies only one utilization equipment.

Concealed: Rendered inaccessible by the structure or finish of the building. Wires in concealed raceways are considered concealed, even though they may become accessible by being withdrawn. [*See* Accessible (as applied to wiring methods).]

Continuous load: A load where the maximum current is expected to continue for 3 h or more.

Controller: A device, or group of devices, which serves to govern, in some predetermined manner, the electric power delivered to the apparatus to which it is connected. [See NEC Section 430-81(*a*)].

Disconnecting means: A device, or group of devices, or other means whereby the conductors of a circuit can be disconnected from their source of supply.

Equipment: A general term including material, fittings, devices, appliances, fixtures, apparatus, and the like used as a part of, or in connection with, an electrical installation.

Feeder: The circuit conductors between the service equipment, or the generator switchboard of an isolated plant, and the branch-circuit overcurrent device.

Ground: A conducting connection, whether intentional or accidental, between an electrical circuit or equipment and earth, or to some conducting body which serves in place of the earth.

Grounded: Connected to earth or to some conducting body which serves in place of the earth.

Grounding conductor: A conductor used to connect equipment or the grounded circuit of a wiring system to a grounding electrode or electrodes.

Grounding conductor, equipment: The conductor used to connect non-current-carrying metal parts of equipment, raceways, and other enclosures to the system grounded conductor at the service and/or the grounding electrode conductor.

Grounding electrode conductor: The conductor used to connect the grounding electrode to the equipment grounding conductor and/or to the grounded conductor of the circuit at the service.

Outlet: A point on the wiring system at which current is taken to supply utilization equipment.

Raceway: Any channel for holding wires, cables, or bus bars which is designed expressly for, and used solely for, this purpose.

Readily accessible: Capable of being reached quickly, for operation, renewal, or inspections, without requiring those to whom ready access is requisite to climb over or remove obstacles or to resort to portable ladders, chairs, etc. [*See* Accessible (as applied to equipment).]

Receptacle: A contact device installed at the outlet for the connection of a single attachment plug. A single receptacle is a single contact device with no other contact device on the same yoke. A multiple receptacle is a single device containing two or more receptacles.

Service: The conductors and equipment for delivering energy from the electricity supply system to the wiring system of the premises served.

Service conductors: The supply conductors which extend from the street main, or from transformers, to the service equipment of the premises supplied.

Service drop: The overhead service conductors from the last pole or other aerial support to and including the splices, if any, connecting to the service-entrance conductors at the building or other structure.

Service-entrance conductors, overhead system: The service conductors between the terminals of the service equipment and a point usually outside the building, clear of building walls, where joined by tap or splice to the service drop.

19-4 IDENTIFICATION OF WIRING MATERIAL

In planning and discussing wiring equipment, as well as in making a list of material needed for a wiring job, it is necessary to know the names and uses of the various items available for wiring systems.

The main parts composing a service entrance, exclusive of the conductors and raceway, are shown in Fig. 19-2. Names of the items and uses are as follows: (1) Service head, used on rigid conduit or electrical metallic tubing to meter socket to avoid entrance of water in raceway and insulate service-entrance cable. (2) Meter socket for holding meter. (3) Sill plate, used to enclose and seal hole for entrance of service cable into building. (4) Service cabinet, provides disconnect for system, and main line and branch-circuit overcurrent protection. (5) Clamp for strapping service-entrance conduit. (6) Ground clamp for connection of grounding wire to water pipe. (7) Split-bolt compression connector for connecting service-entrance conductors at service drop.

Other service equipment include branch-circuit cabinets and safety switches. A service cabinet may

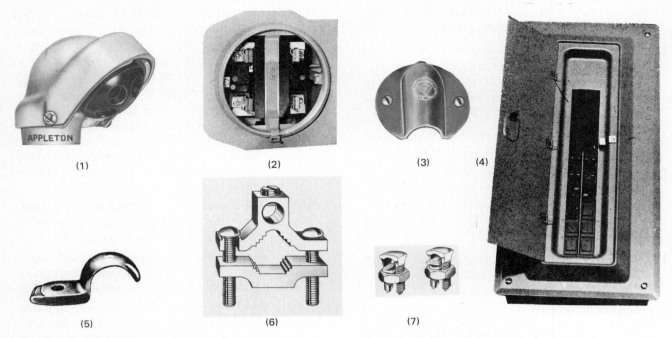

(1) (2) (3) (4)

(5) (6) (7)

Fig. 19-2   Service-entrance equipment.   *(Appleton Electric Co.; General Electric Co.; Ideal Industries, Inc.)*

have a main circuit breaker, four two-pole breakers for ranges, clothes dryers, air conditioners, or other 240-V loads, and single-pole branch circuit breakers for small-appliance circuits and lighting and receptacle circuits.

A branch-circuit cabinet is used to supply floors above or below the floor with the service cabinet. The use of a branch-circuit box, fused or with circuit breakers, gives greater convenience in case of trouble and results in less voltage drop to areas at great distances from the service cabinet.

A safety switch is used anywhere that a fused disconnect is necessary or desirable. Larger safety switches contain cartridge fuses.

Devices commonly used for overcurrent protection are shown in Fig. 19-3. They are (1) single-pole circuit breaker; (2) type S plug fuse, 15 A and less not interchangeable with 16 A and above to 30 A (the only type of plug fuse allowed in new installations); (3) time-delay plug fuse with standard (Edison) base; (4) knife cartridge fuse; (5) ferrule cartridge fuse.

Wiring-equipment items commonly used are shown in Fig. 19-4. They are (1) octagonal outlet box; (2) square outlet box; (3) switch box; (4) lampholder, for mounting directly to ceiling box; (5) cable connector, for connecting cable to box through knockout in box.

A duplex receptacle is shown in (6). Some duplex receptacles have separate connections for each receptacle, one for direct connection to the supply lines, and one for connection through a switch: (7) duplex receptacle cover; (8) switch cover.

There are three types of switches commonly used in wall boxes: (9) single-pole toggle switch, two terminals; (10) three-way toggle switch, three terminals, one identified by dark color.

Miscellaneous wiring devices are shown in Fig. 19-5. They are (1) dryer pigtail assembly; (2) receptacle for connecting and grounding electric ranges (clothes-dryer assemblies are similar but are not interchangeable); (3) weatherproof socket cover; (4) switch for mounting in strap.

19-5   PLANNING A RESIDENTIAL WIRING SYSTEM

In planning a residential wiring system, a simple floor plan of each floor of a house, including the basement,

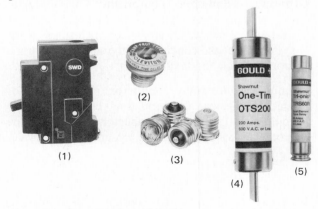

(1)   (2)   (3)   (4)   (5)

Fig. 19-3   Overcurrent devices.   *(Square D Co.; Leviton; Gould)*

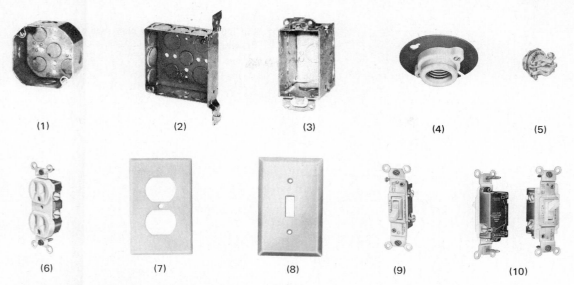

Fig. 19-4 Wiring devices. (*Appleton Electric Co.; Leviton*)

if any, should be drawn. The drawing should be made to a scale of about 1/2 in. = 1 ft. This will allow sufficient space to draw symbols of electrical equipment in their proper locations and, after the plan is complete, to draw the wiring circuits on the same drawing.

19-6 FLOOR PLAN

A floor plan should be as simple as possible. It should contain only the information necessary for the planning of an electrical system. Such necessary information includes location of walls, doors, and windows in the house, the direction the doors open, the location of fixed objects such as wall projections, offsets, stairways, fireplaces, plumbing, heating, and air-conditioning equipment, hot- and cold-air registers, and any other feature of the house that could influence the planning of the electrical system. The size and shape of these objects should be drawn to the scale used in the plan. A sample of a simple floor plan with symbols is shown in Fig. 19-6.

19-7 ELECTRICAL SYMBOLS

After the floor plan is completed, electrical symbols should be drawn in at the location where the equipment represented by the symbols is to be placed. Standard electrical symbols used in residential wiring are shown in Fig. 19-7. In case a symbol cannot be found to cover a special condition, a circle containing a reference mark can be used, and the special condition noted in the key to the symbols on the drawing.

19-8 PLANNING FOR CONVENIENCE AND ADEQUACY

Extreme care should be used in this stage of planning to ensure an adequate, efficient, and convenient electrical system. The initial system will probably be expected to serve the basic needs of the residence for many years. Unless provisions are made for future expansion or revision, any alterations or additions made to the system after it is initially installed will cost several times more than they would cost if included in the original plan.

In addition to the location of lighting and receptacles to serve the specified needs, consideration should be given to special needs and equipment. Multipoint control of lighting and receptacles, the types and location of special outlets and receptacles for entertainment, communication, and security equipment, and the possible use of alternate power sources should receive attention during the planning stage.

Often local electrical codes specify the minimum number, type, and location of *receptacles* (sometimes

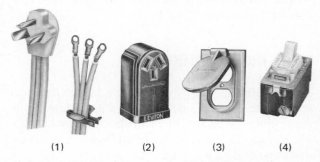

Fig. 19-5 Miscellaneous devices used in residential wiring. (*Leviton*)

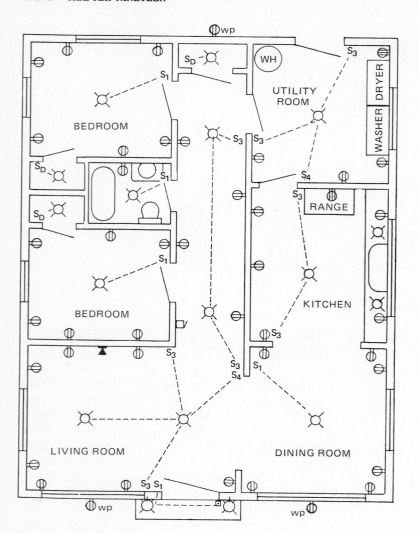

Fig. 19-6 Simple floor plan used as an electrical drawing.

called *convenience outlets* or simply *outlets*). The location of these receptacles is designed to eliminate (or greatly reduce) the need for extension cords to serve table and floor lamps, entertainment equipment (such as radios, television sets and other video equipment, stereo systems, etc.), and small appliances.

Special thought must be given to present and future needs of service to heavy loads and locations that present potential shock hazards (such as bathrooms, laundries, and swimming pools).

19-9 BRANCH CIRCUITS

Although the number of outlets on a branch circuit is not limited by the Code, not more than eight to ten (depending on the load of each) should be included in one circuit. The limiting factor on such circuits is the *overcurrent device rating*.

Special-duty branch circuits, such as those for fixed appliances, may be protected at or less than the maxi-

mum ampacity of the wire sizes used for the circuit.

All 120-V circuit loads in a three-wire service panel or branch-circuit panel should be divided as evenly as possible between the neutral and the two outside or "hot" conductors in the panel.

Use of branch-circuit panels should be considered in residences requiring long runs of wire from the service panel. A branch-circuit panel is supplied by a feeder from the main service-entrance panel. Branch-circuit panels afford short branch-circuit runs, reduced voltage drop, and convenience in resetting circuit breakers or replacing fuses.

19-10 RECEPTACLE REQUIREMENTS

Since many hazards are created by poor planning, installation, and use of receptacles, the Code covers these matters in detail in an effort to promote convenience and safety. While the Code is not meant to be a design manual, special attention should be given to

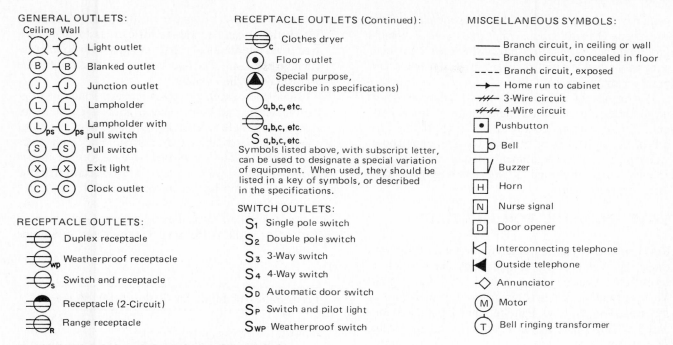

GENERAL OUTLETS:

Ceiling Wall

Light outlet	
B B	Blanked outlet
J J	Junction outlet
L L	Lampholder
L_{ps} L_{ps}	Lampholder with pull switch
S S	Pull switch
X X	Exit light
C C	Clock outlet

RECEPTACLE OUTLETS:

Duplex receptacle

Weatherproof receptacle

Switch and receptacle

Receptacle (2-Circuit)

Range receptacle

RECEPTACLE OUTLETS (Continued):

Clothes dryer

Floor outlet

Special purpose, (describe in specifications)

$\bigcirc$a,b,c, etc.

a,b,c, etc.

S a,b,c, etc

Symbols listed above, with subscript letter, can be used to designate a special variation of equipment. When used, they should be listed in a key of symbols, or described in the specifications.

SWITCH OUTLETS:

S₁ Single pole switch

S₂ Double pole switch

S₃ 3-Way switch

S₄ 4-Way switch

S_D Automatic door switch

S_P Switch and pilot light

S_{WP} Weatherproof switch

MISCELLANEOUS SYMBOLS:

Branch circuit, in ceiling or wall

Branch circuit, concealed in floor

Branch circuit, exposed

Home run to cabinet

3-Wire circuit

4-Wire circuit

Pushbutton

Bell

Buzzer

Horn

Nurse signal

Door opener

Interconnecting telephone

Outside telephone

Annunciator

Motor

Bell ringing transformer

Fig. 19-7 Electrical symbols used on architectural plans.

the requirements of the Code in these matters. Articles 210 and 220 of the Code cover these requirements.

According to the Code, receptacle outlets shall be installed in every kitchen, family room, dining room, breakfast room, living room, parlor, library, den, sunroom, recreation room, and bedroom so that no point along the floor line of usable wall space is more than 6 ft from a receptacle. Included are any usable wall space 2 ft wide or more, wall space occupied by sliding panels in exterior walls, and fixed room dividers such as free-standing bar-counters. This means that a floor lamp equipped with a standard 6-ft cord can reach a receptacle from any point along usable wall spaces in these rooms.

In kitchen and dining room areas a receptacle outlet shall be installed at each counter space wider than 12 in. Counter-top spaces separated by range tops, refrigerators, or sinks shall be considered as separate counter-top spaces. Receptacles rendered inaccessible by the installation of stationary appliances will not be considered as these required outlets.

Receptacle outlets shall, insofar as practicable, be spaced equal distances apart. Receptacle outlets in floors shall not be counted as part of the required number of receptacle outlets unless located close to the wall. At least one wall receptacle outlet shall be installed in the bathroom adjacent to the basin location. Bathroom receptacles must have ground-fault circuit interrupters (GFCI).

Outlets in other sections of the dwelling for special appliances such as laundry equipment shall be placed within 6 ft of the intended location of the appliance.

At least two small-appliance receptacle circuits are required by the Code. These circuits shall be wired with No. 12 or larger wire and overcurrent protection at 20 A. They can contain nothing but receptacles and are the only receptacles to be installed in the kitchen, pantry, family room, dining room, and breakfast room. At least one 20-A branch circuit shall be provided for the laundry receptacle. Kitchen appliance receptacles shall be served by at least two appliance circuits, but these circuits can continue to other rooms listed above. All receptacles installed on 15- or 20-A branch circuits shall be of the grounding type. All outdoor residential 120-V, 15- or 20-A receptacles shall have ground-fault protection.

19-11 GROUNDING RECEPTACLES

A grounding receptacle can be grounded by mounting it in a box that is grounded by conduit or armored cable and connecting its grounding terminal with a grounding wire, bare or green-insulated. It shall not be grounded by use of the neutral or grounded circuit conductor of the wiring system.

Where two or more grounding conductors enter a box, they shall be so connected that electrical continuity will not be interrupted if the receptacle or device is removed. Metallic boxes shall be grounded. Most wall boxes contain a hole for a 10-32 screw or an approved clip for the grounding wire to be used.

If a grounding circuit enters and continues from a box, the two grounding wires can be connected together with a small split-bolt or other means, and the end of one grounding wire connected to the receptacle grounding screw, and the end of the other grounding wire connected to the box by a clip, or under a screw used for that purpose only.

For extensions only in existing installations which do not have a grounding conductor in the branch circuit, the grounding conductor of a grounding-type receptacle outlet may be grounded to a grounded cold-water pipe near the equipment. If it is impractical to reach a source of ground, a nongrounding-type receptacle shall be used.

19-12 PLANNING FOR LIGHTING

Residential lighting relies heavily on table, floor, and other types of portable lamps. Thus the location of receptacles is an important part of any lighting plan. Modern lighting schemes also make extensive use of recessed fixtures, track lighting, chandeliers, and louvered ceiling lights using mixtures of incandescent, fluorescent, and high-intensity discharge lamps. Lighting control includes the common on-off wall switch, three- and four-way switches, low-voltage multipoint switching, dimmers, automatic computer-controlled switching for security and safety, and many others.

The wide range of light choices and control methods means that lighting systems must be carefully planned beyond merely locating fixtures, switches, and receptacles in each area. Styles and tastes in lighting change over the years just as they do in clothing and cars. In the case of lighting, the expectation of change requires an extremely flexible electrical system.

19-13 LIGHT INTENSITY

The main consideration in lighting is providing sufficient light in the proper places. Intensity of light is measured in *footcandles* (*fc*) with a footcandle meter.

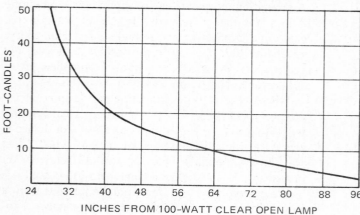

A footcandle is the intensity of light one foot from an international candle. The specifications for composition of an international candle comply with an international agreement. The international metric unit for the footcandle is the *candela*.

Some photographic exposure meters contain a footcandle scale, and some have a conversion table for determining light intensities. If a means of measurement is not available, the graph in Fig. 19-8 gives approximate light intensities at distances in inches from a 100-W clear open light bulb. According to the graph, this lamp has an intensity of 50 fc at 28 in., 40 fc at 30 in., 30 fc at 34 in., 20 fc at 42 in., etc.

Approximate footcandles of light generally needed for seeing while performing various tasks in a home follow:

Footcandles of Light for Tasks in the Home

	Footcandles
Sewing, bookkeeping, detail drawing	50–60
Desk work, fine reading, needlework	40–50
Rough drawing, reading	30–40
Casual desk work, intermittent reading, cooking	20–30
Laundry, rough work	10–20
TV viewing, dining, hallways, stairs	5–10

19-14 PLANNING FOR HEAVY APPLIANCES

After the general and appliance receptacles are included in the plan, the receptacles for heavy appliances, usually requiring individual 120/240-V or 240-V circuits, such as an electric range, clothes dryer, 1-hp and larger air conditioners, space heaters, and so forth, should be located in the plan.

To promote safety in the use of receptacles, the National Electrical Manufacturers Association (NEMA) has adopted a system of assigning individual

Fig. 19-8 Footcandles of light at different distances from a bare 100-W lamp.

configurations for receptables and plugs. Each receptacle is assigned a configuration based on its voltage and current rating. Thus, it would be difficult to mistakenly interchange a plug with a wrong receptacle, since a receptacle will not receive a plug designed for a different rating.

Configurations for most of the commonly used receptacles in a residence are illustrated in Fig. 19-9. The terminals marked W are for connection to the white or neutral wire, terminals marked G are for connection to the green or bare grounding wire. The letters X and Y need not be used unless a certain order of connections is desirable.

19-15 WIRE SIZES

Wire sizes are determined by the load in amperes, insulation, and length of the wire run. Table 3-1 can be used for determining wire sizes for circuits up to about 100 ft in length, and the wire size formula in Sec. 3-12 can be used for determining wire sizes for other circuits to avoid excessive voltage drop.

All branch circuits in a wiring job should be No. 12 or larger. No. 14 can be used, but its overcurrent protection shall not be more than 15 A; No. 12 can be protected at up to 20 A.

If a smaller wire than the regular circuit wire is used in a circuit, such as in switch loops, the size of the fuse or circuit breaker for that circuit is determined by the ampacity of the smaller wire.

19-16 CALCULATING SIZE OF SERVICE ENTRANCE

The minimum size of the service-entrance conductors is determined by the total load, subject to adjustments under certain conditions. These conditions are covered in Arts. 215 and 220 of the Code, which also offer several optional methods of determining the service load; but if the net connected load is calculated and exceeds 10 kW, the service shall be a minimum of 100-A three-wire (NEC 230-41). The Code recommends that 100-A three-wire service be minimum for any individual residence.

Article 220, Part C, of the Code offers a simple optional method of calculating the required minimum size of service equipment. This method involves separating the air conditioning and space heating loads from the rest of the loads in the dwelling. Only the largest air conditioning or space heating load is considered in the optional calculation.

The procedure is as follows. The connected air conditioning load is determined. Next, any central space

heating load is determined. For space heating only 65 percent of the connected load is considered. If there are fewer than four separately controlled electric space heaters, their total connected load is multiplied by 65 percent. If there are four or more separately controlled electric space heaters, their total connected load must be considered. Only the largest of the above loads is to be used in the optional calculation.

For example, if a 12-A, 230-V air conditioner and five separately controlled space heaters, each rated at 1800 W, are connected, only the larger of the two loads (air conditioner or space heaters) is used to compute the feeder size. In this case the space heater load, taken in its entirety, is $5 \times 1800 = 9000$ W or 9 kW. The air conditioner load is $12 \times 230 = 2760$ W or 2.76 kW. The larger of the two is the 9-kW space heater load, and for the sake of the optional calculation, the air conditioner load is not considered. If only three space heaters had been used, the total load would have been $1800 \times 3 = 5400$ W, but only 65 percent would be considered, or $5400 \times 0.65 = 3510$ W.

The balance of the feeder load calculations involves totaling all other loads. These "all other" loads consist of the 20-A small-appliance branch circuits, calculated at 1500 W each; a general-lighting and general-use receptacle load of 3 W/ft^2; the nameplate rating of all fixed appliances such as ranges, ovens, dryers, and the like; and the nameplate rating (in kilovoltamperes) of all motors and other low-power-factor loads.

Example 1(b) in Chap. 9 of the NEC gives the step-by-step procedure used in the optional calculation. The dwelling in this example should be served with a minimum 100-A service.

In selecting service-entrance feeders and conduit, NEC tables show No. 1 TW rated at 110 A; three No. 1 TW wires require 1¼-in. conduit.

To determine the number and type of branch circuits required in the panel, the 120- and 240-V circuits are classified by the nature of the load and Code regulations.

The Code recommends one 120-V lighting-receptacle circuit for each 500 ft^2 of usable floor space and requires at least two 20-A small-appliance circuits and at least one 20-A laundry circuit. Since the total ratings of fixed appliances on 15- and 20-A lighting-receptacle circuits shall not exceed 50 percent of the circuit rating (NEC 120-23-a), one 120-V circuit should be provided for each 120-V fixed appliance. One 240-V circuit should be provided for each 240-V fixed appliance, also.

Accordingly, the dwelling under consideration would require a minimum of seven 120-V circuits as follows: Two for small-appliance circuits, one for the

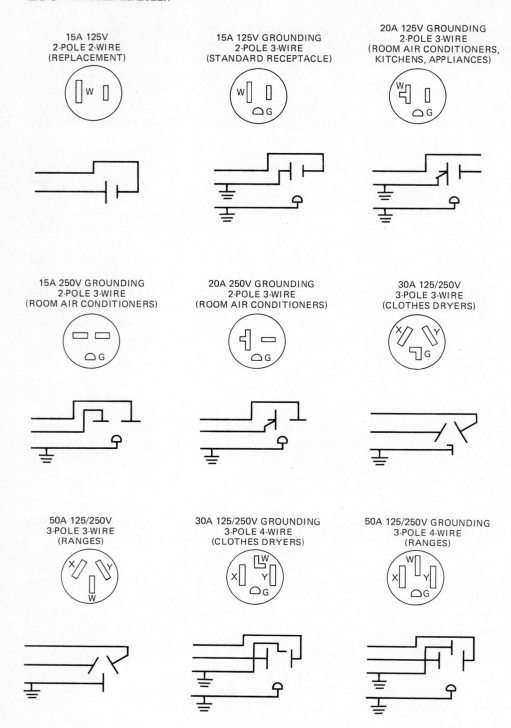

15A 125V
2-POLE 2-WIRE
(REPLACEMENT)

15A 125V GROUNDING
2-POLE 3-WIRE
(STANDARD RECEPTACLE)

20A 125V GROUNDING
2-POLE 3-WIRE
(ROOM AIR CONDITIONERS,
KITCHENS, APPLIANCES)

15A 250V GROUNDING
2-POLE 3-WIRE
(ROOM AIR CONDITIONERS)

20A 250V GROUNDING
2-POLE 3-WIRE
(ROOM AIR CONDITIONERS)

30A 125/250V
3-POLE 3-WIRE
(CLOTHES DRYERS)

50A 125/250V
3-POLE 3-WIRE
(RANGES)

30A 125/250V GROUNDING
3-POLE 4-WIRE
(CLOTHES DRYERS)

50A 125/250V GROUNDING
3-POLE 4-WIRE
(RANGES)

Fig. 19-9 Standard NEMA receptacle configurations.

laundry circuit, one for the dishwasher, and three for lighting-receptacle circuits. The range, water heater, clothes dryer, and air conditioner would require one 240-V circuit each. The electric space heaters for five rooms would have to be provided for, and these circuits would depend on the heaters' ratings.

The service equipment can be of the circuit-breaker type with a 100-A main breaker.

19-17 LOCATION OF SERVICE EQUIPMENT

There are several important factors to be considered in locating the service cabinet. Three of the chief determining factors are accessibility, load center, and location of the service-entrance conductors.

The service cabinet should be located in a place

readily accessible at all times, easily found, and where nothing will have to be moved to get to it in case of fire, tripped breakers (or blown fuses), short circuits, or other emergencies.

It should be located as near as practicable to the center of its load, have the shortest distances possible between it and electrical equipment that requires large amounts of power, and be as near as practicable to the service entrance. The location of the service entrance is determined in consultation with a representative of the utility company. Usually the customer's desires and needs are duly considered, but the utility representative will make the final decision on the location of the point of attachment of the service drop to the building.

In some locations, such as suburban and farm buildings, it is desirable to use a yardpole for electrical service. Local codes sometimes require all service to be underground. In such cases service-entrance conductors originate from pad-mounted or below-ground distribution transformers.

19-18 INSTALLATION OF SERVICE ENTRANCES

A service entrance can be installed in conduit or cable, as shown in Fig. 19-10(a), (b), and (c). Some local codes require rigid metallic conduit. Where rigid conduit is installed, it must be strapped with galvanized straps and screws. The service head screws into one end of the conduit. The other end of the conduit screws into the top of the meter socket. Conduit screws into the bottom of the meter socket and into a service-entrance ell, as shown in Fig. 19-10(b). This ell is equipped with a waterproof cover for access outside the house. Conduit is installed from the ell to the service cabinet.

After the conduit is installed, individual conductors are pulled into the conduit. For 120/240-V service, two black wires for the hot legs and one white wire for the neutral are used. About 3 ft of wire should extend from the service head for connection to the service drop wires. This connection is usually made by the

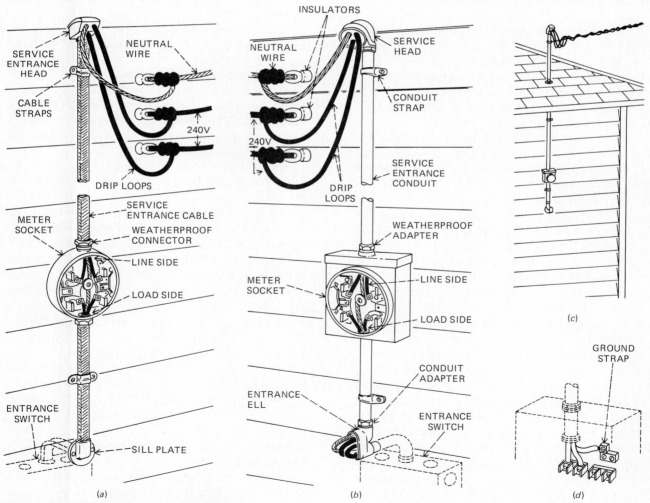

Fig. 19-10 Typical service-entrance installation.

utility company. The two black wires connect to the top outside terminals of the meter socket terminals, and the neutral wire connects to the grounding screw. This connection grounds the meter socket.

The load wires to the service cabinet connect to the two lower outside meter socket terminals, and the neutral wire connects to the grounding screw.

The load wires are pulled down in the conduit and out at the service-entrance ell. They are then turned back and pulled through the conduit to the service cabinet. The connections to the main fuse block or circuit breakers in the service cabinet are shown in Fig 19-10(*d*).

Where service-entrance cable is used for an installation, a cable service head is installed at the top, as shown in Fig. 19-10(*a*), and the cable is connected to the meter socket with a watertight connector. The opening in the building where the cable enters is sealed to prevent water from entering.

19-19 SERVICE GROUNDING

After the service entrance is installed, one end of a grounding wire is connected to any one of three points of the system. It can be connected at the service head to the neutral wire, in the meter socket to the grounding screw, or in the service cabinet or panel to a grounding lug. The other end of the grounding conductor must be connected to the building's grounding system. In the past this simply meant connecting to a cold-water pipe. Widespread use of nonmetallic piping in modern homes and the poor performance of metal water pipes as an effective grounding medium have altered the requirements for grounding systems. Article 250, Part H, of the Code covers grounding electrode systems. The expanded grounding system requires that all major metallic elements of a building be bonded together. This includes metal water pipe that has at least 10 ft of direct contact with earth, the metal frame of the building, a concrete-encased electrode consisting of at least 20 ft of conductor (steel reinforcing bars included), and a ground ring circling the building.

If none of these exist, a metal gas pipe can be used, provided the gas supplier and the local Code authorities approve. Other underground systems or structures may also be used including underground metal tanks and well casings. In any case, a cold-water pipe alone is not acceptable. Supplemental grounding, such as by electrodes driven into the ground, must be provided. These additional grounding points are called *made electrodes* by the Code (Sec. 250-83 Made and Other Electrodes). When grounding to a water pipe, the connection should be made on the street side (the un-

metered side) of the pipe. Whether or not the water pipe qualifies as the system ground, in residential installations the interior metal water pipes must be connected to either the neutral conductor at the service panel or a ground electrode.

19-20 THEORY OF GROUNDING

The subject of grounding has been debated since the beginning of the generation and use of electricity. Grounding is the practice of connecting one side of an electric circuit to the ground, or earth, for protection against *high voltages* and *lightning*.

There are two types of grounds: (1) intentional, and (2) unintentional, or accidental. In this discussion the intentional ground will be called ground, or grounding, and the unintentional ground will be called accidental ground.

Electric distribution lines, extending great distances in the open and entering a building, are subject to hazards from lightning, and they subject the building and its occupants to these hazards. Also, in case of certain accidents, a building and its occupants are subjected to the hazards of the high-voltages of the building's electrical system.

To minimize danger from these hazards, the high-voltage distribution lines are protected by surge (lightning) arresters, and the system is grounded at the transformer. The low-voltage system is *grounded* at the *transformer* and at the *customer's service entrance.*

With this method of grounding, if by accident the high-voltage distribution lines came in contact with the low-voltage lines, as by a fallen tree limb or failure of insulation in a transformer, current flow between the lines and ground would open a cutout in the high-voltage line and deenergize the system, removing the hazard of high voltage. If the system were struck by lightning, the lightning would be bypassed to earth through the surge protection and the grounding system.

Although this system of grounding protects property and persons against high distribution voltages and lightning, it creates possibilities of hazards on the low-voltage or secondary side of a system in the case of an accidental ground. To minimize these hazards, a system of grounding must be used on the secondary system (Fig. 19-11). It can be seen that grounding has nothing to do with the operation of electrical equipment. Its sole purpose is the protection of life and property.

Electrical equipment operates equally well with or without grounding. This fact often leads to carelessness or neglect of the all-important matter of proper

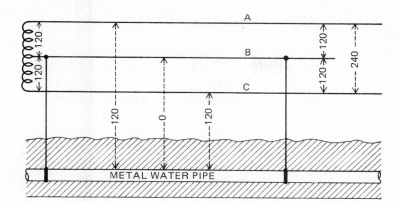

Fig. 19-11 A grounded three-wire secondary system showing voltages between lines and between lines and earth.

and sufficient grounding on the premises. *Improper grounding is an invitation to disaster.*

If a hot wire serving a 120/240-V, three-wire electric range, or any such piece of electrical equipment, accidentally comes in contact with its frame, it creates an accidental ground, and if the frame is properly grounded, a 120-V short circuit will be formed. This short circuit will allow an excessive current to flow from the hot wire through the range frame and ground and back to the neutral wire at the service entrance. The excessive current on the line will blow a fuse or open a circuit breaker, which will deenergize the line.

If the frame of the range is not grounded but contains an accidental ground, there will be a *voltage of 120 V between it and any grounded object,* such as water or gas pipes, water faucets, gas heaters, radiators, space heaters, air-conditioning or heating ducts, etc., and if a person touches the ungrounded frame and any grounded object, he or she will receive a 120-V shock. Such a shock through certain parts of the body *can result in death.* Accordingly, the Code requires the frames of all permanently installed electrical equipment, and of some portable appliances, to be grounded.

Range wall ovens, counter-top cooking units, and clothes-dryer frames can be grounded with the uninsulated neutral wire in service-entrance cable or to the insulated grounded circuit conductor, but these are the only cases in which a current-carrying neutral wire can be used for grounding (NEC 250-60). The grounding wire in any wire assembly must be bare or insulated with green insulation (NEC 250-59*b*).

19-21 GROUNDING PORTABLE EQUIPMENT

The frames of portable equipment are grounded by use of a three-wire cord and a grounding receptacle. A grounding receptacle is a receptacle with a grounded contact. The grounding contact of the receptacle is grounded to the strap supporting the receptacle. It also has a grounding terminal screw. This screw is usually identified by a dark or green color.

When the receptacle is mounted in a box wired by conduit or armored cable, it is automatically grounded by the supporting strap, but if the box is wired with nonmetallic cable, the cable must contain a grounding wire, *bare or green-insulated,* for connection to the grounding terminal of the receptacle and to the box. The identified circuit neutral wire cannot be used for grounding.

19-22 OUTLET AND JUNCTION BOXES

Cable entering metal boxes shall be secured to the box. Built-in clamps in a box, or connectors, are used for this purpose. The sheathing of cable should extend about ½ in. inside the box from the clamp or connector. The insulating bushing on the end of armored cable *shall be visible* (NEC 333-9). Junction boxes shall be *accessible at all times* without removing any part of the building (NEC 370-19).

To allow for sufficient working room, the maximum number of wires entering a box is limited by the Code [NEC 370-6 and Table 370-6(*a*)]. This table applies to bare boxes with no cable clamps, hickeys, or fixture studs. A deduction of conductors is to be made for each of the following conditions: one for one or more cable clamps, hickeys, or fixture studs; one for each strap containing one or more devices (switches, receptacles); and one for one or more grounding conductors entering the box. A conductor running through a box and not spliced is counted as one.

For boxes or combinations of boxes not shown in the Code table, free space for each conductor entering, of sizes shown, shall be as follows: No. 14, 2 in.[3]; No. 12, 2.25 in.[3]; No. 10, 2.5 in.[3]; No. 8, 3 in.[3]; No. 6, 5 in.[3] [NEC Table 370-6(*b*)]. Thus, a 2 × 3 × 4-in. box, containing 24 in.[3], could have a maximum of 12 No. 14 wires entering it, which could be 6 two-wire cables, 4 three-wire cables, 2 three-wire and 3 two-wire cables, or any combination equal to 12.

19-23 SWITCHES

No switch or circuit breaker shall disconnect the grounded conductor of a circuit unless it *simultaneously disconnects all ungrounded conductors*. Knife switches shall be placed so that gravity will not tend to close them and shall be connected so the blades are dead in the open position.

Toggle switches should be mounted so the toggle is down when in the "off" position, except three-way and four-way switches which do not have an "off" position.

AC General-Use Snap Switch This is a general-use switch, marked AC, following its electrical rating, suitable for ac circuits only, which can be loaded to its ampere rating at its rated voltage on resistive and inductive loads, including tungsten filament lamps, but at only 80 percent of its rating with motors. This means the switch can be used in motor circuits only if it has a rating of 125 percent of the full-load current of the motor.

AC/DC General-Use Snap Switch This is an ac/dc switch, which can be loaded to its ampere rating with resistive loads, except tungsten lamps, or one-half its rating with motor loads; and if T-rated following its electrical rating, it can be loaded to its full rating with tungsten filament lamps.

In wiring switches, if the power wires enter the switch enclosure first and then the fixture or equipment enclosure, the switch is connected in the hot or ungrounded line. If the power wires enter the equipment enclosure first and continue to the switch enclosure, the switch is in a switch loop. In a switch loop, switching is done in the "hot," or ungrounded, line, but if a cable contains a white wire, this wire can be used in the hot side if "connections are so made that the unidentified conductor is the return conductor from the switch to the outlet" (NEC 200-7—Exc. 2). This means that the white identified wire is connected from the power to the switch, and the black wire is connected from the switch to the outlet. This is the only time a white wire is permitted to be used as an ungrounded circuit wire, except when the white wire is painted black at outlets where visible and accessible (NEC 200-7—Exc. 1).

19-24 WIRES AND CABLES

Wires and cables are classed by the Code according to their insulation. The classifications are designated by type letters. Usage of wires and cables, as to ampacity, dry or wet locations, surrounding conditions, etc., is determined by the type of insulation (Fig. 19-12).

Usually, Code-specified letters indicate the kind of insulation or its usage: type R is rubber, T is thermoplastic, A is asbestos. Letters following the first letters usually indicate the type of insulation: H, heat-resistant; W, water- or moisture-resistant; C, corrosion-resistant; L, lead-covered. Thus type TW insulation is moisture-resistant thermoplastic; RH is heat-resistant rubber.

A complete listing of conductor types, insulations, and applications is given in NEC Tables 310-2(*a*) and 310-2(*b*). Chapter 20 covers wire and cable in detail.

19-25 CABLE WIRING METHODS

The two most commonly used types of wiring systems in residences (except in restricted areas of some cities where conduit is required) are nonmetallic (NM) sheathed cable and armored cable (AC). These two types of cables have practically the same usage, open or concealed in normally dry locations, and the wiring methods are practically the same (NEC Arts. 333 and 336). The differences in uses are that AC cable can be plastered over in masonry or plaster, while NM cable cannot be plastered over. Armored cable contains armor and a grounding wire which automatically grounds equipment, while NM is made with or without a grounding wire.

19-26 ELECTRICAL CONNECTIONS

An electrical connection is more than a mechanical connection—it is a union of two or more parts through which electricity flows. A connection can have good mechanical properties but poor electrical properties. A good connection must have sufficient contact to conduct the desired current without overheating, and it must be mechanically strong to withstand vibration, strains, expansion and contraction, and damaging effects of the elements, corrosion, and time. So, two conductors simply joined do not qualify for an electrical connection.

The number of electrical connections in an average motor, its armature, and controls can exceed several hundred, and failure of any one of these can cause complete stoppage of a piece of equipment. An aver-

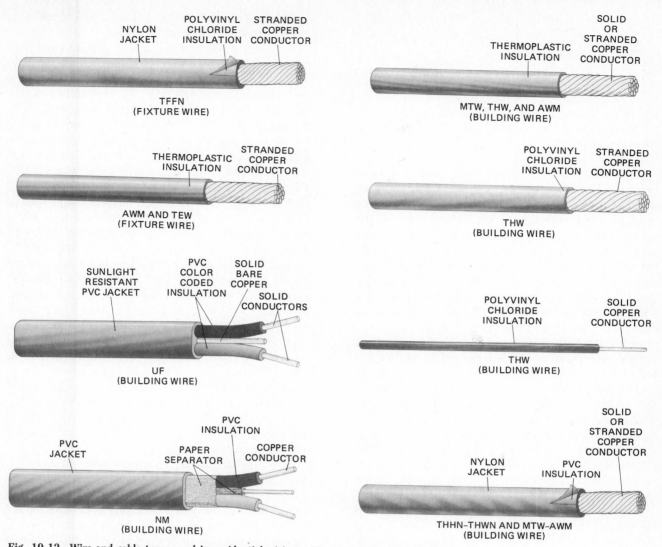

Fig. 19-12 Wire and cable types used in residential wiring. (*General Cable Company*)

age house wiring system contains a thousand or more connections, and therefore a thousand or more possible sources of trouble unless the connections are properly made. A good electrical connection can be made with care and with knowledge of a few facts about connections.

Electrical connections are made by mechanical compression, which includes screwing, clamping, using split bolts and wire nuts, crimping, and the like.

The right and wrong ways of turning a wire for fastening under a screw are shown in Fig. 19-13. If the wrong way is used, the wire will be forced out from under the screw as it is turned. A screw should be loosened and tightened several times as it is tightened to produce a scouring effect which removes burrs and oxides on the wire and gives a clean, secure connection.

Solderless connectors, shown in Fig. 19-14, are used for small and large wires. Figure 19-14(*a*) shows a type of solderless connector for small wires. Wires are prepared and twisted clockwise [Fig. 19-14(*b*)], and the connector is screwed clockwise over them. The internal thread in the connector is tapered, and it tightens as it is screwed on. A split-bolt connector for connecting large wires, such as service conductors, is shown at Fig. 19-14(*c*).

19-27 WIRING CIRCUITS WITH CABLE

Typical methods used in cable wiring for various common conditions are illustrated in the following figures. A simple lighting circuit with a switch loop is shown in Fig. 19-15(*a*).

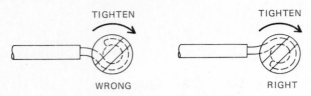

Fig. 19-13 Turning the screw clockwise will tighten it and turn the wire tighter around the screw if the wire is looped the correct way.

The power cable enters the lighting outlet box from the right. The black wire of the power cable connects to the white wire in the switch loop cable. With this black-to-white connection, the white wire in the switch loop can be used in the hot side of the line; otherwise the exposed ends of the white wire would have to be painted black. The black wire returns from the switch and connects to the black wire of the fixture. The white wire from the fixture connects to the white neutral wire of the power cable. Connections in the box are made with solderless connectors.

In Fig. 19-15(b) two fixtures are shown controlled individually by two switches ganged in one switch location. A two-wire power cable enters the first outlet box from the left, and three-wire cable is used in the remainder of the installation. The black power wire connects to a black wire of a three-wire cable to the second box; this wire connects to a black wire in the three-wire cable to the ganged switch box where the black wire connects to one side of both switches. A red wire in this cable connects to one side of the switch at the left, and the other end of the red wire connects to a red wire in the fixture box to continue to the black wire of the next fixture. The white neutral wire from this fixture connects to the neutral wire of the power cable. The switch at the right has a white wire painted black connected to it, and this wire connects to the black wire of the first fixture from the

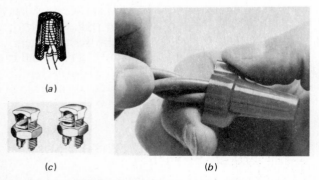

Fig. 19-14 Common solderless connector or wire nut.
(a) Cutaway view of connector with three conductors.
(b) Method of joining conductors with a solderless connector. (*Ideal Industries, Inc.*) **(c)** Split-nut connector for large conductors. (*Ideal Industries, Inc.*)

switches. The white wire of this fixture connects to the white neutral wire in this box.

A typical cable wiring job for a fixture controlled with a single-pole switch is illustrated in Fig. 19-16(a). The power cable enters the outlet box at the top and continues through and out of the left of the box. The white wire of the switch loop connects to the black power wire. The switch is connected in the switch loop, and the black wire from it connects to the black wire of the light. The white wire of the light connects to the white neutral wire in the power cable. Connections are made with solderless connectors.

A hot receptacle connected beyond a light and its switch in a cable job using two- and three-wire cables with grounding wires is illustrated in Fig. 19-16(b). Power enters the light outlet box from the right by means of a two-wire cable. The black wire of the power cable connects to the black wire of a three-wire cable which runs from the light box to the switch box, and the black wire connects to one side of the switch and continues in a two-wire cable to the receptacle. The white wire from the receptacle connects in the switch box to the white wire of the three-wire cable which connects to the white wire of the power cable in the light box. The light circuit from the switch is a red wire which connects to the black wire of the light. The white wire of the light connects to the white neutral of the power cable. The grounding wire connects to the receptacle grounding screw.

Power cable entering the switch box first and serving a switched light and a pull-chain light is shown in Fig. 19-17(a). This switching system does not constitute a switch loop, since switching is done in the hot line before it reaches the light outlet box. Power is brought to one side of the switch by the black wire of the power cable, which continues to the pull-chain light. The circuit through the switch to the first light is a red wire in a three-wire cable. The white neutral wire of each light connects to the white neutral wires in the cables.

A method of ganging a hot receptacle and light switch in one box is shown in Fig. 19-17(b). A two-wire power cable enters the light outlet box from the right. A three-wire cable runs from the light box to the ganged box, the black wire carrying power to the switch and receptacle; the red wire makes the circuit from the switch to the light, and the white wire completes the circuit from the receptacle to the neutral power line. The grounding wire connects to the receptacle grounding screw.

19-28 WIRING A BUILDING

The electrician's knowledge and judgment are sought in the planning of a safe, efficient, and adequate wir-

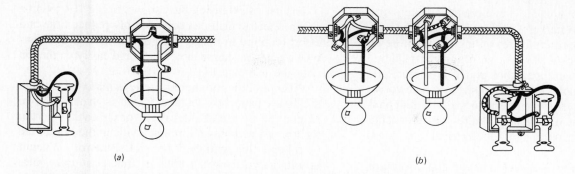

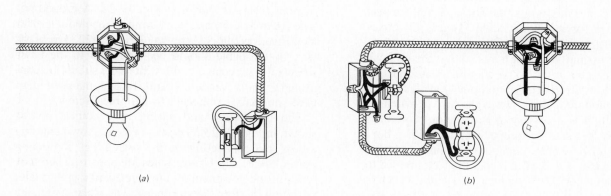

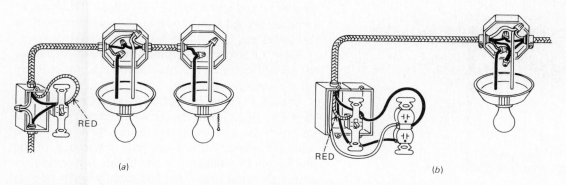

Fig. 19-15 Wiring lamps and switches. (*a*) A single lamp and switch. (*b*) Two lamps each controlled by its own switch.

Fig. 19-16 Cable wiring of fixtures. (*a*) Cable passes through fixture outlet box. (*b*) Receptacle wired beyond fixture and switch. The grounding connection has been omitted for clarity.

Fig. 19-17 Cable wiring of fixtures. (*a*) Wiring a pull-chain fixture. (*b*) Ganged switch and receptacle. The grounding connection has been omitted for clarity.

ing system for a home. The electrician enters the building with an efficiently prepared plan for the wiring system. After carefully reviewing the plan, he or she is ready to go to work in an organized fashion.

An efficiently equipped tool pouch is a prerequisite. The usual tools include two sizes of screwdrivers, a pair of lineman's pliers, a pair of diagonal cutting pliers, an electrician's knife, an adjustable wrench, a cable stripper, a roll of electrical tape, and a claw hammer.

The electrician uses a stick and red chalk to mark the location and height of switch and receptacle boxes. The stick is marked 16 in. for the height from the floor to the center of receptacle boxes and 50 in. for the height from the floor to the center of switch boxes. Switch boxes on walls to be tiled should be located above or below the location of the tile cap. The type of switch needed can be marked at the location.

After all box locations are marked, the electrician mounts all boxes throughout the house. There are several methods for mounting boxes. Nearly all boxes contain holes in the sides, at the top, and bottom to take sixteen-penny common nails so that the box can be nailed directly to the house framing. Care must be used to allow clearance for interior finish, such as door and window facings, when the box is installed. Occasionally it is necessary to nail one or two blocks on the side of a stud for proper clearance for a box.

A close-up view of two methods of mounting boxes is shown in Fig. 19-18. The box at the left is nailed to a stud. The center box is screwed to a board which is mounted between studs. Both methods afford an inexpensive solid mount. When box locations do not occur at a stud, adjustable box supports between studs can be used.

After all wall and ceiling boxes are mounted, the electrician determines routes for wires or cables and bores holes in the framing necessary for the wiring. Holes should be bored in the center of framing or no closer than 2 in. from the nearest edge. An electric drill or a hand brace and bit can be used.

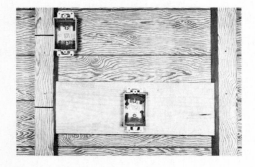

Fig. 19-18 Methods of mounting wall boxes.

The use of extension bits, shown in Fig. 19-19, in a 1/2 in. electric drill can considerably reduce time and effort required in boring a house for wiring. The flat-type bit can be sharpened quickly and easily on the job with the use of a file.

When all holes are bored, cable is threaded through the holes from box to box. Cable is usually drawn from a roll by use of a de-reeler, or the cable is left in its box and withdrawn through a hole in the side of the box, beginning with the inner end of the roll. When it is withdrawn from the box, it comes out in a spiral form. It can be straightened and made flat simply by drawing it through one hand from the box to the free end and not letting it twist. Cable shall be strapped at intervals of 4 1/2 ft or less and 12 in. or less from each box.

A completely wired wall-receptacle box is shown in Fig. 19-20. All receptacle circuits are wired with cable containing a grounding wire. A cable with grounding wire with all wires labeled and with identification colors is shown in Fig. 19-21. A grounding-type receptacle with screw terminals labeled and with identification colors is shown in Fig. 19-22. The black wire of the cable should be connected to the brass screw of the receptacle, the white wire should be connected to the silver screw, and, finally, the grounding wire should be connected to the green screw.

The installation and wiring of a ceiling box is shown in Fig. 19-23. A power cable is shown entering a box, and connections to the switch leg, continuing power cable, and fixture wires are shown. The black wires of the power cables are connected together with a white wire from the switch leg. The black switch leg wire is a fixture wire. The two white wires of the power cables connect together with a white fixture wire. The fixture wires connect to the black and white leads of the fixture. Wire nuts or other solderless connectors are used for the connections.

Where more than one switch is needed in the same location, boxes are ganged and wired at that location. Three switch boxes, two for S_1 or single-pole switches, and one for a S_3 or three-way switch, are shown ganged and wired in Fig. 19-24. The types of switches needed at this location are shown marked on the sheathing.

After all boxes are mounted and wired, including the service-entrance box and cabinet, the job is ready for a rough-in inspection.

Inspectors know the type of work characteristic of their electricians, and the job of an electrician who is known to be careless in his or her work receives a thorough inspection and, usually, a rejection of the job for minor violations of the Code. A rejection requires a second inspection and usually a separate inspection fee. A careful electrician and inspector al-

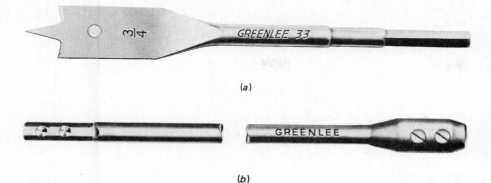

Fig. 19-19 Boring tools.
(*a*) Wood-boring bit. (*b*) Power bit
extension. (*Greenlee Tool Co., a
subsidiary of Ex-Cell-O.*)

ways work toward the same objective—a safe, convenient, and adequate electrical system.

One item that usually receives the special attention of the inspector is the installation of the grounding wire. A galvanized grounding clamp is recommended for connection to a galvanized water pipe, and a copper or bronze clamp is recommended for connection to a copper ground rod. A typical method of grounding to a cold-water pipe is shown in Fig. 19-25.

Once the job receives an "OK" rough-in tag, the electrician returns to finish the job after the carpenters and plasterers have completed the wall and ceiling finish. After all lighting fixtures are installed and a final check of the entire system is made by the electrician, the inspector is called for final inspection of the job.

A neat and orderly service cabinet reflects good planning and workmanship in the entire system. A typical cabinet may contain eight 120-V circuits with four 120-V breakers in reserve for possible future loads and spaces for installation of four more 120-V breakers if needed. It also may contain one 240-V circuit and space for another 240-V circuit if needed later. The inspector places an OK tag on the cabinet for this job. The inspector is always as anxious as anyone to OK a completed job. He or she is an important member of a team that strives to provide safe, dependable, and efficient electrical systems.

19-29 INSTALLATION OF LIGHTING FIXTURES

Various methods are used in the installation of ceiling and wall lighting fixtures. Some of the most commonly used methods are shown in Fig. 19-26.

A fixture weighing more than 6 lb (2.72 kg) or exceeding 16 in. (406 mm) in any dimension shall not be supported by the screw shell of a lampholder. A fixture weighing more than 50 lb (22.7 kg) shall be supported independently of the outlet box (NEC 410-15 and 16). This can be done by anchoring a pipe to the framing above the outlet box, the pipe being extended into the box through a knockout in the bottom for attachment of the fixture.

In Fig. 19-26(*a*) a threaded hanger link is screwed on a fixture stud in the box. This fixture stud is a part of the assembly of the hanger bar. In case a hanger bar is not used, a crowfoot fixture stud is mounted in the box with stove bolts in holes provided in the bottom of the box. A stud is screwed into the link, and the chaindrop lighting fixture is screwed on the lower end of the stud. The electrical connections are made with solderless connectors, and the canopy is raised to contact the ceiling and is secured with a knurled locknut.

In Fig. 19-26(*b*) a strap is secured to a fixture stud with a locknut, electrical connections are made with

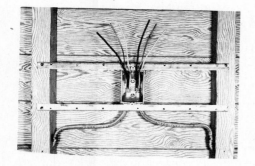

Fig. 19-20 Completely wired receptacle box.

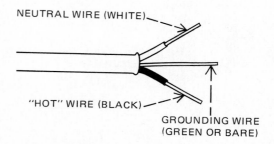

Fig. 19-21 Nonmetallic sheathed cable with grounding wire.

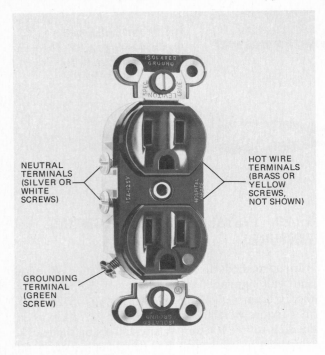

Fig. 19-22 Grounding receptacle.

solderless connectors, and the fixture is mounted on screws in the strap. In Fig. 19-26(c) a strap is secured to the ears of the outlet box with screws, electrical connections are made with solderless connectors, and the fixture is mounted by screws in the strap.

Installation of pan-type fixtures is shown in Fig. 19-27. At Fig. 19-27(a) a coupling is screwed on a fixture stud and a stud screwed into the coupling; electrical connections are made, and the fixture is raised to contact the ceiling and secured by a washer and locknut on the stud which is adjusted to extend through a hole in the pan.

In Fig. 19-27(b) a strap is secured to the ears of the outlet box, electrical connections are made, and a stud is screwed into the center of the strap. The fixture is

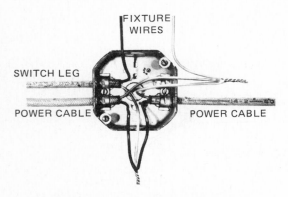

Fig. 19-23 Wiring a ceiling lighting outlet box requiring a switch leg.

Fig. 19-24 Three ganged switch boxes.

raised to contact the ceiling and secured by a washer and nut on the stud, which is adjusted to extend through a hole into the pan.

Installation of a wall-mounted fixture is similar to that of a ceiling-mounted fixture. A strap is mounted by screws to the ears of a wall box, electrical connections are made, and a stud is screwed into a threaded hole in the strap. The fixture is placed against the wall and secured with a knurled nut screwed on the stud, which is adjusted to extend through a hole in the pan of the fixture.

19-30 THREE- AND FOUR-WAY SWITCHES

A *three-way switch* is a single-pole double-throw switch which in its simplest form is shown in Fig. 19-28. There is no open or off position in a three-way switch.

Lights, or any load, can be controlled from two different locations, such as at the bottom and top of a stairway or at the house and in the garage, by the use of two three-way switches.

There are two Code regulations pertaining to three- and four-way switches: All switching shall be done in the unidentified line, that is, the hot line (NEC 380-2), and the wire from the last switch in a combination to the load shall be an unidentified wire, which is any color except white (NEC 200-7—Exc. 2).

Principles of a three-way switch are illustrated in Fig. 19-29. The switch has four contact points inside, which are shown numbered in the illustration. Contact points 1 and 3 are strapped together and are connected to a terminal. This terminal is always identified with a black or dark-brown screw. Contacts 2 and 4 are connected to terminals containing brass- or silver-colored screws. In an ordinary connection the black screw terminal of one switch is connected to the supply, and the black terminal of the second switch is connected to

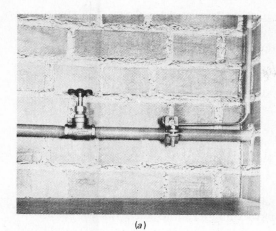

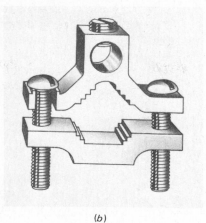

(a)

(b)

Fig. 19-25 Typical grounding installation. (a) Connection is made ahead of the valve. (b) Typical grounding clamp. (*Ideal Industries, Inc.*)

the load. The brass screw terminals are used to connect the two switches together by traveler wires.

The switch blades in the illustrations are shown by shaded lines. In one of their positions, a circuit is made from 1–3 to 4, as shown in Fig. 19-29(a). When switched to the other position, a circuit is made from 1–3 to 2, as shown in Fig. 19-29(b). When switched, they merely transfer the circuit from the bottom line to the top line, or vice versa, as shown in the two illustrations. However, these switches are actually connected to only one side of the circuit lines (the hot leg).

Two three-way switches wired ahead of a light are shown in Fig. 19-30(a). Beginning at the left, the circuit in the hot line is through the left switch, the top traveler to the switch, and through it to the light at the right, through the light to the neutral side of the circuit. In actual wiring, a two-wire cable is used to the first switch, a three-wire cable is used between switches, and a two-wire cable is used from the last switch to the light.

The arrangement in Fig. 19-30(b) shows the light between the switches. This arrangement could be used for a light in a room with a switch at each of two doors, with the power entering a switch. The order of the circuit is the same as in Fig. 19-30(a). A two-wire power cable is used to one switch; a three-wire cable

is used from this switch to the light and the other switch.

In Fig. 19-30(c) the power goes to the light outlet box first and then to the switches, with a two-wire cable to the light and first switch and a three-wire cable between switches.

A pictorial illustration of an actual cable-wired three-way installation is shown in Fig. 19-31.

In Fig. 19-32, the power enters the light outlet box first, then the switches and light. This arrangement is often used in house wiring for lighting in a room controlled at two doors.

A garage light, controlled from the house or the garage, with a receptacle in the garage that can be made hot regardless of whether the light is on or off, is shown in Fig. 19-33. This receptacle circuit is another interesting application of a three-way switch. It will be noticed that this three-way switch can select the hot wire of the travelers for the receptacle. This system requires only a three-wire cable with a receptacle grounding wire from the house to the garage. The grounding wire is shown by dashed lines.

Another interesting connection of two three-way switches is shown in Fig. 19-34. Using a four-wire cable between switches, this system affords two-point control of a light at the house and a light at the garage and also affords a permanently hot receptacle in the

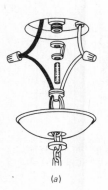

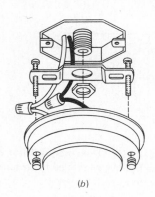

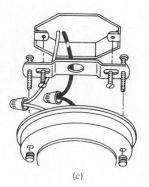

(a)

(b)

(c)

Fig. 19-26 Methods of installing ceiling fixtures. (a) Method used for large or heavy fixtures. (b) Attaching fixture to box stud. (c) Attaching fixture to ears on ceiling box.

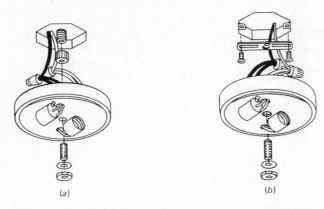

Fig. 19-27 Installing pan-type fixtures. (a) Using a stud to secure the pan. (b) Using a strap to secure the pan to the ears of the ceiling box.

garage. Since all switching is done in the hot line, this circuit complies with the Code. The receptacle grounding wire is shown by a dashed line.

A *four-way switch* is actually a double-pole double-throw switch (also called a *reversing switch*). In one position, it closes two circuits in one direction, and, when switched, it closes two circuits in another direction. There is no open or off position. This is illustrated in Fig. 19-35. In the position shown in Fig. 19-35(a), the circuits are from 1 to 2 for one circuit, and from 4 to 3 for another circuit. When the switch is operated, as shown in Fig. 19-35(b), it makes a circuit from 1 to 3, and another circuit from 4 to 2.

It will be noticed that a four-way switch simply transfers a circuit from one line to another. The lines are crossed at the bottom of the switches in the illustration. This is necessary in some cases, but in most switches the cross is made inside the switch and the lines are not crossed outside.

One or more four-way switches can be connected in the travelers between three-way switches for multipoint control. A combination of three- and four-way switches connected for switching in one side of a line is shown in Fig. 19-36. The shaded lines in the switches represent the position of switch blades and show the circuits through the switches. All switching is done in the hot line. The fixture, as in the three-way combinations previously shown, can be connected out of any of the boxes. Many interesting solutions to control problems can be devised by the use of three- and four-way switches.

19-31 LOW-VOLTAGE WIRING

Although three-way and four-way switching is popular for multipoint switching of lighting, when more than four switching locations are involved, the use of low-voltage systems becomes not only economically feasible but also electrically more convenient.

A low-voltage system consists of a step-down transformer, a rectifier, low-voltage relays, and momentary contact switches. A typical system is shown in Fig. 19-37. In this system four switching locations operate a bank of lights. A single relay controls all six fixtures. Since the relays operate on direct current, a rectifier is wired immediately after the transformer secondary. The switches are rocker-type momentary contact. After the switch is pressed, it returns to a neutral position. If a lamp is burned out or if the switch is located remotely from the lamp, there would be no way of knowing whether the circuit was turned on. To solve this problem, switches are available that are equipped with pilot lights. These switches glow when the on button is pushed and continue to glow until the off button is pushed. Pilot light switches are connected to pilot relays.

To obtain a completely flexible switching system, each fixture can be wired through its own relay. Low-voltage systems usually are available with numerous accessories that permit master control of complete banks of lights or selective fixtures from a single remote location.

The flexibility of low-voltage systems comes from the fact that additional switches and fixtures may be added simply by paralleling the new switches as shown in Fig. 19-38.

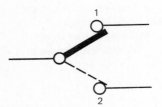

Fig. 19-28 A simple single-pole double-throw switch.

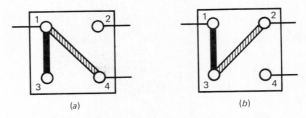

Fig. 19-29 Three-way switch operation. (a) In one position the line connected to terminal 4 is energized. (b) In the other position terminal 2 is energized.

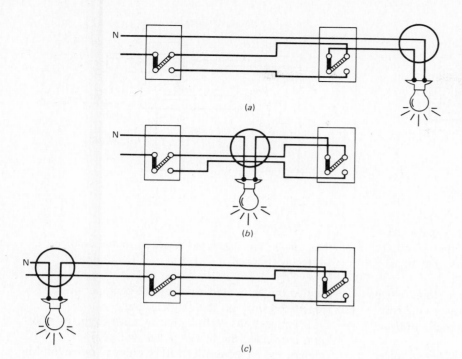

(a)

(b)

(c)

Fig. 19-30 Using a three-way switch to control a lighting fixture from two locations. (*a*) The switches are wired ahead of the fixture. (*b*) The fixture is between the two switches. (*c*) The fixture is wired ahead of the two switches.

19-32 INSTALLATION OF RANGES AND CLOTHES DRYERS

Electric ranges and clothes dryers are usually installed simply by providing a suitable receptacle for them. Ranges usually require 50-A receptacles, and regular clothes dryers require 30-A receptacles. High-speed dryers require 50-A receptacles.

A typical range receptacle installation is shown in Fig. 19-39. This circuit originates at the service cabinet. Ranges usually require 120/240 V for operation. Service-entrance cable can be used for ranges and clothes dryers. The uninsulated neutral wire can be used for grounding. Connection to the range of a three-wire assembly with a male plug is shown. The center connection is for the neutral wire since this terminal is grounded to the frame of the range. This grounds the range.

19-33 INSTALLATION OF WATER HEATERS

Water heaters usually require a 240-V circuit, with a means of disconnect. In a one-family residence, the main service switch or circuit breaker may be used as the disconnect. If the circuit is to be connected to unprotected terminals in the service cabinet, protection can be provided by use of a fused safety switch or circuit breaker, as shown in Fig. 19-40. The frame of the water heater is grounded when the water pipes are connected to it. A temperature-limiting instrument sensitive to water temperature is required by the Code (NEC 422-14). This Code section also requires that the heater be marked to require a temperature and pressure-relief valve.

Many utilities offer a special off-peak rate for electric water heaters. This requires installation of a regu-

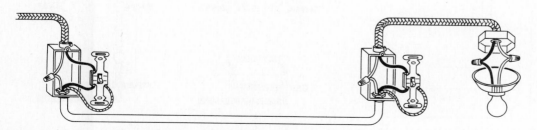

Fig. 19-31 Cable wiring a three-way switch installation.

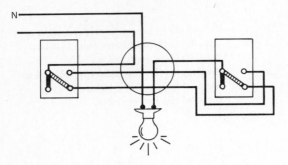

Fig. 19-32 Controlling a room fixture from two doors.

lar meter and an off-peak meter. Dual meter sockets are available for regular and off-peak meters and are recommended for this type installation, but regular individual meter sockets can be used.

Off-peak meters are equipped with a clock which disconnects the meter during peak power-consumption periods of a 24-h day but allows heating during off-peak periods at the special rate.

Generally, there are two ways that off-peak meter systems can be connected. One way will allow heating only during off-peak periods at off-peak rates. Another way allows heating on off-peak periods at off-peak rates and heating also during peak periods at regular rates.

Off-Peak Heating Connections for a typical heater for heating during off-peak hours only are shown in Fig. 19-41. Connection at the regular meter is made to the two top terminals, ahead of the meter, so that power to the off-peak meter does not register on the regular meter. The off-peak meter clock opens the circuit to the heater during peak periods. All power con-

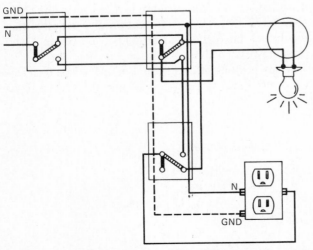

Fig. 19-33 Individual control of a garage fixture and receptacle from two locations.

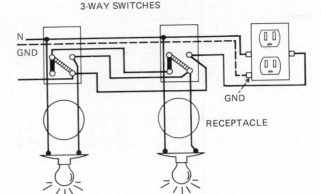

3-WAY SWITCHES

FIXTURE   FIXTURE

Fig. 19-34 Two-fixture and receptacle control using three-way switches.

sumed by the heater is during off-peak periods and is registered only on the off-peak meter.

If connections are incorrectly made to the two bottom terminals, the heater power will register on both meters and the consumer will be charged twice for this power. The neutral line is connected through both meters and the switch, which grounds this equipment. The water pipes to the heater ground the heater.

In operation, during off-peak periods, the top double-throw thermostat is up, making a circuit to the top heating element which heats the water in the top of the tank until the thermostat snaps down. This disconnects the top element and makes a circuit through the bottom thermostat (which is normally in the closed position) to the bottom element. The bottom element heats the water in the bottom of the tank until its thermostat disconnects it.

Off-Peak and Emergency Heating A connection that affords off-peak heating at off-peak rates, and emergency heating at regular rates, is shown in Fig. 19-42. Connection is made to the bottom terminals of the regular meter. The off-peak meter registers only that power that is used by the heater during off-peak, and this is subtracted from the reading of the regular meter and charged at off-peak rates when the power bill is

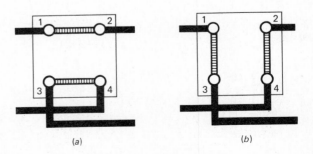

(a)   (b)

Fig. 19-35 Operation of a four-way switch.

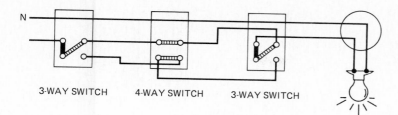

Fig. 19-36 A combination of two three-way switches and one four-way switch used to control a load from three locations.

3-WAY SWITCH   4-WAY SWITCH   3-WAY SWITCH

computed. The remainder of the regular meter reading is charged for at regular rates. The disconnection by the clock during peak periods is made between the top and bottom terminals at the right of the off-peak meter.

A connection to the top right terminal of the off-peak meter allows power to bypass the meter to the heater during peak periods, and this power registers only on the regular meter and is paid for at the regular rate, since it does not register on the off-peak meter to be subtracted from the regular meter.

In operation during off-peak the clock makes a circuit to the bottom heating element, but it cannot heat until the top element heats the water in the top and operates its thermostat. Current to the top element bypasses the clock switch by the connection to the top terminal of the meter and heats the water in the top of the tank until the top thermostat snaps down. This action disconnects the top element and makes a circuit to the bottom element through a normally closed bottom thermostat. This element heats through both meters and raises the water temperature to maximum value throughout the tank for storage before its thermostat opens. Cold water entering the tank at the bottom operates the bottom thermostat for heat as water is used.

During peak periods the circuit to the bottom element is disconnected by the clock, and the top element heats the water in the top of the tank through the regular meter only.

In this particular connection the neutral wire does not always come to the switch, so that the switch must be grounded by other means.

19-34 DOORBELLS OR CHIMES

A signal or bell transformer is used to supply current for the operation of doorbells or chimes. When signal transformers are used, Class 2 wiring can be used (NEC Art. 725). Connections of a one-door or a two-door system are shown in Fig. 19-43. For a one-door system, current flows from the transformer to the pushbutton; when the pushbutton is closed, current flows to the bell, through the bell and back to the transformer.

A combination bell and buzzer (one signal for the front door and one for the back door), using a transformer, is shown in Fig. 19-44(a). When the front-door pushbutton is closed, current flows from the transformer through the switch to the bell, and back to the transformer. The back-door circuit is from the

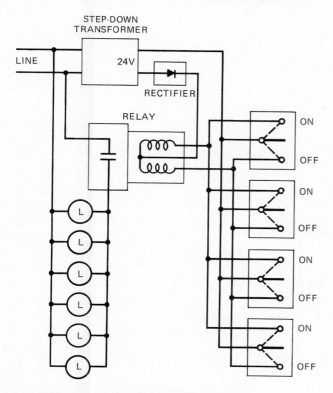

Fig. 19-37 Low-voltage lighting-control system.

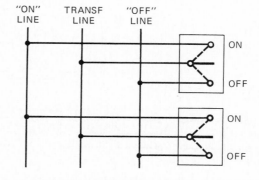

Fig. 19-38 Adding additional switches to a low-voltage system.

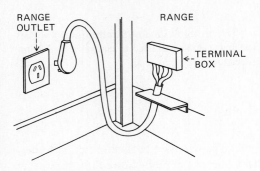

Fig. 19-39 Typical range outlet and plug installation.

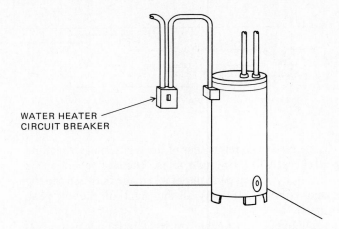

Fig. 19-40 Wiring an electric water heater through a circuit breaker at the heater.

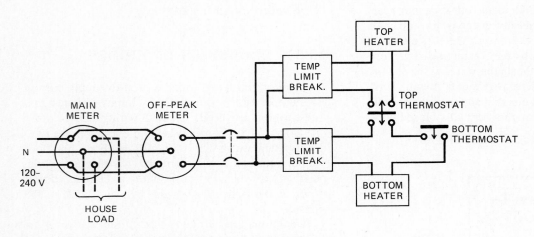

Fig. 19-41 Connections for off-peak water heating only.

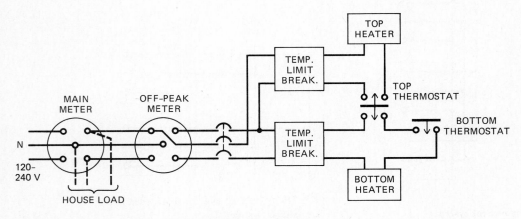

Fig. 19-42 Connections for off-peak and emergency water heating.

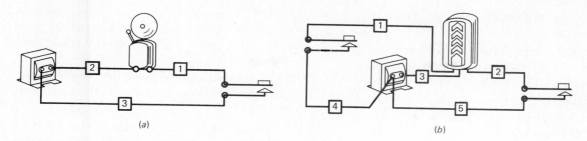

19-43 Doorbell wiring. (*a*) One-door system. (*b*) Two-door system.

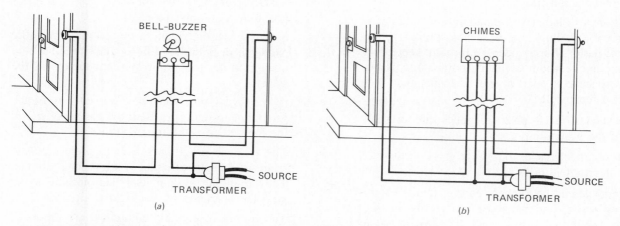

Fig. 19-44 Door signal systems. (*a*) Front doorbell and back door buzzer signals. (*b*) Four-tone front-door and one-tone back-door chimes.

transformer, through the pushbutton and to the buzzer, and return to the transformer. One line is common to both signal devices.

A chimes system using a transformer sounds one note from the back door and four notes from the front door [Fig. 19-44(*b*)].

19-35 TESTING A WIRING JOB

When the rough-in operations are completed, the system should be tested for grounds and short circuits before the inspector is called, and it should be tested again before final inspection. Occasionally, a nail is

driven into a cable by a carpenter during the installation of the interior finish of a building. A direct-reading ohmmeter or a battery and bell are suitable test instruments. To prepare for the test, close all switch loops (if switches have not been wired, twist the switch wires together), *disconnect any load,* such as signal or control transformers, and be certain that no bare wire ends are in contact with a box.

To test, connect a test lead to the neutral bar in the service cabinet and touch the other test lead to each branch-circuit wire where it is connected to the overload device and each ungrounded service-entrance wire. If a reading or signal results, the circuit under test *has a short circuit or a ground.*

SUMMARY

1. A well-planned and wired electrical system will serve efficiently and safely and provide for additional future loads.

2. The National Electrical Code provides standards for the planning and installation of reasonably safe electrical systems.

3. Electricians should be acquainted with local codes before doing any electrical work.

4. An efficient electrician should be skilled in the use of the tools of the trade, know and use electrical trade terms correctly, and be able to identify and specify the wiring materials necessary for the job.

5. Before starting a residential wiring job, the electrician should have a floor plan drawn to scale.

6. Care should be taken to ensure that the electrical system will be adequate and convenient.

7. The number and loading of branch circuits should be carefully designed.

8. In single-phase, three-wire systems, the 120-V loads should be balanced as evenly as possible between each of the hot legs and the neutral.

9. Separate branch-circuit panels fed from the service cabinet should be considered for large residences. These panels could allow shorter branch-circuit runs, reduced voltage drop, and a more convenient physical location for resetting breakers or replacing fuses.

10. Receptacles shall be installed in every kitchen, family room, dining room, breakfast room, living room, parlor, library, den, sun-room, recreation room, and bedroom so that no point along usable wall space 2 ft wide or more is more than 6 ft from a receptacle.

11. Receptacles on 15- or 20-A branch circuits shall be of the grounding type.

12. Bathroom receptacles must have ground-fault circuit interrupters (GFCI).

13. Receptacles in the kitchen, pantry, family room, dining room, and breakfast room must be fed from small-appliance branch circuits rated at 20-A each and wired with No. 12 wire. At least two such circuits are required in each residence.

14. A grounding receptacle can be grounded by mounting it in a box that is grounded and then connecting a jumper between the receptacle-grounding terminal and the box.

15. Location of receptacles and outlets are an important part of lighting design since residential lighting depends heavily on table, floor, and other types of portable lamps.

16. Residential lighting sources are generally incandescent, fluorescent, and high-intensity discharge lamps.

17. The basic unit of light intensity is the footcandle. The metric unit for this quantity is the candela.

18. Heavy appliances such as electric ranges, clothes dryers, central air conditioners, space heaters, and water heaters usually require 120/240-V or 240-V circuits and special receptacles.

19. Wire sizes are determined by the load in amperes, insulation requirements, and length of run.

20. Although No. 14 wire can be used for branch-circuit wiring with 15-A fuse or circuit breaker protection, No. 12 branch-circuit wiring is preferred.

21. The National Electrical Code offers methods for calculating the size of service feeders.

22. The three chief factors in determining the location of the service cabinet are accessibility, load center, and location of the service-entrance conductors.

23. The electric utility makes the final determination of where the service entrance to the building is located.

24. Service entrance can be made by conduit or cable. Some local codes require rigid metallic conduit.

25. Service grounding may not rely solely on connection to a cold-water pipe.

26. Service grounding requires that all major metallic elements of a building be bonded together and connected to the grounding conductor of the system.

27. A metal gas pipe may be used as part of the grounding system, provided the gas supplier and the local code authorities approve.

28. The purpose of grounding is to protect people from the dangers of lethal voltages and currents. The grounding system itself has nothing to do with the operation of electrical equipment.

29. The frames of portable equipment are grounded by use of three-wire cords and a grounding receptacle.

30. The maximum number of conductors entering an outlet or junction box is governed by the size of the wires, the number of accessories in the box such as clamps, hickeys, and studs, the number of devices (switches and receptacles), and the interior volume of the box.

31. Two types of toggle switches in general use are the ac general-use snap switch and the ac/dc general-use snap switch.

32. A T-rated switch may be loaded to its full rating with tungsten filament lamps.

33. The neutral wire must not be switched unless all hot wires are switched simultaneously.

34. Wires and cables are classified according to their insulation: type R is rubber; type T is thermoplastic; type A is asbestos.

35. In wiring a house, the electrician's knowledge and judgment are sought in planning a safe, efficient, and adequate electrical system.

36. A system should be checked for grounds and short circuits before the rough-in inspection is made.

37. Once a job receives an OK rough-in tag, the electrician returns to finish the job after coordinating the work with the other trades.

38. Large or heavy lighting fixtures must not be supported by the screw shell of a lampholder.

39. Three-way switches are used to control lighting from two locations.

40. Four-way switches are used in conjunction with three-way switches to control lighting from three or more locations.

41. For switching lighting from four or more locations, consideration should be given to the use of low-voltage systems. These systems consist of a step-down transformer, low-voltage relays, a rectifier, and momentary contact rocker switches.

42. Electric ranges and clothes dryer are usually installed simply by providing a suitable receptacle.

Ranges usually require a 50-A receptacle; dryers usually require a 30-A receptacle. Service is generally 120/240 V, three-wire.

43. Electric water heaters usually require a 240-V circuit. Some heaters are wired to provide off-peak metering.

44. Doorbells and chimes generally require a low voltage of 12 to 24 V. A stepdown signal or bell transformer can be mounted directly on a ceiling or wall box.

45. After the wiring job has been completed and before final inspection, the system should be tested for grounds and short circuits.

QUESTIONS

19-1 What are the disadvantages and dangers of a poorly planned and wired residential electrical system?

19-2 Define the following terms: (a) ampacity; (b) branch circuit; (c) feeder; (d) grounding conductor; (e) service conductors.

19-3 What is the only type of plug fuse allowed by the NEC? What is a safety feature of this type of fuse?

19-4 What are some of the building features a floor plan should show if it is to be used for planning a residential wiring system?

19-5 What is the maximum number of receptacles in one circuit permitted by the NEC?

19-6 Why should the loads of the 120-V circuits in a 120/240-V branch circuit panel be divided as evenly as possible between each of the two hot legs and the neutral?

19-7 What are some of the advantages of using a separate branch-circuit panel?

19-8 True or false: The maximum distance allowed by the NEC between general-purpose receptacles mounted along a wall is 10 ft. Explain your answer in terms of the NEC section covering this type of installation.

19-9 What type of protection must bathroom receptacles have in addition to branch circuit protection?

19-10 What is the overcurrent protection required of small-appliance branch circuits in kitchens?

19-11 What are the three most common light sources in homes?

19-12 What unit is used to specify lighting intensity?

19-13 What is the preferred minimum wire size for branch circuit conductors? What is the maximum overcurrent protection allowed for this conductor?

19-14 The optional method of NEC Article 220, Part C, is used to calculate the minimum service feeder size for a single-family house. The total load (air conditioning and "all other") was found to be 40 kW. What is the calculated load on which the feeder size will be based, in terms of kilowatts? If the electrical service is 120/240-V three-wire, what is the balanced current in each hot leg? What is the minimum size feeder required to handle the current, assuming THW copper wire is used?

19-15 Using the Code recommendation, how many 120-V circuits should be provided for a residence containing 2000 ft^2 of floor space if the home has a kitchen, bathroom, and laundry room?

19-16 Why is an adequate grounding system essential in a home?

19-17 How is the metal case of a portable electric drill grounded?

19-18 What is the minimum volume required for a metal switch box containing one switch and three No. 12 THW conductors?

19-19 Describe the difference between an ac general-use snap switch and an ac/dc general-use snap switch.

19-20 Are T-rated switches permitted in fluorescent lighting circuits? Explain your answer.

19-21 What do each of the following letters denote in wire designations: (**a**) T; (**b**) N; (**c**) W; (**d**) R; (**e**) H?

19-22 Describe three methods of securing a good mechanical and electrical connection of two wires.

19-23 In what type of wiring job is a three-way switch used?

19-24 In what type of wiring job is a four-way switch used?

19-25 Show how you would control a light from four separate locations using three-way and four-way switches.

19-26 What is a low-voltage lighting control system?

19-27 What is the purpose of a rectifier in a low-voltage system?

19-28 Show how you would control a light from four separate locations using a low-voltage system.

19-29 What is the advantage of off-peak metering for an electric hot water heater?

19-30 What type of test should be performed before the final inspection?

20 INDUSTRIAL WIRING

Industrial wiring refers to electrical wiring in other than residential installations. Nearly all such wiring is contained in raceways installed in a building during construction. Conductors may consist of building wire, cable systems, and busways or combinations of two or three of them. Service-entrance conductors may be brought in by overhead or underground feeders. Motors are an important part of any industrial electrical load.

20-1 RACEWAYS

Raceways are used to provide mechanical protection, easy access for rewiring, support, and a means of identification, among many other uses. Raceways may be made of metal or plastics and include such components as conduit, ducts, wireways, troughs, and metal trays. The choice of the best type of raceway to use in a particular application depends on a number of factors including flexibility, meaning the ability to accommodate expanding or changing systems; safety, both to humans likely to be exposed to the system and to equipment; cost; reliability and endurance; availability and ease of installation; and adaptability to the environment to which it will be exposed.

20-2 RIGID CONDUIT

Rigid conduit shown in Fig. 20-1 is a pipe specially manufactured and finished to allow electrical wiring to be run through it. Although it may resemble water pipe in general appearance, it is not the same as water pipe, and water pipe must not be used to run wires. Metal conduit is made of softer steel than water pipe to permit easier bending, as in Fig. 20-2(a). In addition, conduit is finished internally to provide a smooth surface for pulling wires [Fig. 20-2(b)].

Rigid metal conduit is usually specified as steel or aluminum, though it is available in other metals including wrought iron and copper alloys for special applications. Steel conduit may be galvanized (zinc coated) or enamel coated. If the conduit is to be exposed to severe corrosive environments, it may be coated with polyvinyl chloride (PVC), a plastic material.

Rigid metal conduit is available in the same trade sizes as water pipe. The trade sizes are only nominal since the internal and external diameters are somewhat different, as shown in Fig. 20-3. Standard lengths of conduit may be purchased with tapered pipe threads cut at each end. However, conduit may be cut, reamed, and threaded on the job with standard plumbing or pipe-fitting tools. Some threading dies are equipped to release the taper guides to allow the cutting of a straight thread. A straight thread provides a better means of connection to panels or boxes with the use of bushings and locknuts. Straight threads or running threads are not permitted by the Code for connections at couplings (NEC 346-9b). Rigid metal conduit provides the maximum mechanical protection for wires, and it is an excellent protection against shock hazards since it makes a good electrical path to ground.

Fig. 20-1 Rigid conduit. (*Wheatland Tube Co.*)

Article 346, Rigid Metal Conduit, of the NEC covers the use, installation, and construction specifications for this type of raceway.

20-3 INTERMEDIATE METAL CONDUIT

Intermediate metal conduit (IMC) is lighter weight than rigid metal conduit but still thick enough to take a tapered pipe thread. Trade sizes for IMC are 1/2 to 4 in. As with rigid conduit, standard lengths are available commercially (usually 10 ft) with threads at both ends. Field cutting, reaming, and threading are permitted, but running or straight threads are not permitted by the Code for coupling connections.

Although the outside diameter for a given trade size of IMC is the same as that for rigid conduit, the inside diameter of IMC is greater. This results in the lighter weight of IMC but it also allows easier cable pulling. When properly coated with a protective material, IMC may be installed in concrete or underground.

Article 345, Intermediate Metal Conduit, of the NEC covers the definition and uses of IMC and its installation and construction specifications.

20-4 ELECTRICAL METALLIC TUBING

Electrical metallic tubing (also called EMT or thin wall) is a lightweight steel pipe or tube too thin to be threaded (Fig. 20-4). Although EMT provides some mechanical protection for wiring and a fair conducting path to ground, the Code limits the use of EMT to installations that do not present the possibility of severe damage, either physical or corrosive.

Article 348, Electrical Metallic Tubing, of the NEC covers the use, installation, and construction specifications of EMT.

20-5 RIGID NONMETALLIC CONDUIT

Nonmetallic conduit is manufactured from many materials including PVC, fiber, asbestos cement, fiberglass, epoxy, and high-density polyethylene. Conduits made from these materials are used where conditions of extreme corrosion, rust, electrolysis (an electrochemical process), or physical damage prevent the use of metallic conduit. Most nonmetallic conduit

is lighter than the equivalent metallic conduit. Larger sections can therefore be handled easier. Nonmetallic conduit can be installed to operate at very high and low temperatures. It can also be used in aboveground installations, 600 V or more, without additional protection, such as concrete encasement.

(a)

(b)

Fig. 20-2 Using rigid metallic conduit. (*a*) Rigid metallic conduit can be bent to shape on the job. (*Greenlee Tool Co., a subsidiary of Ex-Cell-O.*) (*b*) Pulling wire in rigid conduit with a "fish" tape. (*Ideal Industries, Inc.*)

TRADE SIZE	INTERNAL DIAMETER (IN.)	EXTERNAL DIAMETER (IN.)
$\frac{1}{2}$	0.622	0.840
$\frac{3}{4}$	0.824	1.050
1	1.049	1.315
$1\frac{1}{4}$	1.380	1.660
$1\frac{1}{2}$	1.610	1.900
2	2.067	2.375
$2\frac{1}{2}$	2.469	2.875
3	3.068	3.500
$3\frac{1}{2}$	3.548	4.000
4	4.026	4.500
5	5.047	5.563
6	6.065	6.625

Fig. 20-3 Trade sizes of rigid metallic conduit. All dimensions are nominal.

Polyvinyl chloride conduit is often used above ground where metallic conduit might be used (Fig. 20-5). This type of conduit is available in two wall thicknesses specified as Schedule 40 (heavy wall) and Schedule 80 (extra-heavy wall). They are available in the same trade sizes as metallic conduit. Standard lengths and bends are available commercially. Field cutting and reaming of PVC conduit can be done easily. Field bending of PVC conduit is done by heating the conduit with hot air or liquid or in electrically heated boxes. This softens the plastic which may then be bent by hand or by using a special jig. For 2-in. conduit and larger, special care must be taken to prevent the cross-sectional area at the bend from being reduced. This can be done by using special flexible metal inserts, filling the conduit with sand or other material, or plugging the ends to trap the air in the conduit. In the latter method, the conduit is heated, causing the trapped air to increase the pressure within the conduit. This pressure then tends to maintain the cross-sectional area of the section being bent. After being bent, the conduit is cooled by means of a wet sponge or rags. The cooled PVC then resumes its rigid state.

Fig. 20-4 Electric metallic tubing (EMT) is too thin to thread. (*Wheatland Tube Co.*)

Fig. 20-5 Polyvinyl chloride (PVC) conduit. (*CertainTeed Corp.*)

Since nonmetallic conduit cannot serve as a ground conductor, an extra green grounding conductor of the appropriate size must be added for most systems. This conductor is usually run inside the conduit, though in some cases it can be run along and outside the conduit.

Article 347, Rigid Nonmetallic Conduit, of the NEC covers the description, uses, installations, and construction specifications of rigid nonmetallic conduit of all types.

20-6 FLEXIBLE METAL CONDUIT

Flexible metal conduit (also called *flex*) is used to facilitate running conduit around obstructions that would be too time-consuming to bend rigid conduit to fit. Flexible conduit is also used to absorb vibrations or irregular movement of equipment requiring permanent electrical connections (Fig. 20-6). The flexibility is achieved by wrapping metal ribbons spirally and interlocking adjacent spirals to form a continuous sealed tube. The metals commonly used are galvanized steel and aluminum. Internal flexible plastic tubing is sometimes provided as a lining for flex. External coverings of metal braid or PVC plastic are also available. These additional coverings allow flexible metal conduit to be used in applications requiring moistureproof or liquid-tight protection.

Although the metal spiral of flex is continuous and capable of carrying current, its resistance is relatively high and the continuity of the circuit is not positive at the various fittings used. For that reason, a separate green grounding conductor run inside the conduit is necessary.

Article 349, Flexible Metallic Tubing, Art. 350, Flexible Metal Conduit, and Art. 351, Liquid Tight Flexible Conduit, of the NEC cover the definition, uses, construction, and installation of flexible conduits.

20-7 OTHER RACEWAY SYSTEMS

Surface Metal Raceways This type of raceway is used for exposed wiring where the possibility of severe

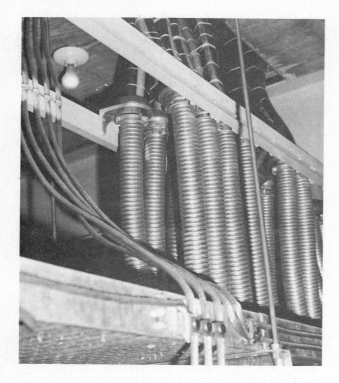

Fig. 20-6 Flexible conduit can bend around obstructions easily.

physical damage is not a problem. Surface raceways are restricted to dry locations and voltages under 300 V.

The principal use of surface raceways is for rewiring or extending existing electrical systems as shown in Fig. 20-7. The raceways include a number of different components that permit right-angle turns, tee connections, and terminations in receptacle and switch boxes and fixtures. Some surface raceway systems permit the location of outlets on continuous strips at regular intervals. Article 353 specifically

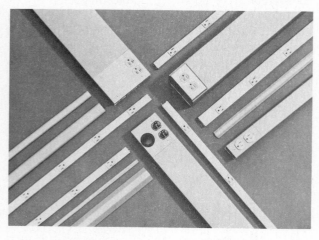

Fig. 20-7 Surface metal raceway. *(Walker)*

Fig. 20-8 Conduits terminating in a square wireway.

covers these multioutlet assemblies. Article 352, Surface Raceways, of the NEC describes the uses and installation of both metal and nonmetallic raceways.

Wireways Commercially available wireways are ducts with square or rectangular cross sections made of sheet metal, as shown in Fig. 20-8. The standard length of each duct is 10 ft. The wiring in the duct is readily accessible through cover plates which make up one of the walls of the duct. The cover plates may be hinged or unhinged, screwed in place, or merely snapped into place. Thus the entire length of duct may be easily accessed for wiring, and wires need only be laid in place without removing the other wires in the duct. Square duct is available in 2 1/2 × 2 1/2, 4 × 4, 6 × 6, and 8 × 8 in. cross sections. Rectangular duct is 4 × 6 in. Nonstandard sizes and lengths may be ordered or made in the field to fit in particular locations.

Wireways cannot be buried, concealed in walls, or exposed to corrosive atmospheres. Raintight ducts are available for use outdoors, though, in general, wireways are mounted exposed indoors.

More elaborate wireway layouts may be built from standard tees, elbows, reducers, adapters, and connectors.

Some of the important advantages of using wireways is that up to 30 current-carrying conductors may be run in the raceway and splices and taps are permitted in the raceway. Wireways may carry systems rated at 600 V or less.

Article 362, Wireways, of the NEC covers the uses and installation of wireways.

Underfloor Raceways There are many systems that may be classified as underfloor raceways, though the NEC differentiates between underfloor raceways, cel-

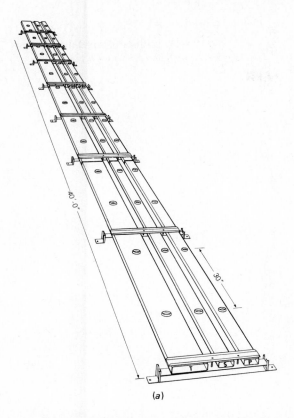

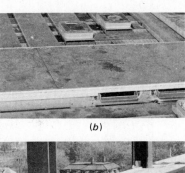

(b)

(c)

(a)

Fig. 20-9 Underfloor raceway systems. (*a*) **A three-duct system for power, telephone, and signal wiring.** (*H. H. Robertson Co.*) (*b*) **A trench duct and connecting ducts installed over a metal cellular floor.** (*c*) **Concrete cellular flooring.**

lular metal floor raceways, and cellular concrete floor raceways. Underfloor raceways, called *underfloor ducts,* consist of separate duct systems buried in the concrete floor or flush with the surface of the floor (Fig. 20-9). Such systems come complete with junction boxes and fittings to provide access along the length of the duct for receptacles and telephone outlets. Some systems that are flush with the floor are provided with continuous cover plates that are screwed into place. When the covers are removed, the entire duct, or trench, is available for wiring. Underfloor duct may consist of single, double, or triple ducts run parallel to provide telephone, signal, and power raceways.

Cellular floor systems, both metal and concrete, are integral parts of a building's structural design. The cells of the cellular floor system may be assigned particular usage such as for power or signal wiring. Header ducts tap into the cells and carry the wiring to the necessary panelboards or boxes.

Article 354, Underfloor Raceways, Art. 356, Cellular Metal Floor Raceways, and Art. 358, Cellular Concrete Floor Raceways, of the NEC cover the uses, installation, and construction specifications of these raceways.

20-8 RACEWAY FITTINGS

Most of the conduit and other raceway systems consist of standard straight lengths as well as factory-built bends of 45° and 90°. However, various types of fittings are available for joining, tapping, and terminating the systems.

Rigid Conduit Fittings Exposed conduit uses fittings to access wires for pulling and splicing. Such fittings are usually cast and contain covers screwed in place. They are usually supplied with inside threads. They are available in standard shapes for particular uses. Straight-through, tees, right- and left-angle ells, and dozens of other shapes allow conduit runs to avoid difficult bends. Types of rigid-conduit fittings are identified by type letters. Type letters usually indicate the use of the fittings. Some commonly used fittings, with type letters, are shown in Fig. 20-10. Beginning at the left, the first fitting is an *LL*. These type letters mean that the fitting is an ell that turns left, viewed from the open side with the longest run up. The second fitting, an *LB*, is an ell that turns back. The third fitting, an *LR*, is an ell that turns right. The fourth fitting is a *T* that tees off from the main run. The *C*

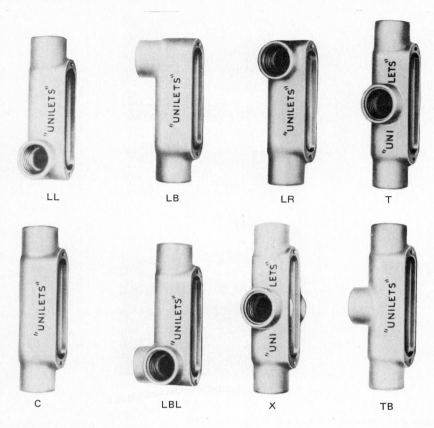

LL   LB   LR   T

C   LBL   X   TB

Fig. 20-10 Fittings for rigid metallic conduit. (*Appleton Electric Co.*)

fitting is for a continuous run. The *LBL* fitting is an ell that turns back and left. The *X* fitting is a cross. The *TB* fitting is a tee that turns back.

Rigid-conduit fittings for service-entrance use are shown in Fig. 20-11. A service-entrance ell to turn into a building is shown in Fig. 20-11(*a*). A service-entrance ell can also be open at the bottom for a grounding conduit and conductor. A service cap is shown in Fig. 20-11(*b*).

Intermediate metal conduit uses the same types of fittings as rigid metal conduit.

Electrical Metallic Tubing Fittings This type of conduit is too thin to support a pipe thread; therefore the various fittings used with EMT are either pressure-fit or clamped in place (Fig. 20-12). In order to make connections or joints in EMT, a fitting is slipped over the end of the tubing. The fitting has an external thread and nut. As the nut is tightened, the fitting squeezes against the tubing. The rest of the exposed thread on the fitting then is used with couplings, locknuts, bushings, and the like. Another type of fitting, with or without external threads, contains a sleeve that slips over the end of the EMT. A crimping tool is then used to crimp the fitting and EMT together.

PVC Conduit Fittings This type of conduit uses fittings that must be cemented in place (Fig. 20-13). A special PVC cement, actually a solvent, is spread over the surface that will accept the fitting. The cement dissolves a thin coating of the PVC. When the two parts are joined, the PVC sets and makes a solid, watertight bond. While this is an advantage of PVC conduit, it is also a disadvantage. This is because once it sets, the joint is permanent and cannot be easily undone. Thus changes in a PVC conduit layout usually require cutting the conduit for new fittings. Old fittings are generally thrown away.

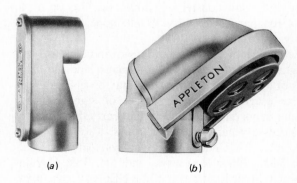

(a)   (b)

Fig. 20-11 Service entrance fittings. (*Appleton Electric Co.*)

Fig. 20-12 Fittings for EMT are either pressure fit or clamped on. (*Appleton Electric Co.*)

20-9 CONDUCTORS

Wires and cables are the basic circuit elements of an electrical system. It is their function to deliver the current to the load. Although the terms *wire* and *cable* are frequently used interchangeably, they should be distinguished from one another. Wire usually refers to building wire which is a single conductor, solid or stranded, made of copper or aluminum with an outside insulation rated at 600 V or less. Except when allowed by the NEC for low-voltage applications, wire is normally run in some type of raceway. Cables are multiple conductor assemblies. The conductors within the assembly are individually insulated, and a separate jacket, braid, or wrapping joins the individual conductors in a single unit. Cable types may be divided into low voltage (under 600 V) or high voltage (over 600 V). Special types of cable in each of these categories are available for particular applications.

20-10 BUILDING WIRE

Table 310-13, Conductor Application and Insulations, of the NEC lists practically all common building wire. The Code states that only "approved" wire can be used. This has been interpreted to mean wire listed by the Underwriters Laboratories (UL).

Copper is the most widely used conductor in building wire. In some localities the electrical code mandates the use of copper wire only. Because of a number of factors, copper makes an excellent electrical conductor. It is relatively abundant, it has one of the highest electrical conductivities, it is very ductile, re-

sistant to a wide range of corrosive materials, and readily accepts solder. In addition, the metallurgical properties of copper make it highly resistant to metal fatigue. It also deforms well under compression and therefore makes positive tight connections at screw terminals and crimped connectors.

Aluminum is often used as the conductor in larger-size wires, No. 6 and above, where it is used as main feeders. Although once popular for branch-circuit wiring in the smaller sizes (No. 14 through 8 AWG), the metal is not as forgiving as copper when less-than-perfect installation practices are performed. Older aluminum metal also has poor metal fatigue properties and can fracture even after moderate flexing.

The principal advantage of aluminum is its light weight. Although the ampacity of aluminum is less than that of copper for a given size, the extra aluminum needed to carry the same amount of current still gives aluminum a weight advantage over copper. Thus labor costs for installation of aluminum wire are less than for the equivalent capacity of copper wire. This cost advantage may be offset by the additional care required to make acceptable terminations. It is extremely important to follow the wire and wiring-device manufacturers' instructions when using aluminum wire. Antioxidants and special screws are often used with aluminum wire to slow the chemical reactions of aluminum and prevent creep due to heat-and-cold temperature cycles.

Wiring devices used with aluminum wire must be rated and marked CO/ALR, indicating that they are suitable for use with aluminum and copper wire.

Copper-clad aluminum conductors consist of solid aluminum conductors coated with a thin layer of copper. The copper does not increase the ampacity of the aluminum wire, but it does provide more reliable terminations. Copper-clad conductors are more expensive than aluminum wire. Most insulations available with copper wire are available with aluminum and copper-clad aluminum wires. Wiring devices marked CO/ALR can also be used with copper-clad wire.

20-11 WIRE INSULATIONS

Most building wires have rubber or thermoplastic insulations. Asbestos in combination with other materials is used for high-temperature installations.

Rubber, natural and synthetic, is available commercially in a range of thicknesses and grades as well as with additional outer jackets for use in locations which would expose the wire to extreme heat, oil, or solvents.

Thermoplastics melt under extreme heat, and other materials must be added to achieve improved insulat-

Fig. 20-13 Fittings for PVC rigid conduit. Cement is used to secure fittings and provide a watertight seal. (*CertainTeed Corp.*)

ing properties. Type THW wire contains an outer layer of PVC. Type THWN replaces the PVC with an outer layer of extruded nylon or polyester. This improves the moisture- and heat-resistant properties while requiring a thinner layer of insulation. The result is that more THWN conductors may be run in a conduit than the comparable size THW. Although the THWN wire may be more expensive, the savings in raceways and labor can offset this extra wire cost.

Type FEP and FEPB wires have an insulation coating of a plastic called fluorinated ethylene propylene (FEP). If an extra outer coating of braid is used, the wire is called FEPB. Type FEPB wire sizes 14 through 8 use a glass braid, while sizes 6 and larger use an asbestos braid.

20-12 ARMORED CABLE

Armored cable is referred to by the Code as types AC, ACT, and ACL. These cables consist of insulated conductors enclosed in corrugated metal tubes, interlocked spiral metal armor, or metal tape wrapping. The metal armor is usually galvanized steel, though aluminum is sometimes used. Inside the armor the insulated conductors are wrapped with paper to reduce chafing against the outer armor. While the metal spiral constitutes a continuous electrical path, its impedance is too high to serve as an effective ground. Therefore a bare solid conductor is run in the armor to provide a grounding conductor.

Special fittings are used to terminate armored cable in boxes and panel boards.

Article 333, Armored Cable, of the NEC covers the use, construction, and installation of this type of cable.

20-13 NONMETALLIC CABLES

Nonmetallic-sheathed cables are assemblies of two or more insulated conductors around which is a plastic sheath. Most nonmetallic cables also include a bare grounding conductor. Type NM cable contains a moisture-resistant flame-retardant outer sheath, while type NMC is also corrosion resistant. While the plastic sheath does provide some protection against physical damage, it can be seriously damaged by sharp objects and punctured by pointed objects. For that reason, nonmetallic cable must be protected by a metal shield if it is embedded in plaster walls. Since this cable is very light and flexible, it is extensively used for general wiring of lighting and receptacle circuits.

Article 336, Nonmetallic-Sheathed Cable, of the NEC covers the construction, use, and installation of type NM and NMC cables.

20-14 MINERAL-INSULATED, METAL-SHEATHED CABLE

Mineral-insulated, metal-sheathed cable (NEC type MI) consists of one or more conductors embedded in a compressed refractory mineral insulation which is enclosed in a liquid-tight and gas-tight continuous copper sheath. The insulation is packed around the conductors so that adequate separation is provided. This type of cable may be used in many hazardous locations, embedded in concrete, exposed to conditions that would not deteriorate the copper sheath, and used outdoors in wet locations.

The outer copper sheath serves as a good conductor for grounding purposes. Because moisture can enter the packed insulation, the exposed ends of the cable must be kept covered. When the cable is terminated, the end of the insulation must be properly sealed. In connecting the conductors of the MI cable, the sheath is first cut away. Without the outer sheath the packed inner insulation cannot protect the conductors. Therefore separate insulation must be used from the sheath to the termination point of each of the conductors.

20-15 SPECIAL-APPLICATION CONDUCTORS

Flat Cables These are assemblies of two or more insulated conductors arranged in flat parallel runs and separated by insulated material. Two types of flat cables are in general use for specific applications (Fig. 20-14).

Type FC flat cable consists of from two to four No. 10 special stranded copper wire. It is meant to be installed in metal surface raceways. Special tap devices are used to connect the cable to wiring devices to supply lighting and small-appliance and power loads.

Article 363, Flat Cable Assemblies, of the NEC covers the uses, construction, and installation of this type of cable.

Type FCC flat conductor cable is specifically designed for installation under carpet squares. The cable consists of three or more copper conductors embedded in parallel in an insulating assembly. Unlike the type FC cables, the conductors in type FCC cable are flat strips of copper which allow for a very low cross section. Type FCC cable actually consists of a complete system of specially designed connectors, terminators,

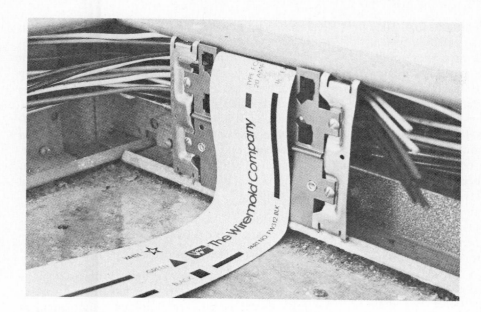

Fig. 20-14 An undercarpet flat cable
fed from a perimeter raceway system.
(*Wiremold.*)

adaptors, boxes, and receptacles. To ensure mechanical protection, metal shields are placed under and over runs of the cable where physical damage may occur. This type of cable is prohibited for use in homes, schools, and hospitals but is widely used in commercial buildings, stores, and offices.

The maximum voltage permitted between ungrounded conductors is 300 V, and individual branch circuits cannot exceed 30 A.

Article 328, Flat Conductor Cable, Type FCC, of the NEC covers the uses, construction, and installation of this type of cable.

Underground Feeder Cable This type of cable (NEC type UF) is designed for direct burial in the earth. It is commonly used as both feeder and branch-circuit wir-

ing. The cable consists of insulated conductors such as THW or RHW or other approved wire. An overall covering of flame-retardant, moisture-, fungus-, and corrosion-resistant insulation adds extra protection. In addition to the insulated circuit conductors in the cable, an insulated or bare grounding conductor may also be included.

Article 339, Underground Feeder and Branch-Circuit Cable, of the NEC covers the use and protection of this type of cable.

20-16 BUSWAYS

Busways are factory-made systems of copper or aluminum conductors designed to carry heavy currents. The conductors can be solid bars, square or rectangu-

Fig. 20-15 Busways.

lar hollow tubes, hollow ovals, or solid I-beams. They can be mounted horizontally or vertically. A typical busway consists of four busses in a three-phase, four-wire system. The four conductors are insulated from one another, and the entire assembly is surrounded by a sheet metal enclosure insulated from the conductors. The enclosure may be ventilated or unventilated.

A variety of busways are available, each designed to serve a particular function as shown in Fig. 20-15. A continuous plug-in busway is used to serve equipment that may be relocated periodically, such as in a woodworking shop. Plug-in busways have regularly spaced openings that permit plugging in switches or circuit breakers. Conduit or flexible cable is then run from these devices to the equipment being served. Switches and circuit breakers are available up to 800 A from some manufacturers.

Trolley busways permit traveling equipment to be continuously connected to a power source. A rolling power-takeoff is in contact with the busway conductors. As the equipment moves, the trolley contact rolls on the conductors. Overhead cranes and hoists use this system.

Industrial lighting fixtures may be supported and supplied by *lighting busways*. This type of busway is hung from the ceiling. The lighting fixtures are then fed from the busway by tapping the required number of conductors. In addition, the fixtures themselves are hung from the busway.

Busways are also used as service-entrance feeders. Busways are available in a wide range of current ratings from 50 to 6000 A.

Article 364, Busways, of the NEC covers the uses, installation, and construction of service-entrance, feeder, and branch-circuit busways.

20-17 SERVICE ENTRANCE

Service entrance is that part of an electrical system extending from the service equipment to the service-supply main feeders. The NEC defines the service equipment as "The necessary equipment, usually consisting of a circuit breaker or switch and fuses, and their accessories, located near the point of entrance of supply conductors to a building or other structure, or an otherwise defined area, and intended to constitute the main control and means of cutoff of the supply."

20-18 SERVICE-ENTRANCE FEEDERS

The NEC requires that the service-entrance feeders be large enough to carry the total load currents that could be drawn in a particular system. While the NEC does

allow certain types of loads to be counted at less than 100 percent (the amount of reduction is called a *demand factor*), it may be wise to use the 100 percent figure. For example, when calculating the lighting load for a storage warehouse, the NEC (Table 220-11) requires that the first 12.5 kW of connected load be calculated at 100 percent when sizing a feeder. However, all *connected* lighting loads over 12.5 kW need be calculated only at a demand factor of 50 percent. This means that if a warehouse has a lighting load of 30 kW, only the following load need be considered when sizing a feeder:

First 12.5 kW at 100% = 12.5 kW
Remaining 17.5 kW at 50% = 8.75 kW
Total load used to size feeder = 21.25 kW

The reasoning, of course, is that it is unlikely that at any given time the entire connected lighting load is going to be turned on. At this point a design judgment must be used to decide whether it is prudent to take advantage of the reduction the Code permits.

Article 230, Services, of the NEC covers the number of services permitted, the type and size of service-entrance conductors, types and installation of service-entrance equipment, and grounding requirements.

20-19 MOTOR WIRING

The NEC sets forth numerous provisions for the installation of electric motors and controllers of all types under nearly all conditions. The contents of the Code as it applies to motors and controllers are important reading for all industrial electricians and technicians.

Article 430 of the NEC pertains to motors, motor circuits, and controllers. Each part of Art. 430, lettered A through L deals with a certain phase of motor installation.

The main factors for consideration in a motor installation are size of branch-circuit conductors and conduit, branch-circuit overcurrent protection, system disconnecting means, motor controller, and motor overcurrent protection.

To illustrate how Code regulations and tables are used in determining the ratings of components in a motor circuit, the parts of Art. 430 and tables involved will be cited in discussing an assumed installation of a three-phase 5-hp 40°C 230-V, continuous-duty squirrel-cage induction motor marked Code letter F and with a service factor of 1.15. The full-load nameplate current is 14 A. This motor is to be started on full voltage with a magnetic starter containing motor-running overcurrent protection devices.

To determine the maximum size of branch-circuit conductors and motor-running overcurrent protection devices and size of conduit, the current requirements of the motor will have to be found.

Current Ratings Current requirements for branch circuits are discussed in Part A of NEC Sec. 430-6, which states that for branch-circuit calculations, current requirements for all types of motors are to be taken from NEC Tables 430-147 through 430-150. Current values for motor-running overload protection calculations are taken from the motor nameplate (NEC Sec. 430-6-*a*). Air-conditioning and refrigeration motors are covered in NEC Art. 440.

For the 5-hp, 230-V, three-phase induction motor, NEC Table 430-150 specifies a full-load current of 15.2 A.

Branch-Circuit Ampacity Part B, Sec. 430-22-*a*, of the NEC requires branch-circuit conductors for a single motor to have an ampacity of not less than 125 percent of the full-load current rating of the motor. For the motor in the above problem

$$\text{Branch circuit ampacity} = 125\% \times 15.2$$
$$= 1.25 \times 15.2$$
$$= 19 \text{ A}$$
$$\text{Use 20 A.}$$

As the ampacity tables in Chap. 3 for THW wire show, No. 12 AWG with an ampacity of 20 A would be acceptable provided only three conductors are run in the conduit. Of course, a voltage-drop calculation should be made to determine that the branch-circuit voltage drop is acceptable according to the Code.

For conduit size, NEC Table 3A shows that up to four No. 12 THW conductors can be placed in 1/2-in. conduit.

Branch-Circuit Overcurrent Protection For branch-circuit overcurrent protection NEC Table 430-152 shows that the maximum rating of an inverse time branch-circuit circuit breaker for a three-phase (polyphase) squirrel-cage motor with Code letter F for full-voltage starting is 250 percent of the full-load motor current, or 175 percent of full-load motor current for dual-element (time-delay) fuses. The full-load current is the value determined from Table 430-150.

For the circuit breaker

$$\text{CB rating} = 250\% \times 15.2 \text{ A}$$
$$= 2.5 \times 15.2$$
$$= 37.5 \text{ A}$$
$$\text{Use 40 A breaker.}$$

For the fuse

$$\text{Fuse size} = 175\% \times 15.2 \text{ A}$$
$$= 1.75 \times 15.2$$
$$= 26.6 \text{ A}$$
$$\text{Use 30 A fuse.}$$

Motor overload protection is provided for in Part C of Art. 430. Part H of NEC Art. 430 provides for disconnecting means for the motor, controller, and branch circuit.

Conductors Supplying Several Motors Conductors supplying two or more continuous-duty motors are required by the Code to have an ampacity equal to the sum of the full-load current ratings of all motors plus 25 percent of the current of the highest-rated motor in the group (Sec. 430-24).

Feeder overcurrent protection for a feeder supplying more than one fixed motor load, with branch-circuit protection based on NEC 430-24, cannot be greater than the highest rating of the branch-circuit protective device in the group, plus the sum of the full-load currents of the other motors, based on Table 430-152 (NEC 430-62).

Problem

Calculate the size of the conductors supplying continuous-duty motors having the following ratings:

System: three-phase, 230 V, 60 Hz
Motors: three, 230 V, three-phase.
Two motors—15 hp each; squirrel-cage induction; full-voltage starting; service factor 1.15; Code letter F.
One motor—30 hp; wound-rotor induction.

Solution

From NEC Table 430-150 the full-load currents of the motors are:
Two 15-hp squirrel-cage motors = 42 A each
One 30-hp wound-rotor motor = 80 A

$$\text{Feeder ampacity} = 42 + 42 + 80 + 80 \times 0.25$$
$$= 184 \text{ A}$$

If THW is used, the wire size would be 000 (3/0).

Code Letters Code letters appear on ac motors 1/2 hp and larger to indicate locked-rotor input of power (starting current) when a motor starts. The value of Code letters in kilovoltamperes per horsepower with locked rotor for motors is given in NEC Table 430-7-*b*. Code letters are used for determining maximum branch-circuit protection by reference to NEC Table 430-152.

SUMMARY

1. Industrial wiring is nearly always contained in raceways.

2. Raceways provide protection for wiring against physical damage.

3. Examples of raceways are conduits, ducts, wireways, wiring troughs, and trays.

4. Rigid metal conduit resembles water pipe, but electrical conduit is made of softer steel to facilitate bending and is finished internally for easier wire and cable pulling.

5. Rigid conduit is available in trade sizes from $\frac{1}{2}$ to 6 in. Internal and external dimensions are somewhat different than the trade sizes.

6. Rigid conduit is available in standard 10-ft lengths, threaded with a standard pipe thread at each end.

7. Rigid metal conduit is a good low-impedance grounding conductor.

8. Intermediate metal conduit (IMC) has a thinner wall than rigid metal conduit but thick enough to take a standard pipe thread. Trade sizes are 1/2 to 4 in. The thinner wall results in a larger internal area, thus allowing more space for conductors.

9. Electrical metallic tubing (EMT) is thin-wall pipe or tubing too thin to be threaded. It is not as good as rigid conduit for use as a grounding conductor.

10. Rigid nonmetallic conduit may be made of polyvinyl chloride (PVC), fiber, asbestos, cement, fiberglass, epoxy, or high-density polyethylene.

11. Polyvinyl chloride conduit is used above ground where metallic conduit is used. The two standard wall thicknesses for PVC conduit are Schedule 40 and Schedule 80.

12. Bending of PVC is facilitated by using a hot box to first soften the conduit.

13. Nonmetallic conduit requires that a separate grounding wire be run with any system using this type of conduit.

14. Flexible metal conduit (or flex) is used to facilitate conduit runs around structural obstructions. It is also used to absorb movement and vibration.

15. Flexible conduit is laid in place first, then wires are pulled through the conduit.

16. Flexible conduit is made of either galvanized steel or aluminum. Because the metal presents a high-impedance path, electrical systems employing flexible conduit usually carry an extra grounding conductor.

17. Surface metal raceways are used exposed, often for rewiring or expanding existing systems.

18. Wireways are completely enclosed duct systems in which wires are placed. A removable cover plate provides access inside the entire duct.

19. Duct shapes may be square or rectangular. The standard length of a duct section is 10 ft. Duct cross sections are $2 1/2 \times 2 1/2$, 4×4, 6×6, and 8×8 in. for square ducts and 4×6 in. for rectangular ducts.

20. Underfloor raceway systems can be classified as underfloor duct, cellular metal floor raceways, or cellular concrete floor raceways.

21. Cellular floor systems, both metal and concrete, are integral parts of a building's structural design, whereas underfloor duct is a separate raceway system installed specifically for electrical and other purposes.

22. Raceways include many fittings and devices designed to facilitate layout, installation, rewiring, and expansion.

23. Fittings for rigid metallic conduit and IMC are made to be screwed together using standard pipe thread.

24. Fittings for EMT are clamped or crimped together.

25. Fittings for PVC conduit must be cemented together. The bond created by the cement is permanent and cannot be easily disassembled.

26. Building wire may be copper or aluminum, stranded or solid. Stranded wire is used in the larger sizes to make bends and pulling easier and to reduce cracking due to metal fatigue.

27. Building wire is usually specified by the type of insulation covering it. The term *approved* wire as used by the NEC has been interpreted to mean wire listed by Underwriters Laboratories (UL).

28. Only wiring devices marked CO/ALR may be used with aluminum wire. These devices can also be used with copper and copper-clad wire.

29. Most building wire uses rubber or thermoplastic insulation. Popular insulation types are RHW, THW, and THWN.

30. Cable consists of more than one conductor, each of which is individually insulated; the entire assembly is then joined with an outer jacket, braid, or wrapping.

31. Nonmetallic cables consist of separate insulated conductors around which is a plastic sheath.

32. Mineral-insulated, metal-sheathed cable consists of conductors embedded in a compressed refractory (clay) material which is enclosed in a liquid-tight and gastight continuous copper sheath.

33. Flat cable (type FC) consists of two or more insulated conductors run in a flat parallel assembly and separated by insulation.

34. Flat conductor cable systems (type FCC) consist of flat strips of copper run parallel and embedded in insulation. This cable is designed specifically for installation under carpeting.

35. Underground feeder cable (type UF) is designed specifically for direct burial in earth.

36. Busways are factory-made systems of aluminum or copper conductors designed for heavy currents and flexible access.

37. Busways may be plug-in type, trolley busways, lighting busways, or service-entrance busways.

38. Service-entrance feeders are those conductors that run from the utility poles or main supply feeders to the service equipment.

39. Motor loads are an important part of industrial wiring systems.

40. Article 430 of the National Electrical Code pertains to motors, motor circuits, and controllers.

QUESTIONS

20-1 Why are raceways used?

20-2 What factors are considered when choosing a suitable raceway?

20-3 How does rigid metal conduit differ from water pipe?

20-4 What is the smallest trade size of rigid metal conduit?

20-5 Can IMC be threaded? If not, how are fittings connected? If so, what type of thread is used?

20-6 Can EMT be used in place of IMC for all types of installations?

20-7 Which is the heavier PVC conduit, Schedule 40 or Schedule 80?

20-8 What are the trade sizes of PVC conduit?

20-9 How is PVC conduit bent?

20-10 Why must a separate grounding conductor be used in most PVC conduit systems?

20-11 How does flexible conduit differ from armored cable?

20-12 What is the principal use of surface raceway?

20-13 What are some advantages of wireways?

20-14 How do cellular metal floor raceways differ from underfloor duct?

20-15 Describe the process for joining a PVC coupling to PVC conduit.

20-16 How are EMT fittings attached to the tubing?

29-17 How does wire differ from cable?

20-18 List some of the reasons why copper is favored over aluminum for branch-circuit wiring.

20-19 What is copper-clad wire?

20-20 What marking indicates that a wiring device is approved for both aluminum and copper wire?

20-21 What do the letters R, T, H, W, and N denote when used to describe wire insulation?

20-22 Does the metal armor of armored cable serve as a low-impedance grounding conductor?

20-23 What are some of the dangers of installing nonmetallic sheathed cable behind walls or imbedded in plaster?

20-24 Why must the ends of mineral-insulated, metal-sheathed cable be sealed?

20-25 How do flat cables (type FC) and flat conductor cable (type FCC) differ?

20-26 What type of busway would be used in a woodworking shop in which power tools are constantly being relocated? Explain your answer.

20-27 Define a service-entrance feeder.

20-28 What article of the NEC covers motor circuits and controllers?

20-29 Must service-entrance feeders be sized to carry the total connected load in a warehouse installation? Explain your answer.

20-30 A 10-hp squirrel-cage motor rated at 208 V, 60 Hz, 3ϕ has a service factor of 1.15 and code letter F. What is the full-load current as given by NEC Table 430-150? What is the minimum feeder size allowed for this motor?

21

MAINTENANCE OF ROTATING EQUIPMENT

When two or more mechanical parts move in contact with each other, friction, heat, and wear result. The efficiency of machines is limited by these factors. The use of lubricants and other antifriction devices and methods have made possible today's highly mechanized industry. Too often lubrication duties are assigned to the most unskilled laborer, often the lowest-paid member of an organization. Lubrication duties should be performed only by specialists or those trained by specialists in the techniques of proper lubrication. Where electrical equipment is concerned, an untrained worker with a grease gun or oilcan can do more damage and cause more trouble than can be remedied by a team of efficient servicers.

21-1 BEARINGS IN ELECTRICAL EQUIPMENT

Rotating electrical equipment such as motors, generators, and rotary converters employ one of two types of bearings to support the rotating element. These two types are *antifriction* and *sleeve bearings*. Antifriction bearings are *ball bearings* or *roller bearings*. Ball bearings are used in nearly all modern rotating electrical equipment. Roller bearings are confined chiefly to slow speeds and heavy loads.

Ball bearings are capable of years of hard service if properly cared for, but the life of a ball bearing *can be reduced to only a few minutes* under improper treatment. It can be damaged in almost unbelievably simple ways which lead to premature failure.

21-2 LIFE EXPECTANCY

Ball bearings are made of specially engineered steel, tempered under carefully controlled conditions, and are machined, ground, and polished to superprecision tolerances. The life expectancy of a ball bearing is determined by its material, degree of hardness, finish, and the treatment given it during its useful life. Premature failure can be caused by mishandling anywhere between its manufacture and installation on the job.

21-3 NOMENCLATURE

A simple ball bearing with names commonly associated with bearings is shown in a cutaway view in Fig. 21-1. It consists chiefly of two grooved steel race rings with a set of steel balls equally spaced and held by a separator between the rings. In a discussion of ball bearings, it is necessary to know the names of the various parts and dimensions.

The two rings are known as the *inner ring* and the *outer ring*. Each ring contains a race for the balls; the inner ring contains the inner ball race, and the outer ring contains the outer ball race. The center opening through the inner ring is the *bore,* or inside diameter of the bearing. The outside diameter is the overall diameter of the outer race. The separator, sometimes called the *cage* or *retainer,* maintains an even spacing beween the balls.

21-4 MEASUREMENTS

Ball bearings are manufactured and measured in millimeters, except in special cases, and are therefore in-

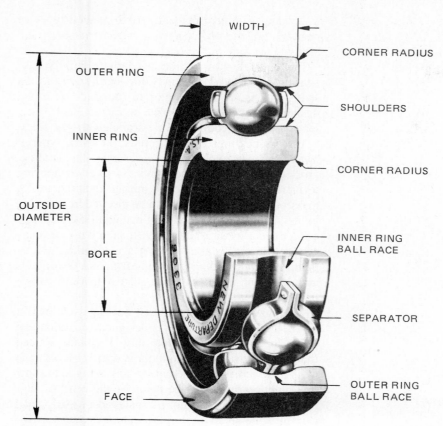

Fig. 21-1 Parts of a ball bearing.

terchangeable throughout the world. This method of measurement presents a problem to the machinist using a micrometer calibrated in decimals of an inch to measure ball bearings and machining shafts and bores for them. Table 21-1 shows the international standard dimensions for inside and outside diameters and width of medium-type bearings in millimeters and equivalents in inches to ten-thousandths of an inch. For example, in the table it will be noticed a size 00 (2/0) bearing has an inside diameter of 10 mm, or

TABLE 21-1
Standard Dimensions of Medium-Type Ball Bearings

Bearing Number	Bore		Outside Diameter		Width	
	mm	in.	mm	in.	mm	in.
00	10	0.3937	35	1.3780	11	0.4331
01	12	0.4724	37	1.4567	12	0.4724
02	15	0.5906	42	1.6535	13	0.5118
03	17	0.6693	47	1.8504	14	0.5512
04	20	0.7874	52	2.0472	15	0.5906
05	25	0.9843	62	2.4409	17	0.6693
06	30	1.1811	72	2.8346	19	0.7480
07	35	1.3780	80	3.1496	21	0.8268
08	40	1.5748	90	3.5433	23	0.9055
09	45	1.7717	100	3.9370	25	0.9843
10	50	1.9685	110	4.3307	27	1.0630
11	55	2.1654	120	4.7244	29	1.1417
12	60	2.3622	130	5.1181	31	1.2205
13	65	2.5591	140	5.5118	33	1.2992
14	70	2.7559	150	5.9055	35	1.3780

0.3937 in., an outside diameter of 35 mm, or 1.3780 in., and a width of 11 mm, or 9.4331 in.

To convert inches to millimeters, multiply the inches by 25.4. To convert millimeters to inches, divide the millimeters by 25.4.

Problem 1

A motor shaft is 0.500 in. in diameter. Find its size in millimeters.

Solution

There are 25.4 mm in 1.000 in. Thus

$$0.500 \text{ in.} \times \frac{25.4 \text{ mm}}{1.000 \text{ in.}} = 12.7 \text{ mm}$$

The shaft is 12.7 mm in diameter.

21-5 SERVICE-DUTY CLASSIFICATIONS

Ball bearings are generally made in three service-duty classifications—*light-, medium-,* and *heavy-duty.* Prefixes and suffixes are added to the bearing numbers, shown in the first column of the table, by the manufacturers to designate their types of bearings. There is no standard in the use of these prefixes and suffixes, and each manufacturer uses its own system.

21-6 TYPES OF BEARINGS

There are several types of ball bearings, each designed for a specific duty. These types should not be interchanged in replacement. Three general types of bearings are *radial, single-row angular-contact,* and *double-row angular-contact.*

Figure 21-1 shows a radial bearing. This type of bearing is designed primarily for loads *radial* to the shaft, that is, at right angles to the shaft. It is not designed to take *heavy axial* or *thrust* loads, i.e., loads in the direction of or parallel with the shaft.

Radial bearings are generally of two types—*loading groove* and *nonloading-groove.* The loading-groove bearing has a groove in its inner and outer ring to facilitate loading the balls during manufacture. A maximum number of balls can be inserted in a bearing in this manner, thus increasing its radial-load capacity but reducing its thrust capacity. It should not be interchanged in an installation with a nonloading-groove bearing. A nonloading-groove bearing has fewer balls and a higher thrust capacity than a loading-groove bearing.

A single-row angular-contact bearing is shown in Fig. 21-2(*a*). This bearing is designed for radial and thrust loads in one direction only. It can be noticed that the outer-ring race shoulder is deeper on one side to take the thrust, and the inner ring is marked to show the direction thrust can be taken by the bearing. It is essential that this bearing be installed properly, and care must be taken not to place *undue pressure* in the *nonthrust direction* in mounting or removal. When used in pairs, they are mounted opposed to each other, either butted together or placed at either end of a shaft. Care must be taken that they are properly mounted with respect to each other.

Double-row angular-contact bearings, illustrated in Fig. 21-2(*b*), are designed to take radial and thrust loads from either direction.

Fig. 21-2 Angular-contact bearings. (*a*) Single row. (*b*) Double row.

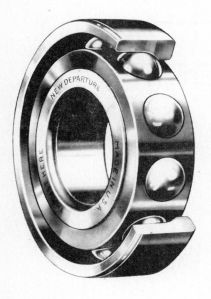

(a)

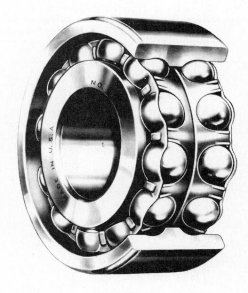

(b)

Fig. 21-3 (*a*) Open-type ball bearing.
(*b*) Shielded ball bearing.

(*a*)   (*b*)

21-7 SEALS AND SHIELDS

The greatest enemies of ball bearings are dirt, friction, and mishandling. Dirt consists of particles out of place. It can be abrasive materials such as dust from an emery grinder, concrete floor, sidewalk, or street. Moisture, acids, and gases in the air are also detrimental to bearings.

A ball bearing must be protected against dirt if normal life is expected. Open-type bearings must depend on a well-designed enclosure for protection from dirt. Other bearings contain shields in the sides that are effective in excluding dirt and retaining the lubricant under normal conditions. Figure 21-3(*a*) shows an open-type bearing, and Fig. 21-3(*b*) a bearing with one shield. The latter is used in motors that have enclosures to protect the bearing from the outside but are open on the inside. Inasmuch as the shield offers protection from dirt inside the motor, the bearing is installed with the shield toward the inside of the motor.

21-8 LUBRICATING PRECAUTIONS

Some bearings contain shields on both sides. These shields are sometimes removable for cleaning and relubrication of the bearing. In a motor, a small amount of lubricant can seep into the bearings between the shield and inner ring when a motor is running. In motors using this method of lubrication, it is essential that care be taken in adding lubricant with a pressure gun, in order to avoid breaking the shields and permitting lubricant to enter the inside of the motor and its windings.

Pure electric motor lubricant is an insulator, but lubricants in a motor attract dirt and absorb moisture, and this, in the presence of atmospheric gases, forms acids that damage the windings and cause short circuits and grounds.

Grease or oil on a commutator glazes the commutator and brushes, causing sparking, pitting, and burning of the bars. This action produces heat, which melts the solder in the lead connections to the bars; centrifugal force throws the solder out, resulting in an open circuit. In turn, an open circuit in an armature can cause rapid destruction of the commutator.

Some bearings are shielded on one side and sealed on the other, as in Fig. 21-4(*a*), and some bearings are sealed on both sides, as in Fig. 21-4(*b*). The latter type is known as a *double-sealed bearing*. Lubricant can be added to the bearing with one seal and shield from the enclosure when the bearing is running, but great care is necessary to prevent breaking the shield and seal.

A bearing with seals on both sides cannot be cleaned and relubricated. It must be replaced if it has an excessively rough feeling.

Occasionally, a new double-sealed bearing will run hotter than normal when it is started the first time. This condition is not dangerous if there is no other cause of the heating. When such a bearing channels its lubricant and cools, it should operate properly thereafter.

SHIELD

(a)

SEAL

(b)

21-9 BALL-BEARING LUBRICANTS

There is very little friction in a ball bearing in operation. Most of the friction is between the balls and separator and at the point of contact of the balls and races. When a loaded ball rolls in the raceway, a slight deformation of the ball and raceway occurs at the point of contact. This deformation process produces a slight rubbing action, which results in friction, making some lubrication necessary.

One of the main functions of a lubricant for ball bearings is protection of highly finished surfaces of the parts of the bearing. A lubricant must dissipate heat in the bearing, prevent corrosion of parts, and protect against water, acid fumes, dirt, or foreign matter of any kind.

Oil or grease is used as a lubricant in ball bearings. Grease is used where it would be difficult to retain free-flowing oil. Most bearings in electrical equipment are therefore lubricated with grease. Grease is made from a high-viscosity mineral oil to which a material is added to give it body and stiffness. Sodium soap or lithium soap is generally used as a base to give the desired body to ball-bearing grease.

Originally grease for lubricating was hog fat or inedible grades of lard, stiffened with fillers such as wax, rosin, or talc. These are not very good lubricants. Grease as referred to in this chapter, however, is more properly specified as *mineral lubricating grease.*

The most commonly used grease is sodium-soap grease. This grease is suitable for a wide range of speeds and temperatures. It affords good protection to surfaces of bearings, and channels easily. Satisfactory operating temperatures range from 30 to 200°F (-1 to 93°C).

Lithium-soap silicone grease is recommended only for extreme temperature and low-load conditions. Operating temperatures are 40 to 400°F (4 to 204°C). Silicones are sometimes used to replace lubricating oils for these temperature extremes, but in general silicones do not have excellent lubricating qualities.

It is always advisable to follow manufacturers' recommendations for lubricating their equipment. However, it is possible to reduce the number of grades, types, and classifications of grease carried in stock by consultation with a lubrication specialist.

Care of lubricants stored in a stockroom is too often neglected. Containers of lubricants should be stored in a dry place of average temperature and kept clean and tightly closed at all times.

21-10 GREASING PRACTICE

Most motors equipped for greasing with a pressure gun contain a pressure-relief plug in the bottom of the enclosure. This plug should be removed before greasing is started.

In greasing with a gun, the nozzle of the gun and the grease fitting on the motor should be thoroughly cleaned to avoid forcing dirt into the enclosure and bearing.

The motor should be running when grease is added, and only one shot inserted at a time, with a few seconds between shots until grease begins to purge through the relief-plug hole. No more grease should be added, and the motor should be allowed to run about five minutes with no evidence of heat apparent before the relief plug is replaced.

21-11 DANGERS OF EXCESSIVE LUBRICATION

Overgreasing is one of the most common causes of bearing failure and destruction of motor windings. A bearing should never be packed more than one-third full of grease. When a bearing is running, the balls must plow a way through the grease; if the bearing is too full, churning of the grease results, which in turn causes friction and heating of the grease. Heat expands the balls, causing them to run tight in the bearing and thus produce more heat. This vicious cycle can continue until the balls get too hot and their original temper or hardness is lost. In this way too much grease can ruin a bearing—sometimes in only a few minutes. In an oil-lubricated bearing, the oil level should not be above the center of the bottom ball. Excessive oil can work its way onto the commutator, slip rings, and windings. The result could be glazed rubbing surfaces, damaged insulation, and potential short circuits.

21-12 CARE OF BEARINGS

Ball bearings should be stored in a clean place free of extreme temperatures or moisture that would cause rusting of the bearing or deterioration of the lubricants.

A new bearing should be kept coated with oil at all times, since its bare finish is highly susceptible to rust and attack by acids in moisture deposited by fingers. A bearing beginning to rust from fingerprints is shown in Fig. 21-5(a).

When it is necessary to unpack a new bearing and there is a delay in installing it, it should be at least protected by keeping it wrapped with oiled paper as shown in Fig. 21-5(b). Preferably, it should be placed back in its box.

21-13 INSTALLATION

The first step in the installation of a ball bearing should be preparation of the part to receive the bearing. In the case of a shaft, it should be the proper size and should be straight, clean, and free of burrs or dents.

A perfect circle is seldom attained in grinding the inner-ring bore of a ball bearing or in machining and grinding a shaft. It is often possible, therefore, to cause a tight bearing to slide easily on a shaft simply by turning the inner ring at the beginning of the bearing seat on the shaft so as to find the position of the least tightness.

The bore of the inner ring of a ball bearing is finish-ground to a certain tolerance. This tolerance for bearings with a bore of 30 mm is generally +0.0 to

(a)   (b)

Fig. 21-5 (a) Ball bearing showing rust beginning to form owing to acids in moisture on fingers. (b) Bearing wrapped in oiled paper as protection.

−0.0004 in. (+0.0 to 0.010 mm). [This means that the bore cannot be oversize but can be 0.0004 in. (0.010 mm) undersize.] The standard tolerance for a shaft for this bearing is ±0.0002 in. (±0.005 mm). Since it is impractical to make bearings and shafts to exact dimensions, occasionally a bearing of a given size may fit perfectly. When a bearing is installed, it should turn freely with no evidence of tightness at any point. If there is evidence of tightness, the bearing will very likely overheat and cause premature failure.

21-14 INSTALLATION PROCEDURES AND EQUIPMENT

If possible, a ball bearing should be installed on a shaft with the use of a hand-operated arbor press and a clean piece of pipe, or tubing, with square ends. By this method, the operator, in the process of mounting the bearing, can "feel" any undue binding, cocking of the bearing, or tight places and can stop and make corrections before damage is done.

In case it is necessary to press a bearing on a shaft, the proper method in using a press is shown in Fig. 21-6. A clean piece of pipe or tubing with square ends is pressing against the inner ring. If a press is not available, or if the use of a press is impractical, the bearing can be driven with a piece of properly prepared pipe, or tubing, and a hammer.

Preparation of the pipe is illustrated in Fig. 21-7. A washer is welded to the lower outside of the pipe to catch dirt that may be jarred loose and tend to fall into

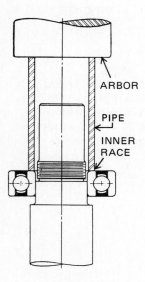

Fig. 21-6 Using an arbor press and a piece of pipe to press a bearing on a shaft.

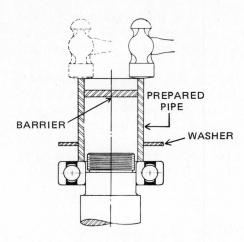

Fig. 21-7 A washer is welded to the pipe used for driving a bearing onto a shaft with a hammer.

the bearing; a barrier has been welded inside the pipe to catch dirt there. If this preparation is impractical, some means of catching dirt should be used. As a substitute, a piece of rag can be stuffed into the pipe, and a rag can be placed around the pipe next to the bearing to catch dirt. In driving, the pipe should be struck with even blows distributed all around the circumference of the pipe.

21-15 HEAT FITTING

A ball bearing can be expanded by heat to install it, but extreme care must be used to avoid ruining the temper of the steel. A ball bearing begins losing its proper temper at around 280°F (138°C). To be safe, a bearing should never be heated above 250°F (121°C). Because the highest rate of expansion of metal is at the beginning of application of heat, overheating a bearing adds comparatively little additional expansion.

A safe way of heating a bearing is to submerge it in clean hot oil for 10 to 20 min. The temperature of the oil should be below 250°F. Sealed bearings should not be heated by this method. The temperature of the oil should be checked regularly with a thermometer.

Another safe method of heating a bearing is to place it over a light bulb (about 200 W) and let the heat from the bulb rise through the inner ring. This is the best method of heating a shielded or sealed bearing. Heating should cease if there is evidence of grease escaping.

Never use an oxyacetylene torch for direct heat on a bearing. However, if a torch is the only means available, the bearing should be placed on spacers on a metal plate, and heat from the torch applied underneath the plate.

The shaft that is to receive the bearing should be well prepared in advance of installation. If a hot bearing is blocked by any form of obstruction before reaching its seat, it will contract and "freeze" to the shaft. This will present a difficult problem in seating it properly or removing it.

21-16 BEARING REMOVAL

In removing ball bearings for cleaning and lubrication, as much caution and care should be used as in installing them. More problems are encountered in safely removing a bearing than in installing it.

The wrong way to remove bearings is illustrated in Fig. 21-8. When a punch or chisel is used in this manner, it is likely to slip and damage the retainer or the seals or shields (if any), or chip the balls or raceways. At best, it can only cock the bearing from side to side, strain the inner race, and damage the shaft. Also, this practice nearly always nicks the shoulder of the shaft. If this method is used, however, all nicks and dents in the shaft and shoulder should be carefully filed before bearing replacement is attempted.

A safe way for removing a bearing from a shaft with use of an arbor press is shown in Fig. 21-9. This method should be used whenever practicable.

Where an arbor press is not available, pullers of the type shown in Fig. 21-10 can be used. Several types of these pullers are available, with various types and sizes of adapters for removal of bearings under most conditions. In using this type of puller on a dc motor or generator armature shaft, care should be taken to avoid damage to the center hole of the shaft. A washer with a small hole should be placed between the end of

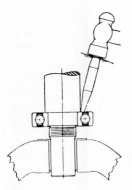

Fig. 21-8 Improper method for removing a bearing. This method often leads to a damaged bearing and shaft.

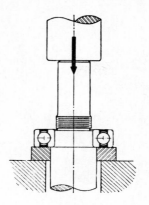

Fig. 21-9 The proper method for removing a bearing from a shaft is with the use of an arbor press.

the shaft and the screw of the puller to avoid damage to the center hole.

Occasionally, the shoulder of a shaft will be too high to allow catching the inner ring with a bearing puller or blocking under it in an arbor press. One possible correction for this condition is to heat the bearing in hot oil, at not over 250°F, and chill the shaft in cold water to shrink it before the bearing cools. Then, drive the bearing with a block of hard wood and a hammer.

Occasionally, a bearing with a hard press fit cannot be removed for replacement without destroying it. One method of removal is to cut the outer ring with an oxyacetylene torch, then heat and expand the inner ring with a torch for removal. If oxyacetylene equipment is not available, the outer ring can be safely broken by wrapping several layers of heavy cloth around it and placing it on an anvil in an upright position, then striking it with a heavy hammer. The cloth eliminates the hazard of flying pieces. A gap can be ground in the inner ring for its removal by use of a portable or bench grinder. An inner ring should not be cut with a torch because of the danger of nicking or warping the shaft.

21-17 BEARING CLEANING

A bearing that is apparently in good condition should be thoroughly cleaned and inspected following removal.

Mineral-spirit solvents, kerosene, unleaded gasoline, and safety naphtha are all good for cleaning bearings. Plant safety rules regarding solvents should be followed in all cases. Bearings coated with a hard crust of oxidized grease or sludge may require overnight soaking in an approved industrial solvent.

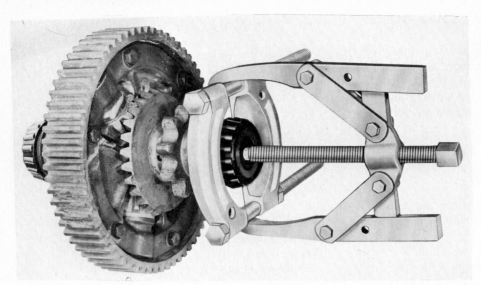

Fig. 21-10 A portable puller used for the removal of a bearing from a shaft. (*Owatonna Tool Co.*)

If a number of open-type bearings are to be cleaned at the same time, they can be placed in a wire basket and agitated in a container of solvent. Shielded or sealed bearings should not be submerged in a solvent; they can be cleaned by wiping them with a cloth moistened with a solvent.

If only one or two bearings are to be cleaned and a wire basket is not available, a stiff 1- or 2-in. paintbrush and a small solvent container can be used satisfactorily. The flared ends of the bristles of the brush should be cut off about 1/2 in. back from the end. Care should be taken to avoid bristles wedging under the balls and breaking off in the bearing.

Following the use of solvents in cleaning, compressed air (preferably filtered) can be used to clear the bearing of the solvent.

Never spin a bearing at high speed with compressed air. It serves no purpose and can actually ruin a good bearing.

21-18 IDENTIFYING BEARING FAILURE

When a ball bearing fails prematurely, the cause of failure should be determined and corrected before a new bearing is installed. The cause of failure can usually be determined by an examination of the bearing that failed. Figure 21-11 shows the inner race and balls of a bearing that failed because of misalignment, which is one of the most common causes of failure. The common causes of misalignment are bent shaft, cocked bearing due to improper mounting, improperly

Fig. 21-11 A failed bearing. (*a*) The inner race and (*b*) the ball of a misaligned bearing.

(a)

(b)

Fig. 21-12 Failure of a ball bearing due to inadequate lubrication.

prepared shaft, dirt or dents on the shaft shoulder, outer ring not true in its bracket, and bracket not at right angles to the shaft.

A bearing that failed because of inadequate lubrication is pictured in Fig. 21-12. The greatest friction in a ball bearing is between the balls and separator. If lubrication is inadequate, this friction will heat and wear the separator. This process darkens the color of the separator. If lubrication continues to be inadequate, the separator will become broken and distorted and the bearing will be ruined, although there may be very little damage to the balls and races, as shown in the illustration.

Failure of a raceway due to poor quality of grease is pictured in Fig. 21-13. When a ball is loaded, it undergoes a certain amount of deformation or flattening at the point of contact with the raceway. This is somewhat similar to an automobile tire in contact with the street. At the point of contact the tire flattens in proportion to the load. In the case of a steel ball, some friction and wear result from the process of deformation. The raceway in this figure shows excessive wear and scoring of its surface due to friction.

A common cause of bearing failure in electrical equipment is stray currents passing through the bearing. The lubricant in the bearing offers resistance to current flow between the balls and races. When the current breaks through this resistance, the arc burns and tempers the metal, causing burned and fused craters to appear in the affected area. A highly magnified view of electrical damage to a raceway is shown in Fig. 21-14.

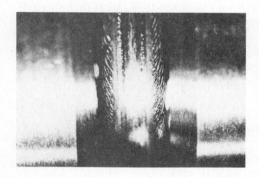

Fig. 21-13 Failure of a raceway due to poor-quality lubricant.

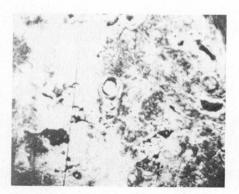

Fig. 21-14 Highly magnified view of a raceway surface showing pitting due to electrical discharge.

Static electricity generated in belt-driven equipment, grounds in the equipment, and arc welding on or near the equipment are common causes of electrical damage to bearings. Arc-welding currents should never be allowed to pass through ball bearings. Static electricity discharges can be eliminated by proper grounding of the rotating member.

Failure of a bearing due to a loose shaft fit can be identified by flaking of the bore and shaft. Particles of flaking metal can get into the bearing and wear the balls and races. In some cases, a drag is produced in the bearing, causing the shaft to turn in the inner ring and wear a groove in the shaft.

An undersized or worn shaft can be restored to proper size by spray-welding, or knurling in a lathe, and machining to required size. Standard bore sizes are shown in Table 21-1. An old bearing should be pressed on the shaft for a trial fit before a good bearing is permanently installed.

A ball bearing will normally fail in time from metal fatigue. Fatigue results under the constant loading and unloading of areas of the races as the balls roll by under the load during the normal life of the bearing. Compression and decompression of the areas gradually weaken the metal in these areas. Eventually, cracks appear and spread, and flaking of the surfaces results. This is the beginning of failure of a bearing at the end of its normal life.

21-19 SLEEVE BEARINGS

Sleeve bearings, sometimes called *plain bearings* or *bushings,* are used in most small and extremely large electric motors and to some extent in other sizes of motors. Sleeve bearings in motors are lubricated with mineral oil; one of several methods of feeding the oil to the bearing is used. Oil of SAE 10 viscosity is generally used in motors up to 5 hp, SAE 20 oil being generally used in large motors. SAE 30-40 is used in extremely large sizes and under higher-temperature

conditions. There are two distinct classifications of sleeve bearings: babbitt and bronze.

Babbitt Bearings   A babbitt bearing usually has a steel, cast-iron, or die-cast body lined with babbitt bearing material. Babbitt bearing material is an alloy of tin, lead, copper, and antimony in various proportions. It is also known as *white metal.*

Babbitt bearing materials are broadly divided into *tin-base* and *lead-base* classifications. The tin-base alloys (tin, copper, and antimony) are harder and stronger and are more suitable for heavy low-speed loads. A commonly used alloy of tin-base babbitt consists of 83-1/3 percent tin, 8-1/3 percent copper, and 8-1/3 percent antimony.

Lead-base babbitt (lead, tin, and antimony) is suitable for higher speeds and lighter loads. A commonly used alloy of lead-base babbitt consists of 75 percent lead, 10 percent tin, and 15 percent antimony.

Bronze Bearings   Bronze is widely used as an alloy bearing material suitable for most bearing applications. A commonly used alloy, SAE 660, consists of 83 percent copper, 7 percent tin, 7 percent lead, and 3 percent zinc. This alloy contains the properties necessary in a good bearing material for antifriction qualities, strength, hardness, ductibility, and machinability.

Sintered graphite bronze is commonly used in small motors for bearings. This material consists of a bronze base with about 40 percent graphite. Graphite bronze is brittle and porous. This type of bearing depends on capillary action to draw oil from outside the bearing through the walls to the inside of the bearing surface for lubrication.

Sintered bearing bronze is made by compressing finely powdered bronze and tin into a mold and heating it below its melting point to form a coherent solid mass. This results in a porous bearing material that can be prelubricated by soaking in oil or lubricated through the walls of the bearing. Figure 21-15 shows

Fig. 21-15 Sintered bronze bearings. The porous nature of this material eliminates the need for oil holes or grooves. (*Eagle-Picher Bearings.*)

examples of sintered bearings. There are no oil grooves in sintered or graphite bronze bearings.

21-20 BRONZE BEARING SHAPES

Bronze bearings are made in many shapes for various applications (Fig. 21-16). A self-aligning bearing is shown in Fig. 21-16(a). This style of bearing is used chiefly in small motors. It is supported in a specially constructed mounting, usually by means of threaded or conical collars, which allow it to move into required alignment. It receives oil through a packing or a cup-retained wick. The bearing in Fig. 21-16(b) is a straight flangeless sleeve generally used in small motors and is lubricated by yarn packing. The bearing in Fig. 21-16(c) is a flanged bearing used in all sizes of motors, especially the larger sizes. It is ring-oiled. A flanged bearing, depending on wool yarn packing for lubrication, is shown in Fig. 21-16(d).

21-21 SLEEVE BEARING LUBRICATION

Sleeve bearings depend on an oil film for lubrication. Under proper or favorable conditions, a rotating shaft, when lubricated with oil, rides or "floats" on a film of oil between it and the bearing. Because there is no metallic contact between the shaft and bearing, there is no friction or wear of these two parts.

Most of the wear on a bearing and shaft occurs during the period of starting. A motor that is started frequently will wear the shaft and bearing more for the same running time than a motor that runs long periods each time it is started.

When a motor is stopped, the oil film, under pressure of the shaft and load, thins in time to a point where metallic contact is made between the shaft and bearing. When the motor is started, this contact causes friction and wear until the oil can establish a protective film.

The steps in the establishment of an oil film are illustrated in Fig. 21-17. In Fig. 21-17(a) the shaft of an armature is shown resting on and in contact with the bearing at the bottom, which is the load area. The load area is the area under pressure and is indicated in the illustration by arrows. When the shaft starts rotation [Fig. 21-17(b)], friction between the bearing and shaft results in wear of these parts. The rotating shaft draws oil into the area of contact, and a film of oil forms and separates the two parts in this area. Because of friction, the shaft has slightly climbed the left side of the bearing, and the load area has changed slightly clockwise. In Fig. 21-17(c) a substantial oil film has been established, and the rotating shaft is "floating" on oil.

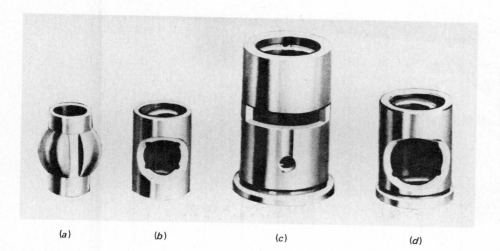

(a)   (b)   (c)   (d)

Fig. 21-16 Bronze bearing shapes. (a) Self-aligning; receives oil through wick or packing. (b) Plain sleeve; oiled through packing. (c) Flanged; oiled through ring. (d) Flanged; oiled through packing. (*Eagle-Picher Bearings.*)

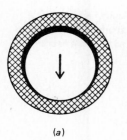
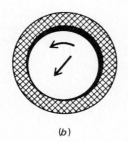
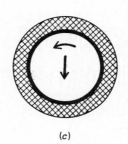

(a)   (b)   (c)

Fig. 21-17 The process of forming an oil film around a shaft in a sleeve bearing.

21-22 LUBRICATING OILS

A good lubricating oil for a specific application must have the proper viscosity to maintain sufficient film to support the load at all speeds, loads, and temperatures. It should have sufficient chemical stability to resist the oxidizing influences of heat and circulation and agitation in warm air. These influences result in increased viscosity and formation of sludge.

Lubricating oil does not wear out. In sealed units, lubricating oil has served for decades without needing replacement. But oil can easily be rendered unfit for service by contamination with dirt, moisture, and other foreign matter and by oxidation.

Manufacturers' recommendations regarding type and viscosity of oil should be followed in all cases unless experience proves otherwise. In the absence of such recommendations, higher-viscosity oil is used for heavy low-speed loads, or high temperatures, while lower-viscosity oils are used for high speeds.

21-23 RING-OILING SYSTEM

In medium- and large-size motors a circulating oil supply is necessary to ensure adequate oil to maintain a sufficient lubricating film and to dissipate heat. This is accomplished in most cases by use of the ring-oiling system. Figure 21-18 illustrates a ring-oiling system and shows the paths for circulating oil.

When the shaft rotates, it turns the ring, and the ring draws oil from the reservoir and deposits it on the shaft. The oil flows from the shaft into the bearing through grooves cut in the inner surface of the bearing.

A typical horizontal-groove pattern is shown on the inner surface of the bearing in Fig. 21-19. The oil is drawn from the groove by the shaft and distributed to all parts of the bearing to form a lubricating film.

Excess oil at the ends of the bearing falls into the circular collection grooves and drains through drain holes in the bearing and end bell to return to the reservoir, as illustrated in Fig. 21-18. Any oil that seeps beyond the outer end of the bearing on the shaft will be thrown by centrifugal force into the open space enclosed by the oil shield and will drain back into the reservoir through the drain holes.

Oil that seeps beyond the bearing on the inner end will collect on the oil slingers and be thrown by centrifugal force into the open space, then drain back into the reservoir.

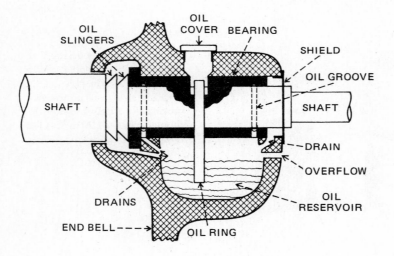

Fig. 21-18 Typical ring-oiled sleeve bearing.

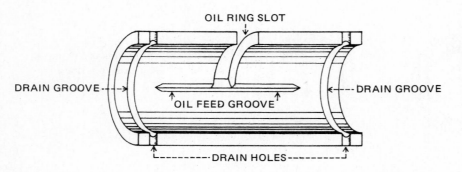

Fig. 21-19 Cutaway view of typical horizontal and circular groove patterns in a general-purpose sleeve bearing.

21-24 YARN-PACKING OILING SYSTEM

A typical yarn-packing oiling system is shown in Fig. 21-20. The bearing in this system has a large hole in the top, exposing a sufficient area of the shaft to the yarn packing to pick up enough oil for lubrication. Oil is drawn from the reservoir by capillary attraction in the wool yarn and fed to the shaft as it rotates.

Some motors use ground packing mixed with a lubricant. This compound is forced under pressure into the bearing cavity and reservoir. A special tool is required to install the compound.

21-25 POROUS-BEARING OILING SYSTEM

The porous-bearing oiling system is an efficient system of oiling extensively used on small motors, such as fans, blowers, and other appliance motors. It is similar to the yarn-packing system, but the bearing is graphite bronze or sintered bronze, without oil holes or grooves (see Fig. 21-15). The oil feeds from yarn packing around the bearing into and through the bearing walls to the shaft as it is needed.

21-26 REPLACING SLEEVE BEARINGS

A worn bearing in an electric motor should be replaced before it can cause severe damage to the motor. The air gap between armature and fields in dc equipment, and between rotor and stator in ac equipment, varies from about 0.015 in. in small motors to 0.050 in. in large motors. If a bearing is allowed to wear too much, it permits the armature core to drag or "pole" on the pole pieces in dc equipment, or stator laminations in ac equipment, and this damages the armature core or ac stator laminations, in some cases, beyond repair.

Armature End Play Before a motor is disassembled, the end play of the armature shaft should be checked and recorded so that correction can be made, if needed, when the motor is reassembled. Measurement can be made by placing a scale along the shaft and against the end bell and moving the shaft in both directions for measurement. To allow for axial expansion of a shaft due to heat, end play should be about 0.003 in. for each inch of shaft length between bearings. If end play is more than this, it can be corrected by installation of sufficient thrust washers. When thrust washers are installed, they should be spaced so the armature core will be centered in the field magnetism to avoid side pull which will wear the thrust washers or bearings.

Removing Sleeve Bearings In removing worn sleeve bearings, care should be used to avoid damaging the end bell and oil rings, if any.

A sleeve bearing can be easily removed by the use of a bearing driver by placing the latter in the bearing and pressing in an arbor press or driving with a hammer.

A bearing driver is illustrated in Fig. 21-21. The small end of the driver should be about 0.010 in. less in diameter than the bore of the bearing, and the body of the driver should be about 0.010 in. less in diameter than the outside of the bearing. These drivers can be purchased or made in a shop on a lathe.

Installing Sleeve Bearings A new bearing can be easily damaged in installing it unless extreme care is taken in the process. Oil grooves, ring slots, and oil holes weaken the body of a bearing and make it liable

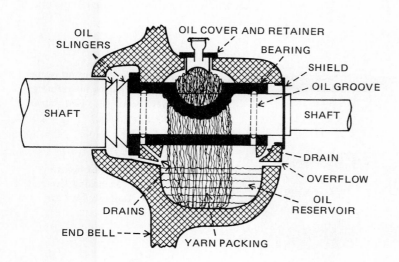

Fig. 21-20 Typical wool yarn-packing oiling system.

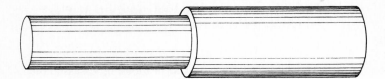

Fig. 21-21 Bearing driver for removing and installing sleeve bearings.

to distortion or collapse. To avoid damage, a bearing must be started straight and in proper alignment with the end bell. It can be pressed in place with a bearing driver, pressed in an arbor press, or driven with a hammer and driver.

Reaming Bearings In most cases after a bearing is installed, it must be reamed to proper size. Clearance between shaft and bearing should be about 0.002 in. for the first inch or fraction of shaft diameter and 0.001 in. for each additional inch in diameter. Thus a 4-in. shaft would require a clearance of 0.005 in.

Reaming of bearings is a process requiring extreme accuracy. Expansion reamers equipped with a pilot guide, as illustrated in Fig. 21-22, should be used with both end bells on the motor to ensure parallelism of the bores of the two bearings for proper alignment. The reamer is adjusted for size by turning the screw at the end of the reamer in or out with a wrench for expansion or contraction. To avoid straining the reamer, not over 0.002 in. of material should be removed at each reaming.

21-27 MAKING SLEEVE BEARINGS

In machining a bearing on a lathe, it is recommended practice to use machined bearing stock, as illustrated in Fig. 21-23, as close to the bearing size as possible.

The outside diameter should be machined to about 0.010 in. oversize, the bore machined to size, and finish cuts to size made outside, without any other

operations that might misalign the stock between these two operations. Then the oil grooves and ring slot (if any) can be cut.

A horizontal groove, such as the one shown in Fig. 21-19, can be cut by turning a properly ground tool one-fourth turn from normal position in a boring bar and scraping the groove in the bearing wall by moving the lathe carriage back and forth by hand.

Supply grooves should not extend closer than 1/2 in. to the oil-collection grooves near the end of the bearing, and the collection grooves should be about 1/4 in. from the end of the bearing.

21-28 CAUSES OF MISALIGNMENT

If an end bell turns freely on the journal of a shaft but binds or locks the shaft when installed on the motor, the cause may be either insufficient thrust or misalignment of the bearings or shaft.

If the shaft is bent, the end bell will wobble when the shaft is turned with the end bell loosely fitted in position. A bent shaft cannot be satisfactorily straightened in all cases. In an emergency, however, a shaft can be straightened by finding the high side of the bend from the center axis in a lathe with the use of a dial indicator, and pressing against the high side in an arbor press.

If bearing wear is excessive and a bent shaft is suspected as being the cause, true conditions can be determined by examining the shaft journal. If it is bent, a polished area will be found on one side of one end of

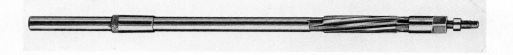

Fig. 21-22 Expansion reamer with pilot guides for reaming sleeve bearings. (*The Martindale Electric Co.*)

Fig. 21-23 Machined bronze bearing stock for making sleeve bearings. (*Eagle-Picher Bearings.*)

the journal and on the opposite side at the other end of the journal.

If the bearing is misaligned, the end bell will wobble when spun on the journal. If this condition is slight, bearing misalignment can be determined by assembling the motor and alternately tapping the end bell loose from the motor and tightening at several places around it, trying the shaft for freedom of movement each time. If a place is found where freedom is afforded the shaft, this place is the high side of the bearing.

A slight reaming with pressure on the reamer toward the high side will correct this condition, provided it is not too severe. If too much reaming is required, replacement of the bearing is the only solution.

If the end bell is cocked because of a dent in the motor frame or end bell, or dirt trapped between the frame and end bell, this condition can be determined by close examination of the end bell fit.

21-29 ELECTRICAL DAMAGE TO BEARINGS

Static electricity from belt-driven equipment, grounded windings, and improper contact of short-circuiting assemblies of repulsion-start induction-run single-phase motors sometimes causes electric currents to flow between a bearing and shaft. When a current punctures the oil film, it pits the bearing and journal. Continued pitting will result in a rough shaft and excessive bearing wear. Proper grounding usually prevents the problems associated with static electricity.

SUMMARY

1. Mechanical parts, moving in contact with each other, produce friction. Friction produces heat and wear. Lubrication minimizes the effects of friction.

2. Antifriction bearings and sleeve bearings are used to support the moving member of rotating electrical equipment.

3. Ball bearings are used more often than other kinds of bearings in modern electrical equipment.

4. Ball bearings are capable of giving years of service if they are properly selected, installed and serviced, but they can be damaged, and this damage can lead to premature failure.

5. The greatest enemies of ball bearings are mishandling and foreign material—dirt, moisture, acids, and gases.

6. New ball bearings should not be unwrapped until they are to be installed. If there is any delay in installation after opening, they should be kept wrapped in oil paper.

7. The first step in installation of a ball bearing is the proper preparation of the part to receive the bearing. Any nicks, burrs, foreign materials, and oxides that can interfere with installation should be removed.

8. Ball bearings are measured in millimeters and are made in sizes according to an international standard.

9. To convert inches to millimeters, multiply inches by 25.4. To convert millimeters to inches, divide millimeters by 25.4.

10. Since there is very little friction in a ball bearing, some of the chief functions of the lubricant, besides lubrication, are protection of the fine finishes of the balls and races against moisture, acids, and gases and dissipation of heat from the bearing.

11. Grease used for electric motor lubrication is properly designated as mineral lubricating grease.

12. Sodium-soap base greases are most commonly used for general-purpose ball bearing lubrication, and lithium-soap silicone-base greases are generally used for low-load, high-temperature operations.

13. Overgreasing is a common cause of ball-bearing and electric motor winding failures. Overgreasing can cause a bearing to heat beyond safe temperatures and destroy its proper temper. Excessive grease gets into motor windings and causes short circuits and grounds.

14. Mineral spirits, solvents, kerosene, and unleaded gasoline are solvents commonly used for cleaning ball bearings.

15. In case of premature failure of a ball bearing, the cause should be determined and eliminated before a new bearing is installed.

16. Sleeve bearings, commonly called plain bearings or bushings, are used chiefly in small and extra-large electric motors.

17. Sleeve bearings are lubricated by a film of mineral oil that forms between the journal and bear-

ing and prevents metal-to-metal contact and friction.

18. Wool yarn is used to conduct oil from a reservoir to bearings. Cotton waste is not an efficient packing.

19. A worn bearing causes an unequal air gap in a motor, resulting in heating, loss of power, circulating currents in some types of windings, and ''poling''—if severe enough.

20. New sleeve bearings can be damaged during installation by collapsing or distorting them. They should be started in the end bell in exact alignment and carefully pressed to position.

21. Pilot-guided expansion reamers should be used in reaming motor bearings. Not more than 0.002 in. of material should be removed at each reaming.

22. Dirt and contamination are enemies of lubricating oil. Oil does not wear out—it becomes unfit for use only by contamination with foreign matter.

QUESTIONS

21-1   What are the two most common types of bearings used in motors?

21-2   Name the two types of antifriction bearings.

21-3   List the factors that determine the life expectancy of ball bearings.

21-4   Identify the parts of a ball bearing.

21-5   What unit of measurement is used in ball bearings?

21-6   Convert 34.5 mm to inches.

21-7   Convert 0.06 in. to millimeters.

21-8   What type of ball bearing is most suitable for loads at right angle to the shaft?

21-9   What type of bearing is used for thrust loads?

21-10 List the greatest enemies of ball bearings.

21-11 What troubles does dirty grease cause in motor windings?

21-12 What troubles are caused by oil or grease on a commutator?

21-13 What is the most commonly used grease for ball bearings?

21-14 How can overgreasing damage a ball bearing?

21-15 Describe a method of heating a bearing to expand it.

21-16 What is the maximum safe heat to use on a ball bearing?

21-17 What are good solvents for cleaning ball bearings?

21-18 Where does the greatest friction occur in a ball bearing?

21-19 What viscosity of oil is used in sleeve bearing motors up to 5 hp?

21-20 How can oil become unfit for service?

21-21 How is oil fed to a bearing by a ring-oiling system?

21-22 When does most wear occur in a motor sleeve bearing?

21-23 What viscosity of oil is used for high speeds?

21-24 What end-play clearance is recommended for armatures?

21-25 What is the proper clearance between a shaft and its bearing?

22

ELECTRICAL METERS AND TESTING

In testing and troubleshooting electrical work, an electrician is constantly using electrical measuring instruments. For maximum and intelligent use and proper care of these instruments, knowledge of their construction and operating principles is essential.

Modern test instruments fall into two categories: analog and digital. Analog instruments contain a printed scale and a movable pointer that moves across the scale to indicate a value. The instrument is read by noting where on the scale the pointer comes to rest. The actual movement that causes the pointer (or *arrow* or *needle*) to move is electromechanical, though some instruments also use electronic circuits. Analog instruments use movements that in some ways resemble the components of a motor. Digital instruments are all electronic. They contain displays that show values directly. Thus "reading" a digital instrument merely means noting the actual values displayed.

22-1 PERMANENT-MAGNET MOVING-COIL METER

The *D'Arsonval movement,* or *permanent-magnet movement,* contains a stationary permanent magnet and a movable coil to which a pointer is attached. When the coil is energized with direct current it will turn in the magnetic field. The same basic meter, with circuit modifications, is used as an *ammeter* or *voltmeter.*

A permanent-magnet moving-coil movement is illustrated in Fig. 22-1(*a*). It is equipped with a permanent magnet to furnish a strong and constant magnetic field. The pointer is mounted on a frame that is pivoted at each end and is free to turn through part of a revolution. The frame contains a rectangular aluminum form on which a coil of fine wire is wound. Each end of the coil is connected to a spirally wound bronze spring at each end of the frame. The other ends of the springs are connected to stationary terminals.

In operation, direct current is conducted from the terminal through the springs to the coil. For proper direction of movement of the pointer, the positive terminal of the meter must be connected to the positive source of current. All dc permanent-magnet meter terminals are marked for this purpose.

Direct current in the coil produces circular magnetic lines of force around the coil that react with the magnetic field of the permanent magnet, similar to the reaction of armature and field magnetism in a dc motor, to produce torque on the frame and move the pointer. Alternating current will only cause the pointer to vibrate. The design of the coil, springs, air gap, etc., is usually such that the movement of the pointer is practically in proportion to the current through the coil; hence, the calibration marks are spaced evenly throughout the scale.

The two springs are wound opposite to each other and are so adjusted that the tension of one counteracts the tension of the other. The springs are connected to an adjustable piece with a Y slot. A small eccentric stud, projecting from an adjusting screw, fits into the Y slot. Turning the screw moves the Y-slot piece containing the springs to adjust the pointer at zero position.

The adjusting screw can be seen near the center of the nameplate of the complete meter in Fig. 22-1(*b*). To facilitate accuracy in reading, this meter contains a pointer or needle flattened at one end, with a mirror directly under the full length of the scale. To minimize an error in reading, the reader's eyes should be positioned so that the pointer will cover its image in the mirror when the reading is made. Viewing the pointer from an angle on either side will result in an erroneous reading. The apparent displacement of the pointer

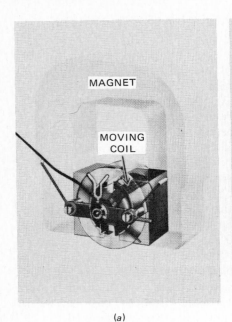

Fig. 22-1 Permanent-magnet moving-coil meter. (a) Cutaway view of magnet and coil. (b) Typical portable meter. (Weston Instruments.)

(a)

(b)

from its true position is known as *parallax*. Use of the mirror image of the pointer eliminates parallax. An extension from the frame opposite the pointer is equipped with a counterbalancing weight resembling a spirally wound spring. This balancing weight is adjusted so the meter, tilted in any position, will read zero. If a meter is dropped or severely jolted, this weight might need readjusting.

In moving through the magnetic field, the aluminum frame cuts magnetic lines of force, and eddy currents are induced in it, which restrains or damps its movements, allowing a steady movement, with no overtravel at the end of the pointer's movement.

Bearings in high-grade meters are carefully designed and manufactured to ensure sensitivity and serviceability, but can be easily damaged by abuse. An electrical meter should be accorded the same consideration and care, or better, as that given a high-grade watch. An incorrect meter can often lead to

inaccurate diagnosis of trouble and expensive operations before the discovery that the meter is defective.

22-2 AMMETERS

A permanent-magnet meter is basically a millivoltmeter. The meter coil can carry only a very small current, usually in the order of 1 mA. In order to measure higher currents, a very low resistance conductor is connected across the moving coil. This conductor, which is actually made of strips of copper, is called a *shunt*. As an ammeter, the millivoltmeter simply measures the voltage drop across the shunt, but the meter scale is marked to indicate amperes. According to Ohm's law, current in amperes is equal to voltage divided by resistance. Thus, if the voltage drop across a known resistance is determined, the current can be calculated.

Ammeter shunts are illustrated in Fig. 22-2. The

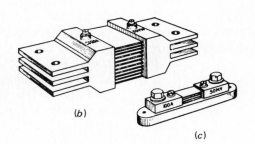

Fig. 22-2 Ammeter shunts used with 50-mV movements. (a) 75 A. (b) 1000 A. (c) 100 A. (Weston Instruments.)

(a)

(b)

(c)

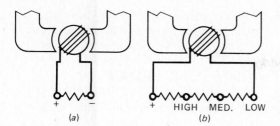

Fig. 22-3 Connections of shunts. (a) Single-range ammeter. (b) Three series shunts provide multirange readings.

leads from the meter connect to the small screws which are shown on top of the shunts. Shunts are usually marked for the size of the millivoltmeter they are designed for and the current range for which they are used.

The proper connection of a shunt for a single-range ammeter is shown in Fig. 22-3(a). The connection of shunts for a multirange ammeter is shown in Fig. 22-3(b).

Ammeters with internal shunts are usually available commercially to 50-A capacity, and shunts are externally connected in meters of larger capacity, because of heating of the shunts in operation. If externally mounted shunts are connected to the meter by leads, the resistance of the lead wire is carefully calculated in designing the resistance for the shunt circuit. In this case, leads should never be shortened, spliced, or replaced with leads of unknown resistance. Replacement leads should be obtained from the manufacturer of the equipment. Standard length of switchboard and panel shunt leads usually is 8 ft.

A formula for calculating the value of a shunt is as follows.

$$R_{sh} = \frac{I_m \times R_m}{I_{fs} - I_m}$$

where R_{sh} = resistance of shunt, Ω
  I_m = maximum current rating of moving coil, A
  R_m = meter resistance, Ω
  I_{fs} = current that will cause full-scale reading with shunt in place, A

Problem 1

A meter with 1000-Ω moving coil has a full-scale reading when 1 mA is flowing. The meter is to be used in a circuit carrying a maximum of 10 A. What is the resistance of the shunt that must be connected across the meter?

Solution

Using the formula

$$R_{sh} = \frac{I_m \times R_m}{I_{fs} - I_m}$$

where I_m = 1 mA = 0.001 A
  R_m = 1000 Ω
  I_{fs} = 10 A
  $R_{sh} = \dfrac{0.001 \times 1000}{10 - 0.001}$
    = 0.1000 Ω

In this case 0.001 A flows through the meter coil and 9.999 A flows through the shunt. The ratio of shunt current to coil current is

$$\frac{9.999 \text{ A}}{0.001 \text{ A}} = 9999$$

$$\approx 10,000$$

Thus the coil resistance is approximately 10,000 times the shunt resistance. As a check 1000 Ω/0.1 = 10,000.

22-3 VOLTMETERS

For measuring voltages above the maximum allowed by the coil, resistance would be added in series with the coil of the millivoltmeter.

In the previous problem, the coil had a resistance of 1000 Ω and a full-scale current of 1 mA. Thus a voltage of $V = I \times R$ = 0.001 A $\times$ 1000 Ω = 1 V caused the full-scale reading. In effect the meter was a 1-V voltmeter. If the meter was to read a full-scale value of 150 V, a series resistor capable of dropping 149 V would have to be added. Since the resistor and coil would form a series circuit, the 0.001-A coil current would have to flow through the series resistor. By use of Ohm's law, the value of this resistor can be calculated.

$$R_{se} = \frac{V_R}{I_M}$$

where R_{se} = series resistor, Ω
  V_R = voltage drop across series resistor, V
  I_m = meter full-scale current, A

Problem 2

The meter in the previous problem is to be used to measure 25 V. What is the value of the series resistor?

Solution

Using the formula

$$R_{se} = \frac{V_R}{I_m}$$

where $V_R = 25 - 1 = 24$ V
$I_m = 0.001$ A
$$R_{se} = \frac{24}{0.001} = 24,000 \ \Omega$$

If the meter scale is printed 0 through 1 with subdivisions representing parts of a volt, the series resistor would extend the value of each subdivision. In the problem above, the new scale would in effect be 0 through 25 V. To obtain the actual voltage to which the pointer indicates, the reading would have to be multiplied by 25. For that reason the series resistor is referred to as a *multiplier resistor*.

A multirange voltmeter can be constructed by connecting a number of multiplier resistors in a manner so that they can be selected for the desired range by connection to the meter's binding posts. A selector switch is used to change the combinations of resistors to obtain the desired range. These meters usually contain a multiscale dial. The resistors used in meters have extremely precise values, and they maintain their values under the usual temperature ranges of the instrument itself.

22-4 OHMMETER

The basic millivoltmeter can be converted to an ohmmeter by adding a voltage source (usually a single dry cell) and series resistors. The scale would be marked for ohms beginning with zero at full scale at the right. The meter coil, dry cell, resistor, and test prods are connected in series. To make adjustments for different battery voltages, a small variable resistor is added in series with the other components. In a voltmeter or ammeter the figures for calibrations increase in value from *left to right*, but in an *ohmmeter* the figures increase in value from *right to left*.

In operation, test prods from the ohmmeter are touched together or shorted, and the rheostat is adjusted to give full-scale deflection of the pointer to zero to indicate zero ohms resistance under these short-circuit conditions. If test resistance is placed across the test prods, it will lessen current flow, and the pointer will move to the left to indicate the value of the resistance on ohms across the test prods.

Ohmmeters are marked in ohms and megohms, with the letter K for 1000 Ω. A megohm (sometimes abbreviated *meg*) is 1,000,000 Ω. The end of the scale is marked *infinity* meaning resistance beyond measurement. Ohmmeters are never used on live circuits.

22-5 WHEATSTONE BRIDGE

For precise resistance measurements in low ranges, a Wheatstone bridge or an electronic meter is commonly used. A Wheatstone bridge consists of a battery, a galvanometer (a sensitive milliammeter), and three precision resistors connected as in the circuit of Fig. 22-4. The unknown resistance to be measured is connected in one of the parallel circuits, and the variable resistor of the bridge is varied. The bridge is balanced when the galvanometer reads zero. The settings of the variable resistor are read and the value of the unknown resistor is calculated using the following formula:

$$R_x = \left(\frac{R_1}{R_2}\right)R_v$$

where R_x = unknown resistor, Ω
R_1 and R_2 = known resistors, Ω
R_v = variable resistance, Ω

Problem 3

A Wheatstone bridge is balanced when the variable-resistor setting is 3.5 Ω. If $R_1 = 50$ Ω and $R_2 = 250$ Ω, what is the value of the unknown resistor?

Solution

Using the formula

$$R_x = \left(\frac{R_1}{R_2}\right)R_v$$
$$= \left(\frac{50}{250}\right)3.5$$
$$= 0.7 \ \Omega$$

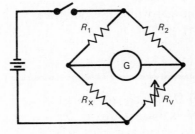

Fig. 22-4 Wheatstone bridge used to measure resistance.

22-6 OHMS PER VOLT

A dc voltmeter usually has the meter resistance in ohms per volt stated on the meter face. This is to indicate the sensitivity of the meter. The total resistance of the meter circuit can be determined by multiplying the ohms per volt by the full-scale voltage of the meter. With this value, the current requirement for full-scale deflection of the pointer can be found to determine the sensitivity of the meter. Current requirement decreases as sensitivity increases.

A 2000-Ω/V voltmeter is usually considered sufficiently sensitive for any general testing and is usually more rugged than more sensitive meters that have up to 20,000-Ω/V ratings. The total calculated resistance of a voltmeter is usually sufficiently accurate to be used as a standard for testing the accuracy of ohmmeters.

Problem 4

Compare the currents needed for full-scale deflection for a 0–150-V meter with 2000 Ω/V and a 0–50-V meter with 5000 Ω/V.

Solution

The 0–150-V meter with 2000 Ω/V has a total resistance of

$$R = 2000 \times 150 = 300,000 \ \Omega$$

The full-scale current is

$$I = \frac{V}{R} = \frac{150}{300,000} = 0.0005 \ \text{A}$$

The 0–50-V meter has a total resistance of

$$R = 5000 \times 50 = 250,000 \ \Omega$$

The full-scale current is

$$I = \frac{V}{R} = \frac{50}{250,000} = 0.0002 \ \text{A}$$

22-7 MOVING-IRON-VANE METERS

Moving-iron-vane (or simply iron-vane) meters operate on the principle of repulsion between like magnetic poles. This principle is demonstrated in Fig. 22-5. Two pieces of soft iron are suspended in a magnetic field produced by a coil. Alternating current in the coil alternates the polarity of the iron pieces.

If one of the iron pieces is made stationary and the other suspended so that it is free to move, as shown in the figure, the free piece will move by repulsion when the coil is energized. A concentric-vane assembly mounted with the coil in the meter is shown in Fig. 22-6. This meter is used for ac or dc measurements.

Since alternating current produces pulsating vibrations and otherwise causes strains in the mechanism, the pointer or needle in this type of meter is trussed to prevent bending. Damping is provided by a light aluminum vane attached to and moved in an air chamber by the pointer shaft.

22-8 ELECTRODYNAMOMETER METERS

These meters operate on the same principle as that of permanent-magnet meters, but field magnetism is furnished by a coil or coils instead of a permanent magnet. Figure 22-7 is a cutaway view of an electrodynamometer mechanism. The moving coil is mounted between two field coils. Current is conducted to the moving coil through the two springs at the top of the pointer shaft.

Electrodynamometer meters operate on alternating

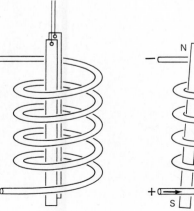
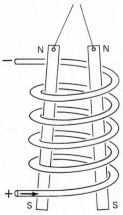
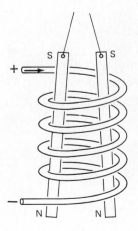

Fig. 22-5 Principles of magnetism applied to two pieces of soft iron. (*a*) Two pieces of iron suspended in a coil. No current is flowing. (*b*) When current flows, the two iron bars are similarly magnetized. (*c*) When the current reverses, the magnetism in both bars reverses equally and the bars remain separated.

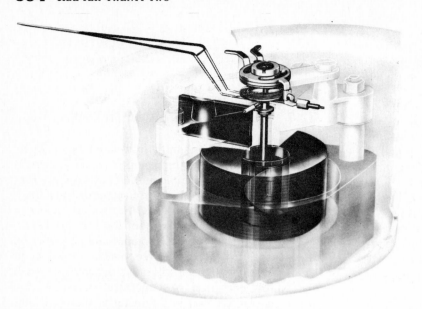

Fig. 22-6 Concentric moving-iron-vane meter movement. (*Weston Instruments.*)

or direct current. By various connections between the field coils, moving coils, and resistors, these meters are used to measure voltage, current, power, power factor, capacity, frequency, etc. Their range can be extended by use of current or potential transformers. When the field coils and moving coil are connected for a milliammeter or millivoltmeter, torque is proportional to the product of the currents, whereas in a permanent-magnet meter torque is proportional to the current.

This type of meter can also be used to measure power. The moving coil can be connected across the line and can act as a voltage coil. The stationary field coils (used instead of the permanent magnets) are connected in series with the load and act as current coils. To be used in this manner the moving coil needs to be wound with small wire, while the field coils need to be wound with few turns of large-size wire.

To measure power, the voltage coil is connected across the load and the current coil is connected in series with the load. Polarity of course must be observed when measuring dc power. When measuring ac power, relative polarity must also be followed as in Fig. 22-8. Note that the terminals marked ± are connected to the same side of the supply. When measuring ac quantities, the wattmeter will read true power.

When measuring total three-phase power, only two wattmeters are required as shown in Fig. 22-9. If the three-phase load is completely balanced, only the power of one phase need be measured using a single meter. The total load will be three times the single meter reading.

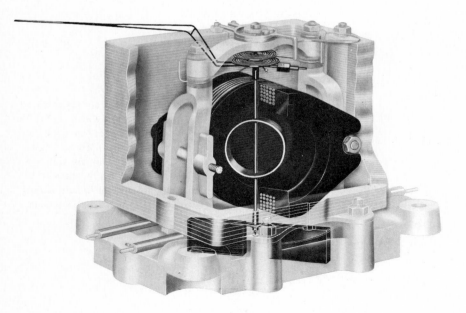

Fig. 22-7 Cutaway view of an electrodynamometer meter movement. (*Weston Instruments.*)

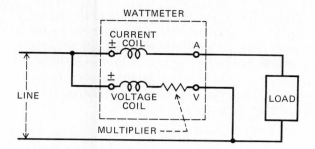

Fig. 22-8 Wattmeter connected to measure power.

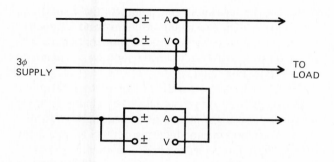

Fig. 22-9 Wattmeters connected to measure total power of a three-phase system.

22-9 METER RECTIFIERS

A permanent-magnet moving-coil meter operates on direct current only, but it can be made suitable for measurements on alternating current by equipping it with a suitable rectifier that converts alternating current to direct current.

A solid-state full-wave meter rectifier is shown in Fig. 22-10(a). A schematic diagram of its connection to a permanent-magnet milliammeter for use on ac circuits is shown in Fig. 22-10(b). Figure 22-10(c) is a schematic diagram of a rectifier connected in series with a multiplier-resistor for ac-voltage measurement.

22-10 MULTIRANGE AC-DC METERS

All the foregoing discussions of permanent-magnet moving-coil meters indicate that several ranges of different conditions can be determined by one basic meter. This is actually done in many multipurpose meters. The meters are commonly known as multimeters (MM) or volt-ohm-milliammeters (VOM).

A meter equipped with shunts, resistances, rectifier, rheostat, and dry cell, capable of measuring ac and dc volts, amperes, and ohms in 26 ranges, is shown in Fig. 22-11. Most multimeters are made for portable handheld field use.

22-11 CLAMP-ON VOLT-AMMETER

A clamp-on volt-ammeter for measuring ac volts and amperes is pictured in Fig. 22-12(a). This meter has a soft-steel split core that can be opened and closed around a current-carrying conductor. A coil is wound around this core to serve as the secondary of a transformer. The coil is connected to the meter circuit itself. When measuring direct current, the alternating magnetism in the conductor induces an alternating current in the coil which is fed to the meter circuit. This alternating current is passed through a rectifier and converted to direct current for the permanent-magnet moving-coil meter. The system is equipped with a selector switch and shunts and a multiple-scale dial for five current ranges.

For use as a voltmeter, the selector switch is set on "volts," and test leads are connected through appropriate holes in the meter and across the circuit to be measured. Several voltage ranges are available for use through tapped resistors in series in the voltage circuit. This type of meter is not suitable for measuring electrical values in the secondary, or rotor, circuit of a wound-rotor (slip-ring) motor, because of the low frequency of the rotor circuit. Moving-iron-vane meters are used for this purpose.

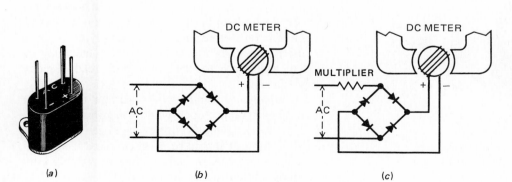

Fig. 22-10 A dc meter with a full-wave rectifier to measure ac quantities. (a) A meter rectifier. (b) A dc meter used to measure ac current. (c) A dc meter with multiplier to measure ac voltage.

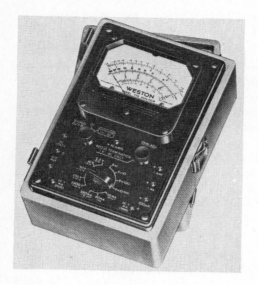

Fig. 22-11 Multimeter with 26 ranges of ac and dc voltage, current, and resistance using a permanent-magnet meter. (*Weston Instruments.*)

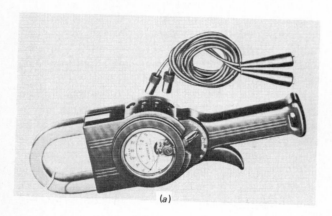

(a)

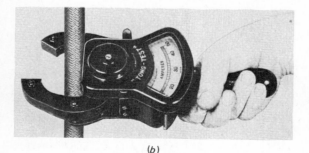

(b)

Fig. 22-12 Clamp-on meters. (*a*) Multirange ac volt-ammeter. (*Weston Instruments.*) (*b*) Moving-iron-vane ammeter. (*The Martindale Electric Co.*)

Clamp-on-type meters are very sensitive to stray magnetic fields. Tests should be made in complete freedom of these fields. Some meters can be made to move full scale in the magnetic field of an electromagnet such as a growler, although not connected to anything. Like other ac instruments, they should be used only within their frequency range.

22-12 MOVING-IRON-VANE CLAMP-ON AMMETER

A moving-iron-vane clamp-on ammeter gathers magnetism from the current-carrying conductors of the circuit being measured and conducts the magnetism into and through the meter mechanism to magnetize the vanes. Because the meter does not contain a winding and no electricity is used in operation, it cannot be burned out [Fig. 22-12(*b*)].

These ammeters operate on alternating or direct current, and their range is changed by changing meter units. A convenient means of measuring low current values is to loop the conductor under test around a jaw of the tongs. The meter reading is divided by the number of loops for actual current values. These meters are also suitable for measuring secondary or rotor currents of wound-rotor motors.

22-13 INSTRUMENT TRANSFORMERS

The ranges of ac meters are extended by current and potential transformers in the manner that dc meter ranges are extended with shunts and resistors.

Instrument transformers afford a high degree of protection in metering electrical values of high-voltage lines. Their use eliminates high voltages at the metering equipment. Also, metering equipment can be more conveniently and centrally located in respect to circuits to be metered. A more detailed discussion of instrument transformers can be found in Chap. 14.

Current Transformers A 5-A ammeter is generally used with a current transformer. The ratio between the number of turns in the primary and secondary windings of the transformer supplies a multiplying factor to be used in reading the meter, or the meter may be scaled according to the ratio and read directly (Fig. 22-13).

The secondary winding of a current transformer should never be left open while the primary is energized, since dangerously high voltages can build up in the secondary. It should be short-circuited if not connected to a meter. A secondary shorting switch is built into most current transformers.

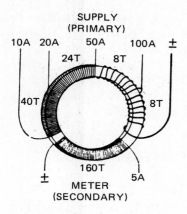

SUPPLY
(PRIMARY)

10A 20A   50A   100A   ±
   24T   8T

40T   8T

      160T   5A
   ±
   METER
(SECONDARY)

Fig. 22-13 Windings and core of a typical current transformer.

Potential Transformers Voltmeters are seldom made with a range beyond 600 V. Potential transformers are usually used with 110- to 150-V voltmeters for measuring voltages above 600 V.

The two primary leads of the transformer are connected across the circuit to be measured, and the secondary leads are connected to the meter. When permanently installed on circuits in excess of 240 V, one of the secondary leads should be solidly grounded.

22-14 DIGITAL METERS

Digital meters are all-electronic instruments using solid-state components such as transistors, diodes, and integrated circuits. Like the analog meters dis-

cussed earlier in this chapter, digital meters usually depend on voltage for their readings.

Digital meters are used to measure voltage, current, and resistance. Digital multimeters are capable of measuring both ac and dc voltage and current as well as resistance.

Digital instruments derive their name from the fact that their displays show specific numbers, or digits.

The heart of the digital display is a seven-segment solid-state component. Figure 22-14(*a*) shows how the seven segments are arranged. With the energizing of one or more of the segments, the digits 0 through 9 can be displayed as in Fig. 22-14(*b*). Two types of seven-segment displays are in common use.

The light-emitting diode (LED) produces a reddish glow when a segment is energized by low-voltage direct current. The glow is bright enough to be seen in the dark. However, when strong direct light falls on the display, the numbers are sometimes difficult to read.

Another common display is the liquid-crystal display (LCD). This type of display shows up dark against a light gray background. Direct light makes the display stand out brighter. However, LCD materials do not produce their own light. Therefore they cannot be read in darkness or dim light without a supplemental light source.

22-15 ACCURACY AND RESOLUTION OF DIGITAL METERS

Since digital meters do not depend on the user's ability to interpret the position of a pointer relative to a

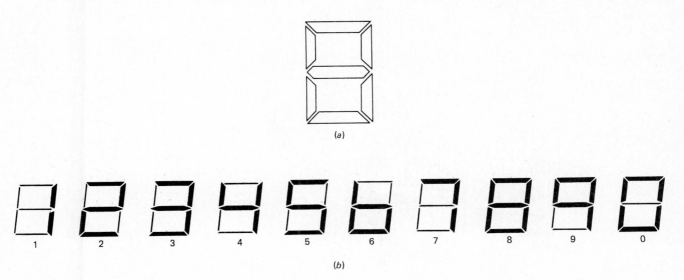

(*a*)

1   2   3   4   5   6   7   8   9   0

(*b*)

Fig. 22-14 A seven-segment display. (*a*) The seven segments. (*b*) The digits from 0 through 9 can be represented with this display.

calibration mark, the circuitry of the meter determines how accurate the reading is. Obviously if only one digit were displayed, the reading would not be very accurate. A two-digit display would be a little better, and three digits better still.

Digital instruments are usually specified in terms of the number of digits in the display. Manufacturers usually specify their meters as 2-1/2 digits, 3-1/2 digits, 4-1/2 digits, and the like. The 1/2 digit refers to the fact that the first digit does not actually indicate the full value. For example, a 4-1/2-digit meter would display a maximum reading of 1999.9 V at the 1000-V range.

Since most digital meters have an overrange of 100 percent, this meter can read another 1000 V. In most cases, however, the meter would not read 2000 V. Digital meters generally have ranges in multiples of 10 and so the next range for this meter would be 10,000 V and the reading would be displayed as 02000 V. In the 10,000-V range the maximum reading would be 19999 V.

22-16 TYPICAL MULTIMETER CIRCUIT

While a discussion of the exact circuitry of a digital multimeter is beyond the scope of this text, the block diagram of Fig. 22-15(a) shows the basic components of such a meter. Figure 22-15(b) shows a 3-1/2-digit digital multimeter. Other common sizes are 2-1/2, 4-1/2, and 5-1/2 digits. The latter two are usually more expensive instruments. Note that this meter measures both dc and ac current. Analog multimeters rarely have ac current ranges.

22-17 SPECIAL FEATURES OF DIGITAL METERS

Digital meters, because of their circuitry, can contain many features not possible with ordinary analog meters. Some are listed below.

Autoranging. Some meters are capable of automatically selecting the best range for a particular measurement.

Polarity indication. On some instruments the plus and minus terminals are labeled. If the test probe is accidentally reversed, a minus sign will flash on with the reading, indicating that the actual polarity is opposite to that of the meter's marking. On the more expensive instruments polarity is set automatically and the readout will indicate the sign. Still

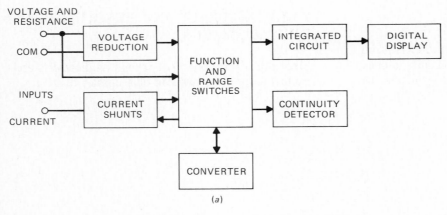

(a)

(b)

Fig. 22-15 A 3-1/2-digit digital multimeter. (*Simpson Electric Co.*)

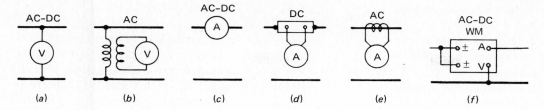

Fig. 22-16 Typical ammeter and voltmeter connections.

other meters contain polarity switches that can reverse connections should the test leads be connected incorrectly.

Power. The electronic circuitry of digital meters requires a power supply. Some meters operate on ac only with internal rectifiers. Others can operate on batteries only. Still others are capable of ac and dc operation.

Overrange indication. An analog meter can be seen to overrange when the pointer slams against the full-scale stop. With digital meters the display will usually indicate when a reading exceeds the range (assuming the meter does not have autoranging). A flashing reading (usually the highest in that range), or a blanked display, or the word *over,* or a number of other indications are common methods of indicating that the reading exceeds the range to which the meter is set.

22-18 TESTING WITH METERS

Analog meters are delicate instruments containing moving parts and delicate bearings made to precision standards. They will serve for years if properly cared for, but they can be ruined in a fraction of a second if improperly handled. A severe jolt can ruin jeweled bearings, bend the needle, and strain the moving mechanism.

If values to be measured are unknown, a meter of sufficient range should first be used that will meet the maximum possible need of the occasion.

Ammeters should always be connected in series with a load—never across the line, since their low resistance offers no protection against burnout. Volt-

meters cannot burn out on a voltage within their range.

Proper connections of ammeters and voltmeters are illustrated in Fig. 22-16. In Fig. 22-16(*a*) a voltmeter is shown across an ac or dc line. In Fig. 22-16(*b*) an ac voltmeter is connected to a potential transformer. In Fig. 22-16(*c*) an ammeter is connected in an ac or dc line. In Fig. 22-16(*d*) a dc ammeter is connected across a shunt in a dc line. In Fig. 22-16(*e*) an ac ammeter is connected to the secondary winding of a current transformer. The line here is the primary winding. In Fig. 22-16(*f*) a wattmeter is used to measure ac or dc power.

22-19 TESTING COILS

A coil can become defective through a short circuit, an open circuit, or a ground—either continuous or intermittent. If a coil is shorted, it will show a lower-than-normal resistance reading. It is advisable for the maintenance department to keep a record of the resistance of all coils under its jurisdiction for comparison purposes in tests.

22-20 VOLTAGE-DROP TESTING

The voltage-drop method can be used to test a series of coils, such as motor field coils. This method is illustrated in Fig. 22-17. A voltmeter is connected across a coil for test as shown in the illustration. With the series of coils energized, the voltage drop across each coil should be the same. If the voltage drop across a coil is less than the remaining coils, it is

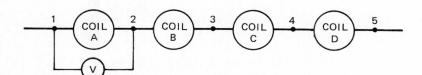

Fig. 22-17 Voltage-drop test for field coils.

shorted, or it is the wrong coil for the set. Intermittent trouble can often be detected by flexing a coil during testing or heating it preparatory to testing.

It is difficult to measure accurately the resistance of a coil of comparatively few turns of a large wire, and in some cases the normal resistance of a coil is not known. In these cases, testing is difficult without proper equipment. A convenient shop-built coil tester is shown in Fig. 22-18.

This tester contains a power coil (120-V ac) on the center leg and a tester coil on each outside leg with an ac milliammeter or millivoltmeter connected in series with them. When the power coil is excited, equal voltage is induced in the two outside tester coils, but they are connected to oppose or balance each other, and no current flows in the meter. If a shorted coil is placed over one outside leg, a flow of induced current in the closed short circuit of the coil produces a magnetic field that counters the cemf in the tester coil on the same leg. This unbalances the voltage between the two tester coils, causing current to flow in the meter circuit and producing a deflection of the meter.

The tester is built on a laminated transformer core stacked to about 2 in.^{2} of cross-sectional area. Because of the varying nature of available core material, this tester will have to be designed on a trial-and-error basis, backed by a knowledge of its operation, and no exact dimensions or data can be given for its construction. Sufficient turns of No. 18 wire are wound in the power coil to allow about 4 A to flow at 120 V. The outside coils are wound with about 40 turns of No. 24 wire, but after the assembly is completed, these turns are adjusted to balance voltages in them so that no deflection of the meter will occur when the power coil is energized.

Care must be taken to keep magnetic materials, such as tools or stray pieces of metal, out of the influence of this tester when it is energized. Too great a change in the magnetic field of one tester coil can result in an overload and burnout of the meter. A coil being tested should be slowly and carefully brought into the field of one tester coil while the tester is energized.

22-21 IMPEDANCE COIL

In testing, it is often necessary to pass a measured or limited current through a device, as in testing the opening of a thermal overload relay. A shop-built adjustable-impedance coil for this purpose is shown in Fig. 22-19. Two L-shaped transformer cores, each 3 in.^{2} in area (one with a coil), are assembled as shown. A guide is bolted to the right end of the coil core. Current flow is regulated by shimming with nonmagnetic material (copper, aluminum, paper, etc.) in the air gap above the coil. An ammeter is connected in series with the 120-V coil for this adjustment. The coil should have sufficient turns of No. 10 wire to limit current flow to about 0.5 A with the core assembled and completely closed.

22-22 KILOWATTHOUR METERS

A dc kilowatthour meter is basically a dc motor driving a train of gears connected to integrating pointers which total the electric energy passing through it. The armature circuit is connected across the line to measure voltage, and the field circuit is in series with the load. Thus the meter computes the product of the volts and amperes of the circuit.

An ac kilowatthour meter contains a potential coil connected across the line to measure voltage and current coils in series with the load to measure current. These coils are mounted in a core with an aluminum disk between them that drives the integrating pointers.

In operation, current in the potential coil produces alternating magnetism in proportion to the voltage that cuts the disk and induces eddy currents in it. The eddy-current magnetic field reacts with the magnetic field produced by the load current coils to produce torque on the disk and drive it. The speed of the meter is in proportion to the product of the in-phase amperes in the current coils times volts across the potential coil.

Permanent magnets, with the disk in their air gaps, restrain movement of the disk and are adjusted toward

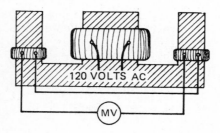

Fig. 22-18 Shop-built coil tester.

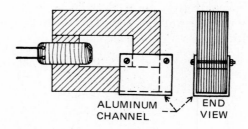

Fig. 22-19 Variable-impedance coil for regulating or limiting ac current.

or away from the center of the disk to afford proper restraint for accurate measurement. In cutting the permanent-magnet fields, eddy currents are induced in the disk, and magnetism of the eddy currents reacts with the permanent-magnet fields to brake the disk.

22-23 READING A KILOWATTHOUR METER

Most kilowatthour meters are constructed with a series of four or five dials on their face. The dial pointer shaft at the right drives the dial pointer shaft adjacent at the left, and this pointer shaft drives the next to the left, etc., resulting in the second and fourth dial from the right turning and reading opposite, or counter-clockwise, to the first and third dials, which turn and read clockwise.

A typical meter face is shown in Fig. 22-20. One reads the dials from left to right, reading each digit the pointer has immediately passed. For example, the meter shown has a pointer between 1 and 2 in the left dial. The reading is 1. The next dial to the right is between 2 and 3. The reading is 2. The next dial to the right (the third from the left) is between 6 and 7. The reading is 6. The right dial would be read as 4. The complete reading is therefore 1264.

If the meters were read once a month and the previous reading of the meter was 0972, a bill for electric energy consumption for a month would be calculated as follows:

Present reading	1264
Previous reading	0972
kWh consumed	292

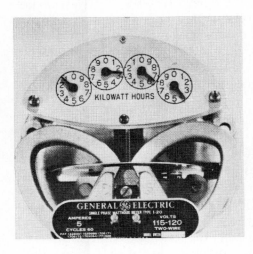

Fig. 22-20 Kilowatthour meter. The thin line just above the nameplate is an edge view of the aluminum disk.

Problem 5

If the next month's reading of the meter was taken from Fig. 22-21, what would be the energy consumption for that month?

Solution

The left dial is just past 1. The reading is 1. The next dial is between 3 and 4. The reading is 3. The next dial appears to be at 8 but notice that the right dial is just past 9. Thus the third dial has not quite reached 8. The reading of the third dial is therefore 7. The complete reading is 1379. Energy consumption is:

Present reading	1379
Previous reading	1264
kWh consumed	115

Problem 6

If the utility company charges for energy consumption according to the following schedule, calculate the monthly charge for 650 kWh of usage.

Service connection (regardless of usage)	$7.50
First 300 kWh	9.5¢ per kWh
Over 300 kWh	11.5¢ per kWh

Solution

Notice that this utility company charges a higher rate as energy consumption rises. This is often done as an energy conservation measure. By charging a higher rate customers are encouraged to limit their consumption of electricity.

The service connection charge is for having service brought to the customer. Thus the first

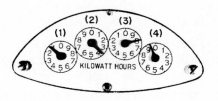

Fig. 22-21 Meter face for Problem 5.

charge for each month is $7.50 regardless of usage. The balance of the charges are computed in the following manner.

Service connection	$ 7.50
First 300 (300 × 0.095 per kWh)	28.50
Over 300 (350 × 0.115 per kWh)	40.25
Total monthly bill	$76.25

SUMMARY

1. Modern test instruments fall into two categories: analog and digital.

2. The permanent-magnet moving-coil meter movement.

3. The principle of operation of a permanent-magnet moving-coil meter is similar to that of a dc motor. Current in the moving coil sets up a magnetic field. The torque developed by the action of this field and the permanent-magnet field moves the meter pointer.

4. The current range of dc meters can be extended by connecting shunts across the moving-coil circuit.

5. The voltage range of dc meters can be extended by connecting multiplier resistors in series with the moving-coil circuit.

6. Ohmmeters contain a dc supply (usually a dry cell) in series with the meter coil and multiplier resistors.

7. The scale of an ohmmeter reads opposite to that of an ammeter or voltmeter. The zero reading of an ohmmeter is full-scale deflection. When the pointer is at the extreme left, the reading is infinity or ∞.

8. The Wheatstone bridge is used to make accurate resistance measurements, particularly of low resistance.

9. The sensitivity of analog meters is usually specified by the unit of ohms per volt.

10. The moving-iron-vane meter contains a stationary coil and two vanes suspended in the magnetic field produced by the coil.

11. Iron-vane meters can be used to make ac as well as dc measurements.

12. Electrodynamometer meters are similar to moving-coil meters except that the permanent-magnet poles are replaced with coil-wound poles.

13. Electrodynamometer meters can be used to measure both ac and dc quantities of voltage, current, power, power factor, capacity, and frequency.

14. When used as a wattmeter to measure power, the stationary field coils are wound with few turns of heavy wire and connected in series with the load. The moving coil is wound with very thin wire and connected across the load.

15. Three-phase power can be measured using two wattmeters. The power of a balanced three-phase system can be measured using a single wattmeter and multiplying the reading of one phase by three.

16. Permanent-magnet moving-coil meters operate on direct current only but can be used to measure quantities of alternating current by connecting a rectifier in series with the moving coil.

17. Meters that can measure both ac and dc quantities of voltage and current as well as resistance are known as multimeters (MMs) or volt-ohm-milliammeters (VOMs).

18. Clamp-on meters are used to measure current and voltage without breaking into the current-carrying conductors.

19. Instrument transformers are used to extend the ranges of ac ammeters and voltmeters. The current transformer uses one of the lines of the system as the primary. The ammeter is connected across the secondary. A potential transformer is connected directly across the lines with the voltmeter connected across the secondary of the transformer.

20. Digital meters contain electronic circuits which measure both alternating and direct current.

21. Digital meter readings are displayed directly by light-emitting diodes (LEDs) or liquid-crystal displays (LCDs).

22. Digital meters are specified by the number of digits they display. Typical displays are 2 1/2, 3 1/2, 4 1/2, and 5 1/2 digits.

23. Features of digital meters include autoranging, polarity indication, overrange indication, and ac/dc operation.

24. Kilowatthour meters are used to measure energy consumption. The difference between two readings taken over a period of time is the actual quantity of energy used during that period.

QUESTIONS

22-1 List the three types of analog meter movements.

22-2 Which of the three movements in question 1 cannot be used directly on alternating current? Why?

22-3 What is another name for the D'Arsonval movement?

22-4 How is parallax prevented in analog meters?

22-5 What device is used to extend the range of a dc ammeter? How is it connected to the circuit and to the meter?

22-6 The device in question 5 is used to extend the range of a dc ammeter from 1.0 mA to 5 A. What is the resistance of the device? The resistance of the meter movement is 1000 Ω.

22-7 What device is used to extend the range of a dc voltmeter from 1 to 150 V? How is it connected to the circuit and to the meter? What is the resistance of the device if the meter resistance is 1000 Ω?

22-8 What does the letter K stand for on an ohmmeter range switch?

22-9 The Wheatstone bridge in Fig. 22-4 is used to measure the value of a resistor. If $R1 = 50\ \Omega$ and $R2 = 100\ \Omega$, what is the value of the unknown resistor if the variable resistor reads 112 Ω when the bridge is balanced?

22-10 Two 150-V voltmeters are being compared to determine which is a more sensitive meter. Meter A had 5000 Ω/V, meter B has a total meter resistance of 750 kΩ. Which is the more sensitive meter? Why?

22-11 What type of meter can be used as a wattmeter?

22-12 Show how two meters can be used to measure the total power of an unbalanced three-wire single-phase system.

22-13 Show how two meters can be used to measure the total power of an unbalanced three-phase four-wire system.

22-14 What device is used to enable dc meters to measure alternating current?

22-15 What are two common names for meters that are capable of measuring both ac and dc current and voltage as well as resistance?

22-16 What type of meter is used to measure current in feeders that cannot be disconnected?

22-17 What are instrument transformers and how are they connected?

22-18 How are readings taken with digital meters?

22-19 What does the specification "3 1/2 digits" mean?

22-20 Explain the autoranging function of a digital meter.

22-21 How is a voltage-drop test used to test motor field coils?

22-22 How does a kilowatthour meter measure electric energy?

22-23 The kilowattmeter in Fig. 22-22(a) was read March 1. On April 1 the meter in Fig. 22-22(b) was read. If the charges for electric energy usage are as given in Problem 6, calculate the March bill.

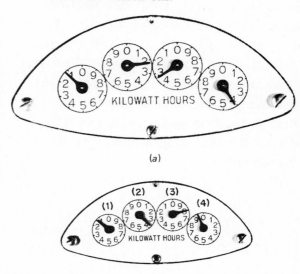

(a)

(b)

Fig. 22-22 Meter faces for question 22-23.

23

SWITCHING LOGIC AND DIGITAL ELECTRONICS

In their simplest forms, controls energize or deenergize circuits. Some controls can regulate speed and reverse motor direction. In more complex applications, controls are used to monitor system operations, interlock different components of the system, and provide sequential operations. Switching logic is a method of analyzing and developing these control circuits. While most control circuits used in industrial electricity involve the mechanical opening and closing of contacts, solid-state electronic devices are replacing magnetic relays in many of the applications. In this chapter emphasis is on two-state devices. The two states are ON (conducting) and OFF (nonconducting). Digital electronics is the study of the behavior of these two-state devices in motor control and other industrial-type circuits.

23-1 SWITCHING LOGIC

Figure 23-1 shows a control circuit containing two NO contactors A and B and a main control coil M all in series. To energize coil M, both coils, A AND B, must be energized. We can write this condition in a short form as

$$A \cdot B = M$$

which resembles a mathematical equation. The dot stands for "AND" so we can read this equation as

The status of A (energized or unenergized) AND the status of B (energized or unenergized) determine whether M is energized or unenergized.

From Fig. 23-1 we can see that M can be energized only if both A AND B are energized.

If we denote an energized circuit with a one, 1, and an unenergized circuit with a zero, one, the four possible conditions of the circuit in Fig. 23-1 can be written as in Fig. 23-2. This figure shows that when A AND B are both unenergized, M is unenergized.

$$0 \cdot 0 = 0$$

If A is unenergized AND B is energized, M is unenergized.

$$0 \cdot 1 = 0$$

If A is energized AND B is unenergized, M is unenergized.

$$1 \cdot 0 = 0$$

If A is energized AND B is energized, M is energized.

$$1 \cdot 1 = 1$$

Figure 23-2 is known as a *truth table* since it gives all the possible conditions of the series circuit containing A, B, and M.

Consider now the control circuit of Fig. 23-3 in which A and B are in parallel. In this circuit, coil M

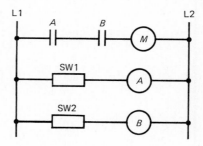

Fig. 23-1 A series control circuit. Both SW1 and SW2 must close to energize coils A and B. This will close the two NO contacts A and B and coil M will be energized.

$A \cdot B = M$

A	B	M
0	0	0
0	1	0
1	0	0
1	1	1

Fig. 23-2 The four possible combinations of contactors A and B in Fig. 23-1.

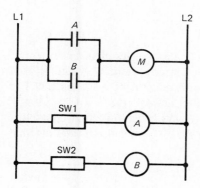

Fig. 23-3 Contactors A and B in parallel with each other and in series with M.

will be energized if either A *or* B (or both) are energized. The short form for this condition is

$$A + B = M$$

The plus sign does *not* denote addition. It stands for "OR." The truth table for this circuit is given in Fig. 23-4.

The symbols and processes just discussed are elements of switching logic. By use of these and similar symbols and equations, a set of rules was developed for analyzing the operation and behavior of control circuits.

$A + B = M$

A	B	M
0	0	0
0	1	1
1	0	1
1	1	1

Fig. 23-4 The four possible combinations of contactors A and B in Fig. 23-3.

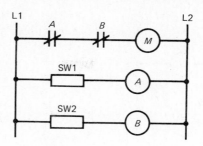

Fig. 23-5 A series circuit with two NC contactors A and B.

NOT Circuits The previous symbols involved NO contacts which closed when their respective operating coils were energized. Figure 23-1 will now be redrawn but contactors A and B will be made normally closed (NC), as in Fig. 23-5. In this circuit, coil M will be energized only if coils A and B are not energized. To indicate NOT, a bar is drawn over the symbol letter. Thus the condition of Fig. 23-5 can be written in equation form as

$$\overline{A} \cdot \overline{B} = M$$

which is read as "A bar AND B bar equals M."

This time the truth table takes the form shown in Fig. 23-6. Note that when $\overline{A} = 1$, $A = 0$. The notation $\overline{A}$ is called the *complement* of A. Thus 1 is the complement of 0 and vice versa.

The equations and the truth tables show an interesting relationship between the AND circuit of Fig. 23-5 and the OR circuit of Fig. 23-3. Compare the truth table of Fig. 23-4 with that of Fig. 23-6. In Fig. 23-6 the complements of A and B ($\overline{A} \cdot \overline{B}$) are combined in an AND circuit. In Fig. 23-4, A and B are combined in an OR circuit. Examine the status of M for each combination of A and B. Notice that each result is the complement of the other. A 1 in Fig. 23-4 becomes a 0 in Fig. 23-6 and vice versa. The general rule, called De Morgan's theorem, is

If an OR circuit is replaced by an AND circuit whose elements are complements of the elements in the OR circuit, the results will be complements of each other.

$\overline{A} \cdot \overline{B} = M$

A	B	M
0	0	1
0	1	0
1	0	0
1	1	0

Fig. 23-6 The NOT truth table for Fig. 23-5.

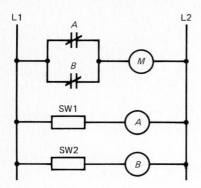

Fig. 23-7 Parallel NOT contactors.

$$\overline{A} + \overline{B} = M$$

A	B	M
0	0	1
0	1	1
1	0	1
1	1	0

Fig. 23-8 Truth table for Fig. 23-7.

This can be written in equation form as

$$\overline{A + B} = \overline{A} \cdot \overline{B}$$

To illustrate this point further, study Fig. 23-7 which has the two NC contactors A and B in parallel. This condition can be written as

$$\overline{A} + \overline{B} = M$$

The truth table representing this equation is in Fig. 23-8. The complement of each of the complements is, respectively, 0, 0, 0, 1. This should be the result of the OR circuit

$$A \cdot B = M$$

Check Fig. 23-2 to verify the fact that $\overline{A \cdot B} = \overline{A} + \overline{B}$ (which is another form of De Morgan's theorem).

23-2 SWITCHING ARITHMETIC

A set of rules and operations can be employed to combine and simplify the equations used to denote different arrangements of contactors. For example, suppose three contactors, A, B, and C, are in series and the group is connected in parallel with three other series contactors, A, B, and D. The equation for such an arrangement would be

$$(A \cdot B \cdot C) + (A \cdot B \cdot D) = Y$$

This equation says that the status of Y in this circuit is determined by contactor A in series with contactor B in series with contactor C and that series combination is in parallel with a series arrangement of another A contactor and another B contactor and contactor D. The actual circuit might look like that of Fig. 23-9. Obviously M will be energized only if $Y = 1$. Let us look at the truth table for this circuit as shown in Fig. 23-10. There are 16 combinations of open and closed contactors A, B, C, and D. Only three result in $Y = 1$ and thus energize M.

However, our original equation can be simplified if A and B are factored out, as would be done in algebra.

$$(A \cdot B \cdot C) + (A \cdot B \cdot D) = (A \cdot B) \cdot (C + D)$$

The new equation indicates a different arrangement of contactors. The series arrangement of A and B is in series with a parallel arrangement of C and D. The circuit for the new equation $(A \cdot B) \cdot (C + D) = Y$ looks like Fig. 23-11. Notice that only four contactors are used rather than the six contactors in the original

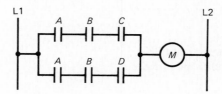

Fig. 23-9 The circuit represented by $(A \cdot B \cdot C) + (A \cdot B \cdot D) = Y$.

$$(A \cdot B \cdot C) + (A \cdot B \cdot D) = Y$$

A	B	C	D	Y
0	0	0	0	0
0	0	0	1	0
0	0	1	0	0
0	0	1	1	0
0	1	0	0	0
0	1	0	1	0
0	1	1	0	0
0	1	1	1	0
1	0	0	0	0
1	0	0	1	0
1	0	1	0	0
1	0	1	1	0
1	1	0	0	0
1	1	0	1	1
1	1	1	0	1
1	1	1	1	1

Fig. 23-10 The truth table for the circuit in Fig. 23-9.

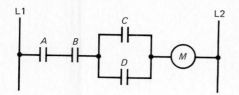

Fig. 23-11 A simplified version of the circuit in Fig. 23-9. The truth table of Fig. 23-10 still applies.

circuit. Yet the two circuits do exactly the same job. The truth table for both circuits would also be exactly the same, so that again only three conditions of the contactors would energize M.

23-3 SWITCHING THEOREMS

The discussions in the previous sections hint at a method for simplifying control circuits merely by applying some basic operations of algebra to circuit equations. But like most mathematical processes certain relationships have to be stated as the basis for building other rules. *Theorems* are statements that can be proved from other established facts or formulas. There are 13 switching theorems that will be used to study and simplify circuits. The proofs of some will be given, and the others are left to the student to prove. The letters A, B, C, and D are used in the theorems to denote contactors (or switches) just as in the previous section.

Theorem 1 $A \cdot (B + C) = A \cdot B + A \cdot C$

Theorem 2 $(A + B) \cdot (C + D) = A \cdot C +$
                 $B \cdot C + A \cdot D + B \cdot D$

Theorem 3 $A + 1 = 1$

Theorem 4 $A \cdot 0 = 0$

Theorem 5 $A \cdot 1 = A$

Theorem 6 $A + 0 = A$

Theorem 7 $A + A = A$

Theorem 8 $A \cdot B + A = A$

Theorem 9 $A \cdot B + (A + B) = A + B$

Theorem 10 $A \cdot B + \overline{B} = A + \overline{B}$

Theorem 11 $A \cdot \overline{B} + B = A + B$

Theorem 12 $A + \overline{A} = 1$

Theorem 13 $A \cdot \overline{A} = 0$

Theorem 3 can be proved by considering what the theorem actually says. A NO contactor, A, is in parallel with a closed circuit. Obviously the circuit will always be closed no matter what the state of A may be.

Similarly Theorem 4 says that the NO contactor A is in series with an open circuit. Obviously an open in a series circuit will remain open no matter what the state of A.

Theorem 10 says that NO contactor A in series with NO contactor B and the combination in parallel with NC contactor B are equal to NO contactor A in parallel with NC contactor B. An examination of each circuit and its associated truth table will show the validity of this theorem. Figure 23-12(a) is the circuit and truth table for $A \cdot B + \overline{B}$. Figure 23-12(b) is the circuit and truth table for $A + \overline{B}$. Notice that the outputs for each are exactly equal.

23-4 APPLICATIONS OF SWITCHING THEOREMS

Let us see how the theorems can help simplify a control circuit. Figure 23-13 is a control diagram consist-

Fig. 23-12 Verification of switching Theorem 10. (a) $A \cdot B + \overline{B}$; (b) $A + \overline{B}$.

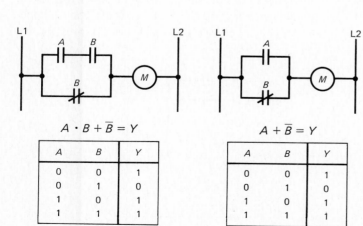

$A \cdot B + \overline{B} = Y$

A	B	Y
0	0	1
0	1	0
1	0	1
1	1	1

(a)

$A + \overline{B} = Y$

A	B	Y
0	0	1
0	1	0
1	0	1
1	1	1

(b)

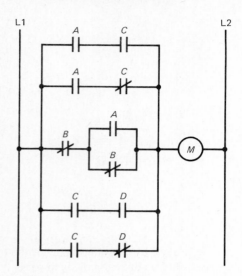

Fig. 23-13 A complex series-parallel control circuit.

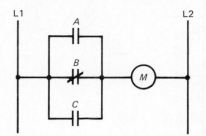

Fig. 23-14 The circuit of Fig. 23-13 has been simplified using switching theorems.

11-contactor, 5-branch circuit could be replaced exactly by three parallel contactors as in Fig. 23-14.

Verification of the equality of the two circuits is left as an exercise for the student.

Problem 1

Suppose a manufacturing line is made up of three conveyor belts, A, B, and C. Belts A and B feed into C. An alarm circuit is required such that if either A or B or both are operating and C stops, the alarm will sound. If either A or B stops and C is running, the alarm will *not* sound. If both A and B stop and C is running, the alarm will sound. It will also not sound if all belts are running or if all are stopped.

Solution

Before drawing the circuit, the conditions of the problem will be recorded in a truth table (Fig. 23-15). Note that the truth table represents all eight conditions of the problem. A "1" indicates a conveyor belt running; a "0" indicates a stopped conveyor belt. Going further, if 1 represents NO

ing of four contactors, A, B, C, and D. The equation for this circuit is

$$A \cdot C + A \cdot \overline{C} + \overline{B} \cdot (A + \overline{B}) + C \cdot D + C \cdot \overline{D} = Y$$

Remember that the bar over the letter indicates NOT, which in the case of the contactors means an NC contactor.

If Y is 1, then M will be energized; if Y is 0, then M will not be energized.

An examination of the switching equation suggests some applications of switching theorems. The equation will be rewritten factoring out some terms and multiplying out others.

$$A \cdot (C + \overline{C}) + \overline{B} \cdot A + \overline{B} \cdot \overline{B} + C \cdot (D + \overline{D}) = Y$$

Applying Theorem 12, $A + \overline{A} = 1$, results in the following simplification of the original equation:

$$A \cdot (1) + \overline{B} \cdot A + \overline{B} + C \cdot (1) = Y$$

And from Theorem 5, $A \cdot 1 = A$, the equation is simplified further to

$$A + \overline{B} \cdot A + \overline{B} + C = Y$$

Notice that either A or $\overline{B}$ can be factored. We will choose A.

$$A \cdot (1 + \overline{B}) + \overline{B} + C = Y$$

which, from Theorem 3, $1 + \overline{B} = 1$, becomes

$$A + \overline{B} + C = Y$$

This is the final reduction. In the process of reducing the equation, we discovered that contactor D served no useful purpose in the circuit and that our original

ROW	A	B	C	Y
1	0	0	0	0
2	0	0	1	1
3	0	1	0	1
4	0	1	1	0
5	1	0	0	1
6	1	0	1	0
7	1	1	0	1
8	1	1	1	0

Fig. 23-15 The truth table for Problem 1.

contacts and NC contacts are represented by 0, each combination of contacts can be drawn as in Fig. 23-16.

For example, row 7 shows *A* and *B* as NO contacts and *C* as an NC contact. Figure 23-16(*g*) represents this condition. In fact, Fig. 23-16(*a*) through (*h*) represents all the combinations of switches required to comply with the original specification. Although each condition shows a *Y* coil, there is actually only one, and all contactors feed through that one control coil.

If we look at Fig. 23-16(*a*) and (*h*), we see that the case of all NO or all NC contactors in a circuit cannot satisfy the specified conditions. If all conveyors are operating, then circuit (*h*) will be open since each of the coils operating *A, B,* and *C* will be energized. Thus the alarm will ring—which is not desired. If none of the conveyors is operating, contacts *A, B,* and *C* will be closed (they are NC) and the alarm will sound again. Circuits (*a*) and (*h*) will therefore be eliminated. Circuits (*b*) through (*g*) must satisfy the conditions of the job. The first condition is that the alarm will sound if

conveyor *A* or *B* or both are operating and *C* is not. With *C* deenergized, circuits (*b*), (*d*), and (*f*) will not sound the alarm. Circuit (*c*) will be complete if *A* is stopped, *B* is running, and *C* is stopped. With the circuit completed, the alarm will sound. Similarly circuit (*e*) will sound the alarm if *A* is running, *B* is stopped, and *C* is stopped. Finally circuit (*g*) will sound the alarm if *A* and *B* are running and *C* is stopped.

The second condition requires that the alarm sound if both *A* and *B* stop and *C* is running. Circuit (*b*) will close sounding the alarm but (*d*) and (*f*) will remain open under these conditions. All other circuits will be open since the NC contactor *C* will open when *C* is running.

The circuit will then resemble Fig. 23-17. The equation for this circuit is

$$\overline{A} \cdot \overline{B} \cdot C + \overline{A} \cdot B \cdot \overline{C} + \overline{A} \cdot B \cdot C + A \cdot \overline{B} \cdot C + A \cdot \overline{B} \cdot \overline{C} + A \cdot B \cdot \overline{C} = Y$$

The circuit of Fig. 23-17 satisfies the conditions of Problem 1, but is it the simplest circuit? Does it use the fewest contactors and the least amount of wiring? These, of course, are important questions to ask before wiring the circuit since material and labor costs are involved and the more parts and wiring there are, the more likely it is for something to go wrong.

There are two common ways of simplifying the solution to Problem 1. The first method involves the switching theorems discussed in Sec. 23-3. The second method involves reducing the truth table in Fig. 23-15.

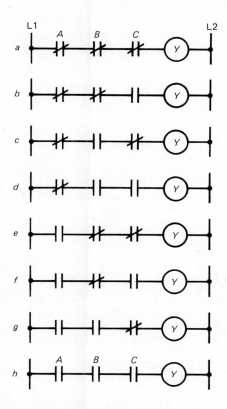

Fig. 23-16 The eight possible circuit combinations satisfying the conditions of Problem 1.

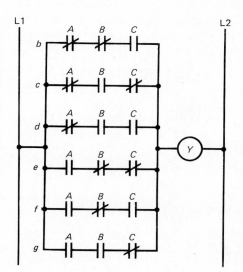

Fig. 23-17 The control circuit for Problem 1 showing the six valid branches.

Using Switching Theorems The 13 theorems listed in Sec. 23-3 can be used to reduce a switching equation to a simpler form. The equation in the solution to Problem 1 will be used to demonstrate this process.

$$\overline{A} \cdot \overline{B} \cdot C + \overline{A} \cdot B \cdot \overline{C} + \overline{A} \cdot B \cdot C + A \cdot \overline{B} \cdot \overline{C}$$
$$+ A \cdot \overline{B} \cdot C + A \cdot B \cdot \overline{C} = Y$$

Factoring like terms (but first we will group them together),

$$\overline{A} \cdot \overline{B} \cdot C + A \cdot \overline{B} \cdot C + \overline{A} \cdot B \cdot \overline{C} + \overline{A} \cdot B \cdot C$$
$$+ A \cdot \overline{B} \cdot \overline{C} + A \cdot B \cdot \overline{C} = Y$$

$$\overline{B} \cdot C \cdot (\overline{A} + A) + \overline{A} \cdot B \cdot (\overline{C} + C)$$
$$+ A \cdot \overline{C} \cdot (\overline{B} + B) = Y$$

By Theorem 12, $\overline{A} + A = 1$, which also means that $\overline{C} + C = 1$ and $\overline{B} + B = 1$:

$$\overline{B} \cdot C + \overline{A} \cdot B + A \cdot \overline{C} = Y$$

Our original switching equation containing 18 contactors and six branches is thus reduced to six contactors and three branches. The circuit represented by this equation is shown in Fig. 23-18.

We can test the circuit against the problem conditions: If C is stopped, branch a will be open, no matter what the condition of belt B. Branches b and c, however, require that both A and B be stopped if the alarm is NOT to sound. Should either A or B or both be running and C stopped, the alarm will sound.

If C is running, branch c will be open. If either A or B is stopped—but not both—neither branch a nor b will close to sound the alarm. However, when both A and B are stopped and C is running, branch a will close and the alarm will sound.

If A, B, and C are either all running or all stopped, none of the branches will close and the alarm will not sound. The reduced circuit therefore satisfies the conditions of the problem.

23-5 TRUTH TABLE SIMPLIFICATION

A truth table can sometimes be condensed by inspection. Notice that the truth table in Fig. 23-15 contains

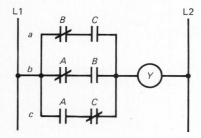

Fig. 23-18 The first step in simplifying the circuit of Fig. 23-17.

	A	B	C	Y
2, 6		0	1	
3, 4	0	1		
5, 7	1		0	

Fig. 23-19 A simplification of the truth table in Fig. 23-15.

eight rows all very similar to one another. Since rows 1 and 8 were eliminated as not satisfying the conditions, only rows 2 through 7 remain. Rows 3 and 4 are the same except for column C; rows 2 and 6 have the same columns except for A. The truth table (Fig. 23-15) will be redrawn combining similar rows but omitting the columns that differ. The new truth table is Fig. 23-19.

Writing the switching equation gives us the following:

$$\underset{\text{rows 2, 6}}{\overline{B} \cdot C} + \underset{\text{rows 3, 4}}{\overline{A} \cdot B} + \underset{\text{rows 5, 7}}{A \cdot \overline{C}} = Y$$

Note that this is exactly the same as the equation arrived at using the simplification-by-switching-theorem method.

While these two methods produced a simplified solution to our original Problem 1, it is by no means the only solution possible. Two other methods for finding solutions to switching problems will be discussed next.

23-6 SUM-OF-THE-PRODUCTS

The *sum-of-the-products* or *minterm* method requires finding all the products of A, B, and C that occur in the original truth table (Fig. 23-15). For example,

Product	Row(s)
$A \cdot B$	7, 8
$\overline{A} \cdot \overline{B}$	1, 2

The complete table of products is given in Fig. 23-20.

Any combination of these products which includes each row just once is a possible solution to Problem 1. Of course, those combinations that include rows 1 and 8 will be eliminated. Checking the combinations in Fig. 23-20 produces the following acceptable combinations:

3, 4	2, 4	4
5, 6	5, 7	5
2, 6	2	6
3, 7	3	7

PRODUCT	ROW(S)	
$A \cdot B$	7, 8	
$\overline{A} \cdot B$	3, 4	*
$A \cdot \overline{B}$	5, 6	*
$\overline{A} \cdot \overline{B}$	1, 2	
$B \cdot C$	4, 8	
$\overline{B} \cdot C$	2, 6	*
$B \cdot \overline{C}$	3, 7	*
$\overline{B} \cdot \overline{C}$	1, 5	
$A \cdot C$	6, 8	
$\overline{A} \cdot C$	2, 4	*
$A \cdot \overline{C}$	5, 7	*
$\overline{A} \cdot \overline{C}$	1, 3	
$\overline{A} \cdot \overline{B} \cdot \overline{C}$	1	
$\overline{A} \cdot \overline{B} \cdot C$	2	*
$\overline{A} \cdot B \cdot \overline{C}$	3	*
$\overline{A} \cdot B \cdot C$	4	*
$A \cdot \overline{B} \cdot \overline{C}$	5	*
$A \cdot \overline{B} \cdot C$	6	*
$A \cdot B \cdot \overline{C}$	7	*
$A \cdot B \cdot C$	8	
A	5, 6, 7, 8	
B	3, 4, 7, 8	
C	2, 4, 6, 8	
$\overline{A}$	1, 2, 3, 4	
$\overline{B}$	1, 2, 5, 6	
$\overline{C}$	1, 3, 5, 7	

Fig. 23-20 Table of products of *A*, *B*, and *C*.

The acceptable combinations are indicated by an asterisk (*) in Fig. 23-20. A natural combination would be 2, 3, 4, 5, 6, 7, but that is the original switching equation from Problem 1. Another would be 5, 6; 3, 7; 2, 4.

$$A \cdot \overline{B} + B \cdot \overline{C} + \overline{A} \cdot C = Y \quad \text{(Fig. 23-21)}$$

Still another combination is 2, 6; 3, 4; 5, 7.

$$\overline{B} \cdot C + \overline{A} \cdot B + A \cdot \overline{C} = Y$$

This will be recognized as the solution obtained using the switching theorem and truth table methods.

Of course, other combinations are possible. Each

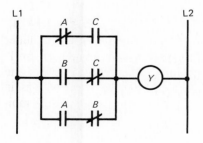

Fig. 23-21 One of the control circuits found using the sum-of-the-products method.

should be tested to see whether it satisfies the conditions of the problem. Since our best circuit contained only three branches and six contactors, any solution that requires more would be rejected. Checking other possible combinations is left as an exercise for the student.

23-7 PRODUCT-OF-THE-SUMS

The *product-of-the-sums* or *maxterm* method can also be used to simplify switching circuits. As with the sum-of-the-products all combinations of sums must be tabulated. Here the De Morgan switching logic theorem can be used. It states that the product of numbers (or letters) is equal to the complement of the sum of the complements of the numbers (or letters). Thus

$$\overline{A \cdot \overline{B}} = \overline{A} + B$$
$$\overline{A \cdot B} = \overline{A} + \overline{B}$$
$$\overline{A \cdot B \cdot C} = \overline{A} + \overline{B} + \overline{C}$$

and so on. This makes it easy to prepare the table of sums using Fig. 23-20. The resultant table of sums is shown in Fig. 23-22. As before, the combinations of sums must include all rows with the exception of 1 and 8. The asterisk indicates the acceptable groups. As with the sum-of-the-products, those combinations

SUM	ROW(S)	
$\overline{A} + \overline{B}$	1, 2	
$A + \overline{B}$	5, 6	*
$\overline{A} + B$	3, 4	*
$A + B$	7, 8	
$\overline{B} + \overline{C}$	1, 5	
$B + \overline{C}$	3, 7	*
$\overline{B} + C$	2, 6	*
$B + C$	4, 8	
$\overline{A} + \overline{C}$	1, 3	
$A + \overline{C}$	5, 7	*
$\overline{A} + C$	2, 4	*
$A + C$	6, 8	
$\overline{A} + \overline{B} + \overline{C}$	1	
$\overline{A} + \overline{B} + C$	2	*
$\overline{A} + B + \overline{C}$	3	*
$\overline{A} + B + C$	4	*
$A + \overline{B} + \overline{C}$	5	*
$A + \overline{B} + C$	6	*
$A + B + \overline{C}$	7	*
$A + B + C$	8	
A	5, 6, 7, 8	
B	3, 4, 7, 8	
C	2, 4, 6, 8	
$\overline{A}$	1, 2, 3, 4	
$\overline{B}$	1, 2, 5, 6	
$\overline{C}$	1, 3, 5, 7	

Fig. 23-22 Table of sums of *A*, *B*, and *C*.

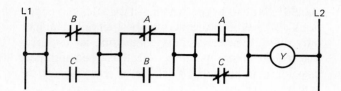

Fig. 23-23 A circuit found using the product-of-the-sums method.

that include each row once are valid solutions to the problem.

For example, 2, 6; 3, 4; 5, 7 is acceptable. The equation is

$$(\overline{B} + C) \cdot (\overline{A} + B) \cdot (A + \overline{C}) = Y$$

which represents the circuit of Fig. 23-23.

Still another possibility is 2, 4; 3, 7; 5, 6, which produces the equation

$$(\overline{A} + C) \cdot (B + \overline{C}) \cdot (A + \overline{B}) = Y$$

and the circuit of Fig. 23-24.

Another possibility of a solution is the use of only rows 1 and 8. This produces

$$(A + B + C) \cdot (\overline{A} + \overline{B} + \overline{C}) = Y$$

This is a circuit in which A, B, and C are in parallel with each other and in series with the parallel combination of $\overline{A}$, $\overline{B}$, and $\overline{C}$. Figure 23-25 shows this circuit.

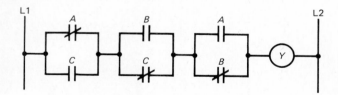

Fig. 23-24 Another product-of-the-sums circuit.

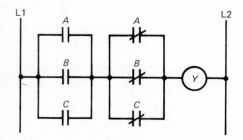

Fig. 23-25 The circuit represented by the equation $(A + B + C) \cdot (\overline{A} + \overline{B} + \overline{C}) = Y$.

23-8 DIGITAL ELECTRONICS

Many electronic devices can act as switches without physically opening and closing a circuit. By applying voltages in a particular range of values and polarity, these devices can change their electrical properties from an excellent conductor to that of an insulator. The specific devices, their characteristics, and applications are discussed in Chap. 24. In this chapter we will discuss their use in switching circuits. Since these devices have just two discrete states or conditions, conducting and nonconducting (or ON and OFF), the principles of binary numbers can be applied.

23-9 BINARY NUMBERS

A mechanical contactor has two states: open or closed. We used 1 to denote a closed contactor and 0 to denote an open contactor. Thus a numbering system and a set of mathematical rules that required only the two numbers (or symbols) 1 and 0 should enable us to record the condition of a contactor and perhaps help us determine the condition of a circuit containing many contactors in series and parallel combinations.

Such a number system does exist. It is called the *binary number system.* Just as in the arithmetic in which we use the digits 0 through 9, so, too, binary arithmetic has a set of rules and operations. As would be expected from a system having just two numbers, the operations and their rules are few and quite simple.

$$1 + 1 = 0 \text{ and carry } 1$$
$$0 + 1 = 1 + 0 = 1$$
$$1 \times 1 = 1$$
$$0 \times 1 = 1 \times 0 = 0$$
$$0 + 0 = 0$$

Most of these rules look familiar from our customary arithmetic. The one strange, or rather new, operation is $1 + 1 = 0$ and carry 1.

The digits 0 and 1 can be combined as multiple-digit numbers, and the arithmetic operations performed on these numbers. For example, the number 11001 can be added to 10011 as follows:

$$
\begin{array}{ccccccl}
⑤ & ④ & ③ & ② & ① & \longleftarrow & \text{column} \\
1 & 1 & 0 & 0 & 1 & & \\
+ \;\; 1 & 0 & 0 & 1 & 1 & & \\
\hline
\end{array}
$$

Starting from column 1,

$$
\begin{array}{r}
1 \\
+ \;\; 1 \\
\hline
0 \text{ carry } 1 \\
\end{array}
$$

Column 2 is

$$0$$
$$+ \ \underline{1}$$
$$1 + 1 \ \text{(carry)} = 0 + 1 \ \text{(carry)}$$

Column 3 is

$$0$$
$$+ \ \underline{0}$$
$$0 + 1 \ \text{(carry)} = 1$$

So for the first three columns the sum looks like this:

```
            ③ ② ①  ⟵  column
      1  1  0  0  1
   +  1  0  0  1  1
      ─────────────
            1  0  0
```

Column 4 produces

$$1$$
$$+ \ \underline{0}$$
$$1$$

Finally column 5 results in the last sum

$$1$$
$$+ \ \underline{1}$$
$$0 + 1 \ \text{(carry) to the sixth column}$$

Thus our example in addition results in the following answer:

```
      1  1  0  0  1
   +  1  0  0  1  1
   ─────────────────
   1  0  1  1  0  0
```

23-10 BINARY-TO-DECIMAL CONVERSION

While binary numbers, themselves, are important in switching and control circuits, the input to, and output from, the control equipment is often stated in terms of our usual number system containing the digits 0 to 9. This number system is called the *decimal system*. Since there are 10 symbols in this system (the 10 digits, of course), we say the system has the base 10. The binary system, on the other hand, has only two symbols, 0 and 1, so we say the binary system has the base 2.

In the decimal system the position of the digit is *weighted* in powers of 10. That is, counting from the right, the digit in that position is multiplied by 10 raised to the power of that position. The first position to the right is called the zero position, and so the digit in that position is multiplied by 10 raised to the 0

6	5	4	3	2	1	0	DIGIT POSITION
10^6	10^5	10^4	10^3	10^2	10^1	10^0	POWER OF 10
3	**2**	**0**	**6**	**1**	**5**	**7**	NUMBER
3×10^6	2×10^5	0	6×10^3	1×10^2	5×10^1	7×10^0	POSITION VALUE

$$3 \times 10^6 + 2 \times 10^5 + 0 + 6 \times 10^3 + 1 \times 10^2 + 5 \times 10^1 + 7 \times 10^0 = 3,206,157$$

Fig. 23-26 The decimal number system.

7	6	5	4	3	2	1	0	DIGIT POSITION
2^7	2^6	2^5	2^4	2^3	2^2	2^1	2^0	BINARY VALUE
128	64	32	16	8	4	2	1	NUMERICAL VALUE
1	**1**	**0**	**1**	**0**	**0**	**1**	**1**	NUMBER
128	64	0	16	0	0	2	1	POSITION VALUE

$$128 + 64 + 0 + 16 + 0 + 0 + 2 + 1 = 211_{10}$$

Fig. 23-27 The binary number system.

power. But 10^0 is equal to 1. So the digit in that first position is merely equal to itself. The next position is 1, and so the digit in that position must be multiplied by 10^1, or 10. This process continues until all positions are calculated and added to give the value of the decimal number. Figure 23-26 shows the value of each digit position in the decimal system up to 1 million. The decimal number shown in the figure is derived as shown from the powers of 10.

Similarly the binary system uses powers of 2. Figure 23-27 shows the positions of the digits. The digit position and position value, however, are shown in the decimal system.

Problem 2

Find the decimal value of the binary number 111.

Solution

The first digit on the right, 1, is multiplied by 2^0, or 1; the second by 2^1, or 2; the third by 2^2, or 4.

(Remember that the answer will be given as a decimal number.)

$$1 \times 1 = 1$$
$$1 \times 2 = 2$$
$$1 \times 4 = \underline{4}$$
$$7$$

Thus 111_2 is equal to 7_{10}. The small numbers to the right of 111 and 7 represent base 2 and base 10, or the binary and decimal systems, respectively.

23-11 DECIMAL-TO-BINARY CONVERSION

A decimal number can be converted to a binary number by successively dividing the decimal number by 2 and recording the remainder at each stage. For example, we will convert 7 to its binary equivalent:

$$7 \div 2 = 3 \text{ remainder } 1$$
$$3 \div 2 = 1 \text{ remainder } 1$$
$$1 \div 2 = 0 \text{ remainder } 1$$

Division stops when the quotient is 0. Thus the binary equivalent of 7 is 111. Unfortunately, this example does not indicate the order of the remainders in the binary numbers as they are obtained. The binary equivalent of 8 will help to solve that problem.

$$8 \div 2 = 4 \text{ remainder } 0$$
$$4 \div 2 = 2 \text{ remainder } 0$$
$$2 \div 2 = 1 \text{ remainder } 0$$
$$1 \div 2 = 0 \text{ remainder } 1$$

Now it appears clear that the first remainder is the first binary digit to the right and each successive remainder takes the next position to the left.

Problem 3

Find the binary equivalent of 117_{10}.

Solution

$117 \div 2 = 58$ remainder 1	2^0 position
$58 \div 2 = 29$ remainder 0	2^1 position
$29 \div 2 = 14$ remainder 1	2^2 position
$14 \div 2 = 7$ remainder 0	2^3 position
$7 \div 2 = 3$ remainder 1	2^4 position
$3 \div 2 = 1$ remainder 1	2^5 position
$1 \div 2 = 0$ remainder 1	2^6 position

The binary equivalent of 117_{10} is therefore 1110101_2.

To check this answer, we need merely multiply each position digit by the power of 2.

$$1 \times 2^6 \text{ or } 64 = 64$$
$$1 \times 2^5 \text{ or } 32 = 32$$
$$1 \times 2^4 \text{ or } 16 = 16$$
$$0 \times 2^3 \text{ or } 8 = 0$$
$$1 \times 2^2 \text{ or } 4 = 4$$
$$0 \times 2^1 \text{ or } 2 = 0$$
$$1 \times 2^0 \text{ or } 1 = \underline{1}$$
$$117_{10}$$

23-12 LOGIC GATES

In previous sections, logic circuits called AND and OR were discussed using contactors and their operating coils. Various combinations of series and parallel contactors and coils were able to open and close circuits according to a programmed sequence of operations. We used switching logic and a set of switching theorems to develop and simplify switching equations involving these contactors.

Similar control can be achieved using electronic components called *semiconductors* or *solid-state devices*. These devices have names such as diodes, transistors, and integrated circuits. In particular, integrated circuits are available that can perform as AND, OR, and other types of *logic circuits*. For this reason these devices are known as *logic gates*. In subsequent sections these logic gates will be described and finally applied to many of the control and switching circuits previously discussed.

23-13 AND GATE

An AND gate has the property of a group of series contactors or switches. The symbol for an AND gate is shown in Fig. 23-28. Although the symbol shows two

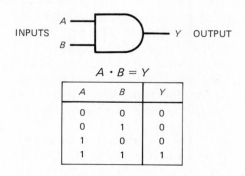

$$A \cdot B = Y$$

A	B	Y
0	0	0
0	1	0
1	0	0
1	1	1

Fig. 23-28 The AND gate symbol and truth table. The symbol represents the function of contactors A and B in Fig. 23-1.

inputs, AND gates can have more. The AND gate operates in an external circuit the same way as the mechanical contactors. However, its internal operation is quite different. Basically an AND gate responds to two levels of voltage, a HI and a LO. The HI could be +5 V dc and the LO could be 0 V; or the HI could be 0 V and the LO −5 V dc.

These voltage levels are determined by the circuit designer, but in terms of switching logic, the HI is a binary 1 while the LO is a binary 0.

Referring to Fig. 23-28, the output Y will be 1 only if A and B are 1. In practical terms this means we would need to apply a voltage (for example, the HI of +5 V) to both inputs in order for the output to be at +5 V.

23-14 OR GATE

An OR gate has the property of a group of contactors or switches connected in parallel. The symbol for an OR gate is shown in Fig. 23-29. Although the symbol shows two inputs, an OR gate can have more. The OR gate operates in an external circuit the same way as the mechanical contactors. Like the AND gate, the OR gate responds to two levels of voltage, a HI and a LO.

Referring to Fig. 23-29, the output Y will be HI (binary 1) if either A or B is HI or both are HI. In practical terms, this means that the output will go HI if any of the inputs are HI.

23-15 EXCLUSIVE OR (XOR) GATE

The OR gate discussed in Sec. 23-14 produced an output (binary 1) under the condition that if one or more (or all) of the inputs were binary 1, the output would be binary 1. Although not generally specified as such,

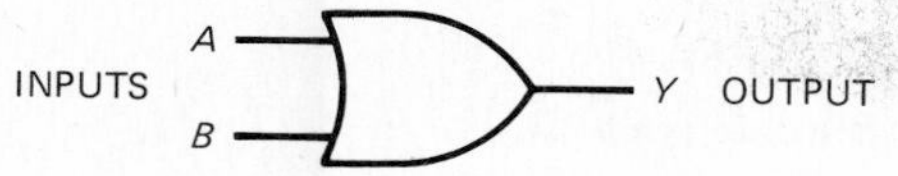

INPUTS $\begin{array}{c} A \\ B \end{array}$) Y OUTPUT

$$A + B = Y$$

A	B	Y
0	0	0
0	1	1
1	0	1
1	1	1

Fig. 23-29 The OR gate symbol and truth table. The symbol represents the function of contactors A and B in Fig. 23-3.

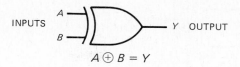

INPUTS $\begin{array}{c} A \\ B \end{array}$) Y OUTPUT

$$A \oplus B = Y$$

Fig. 23-30 The XOR gate symbol and truth table.

this type of OR gate is an *inclusive* OR gate. That is, it includes all combinations of 1.

The XOR gate is called *exclusive* because not all combinations of input 1s produce an output 1. Only an odd number of input 1s will produce a binary 1 in the output. Thus a two-input XOR gate will produce an output 1 if either one of its inputs is 1 but not if they are both 1. On the other hand, a three-input XOR gate will have an output 1 if any one of its inputs is 1 or if all three are 1. If any two of its inputs are 1, then the output will be binary 0.

The symbol for an exclusive OR gate (XOR gate) is shown in Fig. 23-30. The symbol resembles the OR gate with the exception of the extra line at the input. In equations the symbol $\oplus$ is used to differentiate between the XOR operation and the OR operation denoted by $+$. The common use of the XOR gate is in circuits that must distinguish one input from another and act accordingly.

23-16 NOT GATE

The NOT gate is simply a device for inverting an input. In the NOT gate, also called an *inverter*, a single input will produce its complement at the output. Figure 23-31 shows the symbol for the NOT gate. As a logic symbol the triangle is always shown with the output circle, indicating that the output is the complement of the input. For example, an input 1 produces an output 0. Since the purpose of the inverter is merely to change the state of a signal, it consists of just one input line and one output line.

23-17 NAND GATE

The NAND gate can be thought of as an inverted AND gate. The word NAND, in fact, comes from NOT AND.

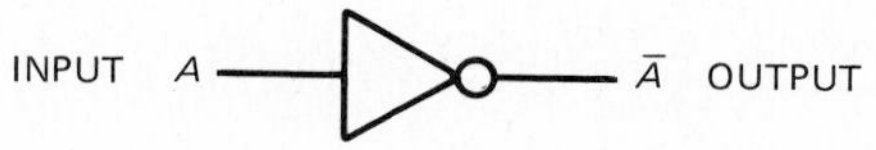

INPUT A ——▷○—— $\bar{A}$ OUTPUT

Fig. 23-31 The NOT gate symbol.

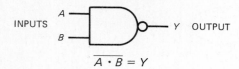

INPUTS
A
B
Y OUTPUT

$$\overline{A \cdot B} = Y$$

Fig. 23-32 The NAND gate symbol and truth table.

The AND gate will produce an output (binary 1) if all inputs are 1s. With a NAND gate, if all inputs are 1, the output is binary 0. By the same token, the AND gate has no output (binary 0) if any of the inputs are 0. The NAND gate, however, will produce a binary 1 if any of the inputs are 0.

The symbol for the NAND gate is shown in Fig. 23-32. Notice that it resembles the AND gate with the exception of the small circle at the output. In logic symbols, the small circle is used to denote inversion or complement. Thus an output that would ordinarily be binary 1 is inverted and becomes binary 0. An output of X becomes $\overline{X}$.

As with other gate symbols, although only two inputs are shown, more than two inputs can be used.

23-18 NOR GATE

The NOR gate is an OR gate which inverts its output. Since the OR gate produces a binary 1 if any input is 1 or if all inputs are 1, the NOR simply inverts the output to binary 0. The symbol for a NOR gate is shown in Fig. 23-33. Although this symbol has only two inputs, NOR gates can have more. The truth table for a two-input NOR gate is given in Fig. 23-34. Compare the

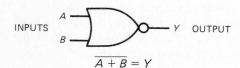

INPUTS
A
B
Y OUTPUT

$$\overline{A + B} = Y$$

Fig. 23-33 The NOR gate symbol.

$$\overline{A + B} = Y$$

A	B	Y
0	0	1
0	1	0
1	0	0
1	1	0

Fig. 23-34 The NOR truth table.

truth table for the NOR gate with the truth table for the OR gate. Does the relationship concerning complements apply here?

23-19 XNOR GATE

The exclusive OR gate (XOR) produced an output binary 1 only when there was an odd number of input 1s. The XNOR is merely an extension of the NOR gate with the same qualification. Remember that in a NOR gate a 1 input in any of the input lines produces a binary 0 in the output. Thus in an XNOR gate the 0 is produced only for an odd number of 1 inputs. The XNOR symbol is shown in Fig. 23-35 and its truth table in Fig. 23-36.

23-20 APPLICATIONS OF LOGIC GATES

Logic gates are building blocks to larger, more complex control and processing operations. For example, a simple binary number adding machine can be made using the logic gates previously discussed. At this point it would be well to review Sec. 23-9 on binary numbers since they will be used extensively in this section. The binary adder we will develop makes use of the following operations:

$$1 + 1 = 0 \text{ with a 1 carry}$$
$$1 + 0 = 1$$
$$0 + 1 = 1$$
$$0 + 0 = 0$$

In this case the + sign has the meaning of addition.

Since numbers are always added two at a time, we need provide only two inputs, A and B. In Fig. 23-37

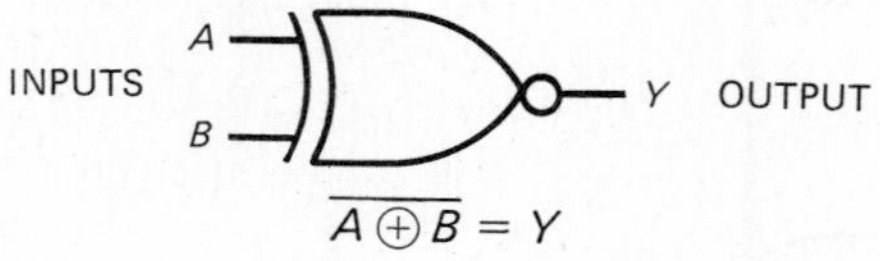

INPUTS
A
B
Y OUTPUT

$$\overline{A \oplus B} = Y$$

Fig. 23-35 The XNOR gate symbol.

$$\overline{A \oplus B} = Y$$

A	B	Y
0	0	1
0	1	0
1	0	0
1	1	1

Fig. 23-36 The XNOR truth table.

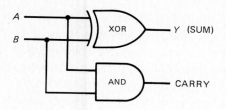

Fig. 23-37 A partial, or half, adder.

two inputs feed into an AND gate and an XOR gate. We will test each of the four additions given above.

First, assume both *A* and *B* have a binary 1 for the sum 1 + 1. Since there are two input 1s, the XOR gate will produce an output 0. However, the AND gate will produce a 1 when both *A* and *B* are 1. Thus there is an output of 1 on the carry line. This checks with the known results of 1 + 1.

Next, assume *A* = 0 and *B* = 1. In that case, *Y* = 1 and the carry line is 0. Remember the AND gate needs a 1 on both inputs to produce a 1.

Similarly, when *A* = 1 and *B* = 0, *Y* will produce a 1 and again there will not be a carry.

Finally, with *A* = 0 and B = 0, *Y* will be 0 and there will be no carry.

We have demonstrated how logic gates can be connected to produce a simple binary number adding machine, but our adding machine is incomplete since it is not equipped to handle the "carry 1" that results from 1 + 1. By connecting another similar pair of gates to our first pair, we can add a carry line to our input. Figure 23-38 shows such an arrangement.

With this new circuit we can add binary numbers containing more than one digit. For example, we will follow the process of adding the two binary numbers previously used in Sec. 23-9.

```
  ⑤  ④  ③  ②  ①   ←—— column
   1   1   0   0   1   ←—— A
+  1   0   0   1   1   ←—— B
```

The numbers in the circles represent the order in which the numbers will be added. In the first opera-

tion *A* = 1 and *B* = 1; there is no carry from a previous operation. The XOR 1 produces an output 0 which goes to XOR 2. The AND 1 produces an output 1 which goes to the OR. At XOR 2 the 0 from XOR 1 and the 0 Carry In produce a *Y* = 0 as the sum. The AND 2 with two 0 inputs produces a 0 as an input to the OR gate. The OR gate, having one input 1, produces a Carry Out of 1. So after the first operation, the addition looks like this:

```
              1        ←—— carry
  ⑤  ④  ③  ②  ①   ←—— column
   1   1   0   0   1   ←—— A
+  1   0   0   1   1   ←—— B
                   0
```

In column 2, *A* = 0, *B* = 1, carry = 1. Therefore XOR 1 produces an output 1, and AND 1 produces an output 0. The XOR 1 output 1 goes to XOR 2. Since XOR 2 has two input 1s (remember the carry 1 from column 1), it produces an output sum of 0. Meanwhile the OR gate, having at least one input 1, produces a carry 1. The answer, after two columns of addition, is

```
           1           ←—— carry
  ⑤  ④  ③  ②  ①   ←—— column
   1   1   0   0   1   ←—— A
+  1   0   0   1   1   ←—— B
               0   0
```

In column 3, *A* = 0, *B* = 0, carry = 1. Thus XOR 1 and AND 1 produce output 0s, and the OR has one input 0. The XOR 1 output 0 goes to XOR 2. The carry 1 and the 0 from XOR 1 produce an XOR 2 output 1 as the sum. With only one input 1, AND 2 has an output 0 which goes to the OR. The OR now has two input 0s and its output is 0, that is, no carry.

```
                       ←—— carry
  ⑤  ④  ③  ②  ①   ←—— column
   1   1   0   0   1   ←—— A
+  1   0   0   1   1   ←—— B
           1   0   0
```

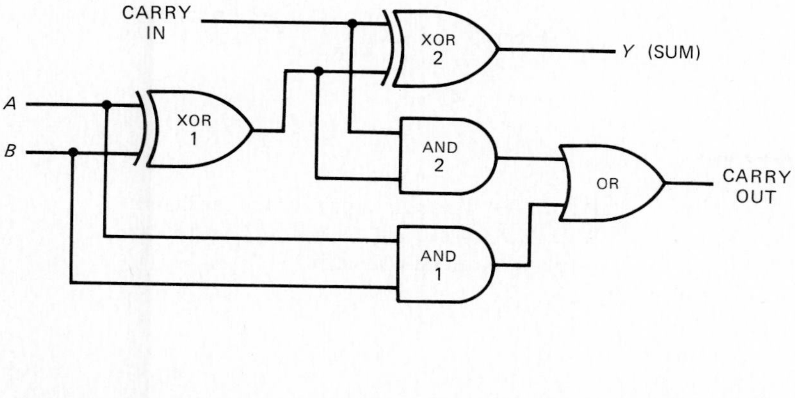

Fig. 23-38 The logic circuit for a full adder for binary numbers.

The student should verify that the sum in column 4 is 1 with 0 carry. Finally, in column 5, $A = 1$, $B = 1$, and carry $= 0$. This addition is similar to column 1. The result is a sum of 0 and a carry of 1. The carry therefore becomes the final digit of the answer.

$$
\begin{array}{r}
1\ 1\ 0\ 0\ 1 \\
+\ \underline{1\ 0\ 0\ 1\ 1} \\
1\ 0\ 1\ 1\ 0\ 0
\end{array}
$$

This, of course, checks with the answer obtained in Sec. 23-9.

A further check can be performed using decimal numbers. Converting 11001 to decimal numbers,

$$
\begin{array}{ccccc}
2^4 & 2^3 & 2^2 & 2^1 & 2^0 \\
1 & 1 & 0 & 0 & 1
\end{array}
$$

Thus we see

$$
\begin{aligned}
2^4 \times 1 &= 16 \\
2^3 \times 1 &= 8 \\
2^2 \times 0 &= 0 \\
2^1 \times 0 &= 0 \\
2^0 \times 1 &= \underline{1} \\
&\ 25
\end{aligned}
$$

Similarly 10011 converted to a decimal number is

$$
\begin{aligned}
2^4 \times 1 &= 16 \\
2^3 \times 0 &= 0 \\
2^2 \times 0 &= 0 \\
2^1 \times 1 &= 2 \\
2^0 \times 1 &= \underline{1} \\
&\ 19
\end{aligned}
$$

Finally 101100 is

$$
\begin{aligned}
2^5 \times 1 &= 32 \\
2^4 \times 0 &= 0 \\
2^3 \times 1 &= 8 \\
2^2 \times 1 &= 4 \\
2^1 \times 0 &= 0 \\
2^0 \times 0 &= \underline{0} \\
&\ 44
\end{aligned}
$$

The addition in decimal numbers is

$$
\begin{array}{r}
25 \\
+\ \underline{19} \\
44
\end{array}
$$

which checks correctly.

Problem 3

Using combinations of the logic gates discussed in this chapter, produce a logic circuit that will satisfy the conditions of Problem 1 in Sec. 23-4.

Solution

In Problem 1 the alarm system was assumed to operate when coil Y was energized. The closing of various contactors completed a circuit that placed coil Y across the line. In a typical installation the coil might be operated by 120 V, though lower voltages are also used. Each of the contactors was also assumed to have been operated by a coil that was energized when the respective belt operated and was deenergized when the belt stopped.

A somewhat similar operation may be assumed when using logic gates. The inputs to the gates are highs and lows. We will assume that a sensor in the belt circuit sends a HI, or +5-V signal, to the control circuit when the belt is running and a LO, or 0-V signal, when the belt is stopped. The input signals to the logic gate determine whether the gate will conduct and produce a HI or not conduct and therefore produce an output LO. We also assume that the +5 V output turns on the alarm circuit. Thus an output 1 means the alarm will sound.

Returning to the problem, the first condition to be satisfied is that with "A or B or both running" and "C stopped" the alarm will sound.

A or B or both running suggests an OR gate; *and C stopped* suggests an AND gate and an inverter, or NOT gate. The need for a NOT gate arises from the fact that when C is stopped (a 0 input) some reaction is expected. We normally associate a 1 with triggering a reaction, though this is certainly not the case in many control circuits. An input 0 can also be used to trigger a reaction.

Figure 23-39 shows a possible arrangement of logic gates that would effect the first condition of the problem. This circuit also satisfies the condition that no alarm sound if all belts are running

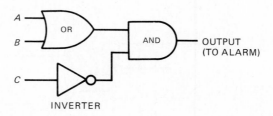

Fig. 23-39 **The logic circuit for the condition: When C is stopped and either A or B or both are running, the alarm will sound.**

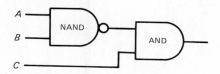

Fig. 23-40 The logic circuit satisfying the condition: When *C* is running and both *A* and *B* are stopped, the alarm will sound.

(that is, $A = B = C = 1$) or all belts are stopped ($A = B = C = 0$).

The second condition of Problem 1 is that with *C* running, the alarm will sound if both *A* and *B* are stopped.

The condition that *A and B* be stopped suggests an AND gate, but since a reaction is expected with NO inputs, a NAND seems more appropriate. The status of *A* and *B and C* running again suggests an AND gate. The logic diagram for this second condition is shown in Fig. 23-40.

Again, this diagram satisfies the condition that no alarm is sounded when all belts are stopped or all belts are running.

The two logic diagrams can not be interconnected to form a single logic circuit that will satisfy all the conditions of the original problem. In Fig. 23-41 AND 1 and AND 2 feed into a common output. An output 1 will turn on the alarm circuit. The truth table for this diagram is given in Fig. 23-15. The student should verify each row with the logic diagram.

As with problems involving contactors, there are a number of possible solutions to most switching and logic problems. Usually the solution requiring the fewest contactors or gates and the least complicated interconnections is the one used. In some cases extra parts or backup systems, called *redundant* systems,

are used to ensure operation. Often extra subsystems are added to monitor the main system. All these require extra components and add to the complexity of the circuits.

23-21 OTHER DIGITAL CIRCUITS

Logic gates are the key elements of any electronic control system, but they are not the only ones. Very often these elements are packaged as individual or integrated digital devices.

Flip-Flops A flip-flop is a circuit, or device, that can maintain a binary state indefinitely until switched, or *toggled,* by a pulse which changes the output to the complementary binary state. For example, the flip-flop output may be set as a binary 1. It will stay at that state until another input pulse changes the output to a binary 0. A *reset-set,* or *R-S,* flip-flop has two outputs; one is a binary 1, the other is 0. Both cannot be 1. There are also two inputs, *R* and *S.* By setting the inputs to 1 or 0, the output states can be made to change in a controlled manner.

Shift Registers A shift register is a device for storing data, usually in the form of a binary number. A series connection of flip-flops, the output of one becoming the input of the next, can store binary 1s or 0s. As the end flip-flop is pulsed, the other flip-flops in the series change their state. A parallel arrangement of flip-flops can also store information, or data, with each flip-flop being pulsed individually. Shift registers are therefore memory devices since they will hold their data until changed.

Counters By connecting flip-flops in series, similar to the shift register described above, a counting process can be achieved. Regular pulses of direct current

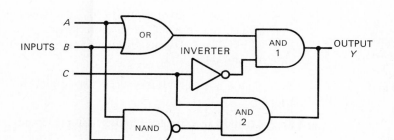

Fig. 23-41 The complete logic diagram satisfying the condition of Problem 3.

called *clock pulses* toggle the action of the flop-flop, changing the state of each one in turn. Depending upon the frequency of the clock pulse, the output of the previous flip-flop is divided by a multiple of 2, 4, 8, etc. As a result of the timed pulses, the state of each flip-flop is advanced to the next binary state—a 1 becomes 0 and a 0 becomes 1.

23-22 LOGIC SYMBOL STANDARDS

The symbols used to represent the various logic gates in this chapter are accepted as standard by the Institute of Electrical and Electronics Engineers (IEEE) and the American National Standards Institute (ANSI). The symbols are used on electronics drawings, especially those involving computers, computing and digital circuits, and consumer electronic equipment.

Equipment used in industrial equipment and installations that require drawings of logic gates and other digital symbols generally use symbols standardized by the National Electrical Manufacturers Association (NEMA). In particular, schematic and logic diagrams of programmable motor controls and other solid-state control equipment employ the NEMA standard symbols. Figure 23-42 is a comparison chart showing the differences in the symbols for similar logic devices.

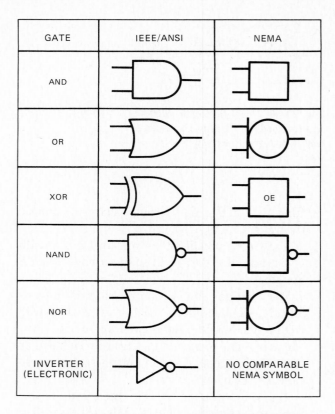

Fig. 23-42 Comparison of IEEE/ANSI and NEMA Standard Distinctive Shape logic symbols.

SUMMARY

1. The operation of contactors, connected together to form control circuits, can be analyzed using switching logic.

2. Two NO contactors, A and B, connected in series can be represented by the equation $A \cdot B = M$.

3. A listing of all combinations of contactors in a control circuit and the state of the output for each combination is known as a truth table.

4. Two NO contactors, A and B, connected in parallel can be represented by the equation
$$A + B = M.$$

5. A bar over a letter, such as $\overline{A}$ *(read "A bar")*, indicates the complement of that letter.

6. The two states of a circuit, energized and unenergized, may be represented by a 1 and 0, respectively. The number 1 is the complement of 0 and vice versa.

7. If $A = 1$, then $\overline{A} = 0$.

8. If A is a NO contactor, $\overline{A}$ is an NC contactor.

9. Switching arithmetic can be used to add and multiply the elements of a switching equation.

10. There are 13 switching theorems that can be used to combine, reduce, and simplify switching equations.

11. Other methods used to simplify switching equations are truth table simplification, sum-of-the-products, and product-of-the-sums.

12. Sum-of-the-products is also known as minterm; product-of-the-sums is also known as maxterm.

13. Digital electronics involves the study of circuits and devices that operate in discrete states. Binary, or two-state, devices are commonly used in digital electronics.

14. The binary number system uses two digits, 1 and 0.

15. Each position in a binary number is represented by a power of 2.

16. Each position in a decimal number (which uses the digits 0 through 9) is represented by a power of 10.

17. Binary numbers may be converted to decimal numbers by adding the value of each power of 10 occupied by a binary 1.

18. A decimal number may be converted to its equivalent binary number by dividing the decimal number successively by 2 and recording the remainder obtained at each step.

19. Logic gates are solid-state electronic devices that combine input signals internally and either pass a signal through as an output or else block the output.

20. The AND gate produces a binary 1 output only if each of its inputs is 1.

21. The OR gate produces a binary 1 output if any, or all, of its inputs are 1.

22. The XOR gate produces a binary 1 output if any *odd* number of inputs is a binary 1. If an even number of inputs are 1, the output will be 0.

23. The NOT gate, or *inverter,* changes the state of its input signal to the complement of that signal at the output of the inverter.

24. The NAND gate is an AND that inverts its output. If all inputs to the NAND gate are 1, the output is 0.

25. The NOR gate is an inverted OR gate. Thus if any, or all, of its inputs are 1, the output is 0.

26. The XNOR gate is an inverted exclusive OR gate. If an odd number of inputs are 1, then the output is 0.

27. A flip-flop is a device that maintains its state indefinitely until switched by a pulse to its complementary state.

28. An *R-S* flip-flop contains two inputs, reset and set, and two outputs, one at binary 1, the other at 0.

29. A shift register is used to store information or data in the form of a binary number.

30. A counter is a series of flip-flops that switch, or toggle, each successive flip-flop to change their state and thus represent an increasing or decreasing binary number.

QUESTIONS

23-1 How would the statement "coil M is energized only if NO contactors A and B are both closed" be written as a switching equation?

23-2 Draw the truth table for $A + B + C = M$.

23-3 Is $\overline{A} + \overline{B} = \overline{A + B}$? Prove your answer by the use of truth tables.

23-4 What is the complement of 1?

23-5 What is the complement of $A \cdot \overline{B}$? Prove your answer by the use of a truth table.

23-6 Coils A, B, and C operate a set of contactors. If any two of the coils are energized, a bell will ring. Show how the contactors should be wired to the bell to satisfy this condition. Show how the bell will be energized.

23-7 Using switching theorems, simplify the following switching equation:

$$\overline{A} \cdot B + A \cdot B \cdot C + \overline{B} \cdot C$$
$$+ B \cdot A + A \cdot C = Y$$

23-8 If A represents a NO contactor, what does $\overline{A}$ represent?

23-9 Simplify the equation

$$(A + B + C) \cdot (A + B + C) = Y$$

using the truth table simplification method.

23-10 What is another name for product-of-the-sums?

23-11 What is another name for sum-of-the-products?

23-12 What does "discrete states" mean in terms of digital electronics?

23-13 What is the value of each position of a binary number?

23-14 In the number 1110_2 what does the 2 represent?

23-15 Convert 101_2 to a decimal number.

23-16 Convert 101_{10} to a binary number.

23-17 Write the equation for a three-input NAND gate.

23-18 Write the equation for a four-input XNOR gate.

23-19 Draw the truth table for the XNOR gate in question 23-18.

23-20 Draw a logic gate circuit that would satisfy the equation

$$\overline{A \cdot B \cdot C} = Y$$

but do not use a NAND gate.

23-21 Draw a logic gate circuit to satisfy the equation

$$A + B + A \cdot C = Y$$

23-22 Draw the truth table for

$$A + B + (B \cdot C) = Y$$

23-23 Draw the truth table for

$$(A + B) + (A \cdot B) = Y$$

23-24 Explain the operation of a reset-set flip-flop.

23-25 In what form does the shift register store data or information?

24

INDUSTRIAL SOLID-STATE DEVICES AND CONTROLS

The majority of the electrical equipment involved in industrial electricity operates on the principles of electromagnetism and mechanical movement. Motors and generators, switches, circuit breakers, transformers, and magnetic starters are a few examples of applications of these principles. There is another class of items that depends wholly on the movement of electrons through a device. The field of study associated with this principle is known as *electronics*. The industrial electrician has worked for years with equipment falling into this classification. Mercury-arc rectifiers, vacuum-tube amplifiers, thyratrons, copper oxide and selenium rectifiers, and photoelectric cells are but a few of the devices common in older industrial installations. In recent years developments in electronics involving the behavior of solid materials (rather than vacuum or electron tubes), miniaturization, and logic circuits have led to a new technology commonly called *solid-state electronics*. This chapter is devoted to applications of this new technology.

24-1 DIODES

A diode is a two-element device ("di" meaning *two* and "ode" referring to *electrode*) that allows current to flow through it in only one direction. Solid-state diodes are made of germanium, silicon, or similar materials that display the properties of both conductors and nonconductors. Such materials are known as *semiconductors*. The molecular structure of the diode is such that when the negative end of the diode is connected to the negative side of a line and the other end is connected to the positive side, current will flow through the diode. When the polarity is reversed (as would be the case if the diode were connected to an ac line), the diode would stop conducting and current would stop flowing. Thus any load connected in series with the diode would have current flowing through it in one direction. In effect, a varying or pulsating dc would be flowing in that circuit (see Fig. 24-1).

Diodes, in fact, are used to change ac to dc. When used for this purpose, diodes are called *rectifiers*. Diodes are also used in circuits to make sure current is flowing in the right direction. In this application the diode is connected in the circuit so that it does not impede the flow of current in the normal direction. Should the polarity somehow reverse itself, the diode would prevent current from flowing.

The symbol for a diode is shown in Fig. 24-2. Since it is essential that the two terminals are connected to the proper polarity, the end of the diode is coded in some fashion to identify the cathode, or minus (−) terminal. Unfortunately there is no universal coding or identification scheme. Some of the more common methods of identification are given in Fig. 24-3.

Diodes are rated according to the maximum current they can carry and the maximum voltage across which they can be connected. If a reverse voltage higher than

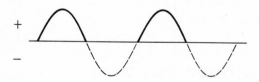

Fig. 24-1 Pulsating dc. Current flows only during the positive half of the ac cycle.

ANODE ————▶|—— CATHODE

Fig. 24-2 Symbol for diode.

383

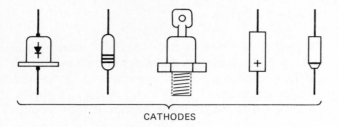

CATHODES

Fig. 24-3 Some common ways of identifying the cathode terminal of a diode. Although the terminal is marked +, the cathode consists of material that has an excess of electrons.

the maximum rating is connected across the diode, the diode material would break down and the diode would short out. In most cases this would permanently damage the diode.

24-2 ZENER DIODE

A certain type of diode called a *zener* is made to handle excessive reverse voltages without permanent damage. This property of the zener diode is used to limit or regulate voltages in a circuit. The breakdown voltage of the zener is a specification of the device. When it is connected in a circuit, the zener blocks the reverse current so long as its breakdown voltage is not exceeded. Should the reverse voltage reach the breakdown point, the zener would act as a short circuit and allow current to flow through another part of the circuit. For this reason, a resistor or other current-limiting device must be used in a zener circuit. This property is also used in a circuit to serve as a switch. In this application the zener is connected in reverse so that current is prevented from flowing until the circuit is turned on by the breakdown voltage of the zener.

The symbol for a zener diode is given in Fig. 24-4. Physically, zener diodes resemble the ordinary diodes described in Sec. 24-1.

24-3 LIGHT-EMITTING DIODES

Certain semiconductor materials, when connected so that they conduct, will emit light at the point where the negative lead of the voltage source touches the

ANODE ——————— CATHODE

Fig. 24-4 Symbol for zener diode.

semiconductor material. Diodes possessing this property are called *light-emitting diodes,* or LEDs (usually pronounced "ell-e-dees" or leds—to rhyme with heads). The semiconductor material used for LEDs is gallium arsenide, though other materials are also used. The semiconductor material and the junction with the voltage source are encased in clear plastic. When the LED is energized, it emits a reddish light. Other semiconductors as well as tinted plastic will produce light of different colors.

The symbol for the LED is shown in Fig. 24-5(*a*). The LED is often used as an indicator lamp or pilot light. The type of LED used for this purpose is shown in Fig. 24-5(*b*). In Chap. 22 the use of the LED for alphabetic and numeric displays was discussed. Figure 24-5(*c*) is a single unit containing seven LEDs embedded in plastic. The pins connecting each of the ends are brought out of the opposite side of the unit. Very often all the anodes or all the cathodes are joined internally and only a single pin is brought out as the anode or cathode connection.

These seven-segment units are usually made so that they can be plugged into a standard socket that accepts other solid-state devices called integrated circuits, or ICs.

The typical LED circuit consists of the diode in series with a resistor and connected to a 5-V dc supply. Some LEDs are manufactured with an internal resistor, while others require an external resistor. The purpose of the resistor is to limit current through the LED. This is because once the LED is fired, that is, begins to conduct, it presents a very low resistance to the circuit. The excessive current that would flow through the LED in that case would damage it. The forward voltage necessary to keep the LED lit is typi-

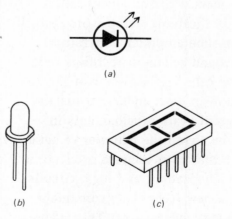

(*a*)

(*b*)   (*c*)

Fig. 24-5 Light-emitting diodes. (*a*) Schematic symbol. (*b*) Single LED. (*c*) LEDs used in a seven-segment display.

cally 1.5 to 1.9 V, while the forward current is 20 to 50 mA. A 150-Ω resistor is frequently used in series with the LED.

24-4 LIQUID CRYSTAL DISPLAY

Another popular type of seven-segment display is the liquid crystal display (LCD). Unlike the LED, the LCD does not produce its own light. Thus to be seen, an LCD must be viewed in an external light. The basis of the LCD is a material called a *nematic fluid*. This type of liquid changes its light-transmitting properties under the influence of an electric current. The common LCD fluid appears black when it conducts. Since the rest of the fluid appears shiny grey (or other light color), the black surface stands out. Each of the segments of a seven-segment display is a separate unit of fluid so that the segments may be energized individually as with the seven-segment LED display.

Since most applications of display require, or are used in, an illuminated area, LCDs have become very popular for electric meter displays, watches, portable computer displays, and the like.

The power requirements of LCDs are much less than those of the LED, thus enabling the use of smaller battery supplies for continuously powered displays such as in watches and meters. The LCD, however, requires an ac voltage to operate, generally a 30-Hz square wave. Therefore additional circuitry is used in the LCD equipment to produce the square wave from the dc-battery supply. An auxiliary lamp is sometimes provided in equipment that must be viewed under dimly lit conditions.

24-5 TRANSISTORS

Since its invention in the early 1950s the transistor has revolutionized the field of electronics. Basically the transistor is a diode to which a third element, or electrode, has been added. Early transistors were made of germanium, but this semiconductor material has been largely replaced by silicon. The word *transistor* comes from the prefix "trans," *to change,* and "resistor." Thus a transistor is a device whose resistance can change. While the applications of transistors are almost limitless, their function falls into two major categories: amplifying and switching.

The symbol for the transistor is shown in Fig. 24-6. Note that there are actually two types of transistors, differentiated from one another by the direction of the arrow in the emitter. The N and P in their designation merely refer to negative (N) and positive (P) polarities. A simplified view of a transistor is shown in Fig.

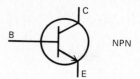

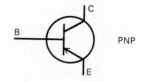

Fig. 24-6 Symbol for transistor. The three leads are *B*ase, *E*mitter, and *C*ollector.

24-7. The line separating the N area from the P area does not actually exist. It is drawn to show more clearly that different parts of the transistor exhibit different electrical properties.

The transistor itself is a tiny crystal of pure silicon (or germanium) to which special impurities have been added in specific areas. The process of adding impurities is called *doping*. One type of impurity will create an excess of free electrons in the region to which it is added. This would become an N, or negative, area. Another type of impurity creates a deficiency of electrons, and this would be a P, or positive, area. Because electrons have been removed, this area is said to have "holes." Basically these holes are positive charges.

The third element of the transistor serves as a valve or gate for the flow of electrons between the emitter and the collector. The operation of the third element determines the function of the transistor.

Amplifying Function An amplifier takes small changes in input power and increases them so that the output power is considerably greater. Transistors are used in radios, TVs, and audio equipment for just

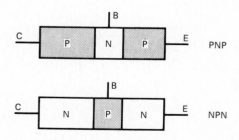

Fig. 24-7 A simplified cross section of a transistor crystal.

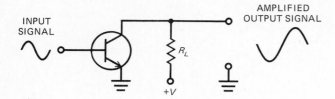

Fig. 24-8 A transistor amplifier.

such a purpose. Industrial control equipment also uses transistors to amplify weak signals from control transducers and other monitoring equipment. The circuit in Fig. 24-8 shows a transistor being used as an amplifier.

Switching Mechanical switches are generally opened and closed by use of manual, magnetic, or pneumatic power. There is movement and contact between two or more parts. The advantage of using a transistor as a switch is that no mechanical movement is involved and the action is as fast as the speed of an electron (which is the speed of light).

The switching action of a transistor is similar in some respects to that of a relay. In a relay, a low voltage and current is used to activate a coil which closes the contacts in a circuit having a much higher voltage and current.

A transistor switching circuit is shown in Fig. 24-9. The transistor used in this circuit is an NPN type. This type is by far the more common of the two transistors used. However in most cases a PNP type can also be used provided the proper polarities of the applied voltages are observed.

Referring to Fig. 24-9, note that the load circuit is from the $+V_{CC}$ terminal, through the collector, base, and emitter of the transistor, to ground. However unless the gate voltage-to-ground is above 0.6 V, the base-emitter junction will resemble an open switch (though physically nothing is actually open) and no current will flow through the load.

When the base voltage exceeds 0.6 V, the transistor will conduct. Current will flow from V_{CC} through the

load, through the transistor, and to ground. At this point there will be a voltage drop across the transistor. As the base voltage is increased, the voltage drop across the transistor decreases until a point in the transistor's characteristics called *saturation* is reached, at which time the entire V_{CC} is dropped across the load. The transistor in this condition acts as a closed switch; that is, it is a short.

24-6 TRANSISTOR SWITCHING CIRCUITS

A switch, whether it is mechanical or electronic, must operate in a reliable positive fashion. When the contacts of a mechanical switch are not touching, there is no path for current to flow and the circuit is open. When the contacts are touching, current has a clear path and the switch is fully closed.

Since there is no actual opening or closing of a transistor switch, its ability to control the flow of current is not as absolute as a mechanical switch. The only positive action takes place when the base voltage reaches the saturation point for permitting the full-load current to flow or when the base voltage is well below 0.6 V for cutoff, or nonconducting. Therefore when a transistor is to be used as a switch, the base circuit is designed so that its voltage is well below 0.6 V, usually below 0.3 V at cutoff and in the saturation area during turn on. The circuit of Fig. 24-10 is designed to operate with this swing in base voltage.

24-7 THYRISTORS

A thyristor is a semiconductor device that acts as a switch. Like a mechanical switch, it has two states: ON (or behaving like a closed switch) and OFF (or behaving like an open switch).

Thyristors are available as either two-terminal or three-terminal devices. Three-terminal thyristors can

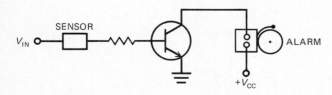

Fig. 24-9 A transistor used as a switch. The sensor in the base circuit will close and the base circuit will be complete. The transistor will be turned on and the alarm will ring.

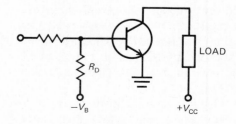

Fig. 24-10 A transistor switching circuit. Resistor R_D prevents premature turning on of the transistor when the input voltage is close to turn on.

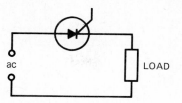

Fig. 24-12 An SCR in series with an ac load.

be switched on by applying the proper voltage to the third terminal, called the *gate*. Two-terminal thyristors remain off until the voltage across them reaches the breakover value. At that point the thyristor is fully on and in effect is shorted.

Unlike other devices capable of switching action, such as transistors, thyristors are able to have only two states, ON and OFF. While transistors have this ON-OFF property, they can also handle a whole range of conducting values. Examples of thyristors used in industrial applications include SCRs, triacs, diacs, and unijunction transistors.

24-8 SILICON-CONTROLLED RECTIFIERS

The silicon-controlled rectifier (commonly abbreviated SCR) is essentially a diode to which a third electrode has been added. Like a diode, the SCR allows current to flow in only one direction. Unlike the usual diode, the third element of the SCR permits turning on or off this forward flow of current. In addition, SCRs are available that can handle very high voltages (in the order of kilovolts) and hundreds of amperes. This makes the SCR extremely useful for a wide range of industrial applications. It is especially useful as a high-power switch that needs only low power to operate.

The schematic symbol for the SCR is shown in Fig. 24-11. Current will not flow between the anode and cathode until there is a burst of current from the gate to the cathode. The gate and cathode form a PN junction similar to that found in transistors; therefore the 0.6 V necessary to produce forward current flow in a silicon device must be present. The gate current necessary to turn on the SCR is a characteristic of the SCR itself, with typical values ranging up to 100 mA. Once the gate current has turned on the normal anode-cathode current, the gate current can drop to zero. The anode-cathode current will continue to flow until its value falls below a minimum value, called a *holding current*. A typical holding current is 10 mA.

If an SCR is connected in series with a load and an ac source as in Fig. 24-12, the SCR can control the average current flowing to the load. The basis of this control is the current flowing in the gate circuit. By

controlling the time it takes the gate voltage to reach the value needed to turn on the SCR, only a portion of the ac current wave will flow to the load.

Once the gate turns the SCR on, it stops controlling current flow in the SCR. Even when the gate voltage falls below the value that turned on the SCR, current will continue to flow. With an ac wave, current is zero at every half cycle, and this will cause the SCR to turn off. When the cycle reverses polarity, the gate circuit goes through its control action again.

Examples of how the SCR controls the average current through a load in an ac circuit can be seen in Fig. 24-13. In Fig. 24-13(*a*) the gate pulse begins at about a quarter of the positive half cycle, or 45°. The SCR begins to conduct at that point and continues to conduct for the rest of the half cycle as shown by the shaded area. Notice that the gate pulse in the meantime has returned to zero. Since the SCR is a rectifier, only forward current will flow. During the reverse half cycle, a gate pulse will not turn on the SCR and no reverse current will flow.

In Fig. 24-13(*b*), the gate pulse occurs late in the half cycle at about 150°. Again the shaded area indicates the current flow through the load. In comparing current flow in the two parts of Fig. 24-13, it can be seen that the average current and voltage decreases as the timing of the gate pulse occurs later in the cycle.

In a dc circuit, the average voltage and current is not varied by the SCR since the supply voltage remains constant with time. Thus at any point in time that the gate pulse occurs, the SCR is turned on at the constant dc value. Figure 24-14 shows the effect of the gate pulse on the load current.

24-9 SCR GATE CONTROL

When a pulse is applied to an SCR but no current flows into the SCR from the gate circuit, the SCR turns on as described in the previous section. The breakover voltage is the value at which current suddenly increases to a high value. If gate current does flow into the SCR, the breakover voltage of the SCR

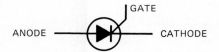

Fig. 24-11 Symbol for an SCR.

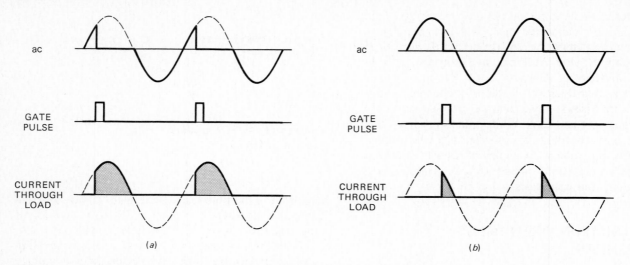

ac

GATE
PULSE

CURRENT
THROUGH
LOAD

(a)

ac

GATE
PULSE

CURRENT
THROUGH
LOAD

(b)

Fig. 24-13 The timing of the gate pulse determines the average current flowing through the load. (a) The gate pulse occurs at 45°. (b) The gate pulse occurs at 150°.

is lowered. Thus changes in gate current will affect the voltage at which the SCR will conduct.

Various types of circuits are used to produce a gate voltage needed to make the SCR conduct. Figure 24-15 is a simple manual circuit in which a switch completes the gate circuit. In this circuit the gate is in series with the load, an ON-OFF switch, and a gate circuit resistor. The gate circuit resistor permits only a few milliamperes to flow in the gate circuit. When the switch is closed, current will flow in the gate circuit and the SCR will turn on. If the switch remains closed, the load will receive the full half-cycle voltage each time the ac wave goes positive. If the gate circuit resistor is replaced by a variable resistor, the gate current can be varied, and therefore the breakover volt-

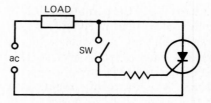

LOAD

ac

SW

Fig. 24-15 A simple gate circuit used to turn on an SCR.

age can be varied. If the breakover voltage is increased, the turn-on point will come later on the ac wave; a lower breakover voltage will produce an earlier turn-on point.

Another gate control circuit is shown in Fig. 24-16. In this circuit the charging time of the capacitor is used to develop the voltage in the gate circuit. The charge time can be varied by changing the value of the variable resistor.

Other circuits can be used to obtain conduction over

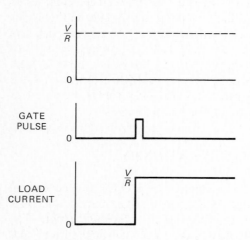

$\frac{V}{R}$

0

GATE
PULSE

0

LOAD
CURRENT

$\frac{V}{R}$

0

Fig. 24-14 The gate pulse turns on a constant dc to the load.

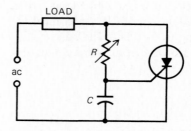

LOAD

ac

R

C

Fig. 24-16 The gate circuit voltage is determined by the charge time of capacitor C.

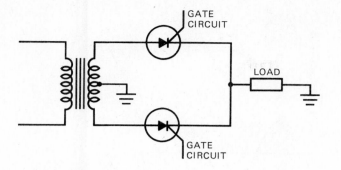

Fig. 24-17 Full-wave rectifier control circuit with two SCRs.

the full ac cycle. The full-wave rectifier circuit of Fig. 24-17 will pass both halves of the ac wave but in the forward current direction only.

While gate control in dc circuits is similar for turn on, the turn-off process is not similar since the dc voltage does not go to zero except when power is turned off. When turning off power to the SCR is not practical, the circuits of Fig. 24-18 can be used. In Fig. 24-18(*a*), a transistor acts as a shorting switch. By triggering the base, the transistor conducts and shunts the load current, thus bypassing the SCR. The SCR current falls below its holding value and it ceases to conduct. The transistor's base circuit is opened and the transistor stops conducting. To avoid destroying the transistor with the high-load current, the transistor is permitted to conduct for only the time necessary to turn off the SCR, which is a matter of milliseconds. In Fig. 24-18(*b*) turn-off action is accomplished by a voltage of opposite polarity applied across the SCR main terminals. In this case, triggering the transistor effectively places a charged capacitor of opposite polarity across the SCR, turning it off. The capacitor also prevents the load current from passing through the transistor. The transistor collector is connected to the dc source through a high resistance, shunting the load and the capacitor.

In both circuits of Fig. 24-18, separate triggering circuits are shown for the SCR gate and the transistor base. In practice these circuits can be combined, but such combination triggering circuits are complex.

Another type of SCR known as a *light-activated SCR* (LASCR) contains a light-sensing element in its housing. The internal junction of the gate and cathode is made on a thin piece of silicon. The top of the LASCR is a small glass lens through which external light can travel and strike the gate-cathode juncture. The light causes current to flow and, at the proper value, trigger the SCR. A voltage on the gate can set the level to which the additional light-activated current can turn on the SCR.

While the LASCR has limited current-handling properties, it can be used to trigger high-power SCRs. The LASCR by itself or in conjunction with another SCR or other thyristor is used to control motors, heaters, digital circuits, and other applications in which the primary triggering means is light.

24-10 TRIACS

The triac is a three-terminal semiconductor similar to an SCR. Since the SCR is a rectifier, it conducts current in only one direction. The triac, however, can conduct in either direction once it is turned on.

The schematic symbol for a triac is shown in Fig. 24-19. Notice that the triac symbol uses two diode symbols facing in opposite directions. The third terminal is the gate. The main terminals are labeled MT1 and MT2 while the gate is labeled G.

The triac is usually connected in series with a load but does not conduct until its breakover voltage is reached. In this respect the triac is an open switch. The gate controls the opening and closing of the switch, that is, the conducting or nonconducting. The general circuit arrangement for a triac is shown in Fig. 24-20.

Triacs can be used to vary the average current going

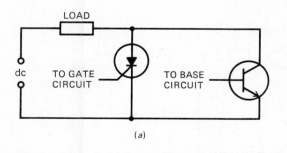

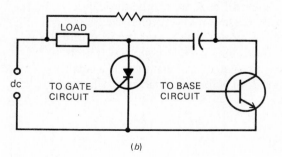

(*a*)　　　　　　　　　　　　　　　　　　(*b*)

Fig. 24-18 Gate control in a dc circuit. (*a*) The transistor acts as a switch and shorts the voltage in the gate circuit. (*b*) A capacitor with opposite polarity is placed across the SCR by the conducting transistor.

Fig. 24-19 Symbol for a triac.

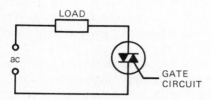

Fig. 24-20 A series circuit with a triac and load connected to an ac source.

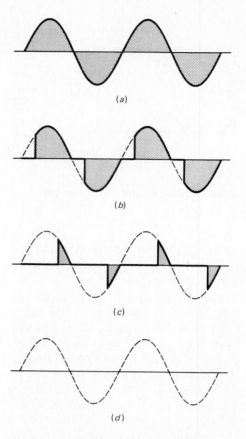

Fig. 24-21 Current waves through a load for different conduction angles. (*a*) The full ac cycle. (*b*) Conduction begins at 45°. (*c*) Conduction begins at 120°. (*d*) No conduction.

to a load. In a lighting circuit, varying the current to a lamp will vary the amount of light emitted by the lamp. Thus the triac can be used as a light-dimming control. In a small motor circuit varying the current will change the speed of the motor. In a heater, the amount of heat can be regulated by varying the current.

If the triac is allowed to conduct over the entire ac cycle, the full current drawn by the device will flow in the circuit. However, if the triac conducts during only part of the cycle, the average current will be less than the effective, or rms, value of current as read on an ammeter.

Figure 24-21 shows the different current waves that result from varying the conduction periods for a typical triac.

The operation of the triac in a typical circuit such as in Fig. 24-22 is as follows. When the switch SW is closed, the capacitor will begin to charge through $R1$ and $R2$. When the voltage across C reaches the trigger voltage for G (about 0.6 to 2.0 V), the triac is turned on and begins to conduct. At this point the triac is almost like a closed switch except that there is a small voltage drop (1 or 2 V) across the triac even when it conducts.

24-11 DIACS

The diac is a two-terminal thyristor used primarily as a switch diode for the triac or SCR. Notice from the symbol of the diac (Fig. 24-23) that it behaves like two diodes pointing in opposite directions. In fact the diac is also known as a *bidirectional trigger diode*.

The diac is usually connected in the gate circuit of the triac or SCR. It does not conduct until the breakover voltage is reached across its terminals in either direction. The breakover voltage is practically equal in both directions. Typical breakover voltages are 30 to 32 V with a less than 1-V difference between the various polarities.

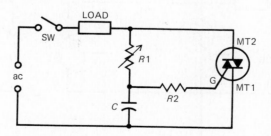

Fig. 24-22 Triggering time of the triac can be varied by use of a variable resistor.

Fig. 24-23 Symbol for a diac.

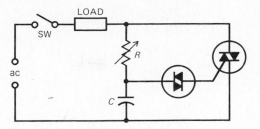

Fig. 24-25 A diac used in the gate circuit of a triac.

The curve of current through the diac versus the voltage across the diac is shown in Fig. 24-24. The curve shows that between the negative breakover voltage and the positive breakover voltage, the diac will present a very high resistance and almost no current will flow through it. Once the voltage goes over the breakover value of $+32$ V or under -32 V, the diac switches on and is in effect short-circuited in the gate circuit of the triac or SCR.

Figure 24-25 is a control circuit containing the diac in the gate circuit of a triac. The circuit operates as follows. When the switch is closed, the capacitor C begins to charge through the load and resistance until the voltage across C reaches the diac's breakover voltage. At that point, the diac conducts, the capacitor discharges, and the triac is turned on. The balance of the 120-V ac half cycle flows through the load and the triac. When the polarity of the ac wave reverses, the capacitor-charging process reverses and the reverse breakdown voltage is reached. Again the diac conducts, the capacitor discharges, and the triac is turned on. The rest of the ac half cycle is therefore applied to the load. By varying the resistance, the charge time is changed and the portion of the ac cycle that is applied to the load is changed.

The circuit described can be used as a dimming switch or a speed controller for small motors. If the variable resistor is replaced by a temperature or pres-

sure sensor or other ON-OFF device, the load can be controlled by changes in temperature or pressure levels, mechanical tripping, and the like. Blowers, fans, heaters, lamps, or other loads can be turned on or off by the triac in this fashion.

24-12 UNIJUNCTION TRANSISTORS

The unijunction transistor (UJT) is a switching device, similar to a bipolar transistor. Unlike the bipolar transistor, the UJT is not used as an amplifier; it behaves like, and is used as, a thyristor.

The symbol for the UJT is shown in Fig. 24-26. A simple circuit containing a UJT as in Fig. 24-27(a) operates as follows. The voltage required at the emitter to produce current through the UJT is called the *peak voltage*. The value of the peak voltage for a specific type of UJT is equal to a percentage of the voltage across the bases plus the forward voltage drop of a silicon junction (0.6 V).

When the voltage between the emitter and base 1 (B1) goes above the peak voltage, the UJT will turn on. A burst of current will then flow from the emitter to base 1. The circuit of Fig. 24-27(a) reaches the peak voltage in the process of charging the capacitor. When switch SW is closed, current will flow from the $+V$ supply, through $R1$ and begin charging C. When the voltage across the capacitor goes just over the peak voltage for the UJT, the UJT will be turned on and in effect the E-to-B1 path will become a short circuit. A burst of current will then flow from the

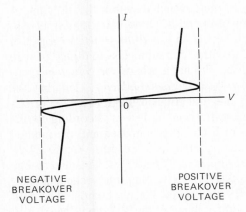

Fig. 24-24 The current (I) − voltage (V) characteristic curve of the diac.

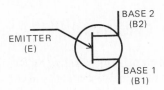

Fig. 24-26 Symbol for a unijunction transistor.

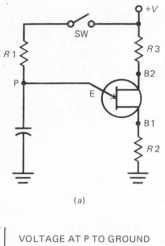

(a)

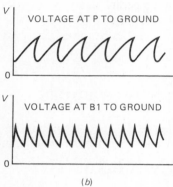

(b)

Fig. 24-27 A UJT oscillator.
(a) Simple circuit. (b) Waveforms
at various points in the circuit.

capacitor through the UJT, through $R2$ to ground. This will discharge the capacitor, lowering the voltage below the peak voltage for the UJT. The UJT will then quickly turn off and conduction will stop.

This charging–rapid discharge characteristic of the UJT makes it useful for producing sharp voltage and current pulses that can be used for triggering other thyristor circuits. The circuit of Fig. 24-27(a) is known as a *relaxation oscillator*. The voltage waveforms at point P and base 1 (B1) are shown in Fig. 24-27(b). If B1 is connected to the gate of a triac or SCR, it can provide the triggering pulses to turn on these devices.

24-13 INTEGRATED CIRCUITS

Solid-state devices, such as transistors, diodes, SCRs, and the like, are formed on extremely small crystals of silicon, germanium, or other materials. By introducing certain impurities in different parts of these materials, various areas are made rich or poor in electrons.

Very tiny wires (leads) are connected to these separate areas and brought out to terminals. The small bits of silicon or other material are then mounted in various types of housing or encased in plastic. These solid-state components are known as *discrete devices*.

An *integrated circuit* device is similar to a discrete device in its basic composition, with a major difference. An integrated circuit combines many discrete components such as transistors and diodes as well as resistors and capacitors directly on a single basic material called a *substrate*.

Integrated circuits, generally referred to as ICs, are usually classified according to the degree of combining discrete components. For example, the logic gates mentioned in Chap. 23 consist of transistors and resistors. A single AND gate or OR gate requires just a few transistors. In terms of integrated circuits it is considered small-scale integration (SSI) if up to 10 gates are made on a single piece of IC material (called a *chip*). Medium-scale integration (MSI) includes up to 99 gates on a single chip. Large-scale integration (LSI) has up to 999 gates on a single chip. Finally, very large scale integration (VLSI) consists of 1000 gates or more on a single chip. While every definition might not agree exactly with these numbers, nevertheless the enormous number of discrete devices combined on a single chip is obvious.

The way in which ICs are manufactured is beyond the scope of this text, but excellent explanations may be found in many of the references listed in Sec. 24-24.

24-14 INTEGRATED CIRCUIT TYPES

Integrated circuits are classified in other ways than according to the number of discrete devices contained on a single chip.

Digital ICs are typically identified by the way the transistors are interconnected within the IC chip. A popular group of digital ICs have their transistors connected in a circuit called *transistor-transistor logic,* or simply TTL. Another common class of digital ICs are the CMOS ICs. The name stands for *complementary symmetry metal oxide semiconductor,* denoting the circuitry and material used to make this type of IC.

Another classification of ICs is according to the package in which the chip is housed and the method used for arranging their leads or pins. Figure 24-28 shows four examples of IC packages. The dual-in-line (DIP) packages are made to plug into sockets or standardized holes in printed circuit boards. The DIP package is used for many of the TTL and CMOS ICs, while the MOS/LSI package generally houses special chips called *microprocessors*.

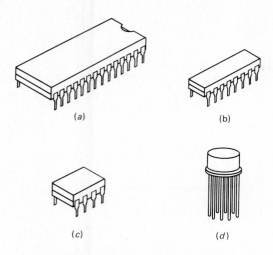

Fig. 24-28 Integrated-circuit package types. (*a*) MOS/LSI. (*b*) Dual-in-line (DIP). (*c*) Mini-DIP. (*d*) TO-5.

24-15 MICROPROCESSORS

A microprocessor is an integrated circuit of large scale or very large scale integration. It is a digital device that is used as the central processing unit (CPU) of a microcomputer. Contained on a single microprocessor chip are hundreds of transistors, resistors, diodes, and capacitors. Their circuitry provides for the basic operations of arithmetic logic (such as discussed in Chap. 23), control, and storage.

Arithmetic logic, which is the operation involved in adding and subtracting binary numbers as well as the manipulation of logic gates (AND, OR, NAND, etc.), is performed in the arithmetic logic unit (ALU). Another part of the microprocessor controls the timing of operations as well as the movement of information or data throughout the microprocessor. A third section of the microprocessor is used for temporary storage of data or information that is being processed.

In order to move data and information throughout the microprocessor, a bus (that is, a conducting path) is connected between each of the major sections of the microprocessor. This path, called a *data bus,* contains as many parallel paths as there are bits in the "words" handled by the ALU. The most common microprocessors used in industry have 8-bit words and 8-bit data busses.

While 8-bit microprocessors are common in industry, there are many different types and classes of microprocessors in use. Some of the more widely used 8-bit microprocessors are the 8080s series, including the 8080 and the 8085 developed by Intel; Motorola's 6800; Zilog's Z80; and MOS Technology's 6502.

Figure 24-29 is a highly magnified view of the Mo-

torola 6800 microprocessor chip. The actual size of this chip is about 1/4 in. square. The black lines around the edges of the chip are the leads that connect to the pins of the microprocessor package.

As modern manufacturing technology advances, 16-bit, 32-bit, and 64-bit microprocessors will become as economical and common as the 8-bit microprocessor. The increase in bit size will lead to faster operations, increased capabilities, and greater and more diversified applications of microprocessors.

24-16 MICROPROCESSOR ARCHITECTURE

A microprocessor can be divided into sections, each of which is interrelated but has specific functions designed into the chip by the engineers and designers who developed the chip. This arrangement of functions is called the *architecture* of the microprocessor. In fact, one microprocessor is distinguished from another by its architecture. The architecture can usually be described or visualized by a *block diagram*. Figure 24-30 is a simplified block diagram of an Intel 8085 8-bit microprocessor. Only the basic lines of interconnection are shown in this diagram.

Arithmetic Section The *accumulator* is a temporary storage place for the quantity that is being operated upon. The quantity itself is in the form of an 8-bit

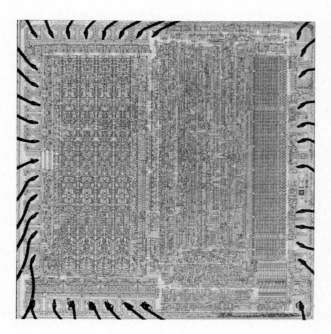

Fig. 24-29 Greatly enlarged view of a Motorola 6800 microprocessor chip. (*Motorola.*)

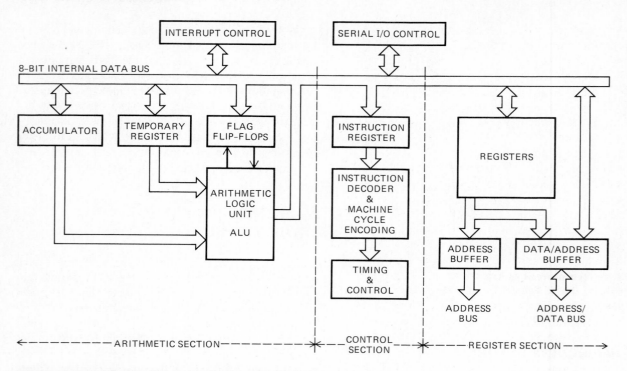

Fig. 24-30 Simplified block diagram of the Intel 8085 microprocessor.

number. The double-ended arrow between the accumulator and bus means information flows both ways between these two parts. Similarly, the *temporary register* holds a quantity from the data bus and then sends it to the ALU. A *flip-flop* is a device that changes its state from 0 to 1 or 1 to 0 when acted upon by an external signal. The *flag flip-flop* is used to indicate the state of certain operations or conditions. For example, if a problem in addition produces a carry, it may be necessary to make this known. We *flag* the fact of the carry by *setting* one of the bits in this unit to 1. There are five flags or bits that can be set (1) or reset (0). The ALU performs the operations themselves (addition, subtraction, and digital logic).

Control Section A microprocessor is designed with a built-in set of instructions that tells the microprocessor what to do and how to do it in processing data and information. Each type of microprocessor has its own instruction set, although a great many similarities in instructions exist among the many different types of microprocessors. The *instruction register* temporarily stores an instruction in the form of an 8-bit number that it receives from the data bus. The *instruction decoder* translates the 8-bit number into a course of action and produces the signals that will initiate and carry through the instructions. The *timing and control* unit provides all the pulsing signals that synchronize

the timing and direction of the quantities being processed.

Register Section Registers are temporary storage locations. In this section are located other processing items. A *program counter,* for example, keeps track of each program step. The program itself is a sequence of operations, designed by the user of the microprocessor. A *stack* is a location in the memory storage unit used to store memory locations. The locations themselves are given numerical addresses. Thus information can be sent or received from a particular location by using its address. A *stack pointer* in the register section is used to store, temporarily, an address where data or an instruction is located. The register section also contains units that lead to the address bus as well as the data bus. The address bus contains 16 parallel paths to the memory locations; eight paths are for addresses, eight are for addresses and data.

24-17 MICROPROCESSOR INSTRUCTION SETS

An *instruction set* is the entire group of orders or directions that are designed into the microprocessor. The instruction set cannot be changed for a particular microprocessor. But the instructions can be arranged in a sequence which, when put into action, will perform operations much more complicated than a single instruction. The instruction set is built into the hardware of the microprocessor itself, but the sequence of these instructions is under the control of the user of the microprocessor and can be changed, enhanced, or even deleted completely. The sequence is called a *program*. Programs are usually distinguished from hardware by being called *software*.

Instructions are often given *mnemonic codes* to indicate what they do. A mnemonic code is merely a set of letters which help us remember what they represent. For example, the code ADD is an instruction that adds one quantity to another; MOV might mean that a quantity is to be moved from one memory address to another, and so forth.

The job of the programmer is to arrange all the instructions necessary to perform an operation, in a step-by-step logical order—in fact, in the exact order—in which the instructions are to be followed.

As an example of their extent—and limitation—the Intel 8080 has 72 instructions while the Zilog Z80A has 158 instructions. Both, however, process 8-bit data.

24-18 MICROPROCESSOR SYSTEMS

Although the microprocessor is a powerful solid-state device, its power is best used when integrated into a system that contains other solid-state devices.

A program designed to perform repeated operations, such as in industrial control systems, may use an integrated circuit in which the program has been permanently set. Such a device, called a *ROM* chip, contains a set program that can be read into the microprocessor system. In fact ROM stands for *Read-Only-Memory,* indicating that its set of program steps may be read by the system but new steps cannot be *written* into the ROM. The memory device to which information can be sent and stored and then later retrieved is called *RAM* which stands for *Random-Access-Memory.* The word *random* indicates that any memory location can be found directly without going through a sequence of locations. When information or data are sent to the RAM, we say information is written in the RAM; retrieval of information is called reading the RAM. The RAM is therefore sometimes called a *read-and-write memory.*

In order to address devices outside the structure of the microprocessor itself, an input/output circuit must be provided.

A typical microprocessor system therefore would appear as shown in Fig. 24-31.

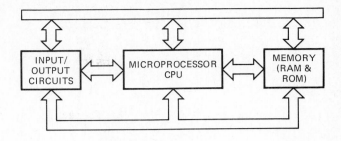

Fig. 24-31 A basic microprocessor system.

24-19 SOLID-STATE MOTOR CONTROLS

Motor controls as discussed in this section involve the direction of rotation, speed, and protection, primarily against overload. Rotation is considered continuous except for jog operations. That is, the shaft is turning with a smooth, constant speed over a period of time. Positional control, that is, movement in small, angular steps, is discussed in Sec. 24-23 on stepping motors. In addition, only power line use of solid-state devices is discussed in this section.

Solid-state devices can be applied to many uses traditionally reserved for magnetic controllers. These include across-the-line and reduced-voltage starters, variable-speed controls, and overload protection.

The high inrush current of large ac motors being started across the line requires the use of very high power solid-state devices in the power lines. Thus, the use of solid-state devices is dependent upon the availability and cost of such devices at high current rating.

Solid-state devices such as thyristors (SCRs, triacs, diacs, and UJTs) are used alone, or in combination, as reduced-voltage starting devices. As discussed in previous sections, by triggering the gate circuit at various points on the ac cycle, the average amount of current delivered to the load, in this case a motor, can be varied.

Because of their very high current ratings SCRs are frequently used for the very large ac motors. However SCRs pass current in only one direction so that their use in an ac circuit would greatly reduce the power available to the motor circuit even when the gate turns on the SCR completely. Therefore SCRs are usually connected in parallel in reverse order as in Fig. 24-32(*a*). Such units are available commercially in a single housing. This arrangement closely resembles a triac which, because of its limited current-handling ability, cannot be used for very large horsepower motors.

One danger of the parallel-reverse SCRs is that a strong transient or surge voltage in the line can turn on the SCR. To prevent this, commercial SCRs are pro-

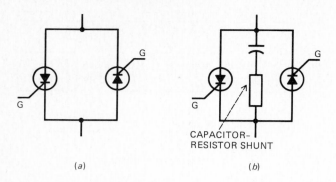

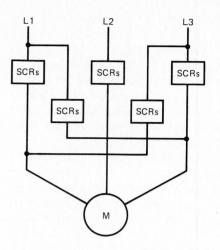

Fig. 24-32 A commercial solid-state control for power line circuits. (*a*) Two SCRs provide full-cycle conduction. (*b*) Surge protection is provided by a shunt capacitor and resistor.

Fig. 24-34 An SCR reversing starter in a three-phase motor circuit.

vided with an integral shunt circuit consisting of a resistor in series with a capacitor, as in Fig. 24-32(*b*). Should a surge occur, the shunt will dissipate the energy and prevent the SCR from turning on.

The operation of the parallel SCRs is similar to that of the triac. During one-half of the ac cycle, the gate will turn on one of the SCRs. At zero voltage both SCRs will turn off. When the second half of the cycle begins in the reverse direction, the gate will turn on the second SCR.

These SCRs are placed directly in the power lines supplying the motor, as in Fig. 24-33, so in effect they replace the main mechanical contacts that close the circuit in a magnetic starter. However, the SCRs have one important advantage over the magnetic starter. The reduced-voltage magnetic starter must drop resistance in steps, thus preventing a smooth,

continuous reduction in voltage. The gate circuit of the SCR, on the other hand, can vary the turn-on point in a smooth progression. Rather than reducing resistance, the SCR merely increases the average current being delivered to the motor. The gate circuits of the SCR use other solid-state circuits to time the turn-on point and measure the point on the ac wave at which to operate.

Three-phase motors are reversed by interchanging any two lines to the motor. Using the two-SCR arrangement, an effective reversing starter can be made. Figure 24-34 shows lines 1 and 3 reversed. The gating of the SCRs is carefully controlled by electronic circuits that turn on and then off the proper sets of SCRs. This type of reversing starter is better able to withstand the stresses that occur on the contacts of mechanical reversing starters.

24-20 SPEED CONTROL

The three-phase squirrel-cage induction motor is the most widely used motor in industrial applications. The speed of such motors, however, is difficult to vary without wasting energy or greatly reducing the torque of the motor. Very often dc motors must be used to effect accurate and reliable speed control.

Variable speed of induction motors is possible using solid-state devices. The speed of the ac motor depends on the voltage and frequency applied to the motor. Motors with more than one constant speed are available, but these are usually multiwinding motors. Reducing the voltage in steps will reduce the speed, but this also reduces the torque and therefore the load-handling ability of the motor. These motors and meth-

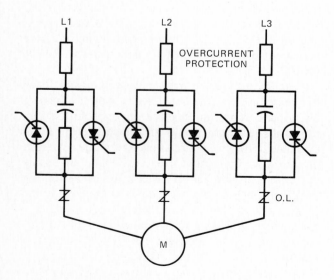

Fig. 24-33 A three-phase reduced-voltage starter using SCRs in the power line.

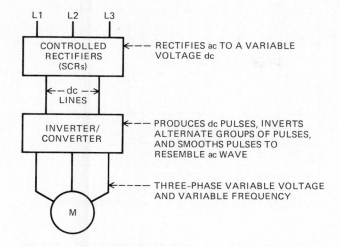

Fig. 24-35 Variable-speed ac motor controller.

ods are not truly variable-speed methods since smooth, wide-ranging speeds are not possible.

Solid-state circuits, however, make it possible to not only vary the frequency but also to reduce the voltage applied to the motor while maintaining the motor's torque.

A simplified diagram of a variable-speed control is shown in Fig. 24-35. The operation of this speed controller is as follows. The three-phase input is rectified to dc using a set of SCRs. By controlling the gate circuit, the dc voltage can be varied. This variable dc is sent into another circuit that converts the steady dc into pulses of controlled duration, then inverts the dc pulse so that, in effect, an ac voltage is produced from the dc pulses as in Fig. 24-36. By varying the pulse width and the magnitude of the dc voltage, the frequency and voltage of the ac output can be varied.

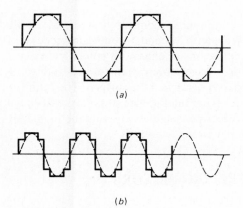

Fig. 24-36 The dc pulses (heavy lines) are inverted and smoothed to resemble an ac wave. (*a*) High voltage, low frequency. (*b*) Low voltage, high frequency.

Additional electronic circuits smooth the stepped or pulsed ac into a wave more closely resembling the sine wave shown.

The advantage of this system is the ability to adjust the speed of the large ac motors over a wide range with savings in energy and without a significant reduction in motor torque.

24-21 MOTOR PROTECTION

Two types of protection are important in motor circuits: protection of the branch-circuit wiring feeding the motor and protection of the motor against overloads and other dangerous operating conditions. Branch-circuit protection of motors is covered elsewhere in this book and relates specifically to NEC Art. 430. The provisions of this article apply to any type of motor starter, whether magnetic or solid-state.

Modern solid-state overload devices are generally designed to monitor many more factors involved in safe motor operation than simply heat rise. Microprocessors form the heart, or rather brain, of these devices. While the exact circuitry of such protective devices is beyond the scope of this book, a listing of the types of critical areas involved is of interest.

1. Motor winding temperatures
2. Motor speed (under and over normal)
3. Loss of load (due to belt snapping, shaft breaking, gear cracking, etc.)
4. Status of loading (over- or underloaded)
5. Level of inrush current and duration
6. Single-phasing
7. Heating effects of jogging, plugging, braking, reversing, etc.
8. Ground faults (leakage current)

These and many more factors can be measured simultaneously by the microprocessor and acted upon as preset by the user or manufacturer. Microprocessors as integral parts of a motor monitoring system can be programmed to react to the various factors as is appropriate to the specific motor to which it is applied.

24-22 PROGRAMMABLE CONTROLS

Basically programmable controls are devices that receive information through an input device, process the information, then produce responses to the content of the information and send the responses out to the equipment and machines that can use the responses.

The National Electrical Manufacturers Association (NEMA) defines a programmable controller as:

> A digitally operating electronic apparatus employing a programmable memory for internal storage of instructions relating to implementation of specific control functions. These include logic, sequencing, timing, counting, and arithmetic, and may control machines or processes via digital or analog input or output modules.

While every manufacturer of programmable controls has its own design approach and objectives, generically the programmable control contains the elements shown in Fig. 24-37.

The microprocessor as the center of activity might be considered the decision-maker of the programmable control. It receives instructions, or a plan of action, in the form of a program. The program may be in the form of a piece of hardware, that is, a plug-in unit containing ROM chips. It may also receive its program from a keyboard, a set of switches or pushbutton, a tape, or by some other means. With the exception of the ROM chips, the programs are stored in memory. The microprocessor can then access memory to obtain its course of action according to the stored program. Visual displays such as a video screen or pilot lights, and audio signals respond to the program so that the operator of the equipment knows the programmable control is in operation and that it is programmed to achieve the required control.

At the other end are the input and output devices. These feed into or are fed by an input/output unit. This unit serves as a traffic manager receiving information from sensing coils, photoelectric cells, transducers, limit switches, and the like. In its output func-

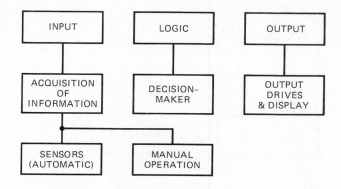

Fig. 24-38 The functions of a programmable control.

tion it sends signals to the gate circuits of thyristors and activates other solid-state devices as well. It can send signals out to starters, turn on (or off) pilot lights, alarms, valves, and similar devices.

Thus the action of a programmable controller may be divided into three basic functions: the input, logic, and output. Figure 24-38 shows these functions and their related actions.

Programmable controls are not unlike the magnetic controls discussed earlier in this text insofar as the input and output devices and functions are concerned. The advantage of the programmable control, however, is that it can be made to learn responses to a whole family of input behaviors—both positive and negative. It can sense trouble and take corrective action, or it can keep operations on an even, positive path. It can do this because it has the ability to read data at very frequent intervals and analyze what the behavior means or what any deviations from behavior mean. It can almost anticipate actions by building into the program a vast catalog of experience which can then be referenced by the controller many times a second.

Modern programmable controls use 8- and 16-bit microprocessors, with 32- and 64-bit microprocessors promising applications of enormous power and range. The application of microprocessor-based industrial controls has barely scratched the surface. The future is likely to see artificial intelligence, robotics, self-diagnostic, and corrective machines entering the industrial scene in overwhelming numbers.

24-23 STEPPING MOTORS

Many mechanical operations involve the controlled movement of levers, wheels, cams, rods, and the like. Industrial robots rely on the manipulation of arms and grips that must be able to move in a straight line in a wide range of angles and to rotate in steps of only a

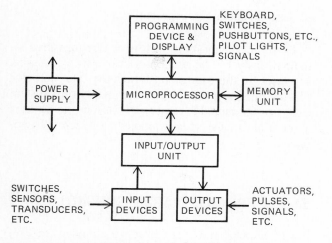

Fig. 24-37 A basic programmable control system.

few degrees. Many of these operations depend on *stepping motors*.

While the stepping motor itself is not a solid-state device, its use and controls do involve such devices. It is therefore appropriate to discuss stepping motors here. Basically the stepping motor is a brushless dc motor. It differs from the usual dc motor in that its movement resembles a jogging action rather than a continuous, smooth rotation of its shaft. If the steps are performed in a rapid sequence, the shaft can have continuous rotation, though the rotor will be in lock-step with the input to the motor.

The basic construction of one type of stepping motor is shown in Fig. 24-39. The stator contains multiple poles, as does the rotor. The rotor poles are permanent magnets arranged with alternate south and north polarity. The motor shown in Fig. 24-39(*a*) has four pairs of poles in the stator and three pairs of poles in the rotor. Since the stator poles are evenly spaced, the poles are separated by 360°/8 = 45°. The rotor has six poles separated 60° apart. Since the two sets are not equal, it is impossible for the poles of the rotor to align exactly with the poles of the stator. To follow the operation of the motor more easily, the poles of the stator are marked in pairs A-A′, B-B′, C-C′, and D-D′. Similarly the poles of the rotor are marked 1-1′, 2-2′, and 3-3′.

If poles A-A′ are energized, the nearest rotor poles of opposite polarity, such as 1-1′, will align themselves. The other four poles of the rotor are then 15° away from the nearest stator pole. Depending on which pair of stator poles are energized next, the rotor will move clockwise or counterclockwise 15°. If the B-B′ coils were energized next, rotor poles 2-2′ would move 15° in a clockwise direction as in Fig. 24-39(*b*). By energizing coils C-C′, D-D′, A-A′, etc., rotor poles 3-3′, 1-1′, 2-2′, etc., will align in

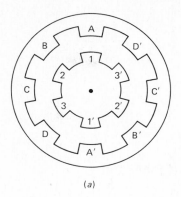

(*a*)

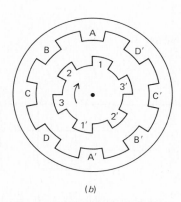

(*b*)

Fig. 24-39 Basic construction of a stepping motor. The stator windings are omitted for clarity. (*a*) The stator and permanent-magnet rotor. (*b*) Stator coils energized for CW stepping.

turn, producing a clockwise rotation in 15° steps. If instead of the A, B, C, D sequence, the coils were energized after coils A-A′, in the order D-D′, C-C′, B-B′, etc., the rotor poles would align in the opposite

(*a*)   (*b*)

Fig. 24-40 A two-stator motor. (*a*) A small stepping motor. (*b*) Exploded view of the same motor. (*Hurst Mfg. Co.*)

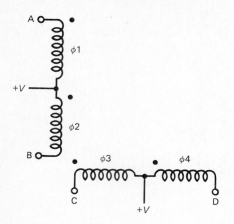

Fig. 24-41 Bifiliar stator winding.

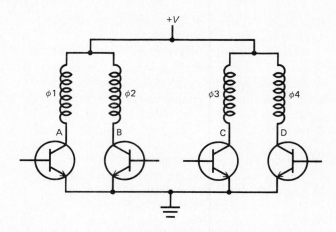

Fig. 24-42 Transistor switching control used to energize stator coils.

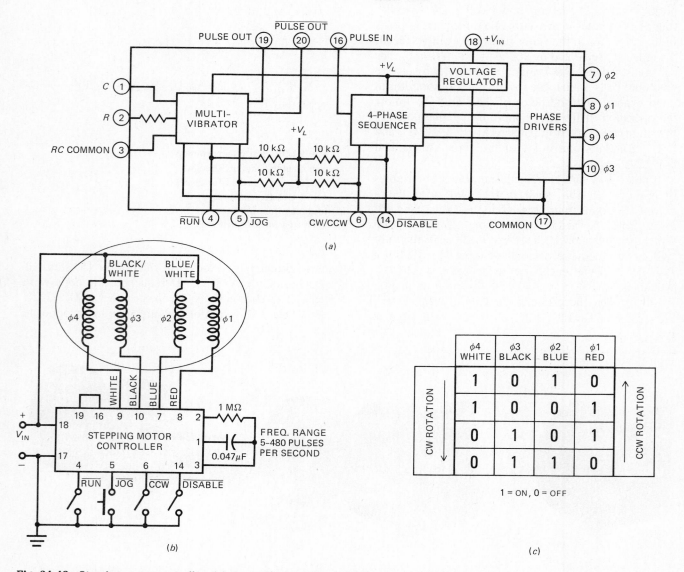

Fig. 24-43 Stepping motor controller. (a) Controller unit. (b) Schematic diagram of motor windings, controller, and external switches. (c) Control table. (*Hurst Mfg. Co.*)

direction. That is, when D-D' are energized, rotor poles 3-3' will rotate to align with D-D', rotating in a counterclockwise direction. By energizing more than one pair of poles, the rotor may step to a position between two poles, thus making it possible to move in 7.5° arcs.

Another type of stepping motor is shown in Fig. 24-40. The exploded view of the motor in Fig. 24-40(b) shows two stators with toothed poles. Two circular coils energize the two sets of teeth in the stator. When assembled, the two stators are offset by one-fourth of the pole pitch, or the angle between adjacent teeth. The rotor in this type of motor has the same number of poles as the teeth in one of the stators. The poles of the rotor are around the circumference of the rotor, alternating north and south poles. When both stators are energized, the rotor will align itself between the two equal stator fields.

The two windings in each stator are known as *bifiliar windings* as shown in Fig. 24-41. Because this motor has four windings, it is referred to as a *four-phase* winding.

The stepping sequence (7.5° and 15° are typical steps) is achieved by energizing the coils in sequence and reversing the magnetic fields in the process. In the bifiliar winding the switching circuit using transistors as switching devices is shown in Fig. 24-42.

An example of a commercial stepping-motor controller is shown in Fig. 24-43(a). As shown in Fig. 24-43(b), this controller contains a logic sequencer which sends a pulse to the motor to provide a single step. The direction of the step, CW or CCW, is determined by the terminal 6 switch. If terminal 6 switch is closed, the stepping motor rotates in the CCW direction; an open switch to terminal 6 produces CW steps.

If terminal 5 switch is closed, a single pulse is sent to the logic sequencer and a "jog" or step is performed. If terminal 4 switch is closed, an ac voltage is sent to the logic sequencer and the stepping motor will rotate continuously, in steps. Closing the switch to terminal 5 during this run operation will not affect the rotation. Closing the switch to terminal 14 will deenergize all four phases of the motor. The switching sequence table of Fig. 24-43(c) shows the order in which two phases are energized while two phases are deenergized whenever the logic sequencer is pulsed through terminal 16.

24-24 BIBLIOGRAPHY

Neither the general theme of this text nor space permits the detailed coverage of basic electronics. A number of excellent texts are available that provide the theory and background in electronics for those students wishing to pursue such study. The following list is by no means exhaustive, but it does represent a cross section of the materials used by students across the country.

Basic Electronics, Bernard Grob, McGraw-Hill Book Company, New York.

Digital Electronics, Roger Tokheim, McGraw-Hill Book Company, New York.

Electronic Circuits and Applications, Bernard Grob, McGraw-Hill Book Company, New York.

Electronic Principles, Albert P. Malvino, McGraw-Hill Book Company, New York.

Electronics: Principles and Applications, Charles A. Schuler, McGraw-Hill Book Company, New York.

SUMMARY

1. A diode is a two-element device that allows current to flow through it in only one direction.

2. Solid-state diodes are commonly made from germanium or silicon crystals.

3. Forward current + to − will flow through a diode from anode to cathode.

4. A zener diode is a special diode that is designed to allow a high reverse current to flow through it without damage.

5. A light-emitting diode (LED) produces light when current flows through it. Seven-segment devices used for digital displays contain small LEDs.

6. A liquid crystal display (LCD) uses a special material called a nematic fluid. The light-transmitting properties of this fluid change when it is placed in an electrical field.

7. The power requirements of LCDs are much less than those of LEDs. However, LEDs produce their own light, while LCDs must be viewed in an external light in order to be seen.

8. Transistors are three-element devices. The elements are called the base, collector, and emitter.

9. Transistors can be used as amplifiers and as switches.

10. Two types of transistors are the NPN and the PNP types.

11. The process of adding impurities to the pure germanium or silicon crystals from which the transistors are made is called doping.

12. A silicon-controlled rectifier (SCR) is a three-element device. In addition to the anode and cathode of a diode, the SCR contains a gate that is able to control the flow of current.

13. Although the gate controls when current will flow, current flows only in the forward direction.

14. The SCR is used as a precisely controlled switch, especially as a power line switch for large ac motors.

15. Once the gate turns on the SCR, it no longer controls current flow. Current will continue to flow in the SCR circuit even if the gate voltage falls below the value that originally triggered the SCR on.

16. The light-activated SCR (LASCR) contains a small lens that allows external light to strike the junction of the gate and the cathode. The light causes current to flow and at the proper value will trigger the SCR on.

17. Thyristors are two- or three-element switching devices. They have only two states: ON or OFF. Examples of thyristors are SCRs, triacs, diacs, and UJTs.

18. Two-element thyristors, such as diacs, conduct when the voltage across them exceeds its breakover value.

19. Three-element thyristors contain a gate which triggers conduction similar to that of the SCR.

20. The triac resembles the SCR except that it can conduct in both the forward and reverse directions.

21. The diac is a two-element thyristor that is used as a switch in the gate circuit of an SCR or triac.

22. A unijunction transistor (UJT) is a three-element thyristor.

23. The UJT is used as a controlled switch or as a relaxation oscillator to produce pulses for other thyristors.

24. A discrete component is a single solid-state device in its own housing. Examples of discrete devices are diodes, transistors, and SCRs.

25. An integrated circuit (IC) contains many discrete devices fabricated on the same base material (called a substrate) and packaged in a single housing.

26. Integrated circuits are classified according to the degree to which the discrete devices are built into the IC. The usual classifications are small-scale integration (SSI), medium-scale integration (MSI), large-scale integration (LSI), and very large scale integration (VLSI).

27. Two popular digital integrated circuits are the TTL and the CMOS.

28. Integrated circuit packages include the MOS/LSI, DIP, MINI DIP, and TO-5 types.

29. The microprocessor is an LSI or VLSI integrated circuit.

30. The microprocessor is used as the central processing unit (CPU) of a microcomputer.

31. Microprocessors are classified according to the size of the words they handle and the number of parallel data lines used. The most widely used microprocessors are 8-bit, but 16-, 32-, and 64-bit microprocessors are available.

32. The three main sections of a microprocessor are the arithmetic section, control section, and register section.

33. The instruction set is designed into the microprocessor. Programs use the instructions for applications of the microprocessor. Programs are called software.

34. Solid-state controls are used to start and stop motors, control their speed and direction of rotation, and protect the motor.

35. Parallel SCRs connected in reverse are used as power line controls.

36. The speed of large squirrel-cage induction motors can be varied over a wide range through the use of variable-voltage, variable-frequency solid-state controls.

37. Microprocessor-based motor protection can monitor the performance of a motor and react to and correct a wide range of motor problems.

38. Programmable controls receive data and information through input devices, process the information, then produce responses to the information and send the responses to machines through output devices.

39. Stepping motors are capable of rotating in very short angular steps. The typical stepping motor consists of a multiple-pole stator and a multiple-pole permanent magnet rotor.

40. Stator poles of stepping motors are energized in a sequence that produces the short, angular movement of the rotor.

41. Steps as short as 7.5° are common in stepping motors.

42. Solid-state controls are used to produce the sequence of steps and the direction of rotation of the rotor of a stepping motor.

QUESTIONS

24-1 What principles are involved in the operation of the majority of industrial electrical equipment?

24-2 List four electronic devices used in older industrial installations.

24-3 What are the two terminals of a diode called?

24-4 Why are rectifiers commonly used in ac circuits?

24-5 What are two main properties on which diodes are rated?

24-6 What voltage is commonly used in LED circuits?

24-7 What type of material is used in the LCD?

24-8 List some advantages and disadvantages of the LCD as compared with the LED.

24-9 What are the three elements of a transistor?

24-10 What two main functions does the transistor perform?

24-11 What are two common materials from which transistors are made? Which is the more common material?

24-12 List two classes of transistors according to whether the elements are positive or negative.

24-13 At what base voltage will the transistor conduct?

24-14 To what class of device does the SCR belong?

24-15 How many elements does a silicon-controlled rectifier have? What are their names?

24-16 How is the SCR made to switch on?

24-17 Why is the usual transistor not considered a thyristor?

24-18 How does the triac differ from the SCR?

24-19 What is a major use of the diac?

24-20 How many elements does a UJT have? Name the elements.

24-21 What are two common applications of the UJT?

24-22 What is the name of the device that contains numerous transistors, diodes, and resistors on a single substrate?

24-23 What do the initials SSI, MSI, LSI, and VLSI stand for?

24-24 What are the three main sections of the microprocessor?

24-25 What is the function of the ALU?

24-26 What is an instruction set of a microprocessor?

24-27 What is the difference between a ROM and a RAM?

24-28 Why must SCRs be connected in parallel-reverse fashion in a motor's power lines?

24-29 What factors affect the speed of ac motors?

24-30 Why are the dc pulses inverted in the variable-speed controller?

24-31 List five critical areas monitored by a microprocessor-based overload device.

24-32 What are the three functional areas of a programmable motor control?

24-33 Describe the basic function of the microprocessor in a programmable control.

24-34 What are the basic parts of a stepping motor?

24-35 Describe the operation of a stepping-motor controller.

APPENDIX: MAKING ELECTRICAL COILS

Making electrical coils is an exacting but interesting job. Good-quality coils, neat in appearance, can be made of the general run of motor windings in the average rewinding shop.

A-1 PRINCIPLES OF COIL MAKING

The main factors in coil making that require special and constant care are the number of turns of wire, wire size, insulation, coil shape, cleanliness, and handling of materials.

The number of turns of wire in a coil is chiefly determined by the design. If the replacement coil is to be used with the same design, the number of turns of wire should be carefully duplicated. The size of the wire in a coil is determined chiefly by the operating current of the coil.

Insulation on a coil should be carefully recorded and duplicated or improved when the coil is made. The *shape of a coil* is an important factor in its quality since a good set of coils can be damaged or ruined if the coils do not fit properly in the space allocated in the equipment.

Cleanliness and proper handling of materials are essential to good coil making. Cleanliness is necessary to avoid inclusion of contaminants such as *oil, grease, dirt, metal chips,* or other *destructive matter* in a coil. Careful handling of materials includes protection against breaks or scratches in wire insulation and inclusion of conducting materials such as metal chips that can cause short circuits. Wire should be handled carefully to avoid bending or kinking, and crossovers in winding should be held to a minimum. Crossovers can cause short circuits, while bends and kinks take up winding room in the slots. Sharp bends in the wire cause the insulating enamel to crack or craze at the bend. Bends and kinks also create high-resistance points which lead to hot spots in the coils.

A-2 WINDING COILS

Form-wound stator and armature coils are usually wound on coil winders with adjustable winding heads like the one shown in Fig. A-1. The six adjustable spindles have removable sleeves carrying eight adjustable spacing spools. The spindles are adjusted for coil size, and the spools adjusted for coil-side dimensions.

A winding head is usually set by the use of a sample coil from the winding to be replaced. In stripping an old winding, one coil should be carefully preserved as a sample.

Three-phase gang-wound coils are shown being wound in Fig. A-1. In gang winding, the wire is not cut between coils that make one pole phase group in a three-phase winding. Winding forms that produce a round-end coil are also used.

In winding, proper tension should be maintained on the wire to keep it straight and to form proper bends at the corners of the coil. Too much tension can damage

405

Fig. A-1 Gang-winding motor coils on an adjustable coil-winder head. (*Crown Industrial Products, Co.*)

wire insulation. Tension can be supplied by drawing the wire through an adjustable fiber friction clamp.

Severe jerking of the wire should be avoided. A square, rectangular, or diamond form jerks the wire at each corner of the form as it turns. Jerking against a heavy spool of wire can damage wire, wire insulation, and equipment. A spring-loaded pulley with a deep groove should be located between the wire spool and tension device to absorb jerking shocks and minimize their effects.

A practical arrangement of a winding system is illustrated in Fig. A-2. A shock-absorbing spring-loaded pulley is shown between the spool of wire and the tension device. After leaving the tension device, the wire is run through a guide to aid in leading the wire to the winding forms.

Before removal of a coil or coils from a form, they are tied with string or small wire in two or more places. The tying is to hold the coil wire in place until the coils are taped or inserted into the winding slots.

A-3 SHOP-MADE FORMS

Occasionally, it is necessary to make special forms for deep, square, or rectangular coils for dc motor fields, transformers, solenoids, or odd-shaped-form coils. Wooden forms can easily be made for these purposes.

Basically, a wood coil form consists of a backplate with a core for the coil and a removable front plate. Slots should be provided for insertion of tie wires around a finished coil. A hole in the front plate near the core can be used to anchor the beginning lead of the coil.

Adhesive tape, placed at intervals around the form before winding, can be taped about the coil after winding for tying instead of using tie wire.

A shop-made form is illustrated in Fig. A-3. This form consists of a front plate and backplate, a core, and suitable clamping and mounting pieces. At Fig. A-3(*a*) is shown a back block for chucking in a lathe or screwing to a faceplate. The core is shown attached

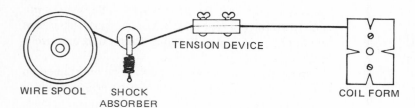

Fig. A-2 A winding system with a shock absorber and tension device.

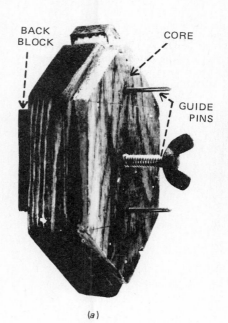

BACK BLOCK

CORE

GUIDE PINS

(a)

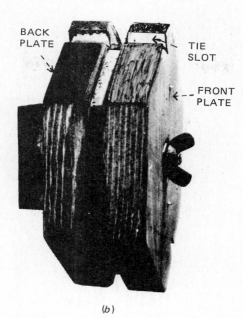

BACK PLATE

TIE SLOT

FRONT PLATE

(b)

Fig. A-3 A shop-made form for winding small coils. (a) Back plate and core with back block for lathe mounting. (b) Assembled form.

to the backplate. Guide pins and a screw serve to align the front plate with the backplate and core. A wing nut is used to hold the front plate to the core.

In Fig. A-3(b) the complete form is shown assembled. Tie slots are shown at both ends of the form for insertion of a string or a small insulated wire to tie a completed coil preparatory to removing it from the form. The top ends of the backplate and front plate and the top front side of the front plate should be painted for quick identification in reassembly of the form and for counting turns when winding. This form can be mounted on a stand and turned with a hand crank.

Coils requiring several thousand turns can be wound in a lathe. In one method the completed form, with the old coil on it, is mounted in the lathe chuck or screwed to a faceplate. The lathe carriage is brought up near the headstock, and its location marked by a position mark on the lathe bed. The lathe is run very slowly in reverse to unwind the old coil with the carriage moving backward, away from the headstock.

After the old coil has been completely unwound, the coil form is insulated and the lathe run forward, winding the new coil until the carriage returns to its original position as indicated by the position mark on the lathe bed. At this point the coil form will have made the same number or revolutions (or turns of wire) as it made in unwinding the coil.

The number of turns in the coil can be calculated by dividing the distance the carriage traveled in inches by the carriage feed in inches per revolution of the chuck. Thus, if the carriage feed is 0.002 in. per revolution

and the carriage travel is 20 in., the number of turns of the chuck is calculated as follows:

$$\text{No. of turns} = \frac{\text{carriage travel, in.}}{\text{carriage feed, in. per rev.}}$$
$$= \frac{20}{0.002} = 10,000$$

A-4 LOOP WINDERS

Large coils and odd-shape coils are conveniently wound on a loop winder. A loop winder with coils, set for four shapes of coils, is shown in Fig. A-4. Coils wound by this method are tied with cord or clipped with narrow strips of sheet metal to hold them in shape when they are removed from the form. They are then varnished and taped before spreading—or taped, spread, varnished, and baked.

A-5 SHAPING COILS

Some coils require shaping after removal from the winder. Shaping is usually done with a coil spreader (Fig. A-5). The coil was originally made in the form of a hairpin loop and was pulled to the shape shown. Knuckles, formed at the ends of the coil in the pulling operation, allow coil ends to turn under the ends of an adjacent coil in a winding. All coils in a set should be identical in shape. A form should never be changed during the making of a set of coils.

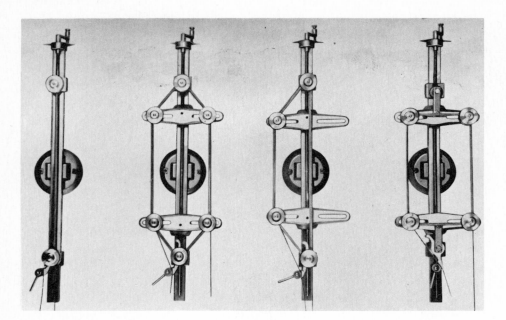

Fig. A-4 Coil loop winders. The four settings shown are used to wind armature and stator coils. (*Armature Coil Equipment, Inc.*)

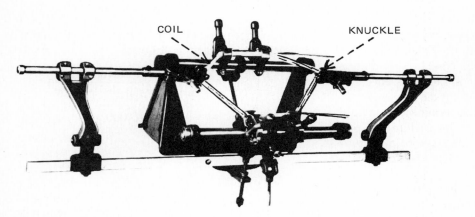

COIL · · · KNUCKLE

Fig. A-5 A coil spreader forming the knuckles at the ends of a coil and shaping the coil sides. (*Armature Coil Equipment, Inc.*)

A-6 TAPING COILS

Large coils and some small coils are taped for insulation, mechanical strength, and protection. Commonly used taping material consists of cotton, linen, glass, asbestos, and a variety of plastics. These materials are either treated or untreated, with or without various types of backing or binders.

Coils are taped by hand or with a taping machine. One form of taping machine is shown in Fig. A-6. Tape is usually half-lapped as it is applied. It is fastened at the ends with one of various kinds of quick-setting adhesives or adhesive-backed tape.

Coils are occasionally made slot-size. Slot-size coils have the slot insulation taped on the coil. Great care must be used in making slot-size coils to ensure a proper fit when the coil is inserted into the slot.

A-7 VARNISHING AND BAKING COILS

Electrical coils require protection against *moisture, dirt, oil,* and *chemicals* before they are installed as a winding. Electrical insulating varnishes, both natural and synthetic, are used to protect coils and windings. Sometimes, coils are varnished only after the winding is completed. When practicable, coils should be varnished *before* and *after* winding.

A good electrical varnish provides an insulating film to protect against moisture, acids, dirt, and oils and to fill pores in insulation and spaces in coils. It bonds all parts of a coil together to give it mechanical strength, reduce the destructive effect of vibration, and increase the ability of the coil to dissipate heat. Varnish provides electrical protection by filling, bonding, and strengthening the windings, as well as improving their appearance.

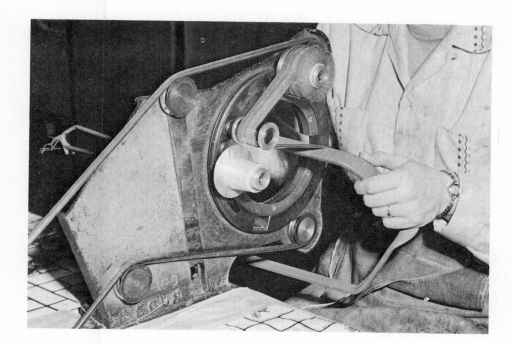

Fig. A-6 A coil-taping machine. (*National Electric Coil, Div. of McGraw-Edison Service.*)

Electrical insulating paints are used to seal, protect, and improve the appearance of windings and electrical parts. These paints are usually applied by brush or spray gun.

A-8 APPLICATION OF ELECTRICAL VARNISHES

When electrical varnish is applied, it should penetrate a winding. Penetration is aided by preheating a winding. This process dries the winding; the heat also thins varnish to allow it to flow freely for good penetration.

Before being varnished, a new winding should be baked at below 212°F (100°C) until thoroughly dry and at increased temperatures up to 275°F (135°C) until it is thoroughly heated. Baking a green coil at over 212°F is likely to create steam in the cells of the insulation and damage it.

Electrical varnishes are of two general types—*air-drying* and *baking*. Air-drying varnishes cure by *contact with air* after their solvents have evaporated. The process of curing is faster when air-drying varnishes are baked. Baking varnishes *will not cure* at normal air temperatures but must be baked at certain critical temperatures. Baking of these varnishes is a *thermosetting* process required for *internal* curing. It is not a "drying-out" operation. To thermoset properly, baking varnishes must be baked at the specific temperatures and for specific times stated in the manufacturer's directions for baking a given varnish. Baking is generally done in an electric oven, although

infrared lamps can be used successfully and, in some cases, can reduce the time of baking.

Four methods are commonly used in applying electrical varnishes—*dipping, spraying, brushing,* and *vacuum-pressure*. A fifth method called the *trickle process* can be used to impregnate certain assembled parts, such as motor armatures. In this process varnish trickles or drips onto a rotating part that has been preheated. Excellent impregnation is achieved, especially when varnishes that are completely free of solvents are used. Although effective, this method is not widely used because of the materials and equipment required.

A-9 VARNISHING BY DIPPING

In varnishing by dipping, coils or windings should be dried first by heating below 212°F, then heated to about 275°F prior to dipping. They are then lowered into a vat or container of varnish and left until bubbling ceases.

Varnish, by displacing air in a winding, causes bubbles to rise from the winding. When bubbling ceases, the winding should be raised from the varnish and allowed to drip until all excess varnish drains from it. A winding left too long in the varnish will cool and will not drip sufficiently to drain off excess varnish.

Armatures should be dipped in a vertical position to allow varnish to penetrate evenly on all sides. They should also be drained vertically so that the varnish drains evenly on all sides. If varnish is allowed to accumulate on one side of an armature, it will cause *imbalance,* requiring a careful rebalancing process.

Spraying and brushing involve the use of an air-spray gun or brush. Air-spraying produces better penetration than brushing. Brushing or spraying is usually done when a winding cannot be dipped or vacuum-varnished.

A-10 VACUUM-PRESSURE VARNISHING

Vacuum-pressure varnishing, which is superior to other commonly used methods of varnishing, is one of the best methods for obtaining thorough penetration of deep windings. Figure A-7 shows a number of coils in a vacuum tank in preparation for varnishing.

Briefly, the vacuum-pressure operation is as follows. Preheated coils are enclosed in a tank. A vacuum of about 28 in. of mercury is drawn on the tank for about 1 h. Varnish is admitted to the tank, and about 50 lb/in.^{2} pressure is applied for 2 h to force varnish into the coils. The coils are then removed and are drained and baked according to recommendations for the varnish.

Fig. A-7 Coils arranged in a vacuum tank before vacuum-pressure varnishing. (*National Electric Coil, Div. of McGraw-Edison Service.*)

INDEX